Differential Equations
with *Mathematica*

Martha L. Abell

James P. Braselton

Department of Mathematics and Computer Science
Georgia Southern University
Statesboro, Georgia

AP PROFESSIONAL

A Division of Harcourt Brace & Company

Boston   San Diego   New York
London   Sydney   Tokyo   Toronto

This book is printed on acid-free paper. ∞

Mathematica is a registered trademark of Wolfram Research, Inc.
Macintosh is a registered trademark of AppleComputer, Inc.

AP PROFESSIONAL
955 Massachusetts Avenue, Cambridge, MA 02139

An Imprint of ACADEMIC PRESS, INC.
A Division of HARCOURT BRACE & COMPANY

United Kingdom Edition published by
ACADEMIC PRESS LIMITED
24–28 Oval Road, London NW1 7DX

Library of Congress Cataloging-in-Publication Data

Abell, Martha L., date.
    Differential equations with Mathematica / Martha L. Abell, James
P. Braselton.
        p.   cm.
    Includes bibliographical references and index.
    ISBN 0-12-041539-9
    1. Differential equations—Data processing. 2. Mathematica
(Computer file) I. Braselton, James P., date. II. Title.
QA371.5.D37A24   1993   93-9134
515'.35'02855369—dc20   CIP

Printed in the United States of America
93 94 95 96 97 EB 9 8 7 6 5 4 3

Contents

Preface

Mathematica's diversity makes it particularly well-suited to performing many of the calculations encountered when solving elementary ordinary differential equations. In some cases, *Mathematica*'s built-in functions can immediately solve a differential equation; in other cases, *Mathematica* can be used to perform the calculations encountered when solving a differential equation. Since one goal of the traditional differential equations course is to introduce the student to basic methods and algorithms and for the student to gain proficiency in them, nearly every topic covered includes typical examples solved by both traditional methods and *Mathematica*. Consequently, we feel that we have addressed one issue frequently encountered when implementing computer assisted instruction. In addition to the above, Differential Equations with *Mathematica* uses *Mathematica* to establish well-known algorithms for solving elementary differential equations.

All *Mathematica* calculations were completed using Versions 2.0 and 2.1 of *Mathematica*. If you are using an earlier or later version of *Mathematica*, your results may not appear in a form identical to those found in the book: some commands in Versions 2.0 and 2.1 are not available in earlier versions of *Mathematica*; in later versions some commands will certainly be changed, new commands added, and obsolete commands removed. In general, however, the text is computer independent. All people who have access to *Mathematica* can use this text with equal success, keeping in mind that results may not physically look identical to those illustrated in the text. In addition, the following conventions are used:

❑ **EXAMPLE** denotes examples primarily completed by traditional methods although *Mathematica* may be used to perform some calculations; and

■ **EXAMPLE** denotes examples primarily completed using *Mathematica* although traditional methods may be illustrated concurrently or used to perform some calculations.

In addition to a comprehensive **Index**, the end of the text includes a **Glossary** of *Mathematica* commands used in the text, and a list of mathematical and *Mathematica* references.

Appropriate uses of Differential Equations with *Mathematica* include:
1) a handbook which addresses some ways to use *Mathematica* for computation of explicit or numerical solutions of a variety of familiar ordinary differential equations; and
2) a supplement for beginning courses in ordinary and/or partial differential equations.

vii

Adequate prerequisites to using this text successfully would include basic familiarity with *Mathematica* and a standard first-year calculus course. In most cases, commands used in the text are briefly discussed when first introduced; a glossary of commands is included at the end of the text.

We have enjoyed working on <u>Differential Equations with *Mathematica*</u> and wish to express our thanks to the folks who have supported us and made our work easier to do. In particular, we'd like to thank our editor, Charles Glaser, and production editor, Elizabeth Tustian, at Academic Press in addition to Brad Horn of Wolfram Research, Inc all of whom have been most helpful during this project.

Martha L. Abell
James P. Braselton
November, 1992

Chapter 1: Introduction to Differential Equations

§1.1 Purpose

The purpose of **Differential Equations with _Mathematica_** is twofold. First, we introduce and briefly discuss in a very standard manner all topics typically covered in an undergraduate course in ordinary differential equations as well as some supplementary topics such as Laplace transforms, Fourier series, and partial differential equations which are not. Second, we illustrate how _Mathematica_ is used to enhance the study of differential equations not only by eliminating the computational difficulties, but also by overcoming the visual limitations associated with the solutions of differential equations. In each chapter, we first briefly present the material in a manner similar to most differential equations texts and then illustrate how _Mathematica_ can be used to solve some typical problems. For example, in **Chapter 2**, we introduce the topic of first-order equations. So as not to trivialize the subject, we do not simply make use of _Mathematica_ commands such as **DSolve** which solve the differential equations for us. Instead, we show how to solve the problems by hand and then show how _Mathematica_ can be used to perform the same solution procedures. In **Chapter 3**, we discuss some applications of first-order equations. In this chapter, since we are experienced and understand the methods of solution covered in **Chapter 2**, we make use of **DSolve** and similar commands to obtain the desired solutions. In doing so, we are able to emphasize the applications themselves as opposed to becoming bogged down in the calculations. You will notice that throughout a majority of **Differential Equations with _Mathematica_** even numbered chapters introduce a particular topic while odd numbered chapters cover some applications of the topic of the previous chapter.

The advantages of using _Mathematica_ in the study of differential equations are numerous, but perhaps the most useful is that of being able to produce the graphics associated with solutions of differential equations. This is particularly beneficial in the discussion of applications since many physical situations are modeled with differential equations. For example, when we solve the problem of the motion of a pendulum, we can actually watch the pendulum move. The same is true for the motion of a mass attached to the end of a spring as well as many other problems. In having this ability, the study of differential equations becomes much more meaningful as well as interesting.

§1.2 Definitions and Concepts

A **differential equation** is an equation which contains the derivative or differentials of one or more dependent variables with respect to one or more independent variables. If the equation contains only ordinary derivatives (of one or more independent variables) with respect to a single independent variable, then the equation is called an **ordinary differential equation**.

❑ **EXAMPLE 1.1**
    Determine which of the following are examples of ordinary differential equations.

1

(a) $\dfrac{dy}{dx} = \dfrac{x^2}{y^2 \cos(y)}$, (b) $\dfrac{dy}{dx} + \dfrac{du}{dx} = u + x^2 y$, (c) $(y-1)\,dx + x\cos(y)\,dy = 1$, (d) $\dfrac{\partial u}{\partial t} = \dfrac{\partial^2 u}{\partial x^2}$,

(e) $x^2 y'' + xy' + \left(x^2 - n^2\right)y = 0$, (f) $\dfrac{\partial^2 u}{\partial t^2} = \dfrac{\partial^2 u}{\partial x^2}$.

Solution:

The equations in parts (a), (b), (c), and (e) are ordinary differential equations. The equations in parts (d) and (f) are not since they contain partial derivatives. Some of these equations are well-known. The equations in parts (d) and (f) are the **heat equation** and the **wave equation**, respectively, and will be solved in **Chapter 12**. Also, the equation in part (e) is **Bessel's equation** which will be solved in **Chapter 6** and discussed in other chapters as well. ∎

If the equation contains partial derivatives of one or more dependent variables, then the equation is called a **partial differential equation**.

❑ **EXAMPLE 1.2**

Determine which of the following are examples of partial differential equations.

(a) $u\dfrac{\partial u}{\partial t} = \dfrac{\partial u}{\partial x}$, (b) $u u_x + u = u_{yy}$, (c) $\dfrac{\partial^2 u}{\partial x^2} + \dfrac{\partial^2 u}{\partial y^2} = 0$, (d) $\dfrac{\partial u}{\partial t} = \dfrac{\partial^2 u}{\partial x^2}$.

Solution:

All of these equations are partial differential equations. In fact, the equation in part (c) is known as **Laplace's equation** and will be discussed in detail in **Chapter 12**. ∎

Differential equations can be categorized into groups of equations which may be solved in similar ways. The first level of classification, distinguishing ordinary and partial differential equations, was discussed above. We extend this classification system with the following definition. The highest derivative in the differential equation is called the **order** of the equation.

❑ **EXAMPLE 1.3**

Determine the order of each of the following differential equations.

(a) $\dfrac{dy}{dx} = \dfrac{x^2}{y^2 \cos(y)}$

(d) $\left(\dfrac{dy}{dx}\right)^4 = y + x$

(b) $u_{xx} + u_{yy} = 0$

(c) $(y-1)\,dx + x\cos(y)\,dy = 1$

(e) $y^3 + \dfrac{dy}{dx} = 1$

Solution:

(a) The order of this equation is first-order since it only includes one first order derivative; (b) This equation is classified as second order since the highest order derivative is of the second order. Hence, Laplace's equation is a second order partial differential equation; (c)This equation is first order since we can solve for dy/dx; (d) This equation is classified as first order since the highest order derivative is the first derivative. Raising that derivative to the fourth power does not affect the order of the equation; (e) Again, we have a first order equation since the highest order derivative is the first derivative. ∎

The next level of classification is based on the following definition:

A **linear** ordinary differential equation (of order n) is of the form

$$a_n(x)\frac{d^n y}{dx^n} + a_{n-1}(x)\frac{d^{n-1} y}{dx^{n-1}} + \cdots + a_2(x)\frac{d^2 y}{dx^2} + a_1(x)\frac{dy}{dx} + a_0(x)y = f(x)$$

where the functions $a_j(x)$, $j = 0, 1, \ldots, n$, and $f(x)$ are given.

If the equation does not meet the requirements of this equation, then the equation is said to be **nonlinear**. A similar classification is followed for partial differential equations.

❑ EXAMPLE 1.4

Determine which of the following differential equations are linear.

(a) $\dfrac{dy}{dx} = x^3$

(b) $\dfrac{d^2 u}{dx^2} + u = e^x$

(c) $(y-1)dx + x\cos(y)dy = 1$

(d) $\dfrac{d^3 y}{dx^3} + y\dfrac{dy}{dx} = x$

(e) $\dfrac{dy}{dx} + x^2 y = x$

(f) $\dfrac{d^2 x}{dt^2} + \sin(x) = 0$

Solution:

(a) This equation is linear since the nonlinear term x^3 is the function f(x) in the general formula above; (b)This equation is also linear. Using u as the dependent variable name does not affect the linearity; (c) Solving for $\dfrac{dy}{dx}$ we have $\dfrac{dy}{dx} = \dfrac{1-y}{x\cos(y)}$. Since the right - hand side of this equation is a function

of y, the equation is nonlinear. (d) The coefficient of the term $\dfrac{dy}{dx}$ is y and, thus, is not a function of

x. Hence, this equation is nonlinear. (e) This equation is linear. The term x^2 is merely the coefficient function. (f) This equation, know as the pendulum equation because it models the motion of a pendulum, is nonlinear since it involves a function of x, the dependent variable in this case. This function is cos(x). ∎

§1.3 Solutions of Differential Equations

When faced with a differential equation, the goal is to determine solutions to the equation. Hence, we state the following definition.

A **solution** of a differential equation on a given interval is a function which is continuous on the interval and has all the necessary derivatives which are present in the differential equation such that when substituted into the equation yields an identity for all values on the interval.

❑ **EXAMPLE 1.5**
Verify that the given function is a solution to the corresponding differential equation.

(a) $\dfrac{dy}{dx} = 3y$, $y(x) = e^{3x}$

(b) $\dfrac{d^2u}{dx^2} + 16u = 0$, $u(x) = \cos(4x)$

(c) $y'' + 2y' + y = 0$, $y(x) = x\,e^{-x}$

Solution:

(a) Differentiating we have $\dfrac{dy}{dx} = 3e^{3x}$. Hence, $\dfrac{dy}{dx} = 3y$.

(b) Two derivatives are required in this case. $\dfrac{du}{dx} = -4\sin(4x)$ and $\dfrac{d^2u}{dx^2} = -4\cos(4x)$. Therefore,

$$\dfrac{d^2u}{dx^2} + 4u = -4\cos(4x) + 4\cos(4x) = 0.$$

(c) Differentiating with the Product Rule we find that $y' = e^{-x} - xe^{-x}$ and $y'' = -2e^{-x} + xe^{-x}$. Hence,

$$y'' + 2y' + y = -2e^{-x} + xe^{-x} + 2\left(e^{-x} - xe^{-x}\right) + xe^{-x} = 0. \quad \blacksquare$$

In the example above, the solution is given as a function of the independent variable. In these cases, the solution is said to be **explicit**. In solving some differential equations an explicit solution cannot be determined. In this case, the solution is said to be **implicit**.

❑ **EXAMPLE 1.6**
Verify that the given implicit solution satisfies the differential equation.

Solution: $2x^2 + y^2 - 2xy + 5x = 0$

Differential Equation: $\dfrac{dy}{dx} = \dfrac{2y - 4x - 5}{2y - 2x}$

Solution:

Using implicit differentiation with the implicit solution to determine $\dfrac{dy}{dx}$, we have

$4x + 2y\dfrac{dy}{dx} - 2\left(y + x\dfrac{dy}{dx}\right) + 5 = 0$, so $\dfrac{dy}{dx} = \dfrac{2y - 4x - 5}{2y - 2x}$. Hence, the given implicit solution

satisfies the differential equation. ■

Most differential equations have more than one solution. We illustrate this property in the following example.

❑ **EXAMPLE 1.7**

Verify that the given solution which depends on an arbitrary constant satisfies the differential equation.

(a) Solution: $y = C \sin x$ (b) Solution: $y = C_1 \sin x + C_2 \cos x$

     Differential Equation: $\dfrac{d^2y}{dx^2} + y = 0$ Differential Equation: $\dfrac{d^2y}{dx^2} + y = 0$

Solution:

(a) Differentiating we obtain $\dfrac{dy}{dx} = C \cos x$ and $\dfrac{d^2y}{dx^2} = -C \sin x$. Therefore,

$\dfrac{d^2y}{dx^2} + y = -C \sin x + C \sin x = 0$;

(b) Similarly, we have $\dfrac{dy}{dx} = C_1 \cos x - C_2 \sin x$ and $\dfrac{d^2y}{dx^2} = -C_1 \sin x - C_2 \cos x$. Substituting into

the differential equation, we have $\dfrac{d^2y}{dx^2} + y = -C_1 \sin x - C_2 \cos x + C_1 \sin x + C_2 \cos x = 0.$ ■

§1.4 Initial and Boundary Value Problems

In many applications of differential equations, we are not only presented a differential equation but one or more auxiliary conditions as well. The number of the conditions typically equals the order of the equation. For example, if we consider the first-order equation which models the exponential growth of a population

$$\frac{dx}{dt} = x$$

where x(t) represents the population at time t. The solution of this equation is $x(t)=ke^x$. Since this solution depends on an arbitrary constant, we call this the **general solution**. However, in a problem such as this, we usually know the initial population. Therefore, we must determine the one solution which satisfies the given **initial condition**. Suppose that x(0)=10, then substitution into the general solution yields k=10. Therefore the solution to the **initial value problem**

$$\begin{cases} \dfrac{dx}{dt} = x \\ x(0) = 10 \end{cases}$$

is $x(t)=10\,e^x$. Notice that this first-order equation requires one auxiliary condition to eliminate the unknown coefficient in the general solution.

❏ EXAMPLE 1.8

The general solution to the first-order equation which determines the velocity of an object of mass m=1 subjected to air resistance equivalent to the instantaneous velocity of the object:

$$\frac{dv}{dt} = 32 - v$$

where v(t) represents the object's velocity is $v(t)=32 + ce^{-t}$. If the initial velocity of the object is v(0)=0, then determine the solution which satisfies this initial condition.

Solution:

Substituting into the general solution, we have v(0)=32 + c=0. Hence, c = – 32, and the solution to the initial value problem is $v(t) =32 + 32e^{-t}$. ∎

Since first-order equations involve a single auxiliary condition which is usually referred to as an initial condition, we distinguish between initial and boundary value problems through the following example.

❏ EXAMPLE 1.9

Consider the second order differential equation which models the motion of a mass with m=1 attached to the end of a spring with spring constant k=1 given by x''+ x = 0, where x(t) represents the distance of the mass from the equilibrium position x =0 at time t. The general solution of this differential equation is x(t) = Acos(t)+ Bsin(t). Since this is a second order equation, we need two auxiliary conditions to determine the unknown constants.
(a) Suppose that the initial position of the mass is x(0)=0 and the initial velocity x'(0)=1. Then this constitutes an initial value, because we have two auxiliary conditions at the same value of t, namely t=0. Determine the solution of this **initial value problem**.
(b) On the other hand, suppose that we know the position at two different values of t such as x(0)=0 and

x(π/2)=−4. Since the conditions are given at different values of t, we call this a **boundary value problem**. Use the given boundary conditions to determine the solution to this problem.

Solution:

(a) Since we need the first derivative of the general solution, we calculate x'(t)=−Asin(t)+ Bcos(t). Now, substitution yields x(0)=A=0 and, thus, x'(0)=B=1. Hence, the solution is x(t)=sin(t).

(b) Substitution into the general solution gives us x(0)=A=0 and x(π/2)= B=−4. Therefore, the solution to the boundary value problem is x(t)=−4sin(t). ∎

Chapter 2: First-Order Ordinary Differential Equations

Mathematica commands used in **Chapter 2** include:

`Apart`	`Evaluate`	`PlotPoints`
`AxesOrigin`	`ExpandAll`	`PlotRange`
`Cancel`	`GraphicsArray`	`PowerExpand`
`Collect`	`ImplicitPlot`	`ReplaceAll`
`ContourPlot`	`Integrate`	`Show`
`Contours`	`NDSolve`	`Simplify`
`ContourShading`	`NIntegrate`	`Solve`
`D`	`Part`	`Table`
`DisplayFunction`	`Partition`	`TableForm`
`DSolve`	`Plot`	

§2.1 Separation of Variables

A differential equation that can be written in the form $g(y)y' = f(x)$ is called a

separable differential equation. Rewriting $g(y)y' = f(x)$ in the form $g(y)\dfrac{dy}{dx} = f(x)$ yields

$g(y)\,dy = f(x)\,dx$ so that $\int g(y)\,dy = \int f(x)\,dx + C$, C a constant.

❏ EXAMPLE 2.1

Show that the equation $\dfrac{dy}{dx} = \dfrac{2y^{1/2} - 2y}{x}$ is separable and solve by separation of variables.

Solution:

The equation $\dfrac{dy}{dx} = \dfrac{2y^{1/2} - 2y}{x}$ is separable since it can be written in the form

$\dfrac{dy}{2y^{1/2} - 2y} = \dfrac{dx}{x}$. To solve the equation, integrate both sides and simplify.

Observe that $\int \dfrac{dy}{2y^{1/2} - 2y} = \int \dfrac{dx}{x} + C_1$ is the same as $\int \dfrac{1}{2y^{1/2}} \dfrac{dy}{\left(1 - y^{1/2}\right)} = \int \dfrac{dx}{x} + C_1$.

To evaluate the integral on the left – hand side, let $u = 1 - y^{1/2}$ so $du = \dfrac{-dy}{2y^{1/2}}$. We then obtain

8

$\int \frac{-du}{u} = \int \frac{dx}{x} + C_1$ so that $-Ln(u) = Ln(x) + C_1$. Recall that $-Ln(u) = Ln\left(\frac{1}{u}\right)$. Then

$\frac{1}{u} = Cx$ $\left(\text{where } C = e^{C_1}\right)$ and resubstituting we obtain that $\frac{1}{1 - y^{1/2}} = Cx$ is the general

solution of the equation $\frac{dy}{dx} = \frac{2y^{1/2} - 2y}{x}$. ■

■ EXAMPLE 2.2

Solve $y' = \frac{dy}{dx} = \frac{x^2 + 8}{\left(x^2 - 5x + 6\right) y^2 \cos(y)}$.

Solution:

We first try to solve the equation with **DSolve** by defining the equation and then entering the correct command. However, in this case **DSolve** is unable to solve this nonlinear equation, so we rewrite the equation

in the form $y^2 \cos(y) dy = \frac{x^2 + 8}{x^2 - 5x + 6} dx$. To solve the equation, we must integrate both the

left- and right-hand sides.

```
▤▢▭▭▭▭▭▭  SeparableEquations  ▭▭▭▭▭🗗▤

In[7]:=
equation=y'[x]==(x^2+8)/
    ((x^2-5x+6) y[x]^2 Cos[y[x]])

Out[7]=
              (8 + x^2) Sec[y[x]]
y'[x] == ─────────────────────────
           (6 - 5 x + x^2) y[x]^2

In[8]:=
DSolve[equation,y[x],x]

    Solve::ifun:
        Warning: Inverse functions are being used by Solve,
        so some solutions may not be found.

Out[8]=
                      (8 + x^2) Sec[y[x]]
DSolve[y'[x] == ─────────────────────────, y[x], x]
                  (6 - 5 x + x^2) y[x]^2
```

In the following, we define **lhs** to be $y^2 \cos(y)$ and **rhs** to be $\frac{x^2 + 8}{x^2 - 5x + 6}$ and then use

Integrate to integrate **lhs** (with respect to y) and **rhs** (with respect to x). We interpret the result to mean that the general solution of the equation $y^2 \cos(y) dy = \dfrac{x^2 + 8}{x^2 - 5x + 6} dx$ is

$$2y \cos(y) + \left(y^2 - 2\right) \sin(y) = x + Ln(x - 3) - 12 Ln(x - 2) + C.$$

```
In[11]:=
lhs=y^2 Cos[y];
rhs=(x^2+8)/(x^2-5x+6);
In[12]:=
Integrate[lhs,y]
Out[12]=
2 y Cos[y] + (-2 + y^2) Sin[y]
In[13]:=
Integrate[rhs,x]
Out[13]=
x + 17 Log[-3 + x] - 12 Log[-2 + x]
```

Note that if the problem had been solved by hand, integration by parts would be used to evaluate

$\int y^2 \cos(y) dy$ and partial fractions would be used to evaluate $\int \dfrac{x^2 + 8}{x^2 - 5x + 6} dx$. In fact, the

Apart command can be used to compute the partial fraction decomposition of $\dfrac{x^2 + 8}{x^2 - 5x + 6}$

which is $1 + \dfrac{17}{x - 3} - \dfrac{12}{x - 2}$.

```
In[14]:=
Apart[(x^2+8)/(x^2-5x+6)]
Out[14]=
        17      12
1 +  ------- - -------
     -3 + x    -2 + x
```
100%

■ EXAMPLE 2.3

Solve the equation $\dfrac{dy}{dx} = \dfrac{x^2}{\sqrt{9 - x^2} \, e^y \cos(y)}$ subject to $y(0) = 0$.

Solution:

Proceeding as in the previous example, we first define the equation and attempt to use **DSolve** to solve the equation.

```
┌──────────────────────── SeparableEquations ───────────────────────┐
│ In[1]:=                                                            │
│ equation=y'[x]==x^2/                                               │
│     (Sqrt[9-x^2] Exp[y[x]] Cos[y[x]])                              │
│                                                                    │
│ Out[1]=                                                            │
│                 x²  Sec[y[x]]                                      │
│ y'[x]  ==  ─────────────────────                                   │
│             E^y[x]  Sqrt[9 - x²]                                   │
│                                                                    │
│ In[2]:=                                                            │
│ DSolve[{equation,y[0]==0},y[x],x]                                  │
│                                                                    │
│     Solve::ifun:                                                   │
│         Warning: Inverse functions are being used by Solve,        │
│         so some solutions may not be found.                        │
│                                                                    │
│ Out[2]=                                                            │
│                      x²  Sec[y[x]]                                 │
│ DSolve[{y'[x]  ==  ─────────────────── ,  y[0] == 0}, y[x], x]     │
│                     E^y[x]  Sqrt[9 - x²]                           │
└────────────────────────────────────────────────────────────────────┘
```

Since **DSolve** is unsuccessful, we rewrite the equation in the form

$$e^y \cos(y)\, dy = \frac{x^2}{\sqrt{9-x^2}}\, dx,\ \text{define } \mathbf{lhs} \text{ to be } e^y \cos(y),\ \mathbf{rhs} \text{ to be } \frac{x^2}{\sqrt{9-x^2}},\ \text{and then use}$$

$\mathbf{Integrate}$ to compute $\int e^y \cos(y)\, dy, \int \dfrac{x^2}{\sqrt{9-x^2}}\, dx$, and name the results $\mathbf{slhs}$ and $\mathbf{srhs}$,

respectively.

```
┌────────────────────────────────────────────────────────────────────┐
│ In[2]:=                                                            │
│ lhs=Exp[y] Cos[y];                                                 │
│ rhs=x^2/Sqrt[9-x^2];                                               │
│                                                                    │
│ In[3]:=                                                            │
│ slhs=Integrate[lhs,y]                                              │
│                                                                    │
│ Out[3]=                                                            │
│  E^y Cos[y]     E^y Sin[y]                                         │
│  ─────────── +  ───────────                                        │
│       2              2                                             │
│                                                                    │
│ In[4]:=                                                            │
│ srhs=Integrate[rhs,x]                                              │
│                                                                    │
│ Out[4]=                                                            │
│   -(x Sqrt[9 - x²])     9 ArcSin[x/3]                              │
│   ─────────────────  +  ──────────────                            │
│          2                    2                                    │
└────────────────────────────────────────────────────────────────────┘
```

We interpret the results to mean that the general solution of the equation is

$$\frac{e^y\left(\cos(y)+\sin(y)\right)}{2}=\frac{9\sin^{-1}\left(\dfrac{x}{3}\right)-x\sqrt{9-x^2}}{2}+c$$ Therefore, to solve the problem we need to find

the value of c for which y(0)=0. We begin by defining **sol** to be the general solution of the equation.

```
In[5]:=
sol=slhs==srhs+c

Out[5]=

E^y Cos[y]   E^y Sin[y]                  x Sqrt[9 - x^2]   9 ArcSin[x/3]
---------- + ----------  == c  -  --------------- + -------------
    2            2                       2                 2
```

and then using **Solve** to find the value of c to make the equation true when x=0 and y=0.

```
In[6]:=
cval=Solve[sol /. y->0 /. x->0,c]

Out[6]=
{{c -> 1/2}}
```

The resulting value of c is $\dfrac{1}{2}$ so we conclude that the desired solution is

$$\frac{e^y\left(\cos(y)+\sin(y)\right)}{2}=\frac{1+9\sin^{-1}\left(\dfrac{x}{3}\right)-x\sqrt{9-x^2}}{2}.$$

```
In[7]:=
solution=sol /. cval[[1]]

Out[7]=

E^y Cos[y]   E^y Sin[y]        1    x Sqrt[9 - x^2]   9 ArcSin[x/3]
---------- + ----------  ==  ---  -  --------------- + -------------
    2            2             2           2                 2
```

The equation **solution** may be graphed with the command **ImplicitPlot** which is contained in the package **ImplicitPlot** which is located in the **Graphics** folder (or directory).

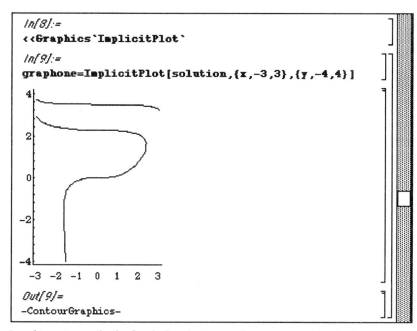

```
In[8]:=
<<Graphics`ImplicitPlot`

In[9]:=
graphone=ImplicitPlot[solution,{x,-3,3},{y,-4,4}]
```

```
Out[9]=
-ContourGraphics-
```

An alternate method of solution is to use the command **NDSolve** to attempt to locate a numerical solution of the problem on a given interval. In this case, we attempt to use **NDSolve** to locate a numerical approximation of the solution for $0 \leq x \leq 2.75$. However, the resulting solution is only valid on the interval [0,2.36544].

```
In[2]:=
altsol=NDSolve[{equation,y[0]==0},y[x],{x,0,2.75}]

      NDSolve::mxst:
         Maximum number of steps reached at the point 2.36544.
Out[2]=
{{y[x] -> InterpolatingFunction[{0., 2.36544}, <>][x]}}
```

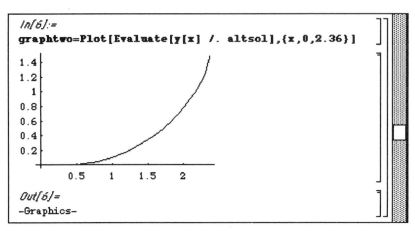

In[6]:=
graphtwo=Plot[Evaluate[y[x] /. altsol],{x,0,2.36}]

Out[6]=
-Graphics-

Finally, both solutions are displayed simultaneously as a **GraphicsArray**. Note the values of x used with each graph when comparing the two plots.

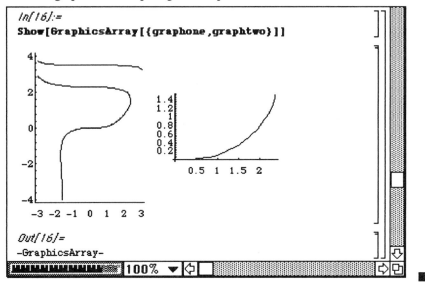

In[16]:=
Show[GraphicsArray[{graphone,graphtwo}]]

Out[16]=
-GraphicsArray-

100%

§2.2 Homogeneous Equations

A differential equation that can be written in the form M(x,y)dx+N(x,y)dy=0 where
$M(tx,ty) = t^n M(x,y)$ and $N(tx,ty) = t^n N(x,y)$ is called a **homogeneous differential**

equation (of degree n).

❑ EXAMPLE 2.4

Show that the equation $\left(x^2 + y\,x\right)dx - y^2dy = 0$ is homogeneous.

Solution:

Let $M(x,y) = x^2 + y\,x$ and $N(x,y) = -y^2$. Since

$$M(tx,ty) = (tx)^2 + (ty)(tx) = t^2\left(x^2 + yx\right) = t^2 M(x,y)$$

and $N(tx,ty) = -t^2y^2 = t^2 N(x,y)$, the equation $\left(x^2 + y\,x\right)dx - y^2dy = 0$ is

homogeneous of degree 2. ∎

Homogeneous equations can be reduced to separable equations by either the substitution y=ux or x=vy.

❑ EXAMPLE 2.5

Solve the equation $\left(x^2 - y^2\right)dx + x\,y\,dy = 0.$

Solution:

In this case, let $M(x,y) = x^2 - y^2$ and $N(x,y) = x\,y$. Then, $M(tx,ty) = t^2 M(x,y)$

and $N(tx,ty) = t^2 N(x,y)$ which means that $\left(x^2 - y^2\right)dx + x\,y\,dy = 0$ is a

homogeneous equation of degree 2.

Assume $x = v\,y$.

Then, $dx = v\,dy + y\,dv$ and

directly substituting into the equation and simplifying yields

$$0 = \left(x^2 - y^2\right)dx + x\,y\,dy = \left(v^2y^2 - y^2\right)(v\,dy + y\,dv) + v\,y\,y\,dy$$

$$= v^3y^2\,dy - y^2\,v\,dy + v^2y^3\,dv - y^3\,dv + vy^2\,dy$$

$$= \left(v^3y^2 - y^2v + vy^2\right)dy + \left(v^2y^3 - y^3\right)dv$$

$$= y^2\,v^3\,dy + y^3\left(v^2 - 1\right)dv. \text{ Dividing}$$

this equation by y^3v^3 yields the separable differential equation $\dfrac{dy}{y} + \dfrac{\left(v^2 - 1\right)dv}{v^3} = 0.$

We solve this equation by putting it in the form $\dfrac{dy}{y} = \dfrac{\left(1-v^2\right)dv}{v^3} = \left(\dfrac{1}{v^3} - \dfrac{1}{v}\right)dv$ and

integrating: $\text{Ln}(y) = \dfrac{-2}{v^2} - \text{Ln}(v) + C_1$ so that $\text{Ln}(vy) = \dfrac{-2}{v^2} + C_1$ and $vy = Ce^{-2/v^2}$

$\left(C = e^{C_1}\right)$. Since $x = vy$, $v = \dfrac{x}{y}$, and resubstituting into the above yields $x = Ce^{-2y^2/x^2}$

is the general solution of the equation $\left(x^2 - y^2\right)dx + xy\,dy = 0$. ■

■ EXAMPLE 2.6

Solve the equation $\left(x^{1/3}y^{2/3} + x\right)dx + \left(x^{2/3}y^{1/3} + y\right)dy = 0$.

Solution:

To show that the equation is homogeneous, we define $M(x,y) = x^{1/3}y^{2/3} + x$ and

$N(x,y) = x^{2/3}y^{1/3} + y$, and then compute $M(tx, ty)$. Note that *Mathematica* does not automatically

compute $(ab)^c = a^c b^c$ so the command **PowerExpand** is used to convert $(tx)^{1/3}(ty)^{2/3}$ to

$tx^{1/3}y^{2/3}$.

```
≡□≡════════ HomogeneousEquations ════════▢≡
In[2]:=
capm[x_,y_]=x^(1/3)y^(2/3)+x;
capn[x_,y_]=x^(2/3)y^(1/3)+y;

In[3]:=
capm[t x,t y]

Out[3]=
                1/3      2/3
t x + (t x)    (t y)

In[4]:=
stepone=PowerExpand[capm[t x, t y]]

Out[4]=
            1/3  2/3
t x + t x    y
```

To see that M(tx,ty)=tM(x,y), the command **Collect** is used to collect together the terms of **stepone** containing the same powers of t.

```
In[5]:=
steptwo=Collect[stepone,t]
Out[5]=
t (x + x^1/3 y^2/3)
```

Then, in the same manner as above, the procedure is repeated to see that N(tx,ty)=tN(x,y). We conclude that the equation is homogeneous of degree one.

```
In[6]:=
stepthree=PowerExpand[capn[t x, t y]]
Out[6]=
t x^2/3 y^1/3 + t y
In[7]:=
stepfour=Collect[stepthree,t]
Out[7]=
t (x^2/3 y^1/3 + y)
```

In this case, the command **DSolve** is used successfully to solve the equation.

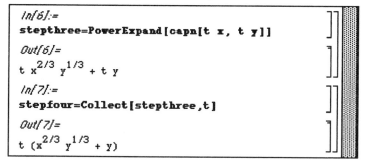

```
In[8]:=
sols=DSolve[
    capm[x,y[x]]+capn[x,y[x]]y'[x]==0,
    y[x],x]
Out[8]=
```
$$\{\{y[x] \rightarrow \frac{((-3\ x^{4/3} + 4\ C[1])^3)^{1/4}}{3^{3/4}}\},$$
$$\{y[x] \rightarrow \frac{I\ ((-3\ x^{4/3} + 4\ C[1])^3)^{1/4}}{3^{3/4}}\},$$
$$\{y[x] \rightarrow -(\frac{((-3\ x^{4/3} + 4\ C[1])^3)^{1/4}}{3^{3/4}})\},$$
$$\{y[x] \rightarrow \frac{-I\ ((-3\ x^{4/3} + 4\ C[1])^3)^{1/4}}{3^{3/4}}\}\}$$

However, we can obtain the same results by repeating the established algorithm to solve the homogeneous equation. We begin by defining **leq** to be the left-hand side of the equation. Note that the symbol Dt[x] corresponds to dx and the symbol Dt[y] corresponds to dy.

```
In[9]:=
Clear[x,y,u]

In[10]:=
leqone=capm[x,y] Dt[x]+capn[x,y]Dt[y]

Out[10]=
(x + x^(1/3) y^(2/3)) Dt[x] + (x^(2/3) y^(1/3) + y) Dt[y]
```

Then set **y=ux**. Note that by the product rule, dy=udx+xdu. Then set **leqtwo** to be the value of **leqone** after all products, including those of the form (ab)^p, are expanded.

```
In[11]:=
y=u x

Out[11]=
u x

In[12]:=
leqtwo=leqone//PowerExpand//ExpandAll

Out[12]=
u^(1/3) x^2 Dt[u] + u x^2 Dt[u] + x Dt[x] +
    u^(2/3) x Dt[x] + u^(4/3) x Dt[x] + u^2 x Dt[x]
```

We then combine together terms that share the factors **u**, **x**, **Dt[u]**, or **Dt[x]** and name the result **leqthree**. Note that we can consider **leqthree** to be equivalent to the mathematical expression $\left(u^{1/3} + u\right)x^2 du + \left(1 + u^{2/3} + u^{4/3} + u^2\right)x\, dx$.

```
In[13]:=
leqthree=Collect[leqtwo,{u,x,Dt[u],Dt[x]}]

Out[13]=
(u^(1/3) + u) x^2 Dt[u] +
    (1 + u^(2/3) + u^(4/3) + u^2) x Dt[x]
```

At this point, we must solve the separable equation $\left(u^{1/3} + u\right)x^2 du + \left(1 + u^{2/3} + u^{4/3} + u^2\right)x\, dx = 0$.

Note that the expression **leqthree**, corresponding to $\left(u^{1/3} + u\right)x^2 du + \left(1 + u^{2/3} + u^{4/3} + u^2\right)x\, dx$,

consists of parts. Namely, the first part of **leqthree**, obtained with the command **leqthree[[1]]**, is **u^(1/3)x^2Dt[u]**; the second part of **leqthree**, obtained with the command **leqthree[[2]]**, is **(1+u^(2/3)+u^(4/3)+u^2)xDt[u]**. Similarly, each part of **leqthree** is composed of subparts. **leqthree[[1,2]]** yields the second part of the first part of **leqthree** which is x^2; **leqthree[[2,1]]** yields the first part of the second part of **leqthree** which is $1+u^{2/3}+u^{4/2}+u^2$.

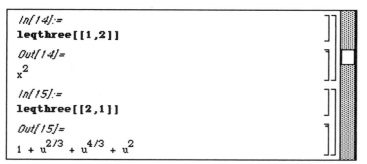

Since the equation we must solve is equivalent to **leqthree**=0, we divide **leqthree** by **leqthree[[1,2]]** and **leqthree[[2,1]]**.

```
leqfour=Cancel[
     Apart[leqthree/
          (leqthree[[1,2]]leqthree[[2,1]])]
          ]
```

Out[16]=

$$\frac{u^{1/3} Dt[u]}{1 + u^{4/3}} + \frac{Dt[x]}{x}$$

The result is equivalent to noticing that the separable equation

$\left(u^{1/3}\right)x^2 du + \left(1 + u^{2/3} + u^{4/3} + u^2\right)x\,dx = 0$ is equivalent to the equation

$\dfrac{u^{1/3}}{1 + u^{4/3}}\,du + \dfrac{1}{x}\,dx = 0$ which is the same as $\dfrac{u^{1/3}}{1 + u^{4/3}}\,du = -\dfrac{1}{x}\,dx.$

Again, noticing that **leqfour** is composed of parts we can avoid unnecessary retyping and compute the

integrals $\int \dfrac{u^{1/3}}{1 + u^{4/3}}\,du$ and $\int \dfrac{1}{x}\,dx$ by observing that $\dfrac{u^{1/3}}{1 + u^{4/3}}$ is the product of the first and second

parts of the first part of **leqfour** and $\dfrac{1}{x}$ is the first part of the second part of **leqfour**.

```
In[17]:=
Clear[x,u,y]

first=Integrate[
    leqfour[[1,1]]leqfour[[1,2]],u]
Out[18]=
3 Log[1 + u^{4/3}]
─────────────
      4
In[19]:=
second=Integrate[leqfour[[2,1]],x]
Out[19]=
Log[x]
```

Hence, the general solution of the differential equation $\dfrac{u^{1/3}}{1+u^{4/3}}\,du + \dfrac{1}{x}\,dx = 0$ is

$$\frac{3\text{Ln}\left(1+u^{4/3}\right)}{4} + \text{Ln}(x) = c, \text{ c any constant}$$

However, since y=ux, u=y/x so we resubstitute for u into the solution and simplify. Finally, we are able to conclude that the general solution of the desired equation is

$$\left[1+\left(\frac{y}{x}\right)^{4/3}\right]^{3/4} = \frac{c}{x}, \text{ c any constant}$$

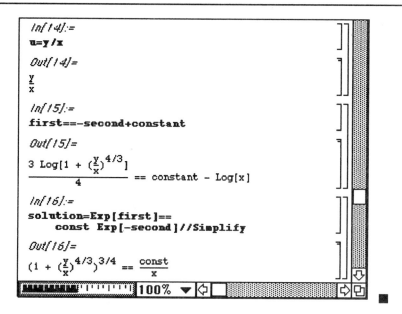

§2.3 Exact Equations

A differential equation that can be written in the form M(x,y)dx+N(x,y)dy=0 where
$\frac{\partial N}{\partial x} = \frac{\partial M}{\partial y}$ is called an **exact differential equation**.

❑ EXAMPLE 2.7

Show that the equation $2xy^3 dx + \left(1 + 3x^2 y^2\right) dy = 0$ is exact and that the equation $x^2 y\, dx + 5xy^2\, dy = 0$
is not exact.

Solution:

Since $\frac{\partial}{\partial y}\left(2xy^3\right) = 6xy^2 = \frac{\partial}{\partial x}\left(1 + 3x^2 y^2\right)$, the equation $2xy^3 dx + \left(1 + 3x^2 y^2\right) dy = 0$ is an exact

equation. On the other hand, the equation $x^2 y\, dx + 5xy^2\, dy = 0$ is not exact since

$\frac{\partial}{\partial y}\left(x^2 y\right) = x^2 \ne 5y^2 = \frac{\partial}{\partial x}\left(5xy^2\right)$. Note, however, that $x^2 y\, dx + 5xy^2\, dy = 0$ is separable. ∎

❑ EXAMPLE 2.8

Solve $\left(2x \sin(y) + 4e^x\right) dx + \left(x^2 \cos(y) - 1\right) dy = 0$ subject to $y(0) = \dfrac{1}{2}$.

Solution:

The equation $\left(2x \sin(y) + 4e^x\right) dx + \left(x^2 \cos(y) - 1\right) dy = 0$ is exact since

$\dfrac{\partial}{\partial y}\left(2x \sin(y) + 4e^x\right) = 2x \cos(y) = \dfrac{\partial}{\partial x}\left(x^2 \cos(y) - 1\right)$. Let $f(x,y)$ be a function

with $\dfrac{\partial f}{\partial x} = 2x \sin(y) + 4e^x$ and $\dfrac{\partial f}{\partial y} = x^2 \cos(y) - 1$. Then, $f(x,y) = \int 2x \sin(y) dx = x^2 \sin(y) + g(y)$.

Since $\dfrac{\partial f}{\partial y} = x^2 \cos(y) - 1$ and $\dfrac{\partial f}{\partial y} = x^2 \cos(y) + g'(y)$, $x^2 \cos(y) - 1 = x^2 \cos(y) + g'(y)$ so

$g'(y) = -1$ and thus $g(y) = -y + C_1$. Therefore $f(x,y) = x^2 \sin(y) - y + C_1$ and $x^2 \sin(y) - y = C$

is the general solution of the equation $\left(2x \sin(y) + 4e^x\right) dx + \left(x^2 \cos(y) - 1\right) dy = 0$.

Since our solution requires that $y(0) = \dfrac{1}{2}$, we obtain that $0^2 \sin\left(\dfrac{1}{2}\right) - \dfrac{1}{2} = C$ so that $C = -\dfrac{1}{2}$.

Therefore, the desired solution is $x^2 \sin(y) - y = -\dfrac{1}{2}$.

In this case, we can graph the resulting solution using the built-in command **ContourPlot** by observing that the level curve of the graph of the function of two variables

$x^2 \sin(y) - y + \dfrac{1}{2}$ corresponding to zero is the graph of the equation $x^2 \sin(y) - y = -\dfrac{1}{2}$.

In this case, the equation is graphed on the square $[-4\pi, 4\pi] \times [-4\pi, 4\pi]$. The option **ContourShading->False** is included to prevent the space between the resulting contours from being shaded; the option **Contours->{0}** specifies that the contour graphed correspond to zero; and the option **PlotPoints->90** is included to assure that 90 points are sampled so that the result is a smoother graph.

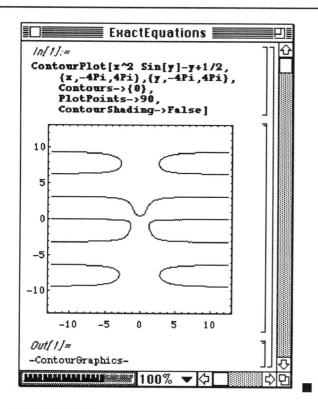

■ EXAMPLE 2.9

Find the general solution of the equation $\left(-1 + e^{xy}\,y + y\cos(x\,y)\right)dx + \left(1 + e^{xy}\,x + x\cos(x\,y)\right)dy = 0.$

Solution:

We begin by defining $m(x,y) = -1 + e^{xy}\,y + y\cos(x\,y),\, n(x,y) = 1 + e^{xy}\,x + x\cos(x\,y)$ and then trying to use **DSolve** to solve the equation.

```
≡□  ══════════════ ExactEquations ══════════════ □▯≡
 In[40]:=
 m[x_,y_]=-1+Exp[x y]y+y Cos[x y];
 n[x_,y_]=1+Exp[x y]x+x Cos[x y];

 In[41]:=
 equation=DSolve[
     m[x,y[x]]+n[x,y[x]] y'[x]==0,y[x],x]

     Solve::ifun:
         Warning: Inverse functions are being used by Solve
             so some solutions may not be found.

 Out[41]=
 DSolve[-1 + E^x y[x] y[x] + Cos[x y[x]] y[x] +
     (1 + E^x y[x] x + x Cos[x y[x]]) y'[x] == 0, y[x],
   x]
```

Since **DSolve** is unsuccessful, we show that the equation is exact since $\dfrac{\partial m}{\partial y} = \dfrac{\partial n}{\partial x}$.

```
 In[42]:=
 D[m[x,y],y]==D[n[x,y],x]

 Out[42]=
 True
```

We use **Integrate** to compute $\int m(x,y)\,dx$ and name the result **stepone**.

```
 In[43]:=
 stepone=Integrate[m[x,y],x]

 Out[43]=
 E^x y - x + Sin[x y]
```

The result means that the desired solution is of the form $e^{xy} - x + \sin(xy) + g(y)$. Therefore, we define **steptwo** to be the derivative of **stepone + g[y]** with respect to y.

```
 In[44]:=
 steptwo=D[stepone+g[y],y]

 Out[44]=
 E^x y x + x Cos[x y] + g'[y]
```

Since $\dfrac{\partial}{\partial y}\big($**stepone + g[y]**$\big) = n(x,y)$, we use the **Solve** command to find the value of g'(y).

```
In[45]:=
stepthree=Solve[steptwo==n[x,y],g'[y]]

Out[45]=
{{g'[y] -> 1}}
```

Therefore $g(y) = y + c$ and the general solution of the equation is $e^{xy} - x + y + \sin(xy) = c$.

```
In[46]:=
stepfour=Integrate[g'[y] /. stepthree[[1]],y]

Out[46]=
y

In[47]:=
solution=stepone+stepfour

Out[47]=
E^x y - x + y + Sin[x y]
```

In this case, we can graph various solutions with the command **ContourPlot** by observing that level of the function $e^{xy} - x + y + \sin(xy)$ correspond to graphs of $e^{xy} - x + y + \sin(xy) = c$ for various values of c.

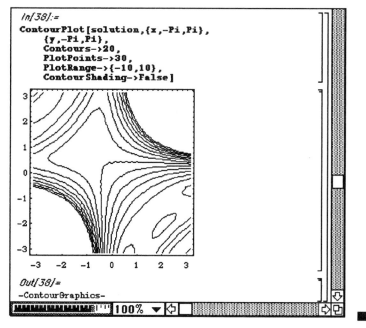

§2.4 Linear Equations

A differential equation that can be written in the form $\dfrac{dy}{dx} + p(x)y = q(x)$ is called a **first – order linear differential equation**.

If $\dfrac{dy}{dx} + p(x)y = q(x)$, then $e^{\int p(x)\,dx}\dfrac{dy}{dx} + e^{\int p(x)\,dx}p(x)y = e^{\int p(x)\,dx}q(x)$. By the Product rule

and Fundamental Theorem of Calculus, $\dfrac{d}{dx}\left(e^{\int p(x)\,dx}y\right) = e^{\int p(x)\,dx}\dfrac{dy}{dx} + e^{\int p(x)\,dx}p(x)y$ so

$\dfrac{d}{dx}\left(e^{\int p(x)\,dx}y\right) = e^{\int p(x)\,dx}q(x)$. Integrating, we obtain $e^{\int p(x)\,dx}y = \int e^{\int p(x)\,dx}q(x)\,dx$ and

dividing by $e^{\int p(x)\,dx}$ yields $y = \dfrac{\int e^{\int p(x)\,dx}q(x)\,dx}{e^{\int p(x)\,dx}}$.

❏ **EXAMPLE 2.10**

Find the general solution of $x\dfrac{dy}{dx} + 3y = x\sin(x)$.

Solution:

Dividing the equation by x yields $\dfrac{dy}{dx} + \dfrac{1}{x}y = \sin(x)$ $\left(\text{where } p(x) = \dfrac{1}{x} \text{ and } q(x) = \sin(x)\right)$. Then,

$e^{\int \frac{1}{x}\,dx} = e^{\text{Ln}\,|x|} = x$, for $x > 0$, and $\dfrac{d}{dx}(xy) = x\dfrac{dy}{dx} + y = x\sin(x)$ so $xy = \int x\sin(x)\,dx$.

Using the Integration by Parts formula, $\int u\,dv = uv - \int v\,du$, with $u = x$ and $dv = \sin(x)$, we obtain

$du = dx$ and $v = -\cos(x)$ so $xy = \int x\sin(x)\,dx = -x\cos(x) + \int \cos(x)\,dx = -x\cos(x) + \sin(x) + C$.

Therefore, the general solution of the equation $x\dfrac{dy}{dx} + 3y = x\sin(x)$ for $x > 0$ is

$y = \dfrac{-x\cos(x) + \sin(x) + C}{x}$.

To graph the solution of the equation for various values of c, we first define **y** (as a function of c) and then create a table of values of **y[c]** for c=−3,−2,−1,0,1,2,3 and name the resulting table **sols**.

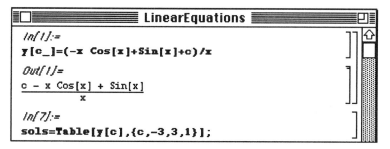

Then **sols** is graphed on the interval $[.01, 5/2\,\pi]$. 0 is avoided since the solutions are undefined when x=0 but the option **AxesOrigin->{0,0}** is included to guarantee that the x- and y-axes intersect at the point (0,0). Notice that the solution corresponding to c=0 is not unbounded, like the other solutions, near x=0.

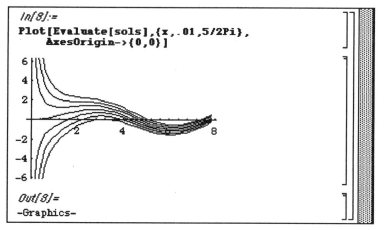

In fact, even though $y[0] = \dfrac{-x\cos(x) + \sin(x)}{x}$ is undefined when $x = 0$,

$$\operatorname*{Lim}_{x\to 0} y[0] = \operatorname*{Lim}_{x\to 0} \frac{-x\cos(x) + \sin(x)}{x} = \operatorname*{Lim}_{x\to 0}\left(-\cos(x) + \frac{\sin(x)}{x}\right) = -1 + 1 = 0.$$ Consequently,

even though *Mathematica* generates appropriate error messages (which are not completely displayed in this case) when the **Plot** command is entered, the resulting graph is correctly displayed.

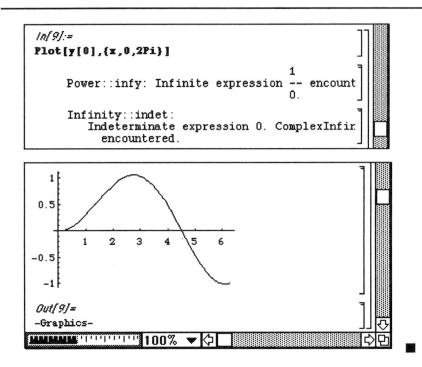

```
In[9]:=
Plot[y[0],{x,0,2Pi}]

                                            1
          Power::infy: Infinite expression -- encount
                                            0.

          Infinity::indet:
               Indeterminate expression 0. ComplexInfir
                    encountered.
```

```
Out[9]=
-Graphics-
```

■ EXAMPLE 2.11

Compare the solutions of $\dfrac{dy}{dx} + y = f(x)$ subject to $y(0) = 0$ where $f(x) = x$, $\sin(x)$, $\cos(x)$, and e^x.

Solution:

Mathematica is able to compute the solution of any first-order linear differential equation
$\dfrac{dy}{dx} + p(x)y = q(x)$ as long as it can compute the integrals $\int p(x)\,dx$ and $\int e^{\int p(x)\,dx} q(x)\,dx$.

LinearEquations

```
In[17]:=
DSolve[y'[x]+p[x] y[x]==q[x],y[x],x]

Out[17]=
{{y[x] ->
                    Integrate[p[x], x]
     C[1] + Integrate[E                q[x], x]
     ───────────────────────────────────────────}}
              Integrate[p[x], x]
             E
```

In this case, $p(x) = 1$ and $q(x) = x$, $\sin(x)$, $\cos(x)$, and e^x.

To compute each solution, the table **funs** is first defined and then the **Table** and **DSolve** commands are used to find the solution of the differential equation
$y' + y = $ **fun s[[i]]** subject to $y(0) = 0$ for $i = 1, 2, 3$, and 4, where **fun s[[i]]** represents the ith element of the list of functions **fun s**.

The resulting list of functions is named **sols** and is expressed in **TableForm**.

```
In[48]:=
funs={x,Sin[x],Cos[x],Exp[x]};

In[50]:=
sols=Table[
        DSolve[{y'[x]+y[x]==funs[[i]],y[0]==0},
        y[x],x],{i,1,4}];
TableForm[sols]

Out[50]//TableForm=

y[x] -> -1 + E^-x + x
y[x] -> 1/(2 E^x) - Cos[x]/2 + Sin[x]/2
y[x] -> -1/(2 E^x) + Cos[x]/2 + Sin[x]/2
y[x] -> -1/(2 E^x) + E^x/2
```

Observe that the first element of **sols** is the list:

{y[x] -> -1+E^{-x}+x} which can be obtained with the command **sols[[1]]**. To evaluate the expression $-1+E^{-x}+x$ for explicit values of x, we must either reenter the expression or extract it from sols. One way of extracting the expression $-1+E^{-x}+x$ from the list sols is to enter

y[x] /. sols[[1]] which replaces **y[x]** by the value $-1+E^{-x}+x$ or to enter **sols[[1,1,2]]**.

To graph each of the explicit solutions in **sols**, we must extract the explicit solutions. One way of extracting the solutions is to create a table of values of **y[x]** where **y[x]** is replaced by the rule specified in the ith element of **sols** as done in the following command. The resulting list of functions is named **toplot** for future use. Alternatively, **sols[[i,1,2]]** explicitly yields the ith function in the list of solutions **sols**.

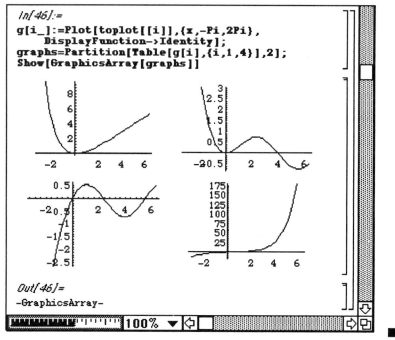

In[28]:=
```
toplot=Table[y[x] /. sols[[i,1]],{i,1,4}]
```
Out[28]=

$$\{-1 + E^{-x} + x, \frac{1}{2 E^x} - \frac{Cos[x]}{2} + \frac{Sin[x]}{2},$$

$$\frac{-1}{2 E^x} + \frac{Cos[x]}{2} + \frac{Sin[x]}{2}, \frac{-1}{2 E^x} + \frac{E^x}{2}\}$$

Finally, each function in **toplot** is graphed on the interval [-π,2π]. The resulting four graphics objects are displayed as a graphics array for easy comparison.

In[46]:=
```
g[i_]:=Plot[toplot[[i]],{x,-Pi,2Pi},
    DisplayFunction->Identity];
graphs=Partition[Table[g[i],{i,1,4}],2];
Show[GraphicsArray[graphs]]
```

Out[46]=
-GraphicsArray-

100% ▼

■ **EXAMPLE 2.12**

Find the general solution of $\dfrac{dy}{dx} - \dfrac{4x}{x^2+1} y = \left(1 + x^2\right)^3 e^x$.

Solution:

In this case, **DSolve** computes the general solution of the equation.

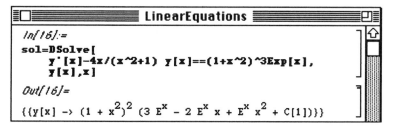

Nevertheless, we can also use the general method to directly construct a solution. We begin by computing $e^{\int \frac{-4x}{x^2+1}\,dx}$ and naming the result **intfac**:

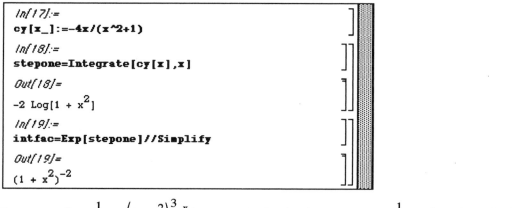

To compute $\int \frac{1}{\left(1+x^2\right)^2}\left(1+x^2\right)^3 e^x dx$, we use the fact that **intfac** $= \frac{1}{\left(1+x^2\right)^2}$ from above; for later use, we name the result **steptwo**. Finally, we are able to construct the general solution

$$\frac{e^x\left(3-2x+x^2\right)+c}{1/\left(1+x^2\right)} = \frac{\texttt{steptwo}+\texttt{c[i]}}{\texttt{intfac}} = \left(1+x^2\right)\left[e^x\left(3-2x+x^2\right)+c\right].$$ We then name the

solution **sol** for later use.

```
In[20]:=
steptwo=Integrate[intfac (1+x^2)^3 Exp[x],x]

Out[20]=
E^x (3 - 2 x + x^2)

In[21]:=
sol=(steptwo+c[i])/intfac // Simplify

Out[21]=
(1 + x^2)^2 (E^x (3 - 2 x + x^2) + c[i])
```

Finally, we will graph the solution for various values of **c[i]**. The following command creates the table of functions obtained by replacing **c[i]** by **i** for i = −2, −1, 0, 1, and 2. The resulting table of functions is named **soltab** and is not displayed since a semicolon is placed at the end of the command.

```
In[22]:=
soltab=Table[sol /. c[i]->i,{i,-2,2}];
```

soltab is graphed on the interval [-1,1] with the commands **Plot** and **Evaluate**.

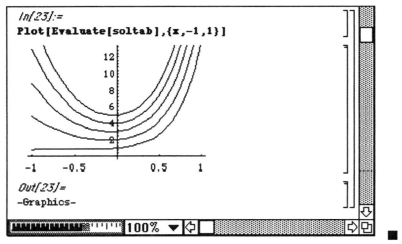

```
In[23]:=
Plot[Evaluate[soltab],{x,-1,1}]



Out[23]=
-Graphics-
```

100% ▼

■ EXAMPLE 2.13

Graph the solution of the equation $\dfrac{dy}{dx} + \dfrac{x}{2 + \sin(x)} y = e^{-x} \cos(x)$ subject to $y(0) = 0$ for $\dfrac{-\pi}{2} \le x \le \dfrac{3\pi}{2}$.

Solution:

Since *Mathematica* cannot compute $\displaystyle\int \dfrac{x\,dx}{2 + \sin(x)}$, **DSolve** cannot be used to produce a general solution of the equation.

```
▤□▤▤▤▤▤▤▤▤▤▤▤ LinearEquations ▤▤▤▤▤▤▤▤▤▤▤▤▤▤◨▤
In/2/:=
Integrate[x/(2+Sin[x]),x]

Out/2/=
Integrate[─────────, x]
          2 + Sin[x]
```

However, since the desired solution is given by the formula

$$y(x) = \frac{\int_0^x e^{\int_0^s \frac{t\,dt}{2+\sin(t)}} e^{-s} \cos(s)\,ds}{e^{\int_0^x \frac{t\,dt}{2+\sin(t)}}}, \text{ the solution can be numerically calculated for any value of } x$$

using the built-in command **NIntegrate**. The following commands define **if[x]**

to be an approximation of $\int_0^x \frac{t\,dt}{2+\sin(t)}$, **num[x]** to be an approximation of $\int_0^x e^{if[t]} e^{-t} \cos(t)\,dt$,

and **sol[x]**, the desired approximation of the solution, to be $\dfrac{\textbf{num[x]}}{\textbf{Exp[if[x]]}}$.

```
In/3/:=
if[x_]:=if[x]=NIntegrate[t/(2+Sin[t]),{t,0,x}]

In/4/:=
num[x_]:=NIntegrate[Exp[if[t]] Exp[-t]Cos[t],
      {t,0,x}]

In/5/:=
sol[x_]:=num[x]/Exp[if[x]]
```

We can then use the **Plot** command to graph **sol[x]** for $\dfrac{-\pi}{2} \le x \le \dfrac{3\pi}{2}$.

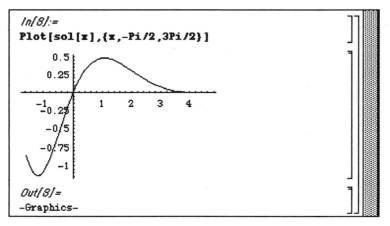

An alternative solution is to use the command **NDSolve** to approximate a solution of the equation on the desired interval. In this case, the resulting solution is expressed as an interpolating function, which has domain [-1.5708,4.71239], and named **altsol**.

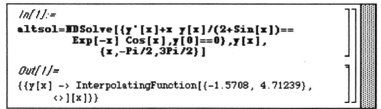

We can then graph the result.

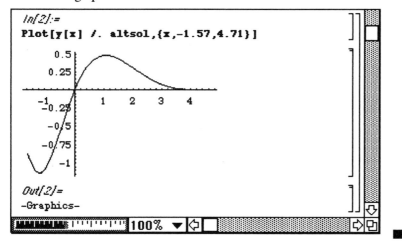

§2.5 Some Special First-Order Differential Equations

A **Bernoulli equation** is a nonlinear equation of the form $y'+p(x)y = q(x)y^n$.

If $n \notin \{0,1\}$, let $w = y^{1-n}$. Then, $\dfrac{dw}{dx} = (1-n)y^{-n}\dfrac{dy}{dx}$ and substituting into the equation

$y'+p(x)y = q(x)y^n$ yields $\dfrac{y^n}{1-n}\dfrac{dw}{dx} + p(x)y^n w = q(x)y^n$ and multiplying by $\dfrac{1-n}{y^n}$ yields

the linear equation $\dfrac{dw}{dx} + (1-n)p(x)w = q(x)$.

■ EXAMPLE 2.14

Solve (a) $y'-2y = \dfrac{2x\sin(3x)}{y}$ and (b) $y'+\dfrac{1}{2}y = \dfrac{-x}{2}\sin(3x)y^3$.

Solution:

Both (a) and (b) are Bernoulli equations with n=3 and –1, respectively. For (a), letting

$w = y^{1-(-1)} = y^2$ changes $y'-2y = \dfrac{2x\sin(3x)}{y}$ into the linear equation $2w'-2w = 2x\sin(3x)$

and dividing by 2 yields $w'-w = x\sin(3x)$. Integrating we obtain $e^{\int -1\,dx} = e^{-x}$ and, computed below,

$\int x e^{-x}\sin(3x)\,dx = \dfrac{(4-5x)\sin(3x) - 3(-1+5x)\cos(3x)}{50e^x}$.

```
┌─■□▤▤▤▤▤▤▤▤ BernoulliExample ▤▤▤▤▤▤□▯▤
│ In[142]:=
│ stepone=Integrate[x Exp[-x] Sin[3x],x]
│
│ Out[142]=
│  -3 (1 + 5 x) Cos[3 x]     (4 - 5 x) Sin[3 x]
│  ─────────────────────  +  ──────────────────
│        50 E^x                    50 E^x
└
```

Therefore the general solution of $w'-w = x\sin(3x)$

is $w(x) = \dfrac{(4-5x)\sin(3x) - 3(-1+5x)\cos(3x)}{50e^x} + ce^{-x} = \big[(4-5x)\sin(3x) - 3(-1+5x)\cos(3x) + c\big]e^{-x}$.

```
In[143]:=
steptwo=stepone+c Exp[-x]
```

```
Out[143]=
 c     3 (1 + 5 x) Cos[3 x]   (4 - 5 x) Sin[3 x]
--- -  --------------------- + -----------------
  x           50 E
 E
```

Finally $y(x) = w(x)^{1/2} = \left|(4 - 5x)\sin(3x) - 3(-1 + 5x)\cos(3x) + c\right|^{1/2} e^{-x/2}$.

```
In[144]:=
solutionone=steptwo^(1/2)//PowerExpand
```

```
Out[144]=
       c     3 (1 + 5 x) Cos[3 x]
Sqrt[ --- -  --------------------- +
        x           50 E
       E
      (4 - 5 x) Sin[3 x]
      ------------------- ]
            50 E
```

For (b), letting $w = y^{1-(-1)} = y^2$ changes $y' - 2y = \dfrac{2x \sin(3x)}{y}$ into the linear equation

$2w' - 2w = 2x \sin(3x)$. Since the general solution of $w' - w = x \sin(3x)$ is

$w(x) = \mathbf{steptwo} = \left|(4 - 5x)\sin(3x) - 3(-1 + 5x)\cos(3x) + c\right|e^{-x}$ from above, we obtain

$y(x) = w(x)^{-1/2} = \left|(4 - 5x)\sin(3x) - 3(-1 + 5x)\cos(3x) + c\right|^{-1/2} e^{x/2}$.

```
In[145]:=
solutiontwo=steptwo^(-1/2)//PowerExpand;
```

```
In[146]:=
cs=Range[-5,7]
```

```
Out[146]=
{-5, -4, -3, -2, -1, 0, 1, 2, 3, 4, 5, 6, 7}
```

```
In[147]:=
tographone=Table[solutionone /. c->cs[[i]],
                 {i,1,13}];
```

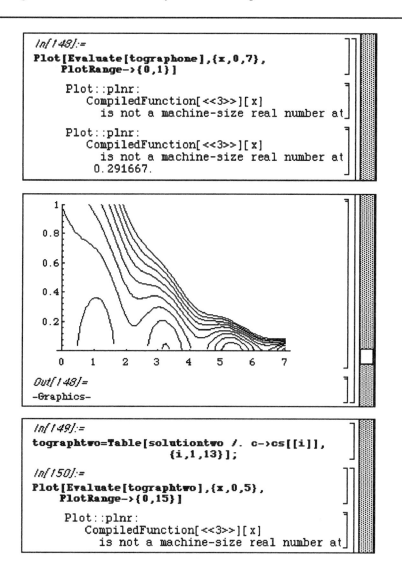

```
In[148]:=
Plot[Evaluate[tographone],{x,0,7},
    PlotRange->{0,1}]

    Plot::plnr:
        CompiledFunction[<<3>>][x]
            is not a machine-size real number at

    Plot::plnr:
        CompiledFunction[<<3>>][x]
            is not a machine-size real number at
            0.291667.
```

```
Out[148]=
-Graphics-
```

```
In[149]:=
tographtwo=Table[solutiontwo /. c->cs[[i]],
                  {i,1,13}];
In[150]:=
Plot[Evaluate[tographtwo],{x,0,5},
    PlotRange->{0,15}]

    Plot::plnr:
        CompiledFunction[<<3>>][x]
            is not a machine-size real number at
```

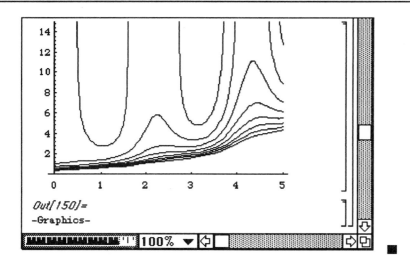

Out[150]=
-Graphics-

Equations of the form $f(x y'-y) = g(y')$ are called **Clairaut equations**.

A general solution of $f(xy'-y) = g(y')$ is given by $f(x c - y) = g(c)$, c any constant.

■ EXAMPLE 2.15

Solve the equation $2(xy'(x) - y(x))^2 = (y'(x))^2 - y'(x)$.

Solution:

We identify the equation as a Clairaut equation with $f(x)=2x^2+1$ and $g(x)=x^2-x$. We begin by defining f and g:

```
ClairautExample
In[75]:=
Clear[f,g,h]
f[x_]=2x^2+1;
g[x_]=x^2-x;
```

and then computing $f(x y'(x) - y(x))$ and $g(y'(x))$ to verify that the equation is a Clairaut equation of this form.

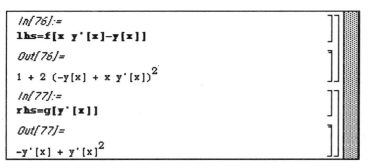

An implicit solution is given by $f(xc - y) = g(c)$, c any constant, which is the equation
$1 + 2(cx - y)^2 = c^2 - c$, computed below and named **implicit**.

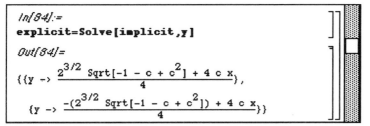

Explicit solutions are calculated by solving the equation **implicit** for y; the resulting list is named
explicit.

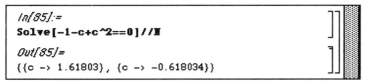

$\sqrt{-1 - c + c^2}$ is not defined when $-1 - c + c^2 < 0$ which is when c is in (approximately) the
interval $(-.62, 1.62)$.

```
In[85]:=
Solve[-1-c+c^2==0]//N
Out[85]=
{{c -> 1.61803}, {c -> -0.618034}}
```

The explicit solutions are extracted from **explicit** with the commands **explicit[[1,1,2]]** and
explicit[[2,1,2]] as shown on the next page:

```
In[87]:=
explicit[[1,1,2]]
explicit[[2,1,2]]
Out[86]=
 3/2               2
2    Sqrt[-1 - c + c ] + 4 c x
------------------------------
              4
Out[87]=
   3/2               2
-(2    Sqrt[-1 - c + c ]) + 4 c x
---------------------------------
                4
```

We will graph the explicit solutions for various values of c. We begin by using **Range** and **Union** to create the list of numbers –5, –4, –3, –2, –1, 2, 4, 6, 8, 10, 12, and 14. Note that none of these numbers lies in the interval (–.62,1.62).

```
In[88]:=
cs=Union[Range[-5,-1],Range[2,14,2]]
Out[88]=
{-5, -4, -3, -2, -1, 2, 4, 6, 8, 10, 12, 14}
```

We then create a table of functions, **tograph**, which consists of replacing c in each explicit solution by the ith element of **cs** for i=1, 2, 3, 4, 5, 6, 7, and 8. Since the resulting list is a list of ordered functions, we remove parentheses with the command **Flatten** so that the result is a list of functions:

```
In[89]:=
tograph=Table[y /. explicit /. c->cs[[i]],
    {i,1,8}]//Flatten
Out[89]=
 Sqrt[232] - 20 x  -Sqrt[232] - 20 x
{----------------, -----------------,
        4                  4
  Sqrt[152] - 16 x  -Sqrt[152] - 16 x
  ----------------, -----------------,
         4                  4
  Sqrt[88] - 12 x  -Sqrt[88] - 12 x
  ---------------, ----------------,
         4                 4
                                      3/2
  Sqrt[40] - 8 x  -Sqrt[40] - 8 x   2    - 4 x
  --------------, ---------------, ----------,
        4                4              4
    3/2         3/2          3/2
  -2    - 4 x  2    + 8 x  -2    + 8 x
  ----------, ----------, -----------,
       4           4           4
  Sqrt[88] + 16 x  -Sqrt[88] + 16 x
  ---------------, ----------------,
         4                 4
  Sqrt[232] + 24 x  -Sqrt[232] + 24 x
  ----------------, -----------------}
         4                  4
```

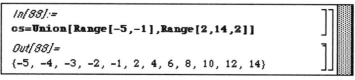

which is then graphed on the interval [–2,2]. In this case, the option **PlotRange->{-15,15}** is used to indicate that the range of y-values displayed is [–15,15].

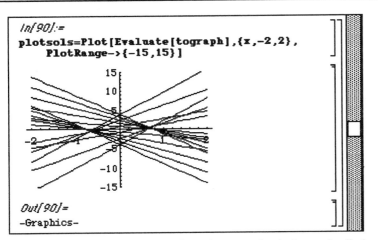

```
In[90]:=
plotsols=Plot[Evaluate[tograph],{x,-2,2},
    PlotRange->{-15,15}]
```

```
Out[90]=
-Graphics-
```

Another solution, not obtainable from the general solution and called a **singular solution**, is obtained by differentiating both **lhs** and **rhs** with respect to x.

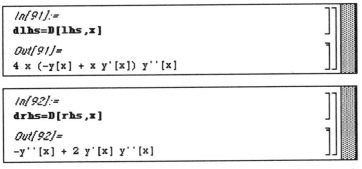

```
In[91]:=
dlhs=D[lhs,x]
```

```
Out[91]=
4 x (-y[x] + x y'[x]) y''[x]
```

```
In[92]:=
drhs=D[rhs,x]
```

```
Out[92]=
-y''[x] + 2 y'[x] y''[x]
```

We obtain the equation $4x\, y''(x)\big(x\, y'(x) - y(x)\big) = y''(x)\big(2y'(x) - 1\big)$. Factoring yields the equation $y''(x)\big(1 - 4x\, y(x) - 2y'(x) + 4x^2 y'(x)\big) = 0$. If $y''(x) \neq 0$ (note that in the general solution obtained above, $y''(x) = 0$), then $1 - 4x\, y(x) - 2y'(x) + 4x^2 y'(x) = 0$. $1 - 4x\, y(x) - 2y'(x) + 4x^2 y'(x)$ is then extracted from **stepone** with **stepone[[1]]**.

```
In[93]:=
stepone=Factor[dlhs-drhs]

Out[93]=

(1 - 4 x y[x] - 2 y'[x] + 4 x² y'[x]) y''[x]

In[94]:=
stepone[[1]]

Out[94]=

1 - 4 x y[x] - 2 y'[x] + 4 x² y'[x]
```

We then use **DSolve** to solve $1 - 4x\,y(x) - 2y'(x) + 4x^2 y'(x) = 0$ and name the resulting list **singular**.

```
In[95]:=
singular=DSolve[stepone[[1]]==0,y[x],x]

Out[95]=

{{y[x] -> x/2 + Sqrt[-1 + 2 x²] C[1]}}
```

The solution $\dfrac{x}{2} + C_1\sqrt{2x^2 - 1}$ is extracted from **singular** with the command **singular[[1, 1, 2]]** and named **sol**.

```
In[96]:=
sol=singular[[1,1,2]]

Out[96]=

x/2 + Sqrt[-1 + 2 x²] C[1]
```

Since **sol** must satisfy the equation $2\big(xy'(x) - y(x)\big)^2 = \big(y'(x)\big)^2 - y'(x)$, we compute $\dfrac{d}{dx}$ **sol** and $\dfrac{d^2}{dx^2}$ **sol**, naming the results **dsol** and **d2sol**, respectively:

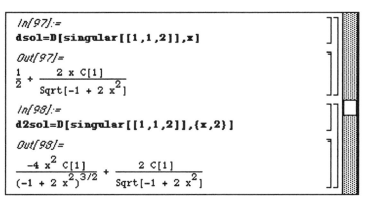

```
In[97]:=
dsol=D[singular[[1,1,2]],x]

Out[97]=
 1     2 x C[1]
 -  + ------------
 2            2
      Sqrt[-1 + 2 x ]

In[98]:=
d2sol=D[singular[[1,1,2]],{x,2}]

Out[98]=
        2
  -4 x  C[1]         2 C[1]
 ------------- + ------------
        2 3/2             2
 (-1 + 2 x )     Sqrt[-1 + 2 x ]
```

and then replace **y[x]** by **sol**, **y'[x]** by **dsol**, and **y''[x]** by **d2sol** in the equation **lhs==rhs**. We then solve the resulting equation, **tosolve**, for **C[1]** and name the resulting list of solutions **roots**.

```
In[99]:=
tosolve=lhs==rhs /. {y[x]->sol,y'[x]->dsol,
            y''[x]->d2sol}

Out[99]=
                -x               2
 1 + 2 Power[--- - Sqrt[-1 + 2 x ] C[1] +
              2

         1     2 x C[1]
     x (- + ------------), 2] ==
         2           2
           Sqrt[-1 + 2 x ]
    1      2 x C[1]       1     2 x C[1]   2
 -(-) - ------------ + (- + ------------)
    2            2      2           2
        Sqrt[-1 + 2 x ]      Sqrt[-1 + 2 x ]

In[100]:=
roots=Solve[tosolve,C[1]]

Out[100]=
            5                   5
 {{C[1] -> Sqrt[-]}, {C[1] -> -Sqrt[-]}}
               8                   8
```

Therefore the singular solutions are $y(x) = \dfrac{x}{2} \pm \sqrt{\dfrac{5}{8}} \sqrt{2x^2 - 1}$.

In the following, we obtain the singular solutions by replacing the constants in the list **singular** by the values obtained in **roots** and name the resulting list **singgraphs**.

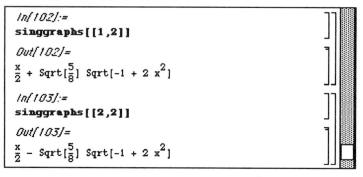

The explicit solutions are extracted from **singgraphs** with the commands **singgraphs[[1,2]]** and **singgraphs[[2,2]]**.

```
In[102]:=
singgraphs[[1,2]]
Out[102]=
x/2 + Sqrt[5/8] Sqrt[-1 + 2 x^2]
In[103]:=
singgraphs[[2,2]]
Out[103]=
x/2 - Sqrt[5/8] Sqrt[-1 + 2 x^2]
```

The singular solutions are not defined when $2x^2 - 1 < 0$ which occurs (approximately) on the interval $(-.708, .708)$.

```
In[104]:=
Solve[-1+2x^2==0]//N
Out[104]=
{{x -> 0.707107}, {x -> -0.707107}}
```

Finally, we graph the singular solutions on the intervals $(.708, 3)$ and $(-3, -.708)$ and then show the resulting graphs simultaneously.

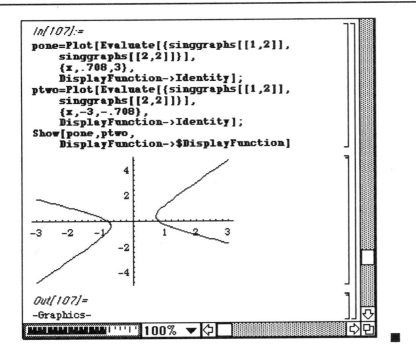

```
In[107]:=
pone=Plot[Evaluate[{singgraphs[[1,2]],
    singgraphs[[2,2]]}],
    {x,.708,3},
    DisplayFunction->Identity];
ptwo=Plot[Evaluate[{singgraphs[[1,2]],
    singgraphs[[2,2]]}],
    {x,-3,-.708},
    DisplayFunction->Identity];
Show[pone,ptwo,
    DisplayFunction->$DisplayFunction]
```

```
Out[107]=
-Graphics-
```

Equations of the form $y = x f(y') + g(y')$ are called **Lagrange equations.**

Let $p = y'(x)$. Then, differentiating $y = x f(y') + g(y')$ with respect to x yields

$y' = x f'(y') y'' + f(y') + g'(y') y''$ and substituting p into the equation yields

$p = x f'(p)\dfrac{dp}{dx} + f(p) + g'(p)\dfrac{dp}{dx} = f(p) + \dfrac{dp}{dx}\left[x f'(p) + g'(p)\right]$. Solving for $\dfrac{dx}{dp}$ yields the linear

equation $\dfrac{dx}{dp} = \dfrac{x f'(p) + g'(p)}{p - f(p)}$ which is equivalent to $\dfrac{dx}{dp} + \dfrac{f'(p)}{f(p) - p} x = \dfrac{g'(p)}{p - f(p)}$.

■ EXAMPLE 2.16

Solve the equation $y = x\left(\dfrac{dy}{dx}\right)^2 + 3\left(\dfrac{dy}{dx}\right)^2 - 2\left(\dfrac{dy}{dx}\right)^3$.

Solution:

We begin by observing that the equation is a Lagrange equation with $f(x)=x^2$ and $g(x)=3x^2-2x^3$ and then defining f and g.

page 46, Chapter 2

Let $p = \dfrac{dy}{dx}$. Then, differentiating $y = x\left(\dfrac{dy}{dx}\right)^2 + 3\left(\dfrac{dy}{dx}\right)^2 - 2\left(\dfrac{dy}{dx}\right)^3$ with respect to x, substituting

$\dfrac{dy}{dx}$ by p, and solving for $\dfrac{dx}{dp}$ yields the equation $\dfrac{dx}{dp} - x\left(\dfrac{f'(p)}{p - f(p)}\right) = \dfrac{g'(p)}{p - f(p)}$. Below we compute

$\dfrac{-f'(p)}{p - f(p)} = \dfrac{-2p}{p - p^2}$, name the result **stepone**, and then compute and simplify $\dfrac{g'(p)}{p - f(p)} = 6$

and name the result **stepthree**.

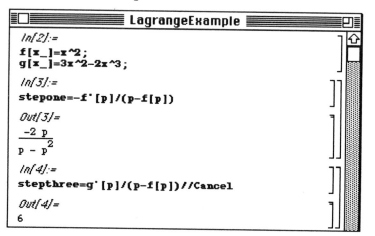

We then compute $e^{\int \mathbf{stepone}\,dp} = \dfrac{1}{(p-1)^2}$, name the result **steptwo**, compute

$\int \mathbf{steptwo\ stepthree}\,dp = \int 6e^{\int \mathbf{stepone}\,dp} = 6p - 6p^2 + 2p^2$, and name the result **stepfour**.

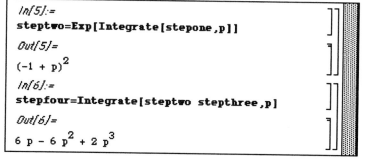

Then the general solution of $\dfrac{dx}{dp} - x\left(\dfrac{f'(p)}{p - f(p)}\right) = \dfrac{dx}{dp} - \dfrac{2p}{p - p^2} x = 6$ is

$$x(p) = \textbf{stepfour} + c\,\textbf{steptwo} = \int e^{\int \frac{2p}{p^2 - p}\,dp}\, 6\,dp + c\!\int \dfrac{2p}{p^2 - p}\,dp$$

$$= c(p - 1)^2 + 6p - 6p^2 + 2p^3.$$

Since $y(p) = x(p)f(p) + g(p)$ we obtain $y(p) = 3p^2 - 2p^3 + p^2\!\left(c(p - 1)^2 + 6p - 6p^2 + 2p^3\right)$.

```
In[7]:=
x[p_]=stepfour+c steptwo
Out[7]=
        2               2       3
c (-1 + p)  + 6 p - 6 p  + 2 p
In[8]:=
y[p_]=x[p] f[p]+g[p]
Out[8]=
     2      3    2            2               2       3
3 p  - 2 p  + p  (c (-1 + p)  + 6 p - 6 p  + 2 p )
```

Below we graph the parametric equations $\begin{cases} x(p) \\ y(p) \end{cases}$ for various values of c by first defining **csone** to be the set of numbers $-10, -8, \ldots, 8, 10$, then defining **tographone** to be a list of the functions $\{x(p),y(p)\}$ in which c is replaced by the ith element of **csone** for i=1, 2, ... ,11, and then finally graphing the list of functions **tographone** with **ParametricPlot**.

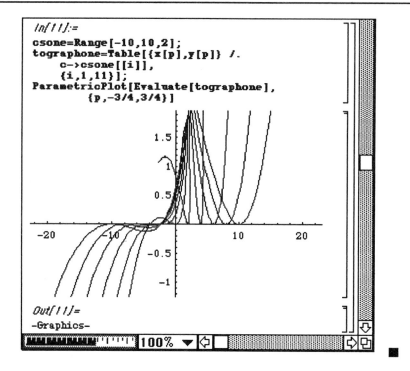

```
In[11]:=
csone=Range[-10,10,2];
tographone=Table[{x[p],y[p]} /.
    c->csone[[i]],
    {i,1,11}];
ParametricPlot[Evaluate[tographone],
    {p,-3/4,3/4}]
```

```
Out[11]=
-Graphics-
```

A **Ricatti equation** is a nonlinear equation of the form $y'+a(x)y^2 + b(x)y + c(x) = 0.$

Let $y(x) = \dfrac{w'(x)}{w(x)}\dfrac{1}{a(x)}$. Then, $y'(x) = \dfrac{w''(x)}{a(x)w(x)} - \dfrac{\left(w'(x)\right)^2}{a(x)\left(w(x)\right)^2} - \dfrac{a'(x)w'(x)}{\left(a(x)\right)^2 w(x)}$ and

substituting into the equation $y'+a(x)y^2 + b(x)y + c(x) = 0$ yields the second - order equation

$$\dfrac{w''(x)}{a(x)w(x)} - \dfrac{\left(w'(x)\right)^2}{a(x)\left(w(x)\right)^2} - \dfrac{a'(x)w'(x)}{\left(a(x)\right)^2 w(x)} + \dfrac{\left(w'(x)\right)^2}{a(x)\left(w(x)\right)^2} + \dfrac{b(x)w'(x)}{a(x)w(x)} + c(x) = 0.$$

Multiplying the equation by $a(x)w(x)$ we obtain

$w''(x) - \dfrac{a'(x)w'(x)}{a(x)} + b(x)w'(x) + a(x)c(x)w(x) = 0$ and simplifying

yields the second - order equation $w'' - \left(\dfrac{a'(x)}{a(x)} - b(x)\right)w' + a(x)c(x)w = 0.$

❏ EXAMPLE 2.17

Convert the Ricatti equation $y'+\left(x^4+x^2+1\right)y^2+\dfrac{2\left(1-x+x^2-2x^3+x^4\right)}{1+x^2+x^4}y+\dfrac{1}{x^4+x^2+1}=0$

to a second-order equation.

Solution:

We begin by identifying $a(x)=x^4+x^2+1$, $b(x)=\dfrac{2\left(1-x+x^2-2x^3+x^4\right)}{1+x^2+x^4}$, and

$c(x)=\dfrac{1}{x^4+x^2+1}$. Letting $y(x)=\dfrac{w'(x)}{w(x)}\dfrac{1}{a(x)}$ yields the second - order equation

$w''-\left(\dfrac{4x^3+2x}{x^4+x^2+1}-\dfrac{2\left(1-x+x^2-2x^3+x^4\right)}{1+x^2+x^4}\right)w'+\left(x^4+x^2+1\right)\dfrac{1}{x^4+x^2+1}w=0$ which simplifies

to $w''-2w'+w=0.$

In **Chapter 4** we will learn how to find an exact solution of $w''-2w'+w=0.$ For now, we use
NDSolve to approximate a solution of

$y'+\left(x^4+x^2+1\right)y^2+\dfrac{2\left(1-x+x^2-2x^3+x^4\right)}{1+x^2+x^4}y+\dfrac{1}{x^4+x^2+1}=0$ subject to $y(0)=0$ on the

interval $\left[\dfrac{-1}{2},1\right].$

We first define the appropriate coefficient functions for this particular Ricatti equation.

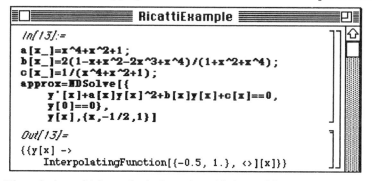

```
                    RicattiExample
In[13]:=
a[x_]=x^4+x^2+1;
b[x_]=2(1-x+x^2-2x^3+x^4)/(1+x^2+x^4);
c[x_]=1/(x^4+x^2+1);
approx=NDSolve[{
    y'[x]+a[x]y[x]^2+b[x]y[x]+c[x]==0,
    y[0]==0},
    y[x],{x,-1/2,1}]
Out[13]=
{{y[x] ->
    InterpolatingFunction[{-0.5, 1.}, <>][x]}}
```

The solution which results is then plotted below with **Plot**.

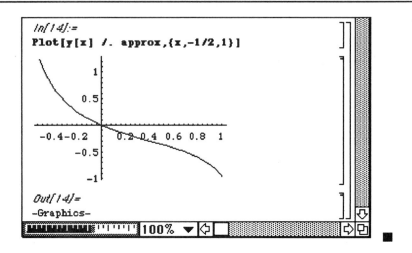

In[14]:=
`Plot[y[x] /. approx,{x,-1/2,1}]`

Out[14]=
-Graphics-

§2.6 Theory of First Order Equations

In order to understand the types of initial value problems which yield a unique solution, the following theorem is stated. At this point, the proof is omitted.

<u>Theorem (Existence and Uniqueness)</u>:
Consider the initial value problem

$y' = f(x,y)$, $y(x_0) = y_0$.

If f and $\dfrac{\partial f}{\partial y}$ are continuous functions on the rectangular region R: a<x<b, c<y<d containing the point

(x_0, y_0), then there exists an interval $|x - x_0| < h$ centered at x_0 on which there exists one and only one solution to the differential equation which satisfies the initial condition.

❑ **EXAMPLE 2.18**
Solve the initial value problem
$$\frac{dy}{dx} = \frac{x}{y}$$
$y(0) = 0.$
Does this result contradict the Existence and Uniqueness Theorem?

Solution:
This equation can be solved by separation of variables which yields the equation ydy = x dx which is easily solved to determine the family of solutions $y^2 - x^2 = C$. Hence, if x = y = 0, then C = 0.

Therefore, two solutions pass through (0,0), $y = x$ and $y = -x$. The problem is also solved with *Mathematica* below.

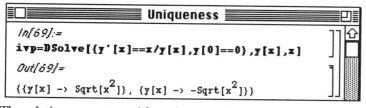

The solutions are extracted from the output list of **ivp** for convenience with the following commands.

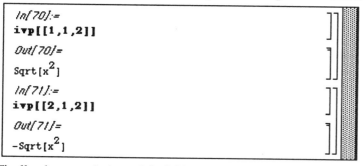

Finally, the two solutions of the initial value problem are plotted.

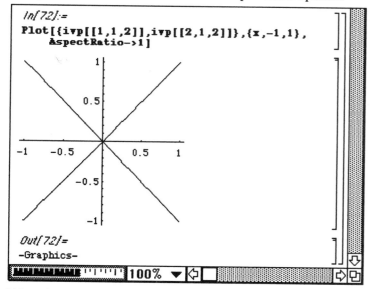

Although more than one solution satisfies this initial value problem, the Existence and Uniqueness

Theorem is not contradicted since the function $\dfrac{x}{y}$ is not continuous at the point $(0,0)$. ∎

■ EXAMPLE 2.19

Verify that the initial value problem

$$\frac{dy}{dx} = y$$

$$y(0) = 1$$

has a unique solution.

Solution:

Notice that in this case, the function $f(x,y) = y$. Hence, both f and $\dfrac{\partial f}{\partial y}$ are continuous on all

regions containing $(0,1)$. Therefore, there exists a unique solution to the differential equation

which also satisfies the initial condition $y(0) = 1$. This is illustrated below with *Mathematica*.

First, the general solution to the differential equation is determined, and several of these curves are plotted.

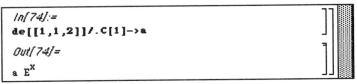

The constant C[1] in the solution is replaced by **a** so that particular values can be substituted for **a** in the formula.

```
In[74]:=
de[[1,1,2]]/.C[1]->a
Out[74]=
a E^x
```

Below, a table of solutions for values of **a** on the interval [0.25,2] using increments of 0.25 is created in **family**.

```
In[75]:=
family=Table[
    de[[1,1,2]]/.C[1]->a,{a,.25,2,.25}]
Out[75]=
{0.25 E^X, 0.5 E^X, 0.75 E^X, 1. E^X, 1.25 E^X, 1.5 E^X,
  1.75 E^X, 2. E^X}
```

The graphs of the functions in **family** are plotted below. Notice that only one of these curves passes through the point (0,1).

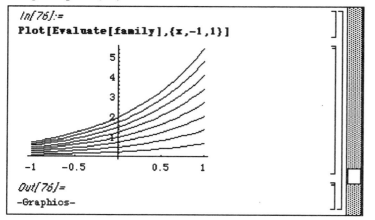

```
In[76]:=
Plot[Evaluate[family],{x,-1,1}]
```

```
Out[76]=
-Graphics-
```

To further verify that this initial value problem has a unique solution, **DSolve** is used to solve the initial value problem.

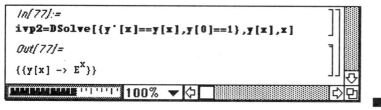

```
In[77]:=
ivp2=DSolve[{y'[x]==y[x],y[0]==1},y[x],x]
Out[77]=
{{y[x] -> E^X}}
```

100%

Chapter 3: Applications of First-Order Ordinary Differential Equations

Mathematica commands used in **Chapter 3** include:

Apart	Evaluate	PlotPoints
AspectRatio	FindRoot	PlotRange
AxesOrigin	GrayLevel	PlotStyle
Cancel	Integrate	PlotVectorField
ContourPlot	Limit	ReplaceAll
Contours	ListPlot	Show
ContourShading	Log	Solve
D	N	Table
DisplayFunction	Part	
DSolve	Plot	

§3.1 Orthogonal Trajectories

Two lines L_1 and L_2, with slopes m_1 and m_2, respectively, are perpendicular if the respective slopes satisfy the relationship $m_1 = \dfrac{-1}{m_2}$. Hence, two curves C_1 and C_2 are orthogonal at a point if the respective tangent lines to the curves at that point are perpendicular.

■ EXAMPLE 3.1

Use the definition of orthogonality to verify that the curves given by $y = x$ and $y = \sqrt{1 - x^2}$ are orthogonal at the point $\left(\dfrac{\sqrt{2}}{2}, \dfrac{\sqrt{2}}{2} \right)$.

Solution:

The derivatives of the functions are given by $y' = 1$ and $y' = \dfrac{-x}{\sqrt{1 - x^2}}$. Substitution of $x = \dfrac{\sqrt{2}}{2}$ into the

second formula for y' yields $\dfrac{-\dfrac{\sqrt{2}}{2}}{\sqrt{1 - \left(\dfrac{\sqrt{2}}{2} \right)^2}} = -1$. Hence, the relationship is verified. The orthogonality is illustrated below with *Mathematica* by plotting these two curves on a small interval centered at $x = \dfrac{\sqrt{2}}{2}$.

First, this particular value of x is entered as **a** for convenience.

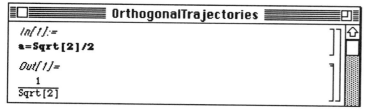

Then, the two curves are plotted on the small interval [a-.05,a+.05] to illustrate that they meet at a right angle.

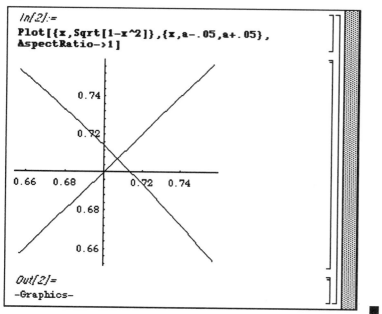

■ EXAMPLE 3.2

Find the family of orthogonal trajectories of the family of ellipses $x^2 - xy + y^2 = c^2$.
Sketch the family and the orthogonal trajectories.

Solution:

We first determine the differential equation satisfied by the family of ellipses. Implicit differentiation

yields $\dfrac{d}{dx}\left(x^2 - xy + y^2 = c^2\right)$ or $2x - y - x\dfrac{dy}{dx} + 2y\dfrac{dy}{dx} = 0.$ Hence, $(2y - x)\dfrac{dy}{dx} = y - 2x,$ so

$\dfrac{dy}{dx} = \dfrac{y - 2x}{2y - x}.$

Below in **sol**, implicit differentiation is carried out with *Mathematica* to yield the desired formula for y'
by differentiating the equation $x^2 - x\,y(x) + \left(y(x)\right)^2$ with respect to x and then solving the resulting
equation for y'(x).

```
In[3]:=
sol=Solve[D[x^2-x y[x]+y[x]^2==0,x],y'[x]]

Out[3]=
{{y'[x] -> -(2 x - y[x]/-x + 2 y[x])}}
```

This formula is extracted with the command **sol[[1,1,2]]**.

```
In[4]:=
sol[[1,1,2]]

Out[4]=
-(2 x - y[x]/-x + 2 y[x])
```

Therefore, the family of orthogonal trajectories satisfies $\dfrac{dy}{dx} = \dfrac{-x + 2y}{2x - y}.$

This is a homogeneous first-order equation which cannot be solved directly with **DSolve** as indicated
below.

First, the right-hand side of the differential equation is defined as **orth** below for use in the next step.
Then, we attempt to solve the equation with **DSolve**. Unfortunately, **DSolve** is unable to solve the
equation.

```
In[5]:=
orth=-1/sol[[1,1,2]]

Out[5]=
-x + 2 y[x]
----------
2 x - y[x]
```

```
In[6]:=
ds=DSolve[y'[x]==orth,y[x],x]
Out[6]=
DSolve[y'[x] == (-x + 2 y[x])/(2 x - y[x]), y[x], x]
```

However, it can be solved by making the substitution $v = \dfrac{y}{x}$. Note that by using the chain rule, we have

$\dfrac{dy}{dx} = x\dfrac{dv}{dx} + v$. By substitution, we have the differential equation involving v: $x\dfrac{dv}{dx} + v = \dfrac{-x + 2y}{2x - y}$ which

is separable. This leads to $\dfrac{2 - v}{v^2 - 1} dv = \dfrac{dx}{x}$. The integral of the left - hand side of the equation is found

below with *Mathematica*.
The symbol Log[x] represents the natural logarithm function, Ln(x).

```
In[10]:=
Integrate[(2-v)/(v^2-1),v]
Out[10]=
Log[-1 + v]   3 Log[1 + v]
----------- - ------------
     2              2
```

Therefore, we have the following equation which must be simplified with logarithmic properties.

$\dfrac{1}{2}\text{Log}[-1 + v] - \dfrac{3}{2}\text{Log}[1 + v] = \text{Log}[x] + C$

Substitution of $v = \dfrac{y}{x}$ and simplification yields:

$\text{Log}\left[\dfrac{y}{x} - 1\right] - \text{Log}\left[\dfrac{y}{x} + 1\right]^3 = 2\text{Log}[x] + \tilde{C}$

$\text{Log}\left[\dfrac{\dfrac{y}{x} - 1}{\left(\dfrac{y}{x} + 1\right)^3}\right] = \text{Log}[x^2] + \tilde{C}.$

Hence, $\dfrac{\dfrac{y}{x} - 1}{\left(\dfrac{y}{x} + 1\right)^3} = Ax^2$ or $\dfrac{x^2(y - x)}{(y + x)^3} = Ax^2$ where $A = $ constant. Therefore, the orthogonal trajectories are given by $\dfrac{y-x}{(y+x)^3} = A$. The family of ellipses and the orthogonal trajectories are

plotted below with *Mathematica*.

The family of orthogonal trajectories is plotted below in **cp1** with **ContourPlot**. Note that we avoid the discontinuity at the origin. The option **ContourShading->False** allows us to view the contour levels more easily. **ContourPlot** merely plots the solutions for different values of the constant A. In addition, the option **Contours** may be used to select particular values of A.

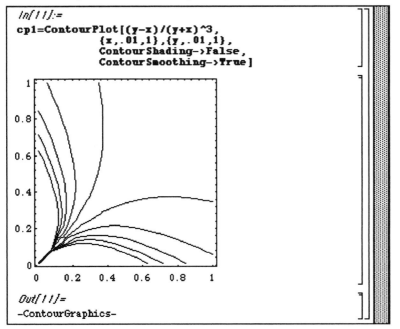

The family of ellipses is also plotted below in **cp2** with **ContourPlot**. Again, the **ContourShading->False** option setting is used so that the contour levels can be seen more easily.

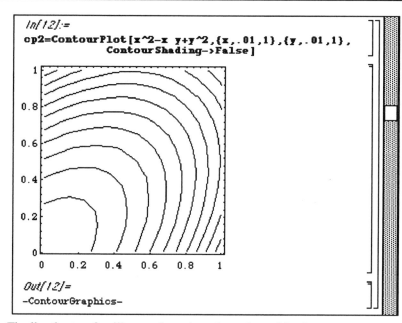

```
In[12]:=
cp2=ContourPlot[x^2-x y+y^2,{x,.01,1},{y,.01,1},
            ContourShading->False]
```

```
Out[12]=
-ContourGraphics-
```

Finally, the two families are then viewed together with **Show**.

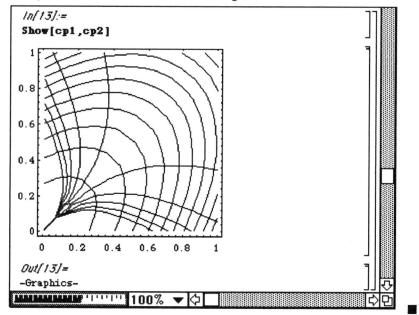

```
In[13]:=
Show[cp1,cp2]
```

```
Out[13]=
-Graphics-
```

§3.2 Direction Fields

The geometrical interpretation of solutions to first-order differential equations of the form

$$\frac{dy}{dx} = f(x,y)$$

is important to the basic understanding of problems of this type. Suppose that a solution to this equation is a function $y=\psi(x)$. Hence, the solution is merely the graph of the function ψ. Therefore, if (x,y) is a point on this graph, the slope of the tangent line is given by f(x,y). A set of short line segments representing the tangent lines can be constructed for a large number of points. This collection of line segments is known as the direction field for the differential equation and provides a great deal of information concerning the behavior of the family of solutions. *Mathematica* is quite useful in overcoming the tedious task of drawing the direction field. The graphics package **PlotField.m** must be opened first in order to take advantage of these useful commands. We illustrate this procedure in the following examples.

■ EXAMPLE 3.3

Find the direction field associated with the differential equation

$$\frac{dy}{dx} = e^{-x} - 2y.$$

Solution:

First, the appropriate *Mathematica* package is loaded. The command **PlotVectorField** can then be employed to plot this field. Notice that the argument of this command is a two-dimensional vector. Hence, the first component is arbitrarily taken to be 1 and the second component is f(x,y) from the differential equation. The direction field is plotted over the rectangular region $[-1,1] \times [-1,1]$.

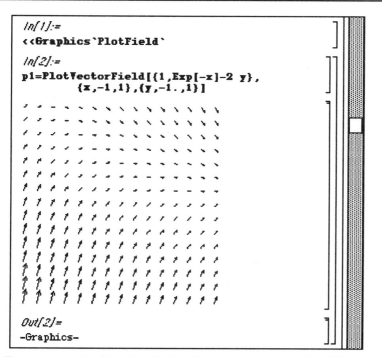

```
In[1]:=
<<Graphics`PlotField`

In[2]:=
p1=PlotVectorField[{1,Exp[-x]-2 y},
       {x,-1,1},{y,-1.,1}]
```

```
Out[2]=
-Graphics-
```

To compare the direction field with the solutions to the differential equation, **DSolve** is used to calculate these solutions. For convenience, the output list is named **sol** so that solutions can be easily extracted with the command **sol[[1,1,2]]**.

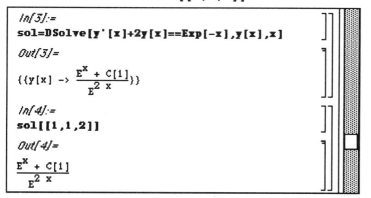

```
In[3]:=
sol=DSolve[y'[x]+2y[x]==Exp[-x],y[x],x]

Out[3]=
               E^x + C[1]
{{y[x] ->  ----------}}
              E^2 x

In[4]:=
sol[[1,1,2]]

Out[4]=
E^x + C[1]
----------
  E^2 x
```

A table of solutions is constructed below by replacing C[1] in **sol[[1,1,2]]** by **a** for **a**=−2, −1.5, −1, −.5, 0, .5, 1, 1.5, and 2 and named **table**.

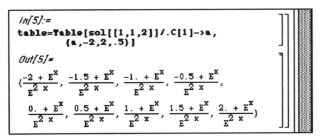

These solutions are plotted below.

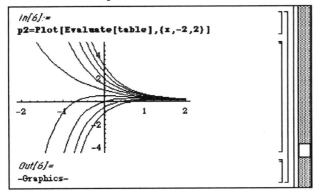

The two plots are shown simultaneously to illustrate that the vectors in the direction field are tangent to the solutions to the differential equation.

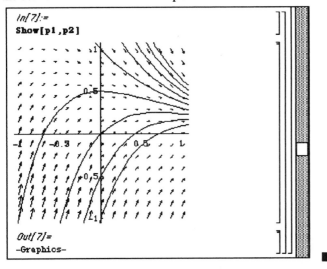

§3.3 Population Growth and Decay

Suppose that the rate at which a population y(t) changes is proportional to the amount present. Mathematically, this statement is represented as the first-order initial value problem

$$\frac{dy}{dt} = ky, \ y(0) = y_0$$

where y_0 is the initial population. If k > 0, then the population is increasing (growth) while the population decreases (decay) if k < 0. Problems of this nature arise in such fields as cell population growth in biology as well as radioactive decay in physics. The model introduced above is known as the Malthus model due to the work of the English clergyman and economist Thomas R. Malthus.

■ EXAMPLE 3.4

The population of the United States was recorded as 5.3 million in 1800. Use the Malthus model to approximate the population if k was experimentally determined to be 0.03. Compare these results to the actual population. Is this a good approximation?

Solution:

For convenience, the Malthus model is solved in general for all values of r and y_0. This enables us to refer to this solution in other problems without solving the differential equation again. The solution **pop** which is a function of t, r, and y_0 is extracted from the output list of **peq** with the command **peq[[1,1,2]]**.

```
≡□         ≡≡≡≡ Population ≡≡≡≡         ▣≡
In[11]:=
peq=DSolve[{y'[t]== r y[t],y[0]==y0},y[t],t]

Out[11]=
{{y[t] -> E^(r t) y0}}

In[12]:=
peq[[1,1,2]]

Out[12]=
E^(r t) y0

In[13]:=
pop[t_,r_,y0_]=peq[[1,1,2]]

Out[13]=
E^(r t) y0
```

This function is then plotted with r = 0.03 and y_0 = 5.3. For comparison, the census figures for the population of the United States for the years 1800-1900 are entered in the list **pdata** and plotted with

ListPlot. The two plots are then displayed with **Show**.

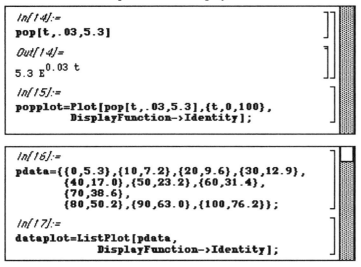

```
In[14]:=
pop[t,.03,5.3]

Out[14]=
5.3 E^0.03 t

In[15]:=
popplot=Plot[pop[t,.03,5.3],{t,0,100},
        DisplayFunction->Identity];
```

```
In[16]:=
pdata={{0,5.3},{10,7.2},{20,9.6},{30,12.9},
       {40,17.0},{50,23.2},{60,31.4},
       {70,38.6},
       {80,50.2},{90,63.0},{100,76.2}};

In[17]:=
dataplot=ListPlot[pdata,
            DisplayFunction->Identity];
```

Notice that the accuracy of the approximation diminishes over time. Hence, another model which better approximates the population is needed.

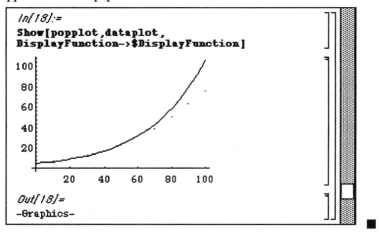

```
In[18]:=
Show[popplot,dataplot,
DisplayFunction->$DisplayFunction]
```

```
Out[18]=
-Graphics-
```

The **logistic equation** (or **Verhulst equation**) is the equation

$y'(t) = (r - a y(t)) y(t)$ subject to the condition $y(0) = y_0$ where r and a are constants.

This equation was first introduced by the Belgian mathematician Pierre Verhulst to study population growth.

■ EXAMPLE 3.5

Use *Mathematica* to determine the general solution of the logistic equation. Approximate the population using $r = 0.03$, $a = 0.0001$, and $y_0 = 5.3$. Compare this result with the census values given in **pdata** in the previous example.

Solution:

This example is solved similarly to the previous example involving the Malthus model in that the general solution is found which depends on the parameters r, a, and y_0. First, **DSolve** is used to determine the general solution of the equation $y'(t) = (r - a y(t)) y(t)$ subject to $y(0) = y_0$ in **de2**.

```
In[19]:=
de2=DSolve[{y'[t]==(r-a y[t]) y[t],
                y[0]==y0},y[t],t]

Out[19]=
               r t
             E    r
{{y[t] ->  ─────────────── }}
                  r t    r
           -a + a E    + ──
                         y0
```

Below, the solution is extracted from the output list of **de2** with the command **de2[[1,1,2]]**:

```
In[20]:=
de2[[1,1,2]]

Out[20]=
       r t
     E    r
─────────────────
        r t    r
-a + a E    + ──
              y0
```

and defined as the function **logistic**.

```
In[21]:=
logistic[r_,a_,y0_,t_]=de2[[1,1,2]]

Out[21]=
       r t
     E    r
─────────────────
        r t    r
-a + a E    + ──
              y0
```

This function is determined for the values of r, a, and y_0 indicated above.

```
In[22]:=
logistic[.03,.0001,5.3,t]

Out[22]=
             0.03 t
    0.03 E
─────────────────────────
                      0.03 t
0.00556038 + 0.0001 E
```

Using **Limit**, we see that the population in this case approaches 300.

```
In[23]:=
Limit[logistic[.03,.0001,5.3,t],t->Infinity]

Out[23]=
300.
```

In order to compare the approximated population obtained with the actual population obtained through the census, we first plot the approximated population below in **popplot2**.

```
In[24]:=
popplot2=Plot[logistic[.03,.0001,5.3,t],
    {t,0,100},
    PlotRange->{0,100},
    DisplayFunction->Identity];
```

Note that when this graph is displayed below with **Show** along with that of the actual population values found in **pdata**, the approximation is more accurate than that obtained with the Malthus model.

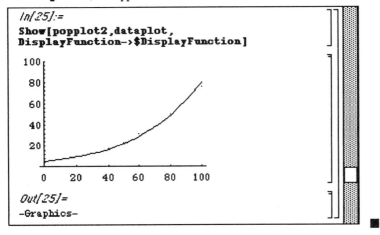

```
In[25]:=
Show[popplot2,dataplot,
DisplayFunction->$DisplayFunction]
```

```
Out[25]=
-Graphics-
```

■ EXAMPLE 3.6

Use *Mathematica* to investigate: (a) the behavior of the solution when the initial population is varied, and (b) the behavior of the solution when values of r and a are varied.

Solution:

(a) First, we investigate the behavior of the curve as the value of y_0 assumes the values of 0.1, 0.2, and 0.4 which are located between 0 and 3. Plots which correspond to these values are created in different levels of gray in **plot1**, **plot2**, and **plot3** below. These three graphs are then displayed simultaneously with **Show**. Notice that the curves which correspond to larger values of y_0 approach the limiting population r/a more quickly.

```
In[26]:=
plot1=Plot[logistic[3,1,.01,t],{t,0,5},
    PlotRange->All,
    AxesOrigin->{0,0},
    DisplayFunction->Identity];
```

```
In[27]:=
plot2=Plot[logistic[3,1,.2,t],{t,0,5},
    PlotRange->All,
    AxesOrigin->{0,0},
    PlotStyle->GrayLevel[.2],
    DisplayFunction->Identity];
```

```
In[28]:=
plot3=Plot[logistic[3,1,.4,t],{t,0,5},
    PlotRange->All,
    AxesOrigin->{0,0},
    PlotStyle->GrayLevel[.4],
    DisplayFunction->Identity];
```

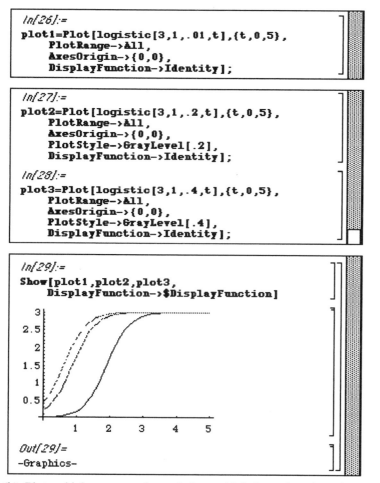

```
In[29]:=
Show[plot1,plot2,plot3,
    DisplayFunction->$DisplayFunction]
```

```
Out[29]=
-Graphics-
```

(b) Plots which correspond to solutions which depend on the values of $r = 3$ and $a = 2$ are considered in this case. The curves that result from the values of $y_0 = 0.5$, 1.5, and 3.0 are plotted simultaneously in varying levels of gray in **plot4** below. Note that each curve approaches the limiting population of 1.5 which is the ratio r/a.

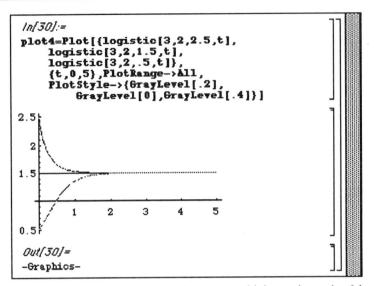

In **plot5** below, the plots for the curves which are determined by r = 3 and k = 0.5 are shown for various values of a. The values considered for a are a = 1, 2, 3. Note that each curve approaches a limit of r/a = 3, 2, 1, respectively.

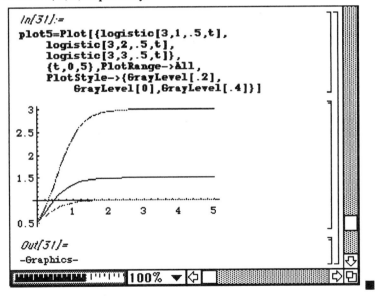

■ EXAMPLE 3.7

Assume that there are initially 10 grams of a radioactive substance. If the half-life of this particular substance is 1 year (half of the substance decays over the first year), then determine the time required for 9 grams of the substance to decay (i.e.,1 gram remain). Also, determine how many grams remain after 6 years.

Solution:

First, the initial value problem, $y'(t) = -ky(t)$ subject to $y(0) = y_0$, is defined and solved in

d1 below. The formula is then extracted and used to define **decay** with the command **decay[t_,k_,y0_]=d1[[1,1,2]]**, a function which describes the amount of substance present at any time t.

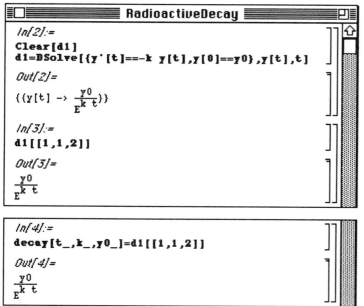

The solution is then plotted for $y_0 = 10$ and $t = 1$ so that the constant of proportionality may be determined. The value of k appears to be near $k = 0.7$.

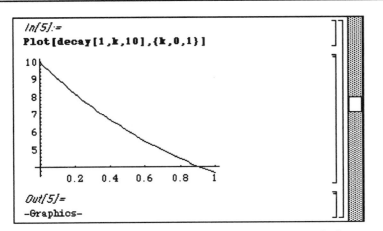

In[5]:=
Plot[decay[1,k,10],{k,0,1}]

Out[5]=
−Graphics−

Hence, this is used as an initial guess with **FindRoot** to obtain a more accurate value in **const**. The k-value is then extracted from the output list of **const** and used to plot the solution as a function of time.

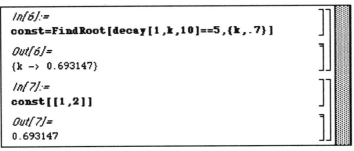

In[6]:=
const=FindRoot[decay[1,k,10]==5,{k,.7}]

Out[6]=
{k -> 0.693147}

In[7]:=
const[[1,2]]

Out[7]=
0.693147

Since we need to determine the value of time when 1 gram remains, we again use the **FindRoot** with an initial guess of t = 4 obtained from the plot. Hence, approximately 3.32193 years are required for 9 grams of the substance to decay. Finally, t = 6 is used as an argument of the function to see that 0.15625 grams of the substance remain after 6 years.

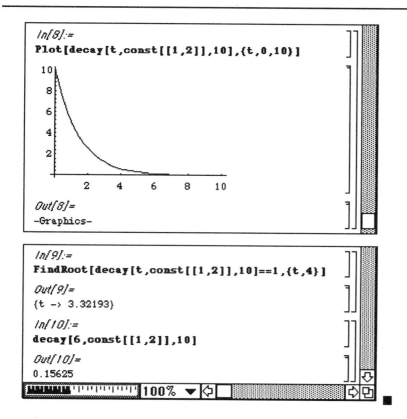

§3.4 Newton's Law of Cooling

Newton's law of cooling states that the rate at which the temperature T(t) changes in a cooling body is proportional to the difference between the temperature of the body and the constant temperature T_s of the surrounding medium. This situation is represented as the first-order initial value problem

$$\frac{dT}{dt} = k(T - T_s),\ T(0) = T_0$$

where T_0 is the initial temperature of the body and k is the constant of proportionality. We investigate problems involving Newton's law of cooling in the following examples.

■ EXAMPLE 3.8

A pie is removed from a 350° oven. In 15 minutes, the pie has a temperature of 150°. Determine the time required to cool the pie to a temperature of 80° so that it may be eaten.

Solution:

As has been the case in the earlier examples, the general solution which depends on the parameters of the problem is determined. Here, the resulting function is called **temp**, the surrounding temperature **temps**, the initial temperature **temp0**, and the constant of proportionality **k**. The solution based on the data indicated in this example is then easily found.

The solution using the parameter values needed for this problem is given below.

```
In[14]:=
temp[75,350,k,15]

Out[14]=
75 + 275/E^(15 k)
```

Since the constant **k** is unknown, it is determined below with **Solve** and called **k1** for convenience. Notice that only one value in the output list of this command is a real number. This number is extracted in the usual manner so that it can used to determine the time at which the pie reaches its desired temperature. This is accomplished with **FindRoot** by, first, plotting the solution to obtain an estimate of the time at which the temperature is 80 degrees and then using this initial approximation with **FindRoot**. Since the value of the function seems to equal 80 near $t = 40$, the initial guess of 40 is used to achieve the more accurate value of $t = 37.1614$. (Note that **FindRoot** could have been used to determine the constant of proportionality above instead of **Solve** in order to avoid the warning messages which result with **Solve**.)

```
In[15]:=
k1=Solve[temp[75,350,k,15]==150,k]//N

    Solve::ifun:
        Warning: Inverse functions are being u
            Solve, so some solutions may not be

Out[15]=
{{k -> 0.0866189}, {k -> 0.0866189 + 0.418879 I},
  {k -> 0.0866189 + 0.837758 I},
  {k -> 0.0866189 + 1.25664 I},
  {k -> 0.0866189 + 1.67552 I},
  {k -> 0.0866189 + 2.0944 I},
  {k -> 0.0866189 + 2.51327 I},
  {k -> 0.0866189 + 2.93215 I},
  {k -> 0.0866189 - 2.93215 I},
  {k -> 0.0866189 - 2.51327 I},
  {k -> 0.0866189 - 2.0944 I},
  {k -> 0.0866189 - 1.67552 I},
  {k -> 0.0866189 - 1.25664 I},
  {k -> 0.0866189 - 0.837758 I},
  {k -> 0.0866189 - 0.418879 I}}
```

This number is extracted in the usual manner so that it can be used to determine the time at which the pie reaches its desired temperature.

```
In[16]:=
k1[[1,1,2]]

Out[16]=
0.0866189
```

This is accomplished with **FindRoot** by, first, plotting the solution to obtain an estimate of the time at which the temperature is 80 degrees and then using this initial approximation with **FindRoot**. Since the value of the function seems to equal 80 near t = 40, the initial guess of 40 is used to achieve the more accurate value of t = 46.264. (Note that **FindRoot** could have been used to determine the constant of proportionality above instead of **Solve** in order to avoid the warning messages which result with **Solve**.)

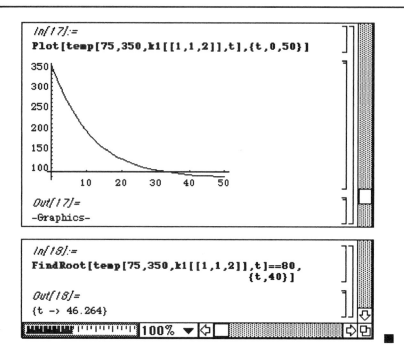

```
In[17]:=
Plot[temp[75,350,k1[[1,1,2]],t],{t,0,50}]
```

```
Out[17]=
-Graphics-
```

```
In[18]:=
FindRoot[temp[75,350,k1[[1,1,2]],t]==80,
                              {t,40}]
```

```
Out[18]=
{t -> 46.264}
```

100% ▼

■ **EXAMPLE 3.9**

In the investigation of a homicide, the time of death is often important. Newton's law of cooling can be used to approximate this time. For example, the normal body temperature of most healthy people is 98.6 degrees. Suppose that at the time that a body is discovered, its temperature is 82 degrees. Two hours later, it is 75 degrees. If the temperature of the surroundings is 65 degrees, what was the time of death?

Solution:

This problem is solved similarly to the previous example. First, the value of the constant of proportionality is found with **Solve**. The real number given in the output list is extracted to obtain a value which is used to plot the solution in order to approximate the time at which the temperature of the body is 98.6.

```
In[8]:=
k2=Solve[temp[65,82,k,3]==75,k]//N

    Solve::ifun:
        Warning: Inverse functions are being ι
            Solve, so some solutions may not be
Out[8]=
{{k -> 0.176876}, {k -> 0.176876 + 2.0944 I},
  {k -> 0.176876 - 2.0944 I}}
```

Since the time at which the body was discovered was taken to be t = 0, the curve must be plotted over negative values of t. This plot reveals that death occurred near t = −3.

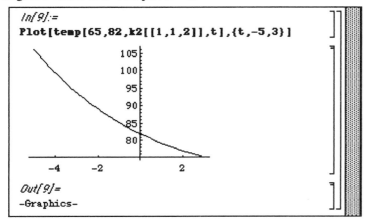

```
In[9]:=
Plot[temp[65,82,k2[[1,1,2]],t],{t,-5,3}]
```

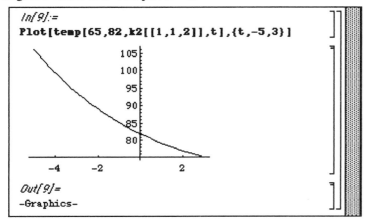

```
Out[9]=
-Graphics-
```

Hence, this is the initial guess used with **FindRoot** to obtain the value of t = −3.85192. We conclude that the person died almost 4 hours prior to being discovered.

```
In[10]:=
FindRoot[temp[65,82,k2[[1,1,2]],t]==98.6,
            {t,-3}]
Out[10]=
{t -> -3.85192}
```

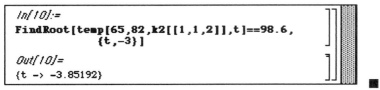

§3.5 Free-Falling Bodies

Newton's second law of motion states that the rate at which the momentum of a body changes with respect to time is equal to the resultant force acting on the body. Since the body's momentum is defined as the product of its mass and velocity, this statement is modeled as

$$\frac{d}{dt}(mv) = F$$

where m and v represent the body's mass and velocity, respectively, and F is the sum of the forces acting on the body. Since m is constant, differentiation leads to the well-known equation

$$m\frac{dv}{dt} = F.$$

If the body is subjected to the force due to gravity, then its velocity is determined by solving the differential equation

$$m\frac{dv}{dt} = mg \text{ or } \frac{dv}{dt} = g$$

where $g \approx 32 \text{ ft/s}^2$ (English system) and 9.8 m/s^2 (metric system). This differential equation is applicable only when the resistive force due to the medium (such as air resistance) is ignored. If this offsetting resistance is considered, the equation becomes

$$m\frac{dv}{dt} = mg - F_R$$

where F_R represents this resistive force. Note that down is assumed to be the positive direction. The resistive force is typically proportional to the body's velocity or the square of its velocity. Hence, the differential equation is linear or nonlinear based on whether or not the resistance of the medium is taken into account.

■ EXAMPLE 3.10

Compare the effects that air resistance has on the velocity of an object of mass m=0.5 which is released with an initial velocity of v_0=0. Consider F_R=16v^2 and F_R=16v.

Solution:

First, after simplification, the differential equation to be solved with F_R=16v^2 is

$$\frac{dv}{dt} = -32 - 32v^2, \ v(0) = 0.$$

We attempt to use **DSolve** to solve the initial value problem. Unfortunately, **DSolve** is unable to yield the result. Therefore, another approach must be taken which involves separation of variables.

```
══════════════ FallingBodies ══════════════
In[2]:=
sol2=DSolve[{v'[t]==g-(c/m)(v[t])^2,v[0]==v0}
                ,v[t],t]

        Solve::tdep:
            The equations appear to involve transce
                functions of the variables in an esse
                non-algebraic way.

Out[2]=

DSolve[{v'[t] == g - c v[t]²/m , v[0] == v0}, v[t],
    t]
```

This separated equation is

$$\frac{1}{1-v^2}\,dv = 32\,dt.$$

Mathematica can be used to aid in the integration by partial fractions as illustrated below:

The solution is found below:

$$\frac{1}{2}\left(\frac{1}{1+v}+\frac{1}{1-v}\right)dv = 32\,dt$$

$$\frac{1}{2}\left(\ln|1+v|-\ln|1-v|\right) = 32t + C_2$$

$$\frac{1}{2}\ln\left|\frac{1+v}{1-v}\right| = 32t + C_2$$

$$\frac{1+v}{1-v} = K_1\exp(64t)$$

$$v = \frac{K_1\exp(64t)-1}{K_1\exp(64t)+1}.$$

Application of the initial condition gives $K_1=1$, so the solution is

$$v(t) = \frac{\exp(64t)-1}{\exp(64t)+1}.$$

This solution is entered below as **rv[t]** and then plotted in **rplot** on the interval [0,.5]. (The display of the plot is suppressed.)

```
In[3]:=
Apart[1/(1-v^2)]

Out[3]=
   -1          1
--------  +  --------
2 (-1 + v)   2 (1 + v)

In[4]:=
rv[t_]:=(Exp[64t]-1)/(Exp[64t]+1)

In[5]:=
rplot=Plot[rv[t],{t,0,.5},PlotRange->All,
DisplayFunction->Identity]

Out[5]=
-Graphics-
```

Next, **DSolve** is used to solve the problem with F_R=16v.

```
In[7]:=
deq=DSolve[{v'[t]==32-32v[t],v[0]==0},v[t],t]

Out[7]=
{{v[t] -> 1 - E^{-32 t}}}
```

The solution is extracted from the output list and also plotted on the interval [0,5] in **plot2**. (The display of this plot is also suppressed.)

```
In[8]:=
deq[[1,1,2]]

Out[8]=
1 - E^{-32 t}

In[9]:=
plot2=Plot[deq[[1,1,2]],{t,0,.5},
        PlotRange->All,
        PlotStyle->GrayLevel[.2],
        DisplayFunction->Identity]

Out[9]=
-Graphics-
```

Both solutions are then displayed together with **Show**. Notice that since the velocity values are on the interval [0,1], more resistance is offered when F_R=16v is used. The graph of this function is the darker of the two curves. In this case, the object reaches its limiting velocity of 1 more quickly than in the case with F_R=16v^2 (the lighter curve).

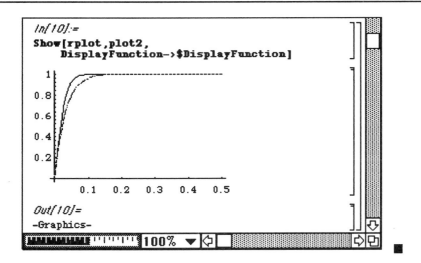

```
In[10]:=
Show[rplot,plot2,
    DisplayFunction->$DisplayFunction]
```

```
Out[10]=
-Graphics-
```

■ EXAMPLE 3.11

Determine the general solution (for the velocity and the position) of the differential equation which models the upward motion of an object of mass m when directed upward with an initial velocity of v_0 from an initial position y_0 assuming that the air resistance equals cv.

Solution:

This initial value problem is solved below with **DSolve**. Notice that the result is named **sol** so that a function, called **velocity**, which depends on the parameters m, c, g, and v_0, may be defined easily.

FallingBodies

```
In[11]:=
sol=DSolve[{v'[t]==-g-(c/m)v[t],v[0]==v0},
            v[t],t]
```

```
Out[11]=
```

$$\{\{v[t] \rightarrow -(\frac{g\ m}{c}) + \frac{\frac{g\ m}{c} + v0}{E^{(c\ t)/m}}\}\}$$

```
In[12]:=
velocity[m_,c_,g_,v0_,t_]=sol[[1,1,2]]
```

```
Out[12]=
```

$$-(\frac{g\ m}{c}) + \frac{\frac{g\ m}{c} + v0}{E^{(c\ t)/m}}$$

For example, the velocity function for the case with

$$m = \frac{1}{128}, \quad c = \frac{1}{160}, \quad g = 32, \quad \text{and} \quad v_0 = 48$$

is found below. Similarly, this function can be employed to investigate numerous situations without the need to solve the differential equation each time.

```
In[13]:=
velocity[1/128,1/160,32,48,t]

Out[13]=
-40 + 88/E^(4 t)/5
```

The position function is easily determined by integrating the velocity function. This is accomplished below with **DSolve** using the initial position y_0. As with the previous case, the output is named **pos** so that the position formula may be extracted from the output list for later use.

```
In[14]:=
pos=DSolve[
     {y'[t]==velocity[m,c,g,v0,t],y[0]==y0},
     y[t],t]

Out[14]=
{{y[t] -> (g m^2)/c^2 - (g m t)/c + (m v0)/c - (m (g m + c v0))/(c^2 E^(c t)/m) +
     y0}}

In[15]:=
position[m_,c_,g_,v0_,y0_,t_]=pos[[1,1,2]]

Out[15]=
(g m^2)/c^2 - (g m t)/c + (m v0)/c - (m (g m + c v0))/(c^2 E^(c t)/m) + y0
```

The position and velocity functions are plotted below using the parameters listed above as well as $y_0 = 0$.

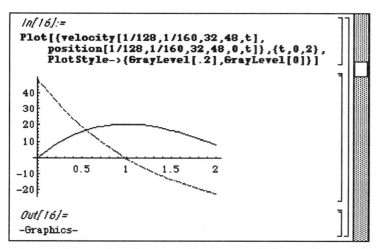

```
In[16]:=
Plot[{velocity[1/128,1/160,32,48,t],
    position[1/128,1/160,32,48,0,t]},{t,0,2},
    PlotStyle->{GrayLevel[.2],GrayLevel[0]}]
```

```
Out[16]=
-Graphics-
```

Solve is then used to determine the time at which the object reaches its maximum height. This time occurs when the derivative of the position is equal to zero. Since warning messages appear, **FindRoot** is then used to illustrate that the derivative of position equals zero at the time at which velocity equals zero for these parameters.

```
In[17]:=
root=Solve[
D[position[1/128,1/160,32,48,0,t],t]==0,t]//N

    Solve::ifun:
        Warning: Inverse functions are being us
        Solve, so some solutions may not be f
Out[17]=
{{t -> 0.985572}, {t -> 0.985572 + 1.5708 I},
  {t -> 0.985572 + 3.14159 I},
  {t -> 0.985572 - 1.5708 I}}
```

```
In[18]:=
FindRoot[velocity[1/128,1/160,32,48,t]==0,
            {t,1}]
Out[18]=
{t -> 0.985572}
```

Below, we compare the effects that varying the initial velocity and position has on the position function. Suppose that we use the same values used earlier for m, c, and g. However, we let $v_0=48$ in one function and $v_0=36$ in the other. We also let $y_0=0$ and $y_0=6$ in these two functions, respectively. Note that the darker curve corresponds to the second function listed below in the **Plot** command.

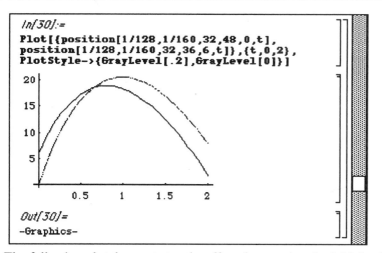

The following plot demonstrates the effect that varying the initial velocity only has on the position function. The values of v_0 used are 48, 64, and 80. The darkest curve corresponds to v_0=48. Notice that as the initial velocity is increased the maximum height obtained by the object is increased as well.

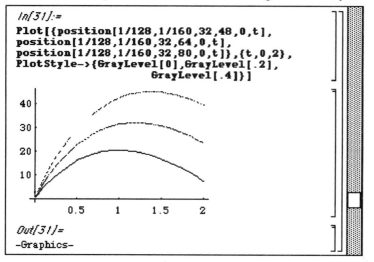

The graph below indicates the effect that varying the initial position and holding all other values constant has on the position function. We use values of 0, 10, and 20 for y_0. Notice that the value of the initial position vertically translates the position function.

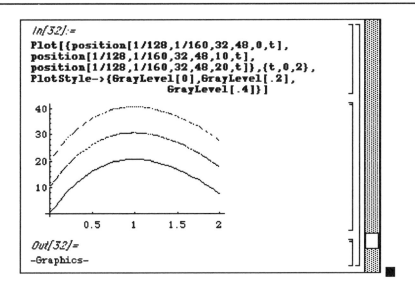

```
In[32]:=
Plot[{position[1/128,1/160,32,48,0,t],
position[1/128,1/160,32,48,10,t],
position[1/128,1/160,32,48,20,t]},{t,0,2},
PlotStyle->{GrayLevel[0],GrayLevel[.2],
                    GrayLevel[.4]}]
```

```
Out[32]=
-Graphics-
```

■ EXAMPLE 3.12

Investigate the motion of the object if $F_R=0$ (i.e., if air resistance is ignored). Solve for the velocity and position as functions of time, initial velocity, and initial position. Compare these results with those of the previous example which involved air resistance. Also, compare the effects that varying the initial velocity and initial position has on these functions.

Solution:

Since we ignore all resistance forces due to the medium, we have the following initial value problem to solve:

$$\frac{dv}{dt} = -g , \; v(0) = v_0.$$

This is done below in **de2**.

```
In[19]:=
de2=DSolve[{v'[t]==-g,v[0]==v0},v[t],t]

Out[19]=
{{v[t] -> -(g t) + v0}}
```

The function **vel2** is then defined by extracting the appropriate formula from the previous output. In order to compare these results to those of the problem in which $F_R=cv$, we consider the following values:

$$m = \frac{1}{128}, \; c = \frac{1}{160}, \; g = 32, \text{ and } v_0 = 48$$

which were also used in Example 3.10. The velocity function **vel2** for the problem involving no air resistance is given below.

```
In[20]:=
vel2[m_,c_,g_,v0_,t_]=de2[[1,1,2]]

Out[20]=
-(g t) + v0

In[21]:=
vel2[1/128,1/160,32,48,t]

Out[21]=
48 - 32 t
```

Next, the position function **pos2** is determined by using **DSolve** to solve the necessary initial value problem. Note that we are simply integrating the velocity function given in **vel2** and applying the initial condition, $y(0)=y_0$.

```
In[22]:=
pos2=DSolve[
     {y'[t]==vel2[m,c,g,v0,t],y[0]==y0},
      y[t],t]

Out[22]=
{{y[t] -> -(g t^2)/2 + t v0 + y0}}
```

The position function **position2** is then defined by extracting the formula from the output above.

```
In[23]:=
position2[m_,c_,g_,v0_,y0_,t_]=pos2[[1,1,2]]

Out[23]=
-(g t^2)/2 + t v0 + y0
```

The position functions for the case with $F_R=0$ and that when $F_R=cv$ are plotted below using the parameter values indicated earlier. Note that the lighter curve is that corresponding to $F_R=cv$.

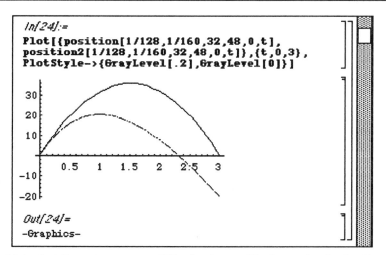

```
In[24]:=
Plot[{position[1/128,1/160,32,48,0,t],
position2[1/128,1/160,32,48,0,t]},{t,0,3},
PlotStyle->{GrayLevel[.2],GrayLevel[0]}]

Out[24]=
-Graphics-
```

Note that these two curves differ in shape. We know that by looking at the formula for the position function in which air resistance is neglected, that the corresponding curve is a parabola. This is verified below. The result in **noairv** represents the t-value at which the object reaches its maximum height, and the result in **noairpos** represents the t-value at which the object reaches the ground. Hence, equal amounts of time (t = 1.5) are necessary for the object to achieve its maximum and then return to its minimum.

```
In[25]:=
noairv=FindRoot[
    vel2[1/128,1/160,32,48,t]==0,{t,1.5}]

Out[25]=
{t -> 1.5}

In[26]:=
noairpos=FindRoot[
    position2[1/128,1/160,32,48,0,t]==0,
                    {t,3}]

Out[26]=
{t -> 3.}
```

On the other hand, the result in **airv** is not one half of the result in **airpos**. Hence, this curve is not a parabola. In fact, more time is required for the object to return to the ground (the horizontal axis) than to move from its initial position to its maximum height.

```
In[27]:=
airv=FindRoot[
    velocity[1/128,1/160,32,48,t]==0,{t,1.5}]
Out[27]=
{t -> 0.985572}

In[28]:=
airpos=FindRoot[
    position[1/128,1/160,32,48,0,t]==0,{t,3}]
Out[28]=
{t -> 2.32028}
```

```
In[29]:=
2 airv[[1,2]] ==airpos[[1,2]]
Out[29]=
False
```

We now combine several of the topics discussed in this section to solve the following problem.

■ EXAMPLE 3.13

An object of mass m = 1 is dropped from a height of 50 feet above the surface of a small pond. While the object is in the air, the force due to air resistance is v. However, when the object is in the pond, it is subjected to a buoyancy force equivalent to 6v. Determine how much time is required for the object to reach a depth of 25 feet in the pond.

Solution:

This problem must be broken into two parts: an initial value problem for the object above the pond, and an initial value problem for the object below the surface of the pond. The initial value problem above the pond's surface is easily found to be

$$\frac{dv}{dt} = 32 - v$$

$v(0) = 0.$

However, to define the initial value problem to find the velocity of the object beneath the pond's surface, the velocity of the object when it reaches the surface must be known. Hence, the velocity of the object above the surface must be determined by solving the initial value problem above. This is done below with *Mathematica*. In order to find the velocity when the object hits the pond's surface the position function of the object must be found by integrating the velocity function. This, too, is done below with **DSolve**. In **d1** and **p1**, below, we obtain the velocity and position functions, respectively, for the object above the pond.

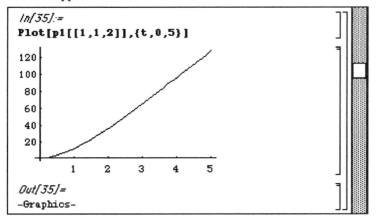

```
In[33]:=
d1=DSolve[{v'[t]==32-v[t],v[0]==0},v[t],t]

Out[33]=
        32
{{v[t] -> 32 - ---}}
              E^t

In[34]:=
p1=DSolve[{y'[t]==d1[[1,1,2]],y[0]==0},
                                     y[t],t]

Out[34]=
              32
{{y[t] -> -32 + --- + 32 t}}
             E^t
```

The position function is plotted below. The value of t such that the object has traveled 50 feet is needed. This time appears to be 2.5 seconds.

```
In[35]:=
Plot[p1[[1,1,2]],{t,0,5}]

120
100
 80
 60
 40
 20

        1    2    3    4    5

Out[35]=
-Graphics-
```

A more accurate value of the time at which the object hits the surface is found below with **FindRoot**. The velocity at this time is then determined by substitution into the velocity function. This result is called **v1** for convenience.

```
In[36]:=
time=FindRoot[p1[[1,1,2]]==50,{t,2.5}]

Out[36]=
{t -> 2.47864}

In[37]:=
v1=d1[[1,1,2]]/.t->time[[1,2]]

Out[37]=
29.3166
```

The initial value problem which determines the velocity of the object beneath the surface of the pond is solved below in **d2** with **DSolve**. Then the position function is found in **p2**.

```
In[38]:=
d2=DSolve[{v'[t]==32-6v[t],v[0]==v1},v[t],t]

Out[38]=
{{v[t] -> 16/3 + 23.9832/E^(6 t)}}

In[39]:=
p2=DSolve[{y'[t]==d2[[1,1,2]],y[0]==0},y[t],t]

Out[39]=
{{y[t] -> 3.99721 - 3.99721/E^(6 t) + 16 t/3}}
```

This position function is then plotted to determine when the object is 25 feet beneath the surface of the pond. This time appears to be near 4 seconds.

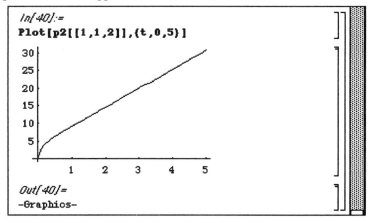

```
In[40]:=
Plot[p2[[1,1,2]],{t,0,5}]
```

```
Out[40]=
-Graphics-
```

FindRoot is used below with an initial guess of 4 to obtain a more accurate approximation of the t-value at which the object is 25 feet beneath the pond's surface. Finally, the time required for the object to reach the pond's surface is added to the time needed for it to travel 25 feet beneath the surface to see that approximately 6.41667 seconds are required for the object to travel from a height of 50 feet above the pond to a depth of 25 feet below the surface..

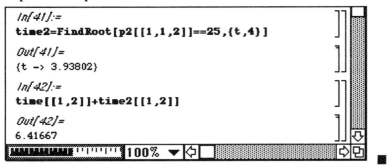

```
In[41]:=
time2=FindRoot[p2[[1,1,2]]==25,{t,4}]

Out[41]=
{t -> 3.93802}

In[42]:=
time[[1,2]]+time2[[1,2]]

Out[42]=
6.41667
```

Chapter 4: Higher Order Differential Equations

Mathematica commands used in **Chapter 4** include:

Coefficient	Exponent	PlotRange
ColumnForm	GraphicsArray	ReplaceAll
ComplexExpand	Integrate	Roots
ContourPlot	LegendreP	Short
Contours	Length	Show
ContourShading	Map	Simplify
D	MatrixForm	Solve
Det	Module	Sum
DisplayFunction	NDSolve	Table
DSolve	NIntegrate	TableForm
Evaluate	Part	Trig
Expand	Plot	Variables

§4.1 Preliminary Definitions and Notation

Definitions:

An ordinary differential equation of the form

$$\sum_{k=0}^{n} a_k(x) y^{(k)}(x) = a_n(x) y^{(n)}(x) + a_{n-1}(x) y^{(n-1)}(x) + \ldots + a_1(x) y'(x) + a_0(x) y(x) = f(x)$$

is called an **nth order ordinary linear differential equation**. If $f(x)$ is identically the zero function, the equation is said to be **homogeneous**; if $f(x)$ is not the zero function, the equation is said to be **nonhomogeneous**; and if the functions $a_i(x)$, $i=0, 1, 2, \ldots, n$ are constants, the equation is said to have **constant coefficients**.

Definition:

Let $f_1(x), f_2(x), f_3(x), \ldots, f_{n-1}(x)$, and $f_n(x)$ be a set of n functions at least $n-1$ times differentiable. Let $S = \{f_1(x), f_2(x), f_3(x), \ldots, f_{n-1}(x), f_n(x)\}$ be a set of n functions. S is **linearly dependent** on an interval I means there are constants $c_1, c_2, \ldots, c_n$, not all zero, so that $\sum_{k=1}^{n} c_k f_k(x) = 0$ for every value of x in the interval I. S is **linearly independent** means that S is not linearly dependent.

❏ EXAMPLE 4.1

Show that $S = \{\cos(2x), \sin(2x), \sin(x)\cos(x)\}$ is linearly dependent.

Solution:

S is linearly dependent since for every value of x, $0 \bullet \cos(2x) + 1 \bullet \sin(2x) - 2 \bullet \sin(x)\cos(x) = 0$. ∎

Definition:

Let $S = \{f_1(x), f_2(x), f_3(x), \ldots, f_{n-1}(x), f_n(x)\}$ be a set of n functions for which each is differentiable at least $n-1$ times. The **Wronskian** of S, denoted by

$W(S) = W(f_1(x), f_2(x), f_3(x), \ldots, f_{n-1}(x), f_n(x))$, is the determinant

$$\begin{vmatrix} f_1(x) & f_2(x) & \cdots & f_n(x) \\ f_1'(x) & f_2'(x) & \cdots & f_n'(x) \\ \vdots & \vdots & \vdots & \vdots \\ f_1^{(n-1)}(x) & f_2^{(n-1)}(x) & \cdots & f_n^{(n-1)}(x) \end{vmatrix}.$$

❏ EXAMPLE 4.2

If $S = \{\sin(x), \cos(x)\}$, compute $W(S)$.

Solution:

$$W(S) = \begin{vmatrix} \sin(x) & \cos(x) \\ \dfrac{d}{dx}(\sin(x)) & \dfrac{d}{dx}(\cos(x)) \end{vmatrix} = \begin{vmatrix} \sin(x) & \cos(x) \\ \cos(x) & -\sin(x) \end{vmatrix} = -\sin^2(x) - \cos^2(x) = -1.$$ ∎

Theorem:

Let $S = \{f_1(x), f_2(x), f_3(x), \ldots, f_{n-1}(x), f_n(x)\}$ be a set of n functions each differentiable at least $n-1$ times on an interval I.

If $W(S) \neq 0$ for at least one value of x in the interval I, S is linearly independent.

❏ EXAMPLE 4.3

Show that $S = \{e^x, x\,e^x, x^2\,e^x\}$ is linearly independent.

Solution:

The Wronskian of S is

$$
W(S) = \begin{vmatrix} e^x & x\,e^x & x^2\,e^x \\ \dfrac{d}{dx}\left(e^x\right) & \dfrac{d}{dx}\left(x\,e^x\right) & \dfrac{d}{dx}\left(x^2\,e^x\right) \\ \dfrac{d^2}{dx^2}\left(e^x\right) & \dfrac{d^2}{dx^2}\left(x\,e^x\right) & \dfrac{d^2}{dx^2}\left(x^2 e^x\right) \end{vmatrix} = \begin{vmatrix} e^x & x\,e^x & x^2\,e^x \\ e^x & (x+1)e^x & \left(x^2+2x\right)e^x \\ e^x & (x+2)e^x & \left(x^2+4x+2\right)e^x \end{vmatrix}
$$

$$
= e^x \begin{vmatrix} (x+1)e^x & \left(x^2+2x\right)e^x \\ (x+2)e^x & \left(x^2+4x+2\right)e^x \end{vmatrix} - x\,e^x \begin{vmatrix} e^x & \left(x^2+2x\right)e^x \\ e^x & \left(x^2+4x+2\right)e^x \end{vmatrix} + x^2 e^x \begin{vmatrix} e^x & (x+1)e^x \\ e^x & (x+2)e^x \end{vmatrix}
$$

$$
= e^x\left[(x+1)\left(x^2+4x+2\right)e^{2x} - (x+2)\left(x^2+2x\right)e^{2x}\right] - x\,e^x\left[\left(x^2+4x+2\right)e^{2x} - \left(x^2+2x\right)e^{2x}\right]
$$

$$
+ x^2 e^x\left[(x+2)\,e^{2x} - (x+1)\,e^{2x}\right]
$$

$$
= 2e^{3x}.
$$

Since the Wronskian of S is not zero, we conclude that S is linearly independent. ∎

■ EXAMPLE 4.4

Determine whether each set is linearly dependent or linear independent:

(a) $\left\{1 - 2\sin^2(x), \cos(2x)\right\}$; and

(b) $\left\{\texttt{LegendreP[1, x]}, \texttt{LegendreP[2, x]}, \texttt{LegendreP[3, x]}\right\}$.

Solution:

For (a) we first define **rowone** to be the list $\left\{1 - 2\sin^2(x), \cos(2x)\right\}$:

```
≡□≡≡≡≡≡≡≡≡≡≡≡ Wronskian ≡≡≡≡≡≡≡≡≡≡□≡
In[44]:=
rowone={1-2Sin[x]^2,Cos[2x]}

Out[44]=
        2
{1 - 2 Sin[x] , Cos[2 x]}
```

and then define **rowtwo** to be the list $\left\{-4\cos(x)\sin(x), -2\sin(2x)\right\}$ obtained by differentiating each element of **rowone**. **matrix** is then defined to be the 2×2 matrix $\{\texttt{rowone}, \texttt{rowone}\}$ and displayed in traditional matrix form with the command **MatrixForm**

```
In[45]:=
rowtwo=D[rowone,x]

Out[45]=
{-4 Cos[x] Sin[x], -2 Sin[2 x]}

In[47]:=
matrix={rowone,rowtwo};
MatrixForm[matrix]

Out[47]//MatrixForm=
        2
1 - 2 Sin[x]          Cos[2 x]

-4 Cos[x] Sin[x]     -2 Sin[2 x]
```

The Wronskian of $\{1 - 2\sin^2(x), \cos(2x)\}$ is the determinant of matrix which is computed

and simplified below. Since the Wronskian of $\{1 - 2\sin^2(x), \cos(2x)\}$ is zero, we can conclude that

$\{1 - 2\sin^2(x), \cos(2x)\}$ is linearly dependent.

```
In[48]:=
wronskian=Det[matrix]

Out[48]=
4 Cos[x] Cos[2 x] Sin[x] - 2 Sin[2 x] +
           2
   4 Sin[x]  Sin[2 x]

In[49]:=
Expand[wronskian,Trig->True]

Out[49]=
0
```

The **Legendre polynomial of order n**, $P_n(x)$, is a solution of the second order differential equation

$$\left(1 - x^2\right)\frac{d^2y}{dx^2} - 2x\frac{dy}{dx} + n(n+1)y = 0 \text{ which satisfies } \int_{-1}^{1} P_n(x)P_m(x)dx = 0 \text{ if } n \neq m.$$

The *Mathematica* command **LegendreP[n,x]** returns the Legendre polynomial of order n.
For (b), we define **row[1]** to be a table of the first three Legendre polynomials $P_1(x)$, $P_2(x)$, and $P_3(x)$,

$x, \dfrac{3x^2 - 1}{2}$, and $\dfrac{5x^3 - 3x}{2}$, respectively.

```
In[50]:=
row[1]=Table[LegendreP[n,x],{n,1,3}]
```

```
Out[50]=
```
$$\{x, \frac{-1 + 3\ x^2}{2}, \frac{-3\ x + 5\ x^3}{2}\}$$

We then define **row[2]** to be the derivative of each term of **row[1]** and **row[3]** to be the derivative of each term of **row[2]**. Note that **row[3]** could also be obtained by computing the second derivative of each term of **row[1]**.

```
In[51]:=
row[2]=D[row[1],x]
```

```
Out[51]=
```
$$\{1, 3\ x, \frac{-3 + 15\ x^2}{2}\}$$

```
In[52]:=
row[3]=D[row[2],x]
```

```
Out[52]=
{0, 3, 15 x}
```

matrix is then defined to be the 3×3 matrix with first row **row[1]**, second row **row[2]** and third row **row[3]** and displayed in matrix form with **MatrixForm**. Then the Wronskian of

$$\left\{ x, \frac{3x^2 - 1}{2}, \frac{5x^3 - 3x}{2} \right\}$$ is the determinant of **matrix**.

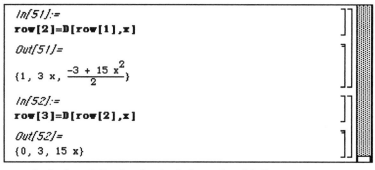

```
In[54]:=
matrix=Table[row[i],{i,1,3}];
MatrixForm[matrix]
```

```
Out[54]//MatrixForm=
```

Since the determinant of **matrix** is not identically zero, we conclude that the set

$$\left\{ x, \frac{3x^2 - 1}{2}, \frac{5x^3 - 3x}{2} \right\}$$ is linearly independent.

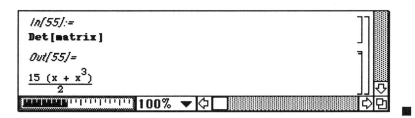

■ EXAMPLE 4.5

In the following example, the command **wronskian** is defined to compute the Wronskian of a list of functions in the variable x. **wronskian[list]** computes the Wronskian of the list **list** by

1. Defining the variables **n**, **r**, and **matrix** local to the function **wronskian**;
2. Defining **n** to be the length of **list**. The <u>length</u> of **list** is the number of elements of **list**;
3. Defining **r[1]** to be **list** and then **r[k]** to be the derivative of **r[k-1]**. **r[1]** corresponds to the first row of the matrix whose determinant will be computed to yield the Wronskian of **list**; **r[k]** corresponds to the kth row of the matrix;
4. Defining **matrix** to be a table of the lists **r[1]**, **r[2]**, ... , **r[n]**; and
5. Computing and simplifying the determinant of **matrix**.

We then use **wronskian** to show that the functions

$1 - 2\sin^2(x)$ and $\cos(2x)$ are linearly dependent while the set of functions $\sin(x)$, $\cos(x)$, and $\sin(2x)$ are linearly independent as is the set of functions $\sin(x), \sin(2x), \sin(3x), \sin(4x)$.

```
In[74]:=
wronskian[{1-2Sin[x]^2,Cos[2x]}]

Out[74]=
0

In[75]:=
wronskian[{Sin[x],Cos[x],Sin[2x]}]

Out[75]=
3 Sin[2 x]

In[76]:=
wronskian[{Sin[x],Sin[2x],Sin[3x],Sin[4x]}]

Out[76]=
                                            135 Cos[6 x]
189 - 315 Cos[2 x] + 180 Cos[4 x] - ———————————— +
                                                 2

              3 Cos[10 x]
15 Cos[8 x] - ————————————
                   2
```

100%

One way to see that $189 - 315\cos(2x) + 180\cos(4x) - \dfrac{135\cos(6x)}{2} + 15\cos(8x) - \dfrac{3\cos(10x)}{2}$ is not the zero

function is to graph it on a suitable interval. Alternatively, since

$189 - 315\cos(2x) + 180\cos(4x) - \dfrac{135\cos(6x)}{2} + 15\cos(8x) - \dfrac{3\cos(10x)}{2}$ has value 768 when $x = \dfrac{\pi}{2}$, it is

not the zero function. ■

A set $S = \left\{ f_1(x), f_2(x), f_3(x), \dots, f_{n-1}(x), f_n(x) \right\}$ of n linearly independent non - trivial solutions

of the nth order linear homogeneous equation

$a_n(x)y^{(n)}(x) + a_{n-1}(x)y^{(n-1)}(x) + \dots + a_1(x)y^{(1)}(x) + a_0(x) = 0$

is called a **fundamental set of solutions** of the equation.

Observe that if $S = \left\{ f_i(x) \right\}_{i=1}^{n}$ is a fundamental set of solutions of

$\displaystyle\sum_{i=0}^{n} a_i(x)y^{(i)}(x) = a_n(x)y^{(n)}(x) + a_{n-1}(x)y^{(n-1)}(x) + \dots + a_1(x)y'(x) + a_0(x)y(x) = 0$ and $\left\{ c_i \right\}_{i=1}^{n}$

is a set of n numbers, then $f(x) = \displaystyle\sum_{i=1}^{n} c_i f_i(x)$ is also a solution of $\displaystyle\sum_{i=0}^{n} a_i(x)y^{(i)}(x) = 0.$

The following two theorems tell us that under reasonable conditions the nth order homogeneous equation

$a_n(x)y^{(n)}(x) + a_{n-1}(x)y^{(n-1)}(x) + \dots + a_1(x)y^{(1)}(x) + a_0(x) = 0$ has a fundamental set of n solutions.

Theorem:
If $a_i(x)$ is continuous on an open interval I for $i = 0, 1, \ldots, n$, then the nth order linear homogeneous

equation $\displaystyle\sum_{i=0}^{n} a_i(x) y^{(i)}(x) = 0$ has a fundamental set of solutions on I.

Theorem:

Any set of $n+1$ solutions of the nth order linear homogeneous equation $\displaystyle\sum_{i=0}^{n} a_i(x) y^{(i)}(x) = 0$ is

linearly dependent.

❏ **EXAMPLE 4.6**

Show that $S = \left\{ e^{-5x}, e^{-x} \right\}$ is a fundamental set of solutions of the equation $y''(x) + 6y'(x) + 5y(x) = 0$.

Solution:

S is linearly independent since $W(S) = \begin{vmatrix} e^{-5x} & e^{-x} \\ -5e^{-5x} & -e^{-x} \end{vmatrix} = -e^{-6x} + 5e^{-6x} = 5e^{-6x} \neq 0$.

Since $\dfrac{d^2}{dx^2}\left(e^{-5x}\right) + 6\dfrac{d}{dx}\left(e^{-5x}\right) + 5e^{-5x} = 25e^{-5x} - 30e^{-5x} + 5e^{-5x} = 0$ and

$\dfrac{d^2}{dx^2}\left(e^{-x}\right) + 6\dfrac{d}{dx}\left(e^{-x}\right) + 5e^{-x} = e^{-x} - 6e^{-x} + 5e^{-x} = 0$, we conclude that S is a fundamental

set of solutions of the equation $y''(x) + 6y'(x) + 5y(x) = 0$. ■

§4.2 Solutions of Homogeneous Equations with Constant Coefficients

If $S = \left\{ f_i(x) \right\}_{i=1}^{n}$ is a fundamental set of solutions of the nth order linear homogeneous equation

$\displaystyle\sum_{i=0}^{n} a_i(x) y^{(i)}(x) = 0$, then a **general solution** of the equation is $f(x) = \displaystyle\sum_{i=1}^{n} c_i f_i(x)$

where $\left\{ c_i \right\}_{i=1}^{n}$ is a set of n arbitrary constants.

❑ **EXAMPLE 4.7**

Show that $y(x) = c_1 e^{-x} \cos(4x) + c_2 e^{-x} \sin(4x)$ is a general solution of

$y'' + 2y' + 17y = 0.$

Solution:

First, $y' = e^{-x}\left(-c_1 \cos(4x) + 4c_2 \cos(4x) - 4c_1 \sin(4x) - c_2 \sin(4x)\right)$ and

$y'' = e^{-x}\left(-15c_1 \cos(4x) - 8c_2 \cos(4x) + 8c_1 \sin(4x) - 15c_2 \sin(4x)\right)$.

Computing and simplifying $y'' + 2y' + 17y$ yields zero.

It is easy to verify that $e^{-x} \cos(4x)$ and $e^{-x} \sin(4x)$ are linearly independent and, consequently,

we can conclude that $y(x)$ is a general solution of $y'' + 2y' + 17y = 0.$

In the following, we use *Mathematica* to graph the solution for various values of c_1 and c_2 by first

defining the general solution y as a function of c_1 and c_2:

```
═════════════════ GeneralSolution ═════════════════
In[18]:=
Clear[y]
y[{c1_,c2_}]=Exp[-x] (c1 Cos[4x]+c2 Sin[4x])

Out[18]=
c1 Cos[4 x] + c2 Sin[4 x]
─────────────────────────
          E^x
```

Then we define a list of ordered pairs which correspond to the values of c_1 and c_2 to be graphed:

```
In[19]:=
vals={{0,1},{1,0},{2,1},{1,-2}};
```

Then, **Map** is used to compute the value of **y[{c1,c2}]** for each element of the list **vals**. The
resulting list of functions is named **funcs**:

```
In[20]:=
funcs=Map[y,vals]

Out[20]=
 Sin[4 x]   Cos[4 x]   2 Cos[4 x] + Sin[4 x]
{────────, ────────, ─────────────────────,
   E^x        E^x            E^x
 Cos[4 x] - 2 Sin[4 x]
 ────────────────────}
         E^x
```

Finally, the list of functions **funcs** is graphed on the interval $[-1,1]$:

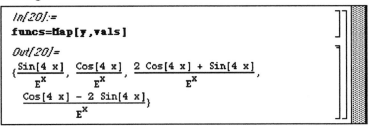

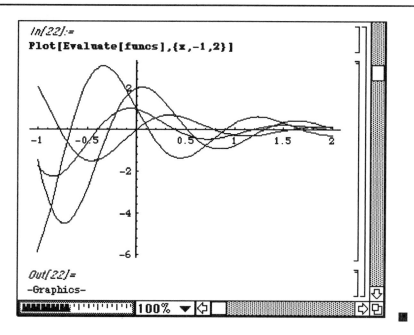

```
In[22]:=
Plot[Evaluate[funcs],{x,-1,2}]
```

```
Out[22]=
-Graphics-
```

■ EXAMPLE 4.8

Show that (a) $y = c_1 e^x \cos(2x) + c_2 e^x \sin(2x)$ is a general solution of $y'' - 2y' + 5y = 0$; and

(b) $y = c_1 e^x + c_2 x e^x + c_3 x^2 e^x$ is a general solution of $y''' - 3y'' + 3y' - y = 0$.

Solution:

For (a) we begin by clearing all prior definitions of y and then defining y. Note that **c[1]** and **c[2]** in the definition of **y** correspond to the constants c_1 and c_2, respectively.

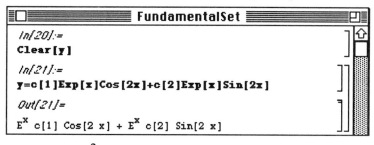

```
              FundamentalSet
In[20]:=
Clear[y]
In[21]:=
y=c[1]Exp[x]Cos[2x]+c[2]Exp[x]Sin[2x]
Out[21]=
 X                 X
E  c[1] Cos[2 x] + E  c[2] Sin[2 x]
```

Then compute $\dfrac{d^2}{dx^2} y - 2\dfrac{d}{dx} y + 5y$ and name the result **stepone**:

```
In[22]:=
stepone=D[y,{x,2}]-2D[y,x]+5y

Out[22]=
-3 E^x c[1] Cos[2 x] + 4 E^x c[2] Cos[2 x] -
  4 E^x c[1] Sin[2 x] - 3 E^x c[2] Sin[2 x] +
  5 (E^x c[1] Cos[2 x] + E^x c[2] Sin[2 x]) -
  2 (E^x c[1] Cos[2 x] + 2 E^x c[2] Cos[2 x] -
     2 E^x c[1] Sin[2 x] + E^x c[2] Sin[2 x])
```

and finally simplify **stepone**. Since the result is 0, we conclude that y is a general solution.

```
In[23]:=
Simplify[stepone]

Out[23]=
0
```

For (b) we proceed in a different manner. First, we define the list **sols**:

```
In[24]:=
sols={Exp[x],x Exp[x],x^2 Exp[x]};

In[25]:=
Clear[y]
```

and then use **Sum** to define **y**:

```
In[26]:=
y[x_]=Sum[c[i]sols[[i]],{i,1,3}]

Out[26]=
E^x c[1] + E^x x c[2] + E^x x^2 c[3]
```

and finally compute and simplify y'''−3y''+3y'−y:

```
In[27]:=
y'''[x]-3y''[x]+3y'[x]-y[x]

Out[27]=
3 E^x c[2] + 6 E^x c[3] + 6 E^x x c[3] +

   3 (E^x c[1] + E^x c[2] + E^x x c[2] + 2 E^x x c[3] +

      E^x x^2 c[3]) - 3

   (E^x c[1] + 2 E^x c[2] + E^x x c[2] + 2 E^x c[3] +

      4 E^x x c[3] + E^x x^2 c[3])

In[28]:=
y'''[x]-3y''[x]+3y'[x]-y[x]//Simplify

Out[28]=
0
```

The equation $a_n m^n + a_{n-1} m^{n-1} + \ldots + a_1 m + a_0 = \sum_{k=0}^{n} a_k m^k = 0$ is called the

characteristic equation of the nth order homogeneous linear differential equation with

constant coefficients $a_n y^{(n)}(x) + a_{n-1} y^{(n-1)}(x) + \ldots + a_1 y'(x) + a_0 y(x) = \sum_{k=0}^{n} a_k y^{(k)}(x) = 0.$

The general solutions of the nth order homogeneous linear differential equation with constant coefficients are determined by the solutions of its characteristic equation.

Let $a y'' + b y' + c y = 0$ be a homogeneous second order equation with constant coefficients and

let m_1 and m_2 be the solutions of the characteristic equation $a m^2 + b m + c = 0.$
   (i) If $m_1 \neq m_2$ and both m_1 and m_2 are real, a general solution of $a y'' + b y' + c y = 0$ is

$y(t) = c_1 e^{m_1 t} + c_2 e^{m_2 t};$
   (ii) If $m_1 = m_2$ and both m_1 and m_2 are real, a general solution of $a y'' + b y' + c y = 0$ is

$y(t) = c_1 e^{m_1 t} + c_2 t e^{m_1 t};$ and
   (iii) If $m_1 = \alpha + i \beta, \beta \neq 0,$ and $m_1 = \overline{m_2},$ a general solution of $a y'' + b y' + c y = 0$ is

$y(t) = c_1 e^{\alpha t} \cos(\beta t) + c_2 e^{\alpha t} \sin(\beta t).$

In (iii) above, $\overline{m_2}$ is the **complex conjugate** of m_2: $\overline{m_2} = \alpha - i \beta = \alpha + i \beta.$

The following three examples illustrate each of the above situations.

❑ **EXAMPLE 4.9**

Solve $y''+3y'-4y = 0$ subject to $y(0) = 1$ and $y'(0) = -1$.

Solution:

The characteristic equation of $y''+3y'-4y = 0$ is $m^2 + 3m - 4 = (m + 4)(m - 1) = 0$. Since the solutions of the characteristic equation are $m = -4$ and $m = 1$, the general solution of $y''+3y'-4y = 0$ is $y(x) = c_1 e^{-4x} + c_2 e^x$. Since $y'(x) = -4c_1 e^{-4x} + c_2 e^x$, applying the initial conditions $y(0) = 1$ and $y'(0) = -1$ yields the system of equations $\begin{cases} c_1 + c_2 = 1 \\ -4c_1 + c_2 = -1 \end{cases}$. Then, $c_1 = \dfrac{2}{5}$ and $c_2 = \dfrac{3}{5}$ so the desired solution is $y(x) = \dfrac{2}{5} e^{-4x} + \dfrac{3}{5} e^x$. ∎

❑ **EXAMPLE 4.10**

Solve $y''+2y'+y = 0$.

Solution:

The characteristic equation of $y''+2y'+y = 0$ is $m^2 + 2m + 1 = (m + 1)^2 = 0$. Since the solution of the characteristic equation is $m = -1$ with multiplicity two, the general solution of $y''+2y'+y = 0$ is $y(x) = c_1 e^{-x} + c_2 x e^{-x}$. ∎

❑ **EXAMPLE 4.11**

Solve $y''+4y'+20y = 0$.

Solution:

The characteristic equation of $y''+4y'+20y = 0$ is $m^2 + 4m + 20 = 0$. Completing the square yields $m^2 + 4m + 20 = (m + 2)^2 + 16 = 0$ so the solutions of the chacteristic equation are $m = -2 \pm 4i$. Note that the quadratic formula can be used to solve $m^2 + 4m + 20 = 0$ as well.

Since the solutions of the characteristic equation are complex conjugates, the general solution of $y''+4y'+20y = 0$ is $y(x) = e^{-2x}(c_1 \cos(4x) + c_2 \sin(4x))$. ∎

More generally, let $a_n m^n + a_{n-1}m^{n-1} + \ldots + a_1 m + a_0 = \displaystyle\sum_{k=0}^{n} a_k m^k = 0$ be the characteristic equation of the nth order homogeneous linear differential equation with real constant coefficients

$$a_n y^{(n)}(x) + a_{n-1}y^{(n-1)}(x) + \ldots + a_1 y'(x) + a_0 y(x) = \sum_{k=0}^{n} a_k y^{(k)}(x) = 0.$$

In the same manner as in the case for a second order homogeneous equation with real constant coefficents, a general solution is also determined by the solutions of the characteristic equation. Instead of stating an exact rule for the numerous situations encountered, we illustrate how a general solution is found in the following examples.

❏ **EXAMPLE 4.12**

Solve $4y''' - 12y'' + 21y' - 26y = 0$ subject to the initial conditions $y(0) = 0$, $y'(0) = -1$, and $y''(0) = 1$.

Solution:

In this case, the characteristic equation is $4m^3 - 12m^2 + 21m - 26 = 0$. Factoring yields the equation $(m - 2)\left(4m^2 - 4m + 13\right) = 0$ and then applying the quadratic formula yields the three roots

$$m = 2 \text{ or } m = \frac{1}{2} + \sqrt{3}\,i \text{ or } m = \frac{1}{2} - \sqrt{3}\,i.$$

In this case, each root has multiplicity one so a general solution of the equation is

$$y(x) = c_1\, e^{2x} + e^{x/2}\left(c_2 \cos\left(\sqrt{3}\,x\right) + c_3 \sin\left(\sqrt{3}\,x\right)\right).$$

To calculate the values of c_1, c_2, and c_3 that satisfy the initial conditions, observe that

$$y(0) = c_1 + c_2 = 0,\ y'(0) = 2c_1 + \frac{1}{2}c_2 + \sqrt{3}\,c_3 = -1,\text{ and } y''(0) = 4c_1 - \frac{11}{4}c_2 + \sqrt{3}\,c_3 = 1$$

yields the linear system of three equations in three unknowns
$$\begin{cases} c_1 + c_2 = 0 \\ 2c_1 + \dfrac{1}{2}c_2 + \sqrt{3}\,c_3 = -1. \\ 4c_1 - \dfrac{11}{4}c_2 + \sqrt{3}\,c_3 = 1 \end{cases}$$

The solution of this system is $c_1 = \dfrac{8}{21}$, $c_2 = \dfrac{-8}{21}$, and $c_3 = \dfrac{-11}{7\sqrt{3}}$. Therefore, the desired solution

is $y(x) = \dfrac{8}{21}\, e^{2x} + e^{x/2}\left(\dfrac{-8}{21}\cos\left(\sqrt{3}\,x\right) + \dfrac{-11}{7\sqrt{3}}\sin\left(\sqrt{3}\,x\right)\right).$ ∎

The command **DSolve** can be used to solve nth order linear homogeneous differential equations with constant coefficients as long as n is smaller than 5. In cases when the roots of the characteristic equation are symbolically complicated, approximations of the roots of the characteristic equation can be computed with the command **NRoots**.

■ **EXAMPLE 4.13**

Find the solution of each problem:

(a) $3y''+2y'-5y = 0$;

(b) $2y''+5y'+5y = 0$ subject to the initial conditions $y(0) = 0$ and $y'(0) = \dfrac{1}{2}$; and

(c) $y''+4y'+4y = 0$ subject to the initial conditions $y(0) = 0$ and $y'(0) = \dfrac{-1}{2}$.

Solution:

(a) is solved using **DSolve**.

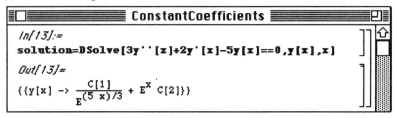

When **DSolve** is used to solve (b), the resulting solution is expressed as a complex exponential. To see that the solution is real, we use **ComplexExpand**.

```
In[18]:=
solution=DSolve[{2y''[x]+5y'[x]+5y[x]==0,
    y[0]==0,y'[0]==1/2},y[x],x]

Out[18]=
{{y[x] -> (-I (-E^((-5 - I Sqrt[15]) x)/4 +
        E^((-5 + I Sqrt[15]) x)/4)) / Sqrt[15]}}
```

Notice that **solution** is a nested list. **solution[[1,1,2]]** yields the second element of the first element of the first element of **solution**. In other words, **solution[[1,1,2]]** yields the expression corresponding to the desired solution. **ComplexExpand** is used to expand **solution[[1,1,2]]** assuming that x is real. The result is clearly a real-valued function.

```
In[19]:=
simp=solution[[1,1,2]]//ComplexExpand

Out[19]=
  2 Sin[Sqrt[15] x]
          4
  ─────────────────
  Sqrt[15] E^((5 x)/4)
```

Finally, the solution is graphed on the interval $[-\pi/2,\pi]$.

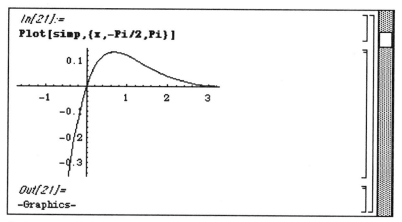

Similarly **DSolve** successfully solves (c).

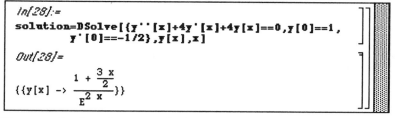

The result is then graphed on the interval [– 1,1]. Note that the command
Plot[solution[[1,1,2]],{x,-1,1}] would produce the same result.

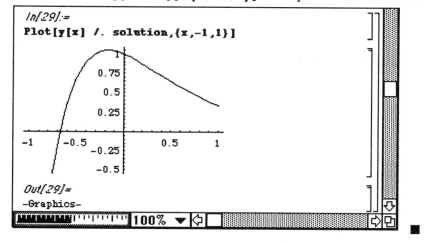

As stated above, in cases when either **DSolve** is slow or does not work, other techniques can frequently be used.

■ EXAMPLE 4.14

Find the general solution of (a) $y'''+3y''+2y'+6y = 0$; and (b) $y^{(4)} - 8y^{(3)} + 30y^{(2)} - 56y^{(1)} + 49y = 0$; and (c) graph the solution of $y'''+3y''+2y'+6y = 0$ subject to the conditions $y(0) = 0$, $y'(0) = 1$, and $y''(0) = -1$ on the interval $[-1,1]$.

Solution:

Using x instead of m, we obtain that

the characteristic equation for $y'''+3y''+2y'+6y = 0$ is $x^3 + 3x^2 + 2x + 6 = 0$. In this case, we use

Solve to solve the equation. Note, however, that either **NRoots** or **Roots** could also be used to solve the equation in this case.

```
▤□▤▤▤▤▤▤▤▤ ConstantCoefficients ▤▤▤▤▤▤▤▤▤
In[1]:=
Solve[x^3+3x^2+2x+6==0]

Out[1]=
{{x -> -3}, {x -> I Sqrt[2]}, {x -> -I Sqrt[2]}}
```

Since the solutions of the characteristic equation are -3, $\sqrt{2}\,i$, and $-\sqrt{2}\,i$, we conclude that the general solution is $y = c_1 e^{-3t} + c_2 \cos(\sqrt{2}\,t) + c_3 \sin(\sqrt{2}\,t)$.

Similarly, the characteristic equation of $y^{(4)} - 8y^{(3)} + 30y^{(2)} - 56y^{(1)} + 49y = 0$ is

$x^4 - 8x^3 + 30x^2 - 56x + 49 = 0$. The solutions are obtained with **Solve**:

```
In[2]:=
Solve[x^4-8x^3+30x^2-56x+49==0]

Out[2]=
{{x -> 4 + 2 I Sqrt[3]}, {x -> 4 - 2 I Sqrt[3]},
          ───────────              ───────────
               2                        2
   {x -> 4 + 2 I Sqrt[3]}, {x -> 4 - 2 I Sqrt[3]}}
          ───────────              ───────────
               2                        2
```

Therefore the general solution of $y^{(4)} - 8y^{(3)} + 30y^{(2)} - 56y^{(1)} + 49y = 0$ is

$y = e^{2t}\left[c_1 \cos(\sqrt{3}\,t) + c_2 \sin(\sqrt{3}\,t)\right] + te^{2t}\left[c_3 \cos(\sqrt{3}\,t) + c_4 \sin(\sqrt{3}\,t)\right]$.

In (a) we found that the general solution of $y'''+3y''+2y'+6y = 0$ is

$y = c_1 e^{-3t} + c_2 \cos(\sqrt{2}\,t) + c_3 \sin(\sqrt{2}\,t)$. Applying the conditions $y(0) = 0$, $y'(0) = 1$, and $y''(0) = -1$

yields the system of equations $\begin{cases} c_1 + c_2 = 0 \\ -3c_1 + \sqrt{2}\,c_3 = 1 \\ 9c_1 - c_2 = -1 \end{cases}$ which has solution $c_1 = \dfrac{-1}{11}$, $c_2 = \dfrac{1}{11}$, and $c_3 = \dfrac{4\sqrt{2}}{11}$.

Therefore the solution to the problem is $y = \dfrac{-1}{11}e^{-3t} + \dfrac{1}{11}\cos\left(\sqrt{2}\,t\right) + \dfrac{4\sqrt{2}}{11}\sin\left(\sqrt{2}\,t\right)$ which can be graphed on the interval [-1,1] with the **Plot** function.

As an alternative we use **NDSolve** to obtain a numerical solution of $y''' + 3y'' + 2y' + 6y = 0$ subject to the conditions $y(0) = 0$, $y'(0) = 1$, and $y''(0) = -1$ on the interval $[-1,1]$.

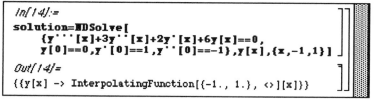

```
In[14]:=
solution=NDSolve[
    {y'''[x]+3y''[x]+2y'[x]+6y[x]==0,
     y[0]==0,y'[0]==1,y''[0]==-1},y[x],{x,-1,1}]

Out[14]=
{{y[x] -> InterpolatingFunction[{-1., 1.}, <>][x]}}
```

and then we graph the result:

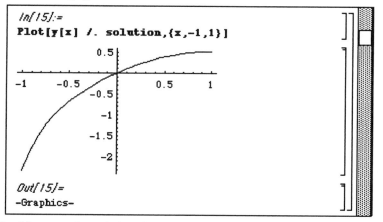

```
In[15]:=
Plot[y[x] /. solution,{x,-1,1}]
```

```
Out[15]=
-Graphics-
```

In this case, the solution found in (a) could be used to obtain an exact solution to the equation. In addition, notice that the result of **NDSolve** is a list, named **solution**. Explicitly, the interpolating function can be obtained with the command **solution[[1,1,2]]**:

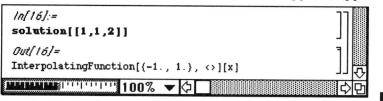

```
In[16]:=
solution[[1,1,2]]

Out[16]=
InterpolatingFunction[{-1., 1.}, <>][x]
```

§4.3   Nonhomogeneous Equations with Constant Coefficients: The Annihilator Method

A **particular solution**, $y_p(x)$, of the differential equation $f\left(x, y'(x), \ldots, y^{(n-1)}(x), y^{(n)}(x)\right) = g(x)$

is a specific function (which contains no arbitrary constants) that satisfies the equation.
The general solution to the nonhomogeneous equation

$$a_n y^{(n)}(x) + a_{n-1} y^{(n-1)}(x) + \ldots + a_1 y'(x) + a_0 y(x) = \sum_{k=0}^{n} a_k y^{(k)}(x) = g(x) \text{ is } y(x) = y_h(x) + y_p(x),$$

where $y_h(x)$ is the general solution of the corresponding homogeneous equation

$$a_n y^{(n)}(x) + a_{n-1} y^{(n-1)}(x) + \ldots + a_1 y'(x) + a_0 y(x) = \sum_{k=0}^{n} a_k y^{(k)}(x) = 0, \text{ and } y_p(x) \text{ is a particular}$$

solution to the nonhomogeneous equation.

Recall that the n - th order derivative of a function y is $D^n y = \dfrac{d^n y}{dx^n}$.

Thus, the linear nth order differential equation with constant coefficients

$$a_n y^{(n)}(x) + a_{n-1} y^{(n-1)}(x) + \ldots + a_1 y'(x) + a_0 y(x) = \sum_{k=0}^{n} a_k y^{(k)}(x) = g(x) \text{ can be expressed in}$$

operator notation

$$a_n y^{(n)}(x) + a_{n-1} y^{(n-1)}(x) + \ldots + a_1 y'(x) + a_0 y(x) = a_n D^n y(x) + a_{n-1} D^{n-1} y(x) + \ldots + a_1 Dy(x) + a_0 y(x)$$

$$= \left(a_n D^n + a_{n-1} D^{n-1} + \ldots + a_1 D + a_0\right) y(x) = g(x)$$

The expression $p(D) = a_n D^n + a_{n-1} D^{n-1} + \ldots + a_1 D + a_0$ is called an **nth order differential operator**.
The differential operator $p(D)$ is said to **annihilate** a function $f(x)$ if $p(D)(f(x))=0$ for all x.

❑ **EXAMPLE 4.15**
   Find a differential operator that annihilates (a) x^2; (b) xe^{-3x}; and (c) $e^x \sin(x)$ and $e^x \cos(x)$.
   Solution:
   The differential operator D^3 annihilates x^2 since $D^3 x^2 = 0$;

the differential operator $(D+3)^2$ annihilates xe^{-3x} since

$$(D+3)^2\left(xe^{-3x}\right) = \left(D^2 + 6D + 9\right)\left(xe^{-3x}\right) = D^2\left(xe^{-3x}\right) + 6D\left(xe^{-3x}\right) + 9\left(xe^{-3x}\right)$$

$$= \left(-6e^{-3x} + 9xe^{-3x}\right) + 6\left(e^{-3x} - 3xe^{-3x}\right) + 9xe^{-3x} = 0; \text{ and}$$

the differential operator $D^2 - 2D + 2$ annihilates both $e^x\sin(x)$ and $e^x\cos(x)$ since

$$\left(D^2 - 2D + 2\right)\left(e^x \sin(x)\right) = D^2\left(e^x \sin(x)\right) - 2D\left(e^x \sin(x)\right) + 2\left(e^x \sin(x)\right)$$

$$= 2e^x \cos(x) - 2\left(e^x \sin(x) + e^x \cos(x)\right) + 2e^x \sin(x) = 0 \text{ and}$$

$$\left(D^2 - 2D + 2\right)\left(e^x \cos(x)\right) = D^2\left(e^x \cos(x)\right) - 2D\left(e^x \cos(x)\right) + 2\left(e^x \cos(x)\right)$$

$$= -2e^x \sin s(x) - 2\left(e^x \cos(x) - e^x \sin(x)\right) + 2e^x \cos(x) = 0. \blacksquare$$

■ EXAMPLE 4.16

Show that (a) $(D+3)^3$ annihilates x^2e^{-3x}; (b) $D^2 - 4D + 13$ annihilates $e^{2x}\cos(3x)$; and

(c) $(D+3)^2\left(D^2 - 4D + 13\right)$ annihilates $x^2e^{-3x} + e^{2x}\cos(3x)$.

Solution:

We begin by defining a function **ann[q,f]** which computes q(D)(f(x)) for the differential operator q. **ann** is defined as follows:

1. The variables **var, exp, c, val**, and **p** are declared local to the function **ann**;
2. **q** is expanded and the result is named **p**;
3. The variable in **p** is named **var**;
4. **exp** is defined to be the degree of **p**;
5. **c[0]** is defined to be the constant term of **p** obtained by evaluating **p** when the value of the variable of **p** is 0;
6. For i greater than zero, **c[i]** is defined to be the coefficient of **exp** in **p**; and

7. The value of $\displaystyle\sum_{i=0}^{exp} c[i]\frac{d^i}{dx^i}f(x)$ is computed which corresponds to the value of q(D)(f(x)).

```
▤▢▤▤▤▤▤▤▤▤ OtherMethods(Annihilator) ▤▤▤▤▤⊡▤
In[1]:=
ann[q_,f_]:=
    Module[{var,exp,c,val,p},

    p=Expand[q];

    var=Variables[p];

    exp=Exponent[p,var[[1]]];

    c[0]=p /. var[[1]]->0;

    c[i_]:=Coefficient[p,var[[1]],i];

    val=Sum[c[i] D[f,{x,i}],{i,0,exp}]

    ]
```

We then use **a n n** to compute $(D+3)^3 \left(x^2 e^{-3x} \right)$ and name the result **o n e**:

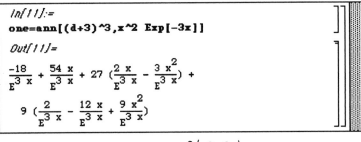

```
In[11]:=
one=ann[(d+3)^3,x^2 Exp[-3x]]

Out[11]=
```
$$\frac{-18}{E^{3\,x}} + \frac{54\,x}{E^{3\,x}} + 27\,(\frac{2\,x}{E^{3\,x}} - \frac{3\,x^2}{E^{3\,x}}) +$$

$$9\,(\frac{2}{E^{3\,x}} - \frac{12\,x}{E^{3\,x}} + \frac{9\,x^2}{E^{3\,x}})$$

Simplifying **o n e** shows that $(D+3)^3 \left(x^2 e^{-3x} \right) = 0$.

```
In[12]:=
Simplify[one]

Out[12]=
0
```

Similarly **a n n** and **S i m p l i f y** are used to compute then simplify $\left(D^2 - 4D + 13 \right) \left(e^{2x} \cos(3x) \right)$

and $(D+3)^3 \left(D^2 - 4D + 13 \right) \left(x^2 e^{-3x} + e^{2x} \cos(3x) \right)$:

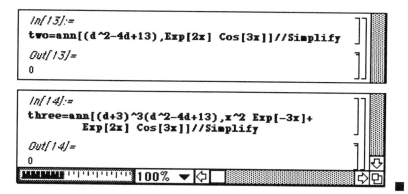

In general

(i) The differential operator D^n annihilates each of the functions $1, x, x^2, \ldots, x^{n-1}$;

(ii) The differential operator $(D - \alpha)^n$ annihilates each of the functions $e^{\alpha x}, x\,e^{\alpha x}, \ldots, x^{n-1}e^{\alpha x}$; and

(iii) The differential operator $\left[D^2 - 2\alpha D + \left(\alpha^2 + \beta^2\right)\right]^n$ annihilates each of the functions $e^{\alpha x} \cos(\beta x)$,

$xe^{\alpha x} \cos(\beta x), \ldots, x^{n-1}e^{\alpha x} \cos(\beta x), e^{\alpha x} \sin(\beta x), xe^{\alpha x} \sin(\beta x), \ldots, x^{n-1}e^{\alpha x} \sin(\beta x)$.

Therefore, the homogeneous linear nth order differential equation with constant coefficients can be expressed as p(D)y=g(x). When g(x) is a function of one of the above forms, another differential operator q(D) which annihilates g(x) can be determined.

Suppose that the differential operator q(D) annihilates g(x). Then applying q(D) to the nonhomogeneous equation yields q(D)p(D)y=q(D)g(x)=0. The form of the particular solution is found by solving the homogeneous equation q(D)p(D)y=0.

❏ **EXAMPLE 4.17**

Solve $y''+2y'-3y = 4e^x - \sin(x)$ subject to the conditions $y(0) = 0$ and $y'(0) = 1$.
Solution:

In operator notation, the equation $y''+2y'-3y = 4e^x - \sin(x)$ is the same as

$$\left(D^2 + 2D - 3\right)y = 4e^x - \sin(x)$$

Notice that by (ii), $(D - 1)$ annihilates $4e^x$ and by (iii), $\left(D^2 + 1\right)$ annihilates $\sin(x)$.

Applying $(D-1)$ and $\left(D^2+1\right)$ to the equation $\left(D^2+2D-3\right)y = 4e^x - \sin(x)$

yields $(D-1)\left(D^2+1\right)\left(D^2+2D-3\right)y = (D-1)\left(D^2+1\right)\left(4e^x - \sin(x)\right) = 0.$

The auxiliary equation of

$(D-1)\left(D^2+1\right)\left(D^2+2D-3\right)y = y^{(5)} + y^{(4)} - 4y^{(3)} + 4y'' - 5y' + 3y = 0$

is $m^5 + m^4 - 4m^3 + 4m^2 - 5m + 3 = (m-1)\left(m^2+1\right)\left(m^2+2m-3\right)$

$$= (m-1)^2\left(m^2+1\right)(m+3) = 0$$

which has roots $m = 1$ (with multiplicity two), $m = \pm i$, and $m = -3$.

First observe that $y_c = c_1 e^x + c_2 e^{-3x}$ is a general solution of the corresponding homogeneous equation $y'' + 2y' - 3y = 0$ and

$y(x) = \underbrace{c_1 e^x + c_2 e^{-3x}}_{y_c} + \underbrace{c_3 x\, e^x + c_4 \cos(x) + c_5 \sin(x)}_{y_p}$ is a general solution of

$y^{(5)} + y^{(4)} - 4y^{(3)} + 4y'' - 5y' + 3y = 0.$ Therefore a particular solution of

$y'' + 2y' - 3y = 4e^x - \sin(x)$ can be found of the form $y_p = c_3 x\, e^x + c_4 \cos(x) + c_5 \sin(x).$

Since $y_p' = c_3 x\, e^x + c_3 e^x - c_4 \sin(x) + c_5 \cos(x)$ and

$y_p'' = c_3 x\, e^x + 2c_3 e^x - c_4 \cos(x) - c_5 \sin(x)$, we obtain

$\left(c_3 x\, e^x + 2c_3 e^x - c_4 \cos(x) - c_5 \sin(x)\right) + 2\left(c_3 x\, e^x + c_3 e^x - c_4 \sin(x) + c_5 \cos(x)\right)$

$\qquad -3\left(c_3 x\, e^x + c_4 \cos(x) + c_5 \sin(x)\right) = 4e^x - \sin(x).$ Simplifying yields the equation

$4c_3 e^x + \left(2c_5 - 4c_4\right)\cos(x) + \left(-2c_4 - 4c_5\right)\sin(x) = 4e^x - \sin(x)$ and equating coefficients

yields the system $\begin{cases} 4c_3 = 4 \\ 2c_5 - 4c_4 = 0 \\ -2c_4 - 4c_5 = -1 \end{cases}$ which has solution $c_3 = 1, c_4 = \dfrac{1}{10}$, and $c_5 = \dfrac{1}{5}.$

Therefore $y_p = x\, e^x + \dfrac{\cos(x)}{10} + \dfrac{\sin(x)}{5}$ and a general solution of $y'' + 2y' - 3y = 4e^x - x\sin(x)$

is $y(x) = c_1 e^x + c_2 e^{-3x} + x\, e^x + \dfrac{\cos(x)}{10} + \dfrac{\sin(x)}{5}.$

Since $y(0) = \dfrac{1}{10} + c_1 + c_2 = 0$ and $y'(0) = \dfrac{6}{5} + c_1 - 3c_2 = 1$, $c_1 = \dfrac{-1}{8}$ and $c_2 = \dfrac{1}{40}$ and thus the desired

solution is $y(x) = \dfrac{-e^x}{8} + \dfrac{e^{-3x}}{40} + x\,e^x + \dfrac{\cos(x)}{10} + \dfrac{\sin(x)}{5}$. ∎

■ EXAMPLE 4.18

Solve $y'''(x) - y''(x) - 7y'(x) + 15y(x) = x^2 e^{-3x} + e^{2x}\cos(3x)$.

Solution:

In this case, we first use **DSolve** to compute the general solution of the equation and name the result **gensol**.

```
▤▢▭▭▭▭▭▭ OtherMethods(Annihilator) ▭▭▭▭▭ ▯▤

In[20]:=
gensol=DSolve[y'''[x]-y''[x]-7y'[x]+15y[x]==
        x^2 Exp[-3x]+Exp[2x] Cos[3x],
        y[x],x]

Out[20]=
{{y[x] ->  37 x         5 x^2        x^3      C[1]
          ---------  + ---------  + ------- + ----- +
          4394 E^3 x   338 E^3 x    78 E^3 x  E^3 x

  E^2 x C[3] Cos[x] -  5 E^2 x Cos[3 x]
                       ---------------- -
                             272

  E^2 x C[2] Sin[x] -  3 E^2 x Sin[3 x]
                       ---------------- }}
                             272
```

Alternatively, we use *Mathematica* to help construct a solution via the annihilator method.
In operator notation, the equation $y'''(x) - y''(x) - 7y'(x) + 15y(x) = x^2 e^{-3x} + e^{2x}\cos(3x)$ can be

written as $\left(D^3 - D^2 - 7D + 15\right)y(x) = x^2 e^{-3x} + e^{2x}\cos(3x)$. Since $(D+3)^3$ annihilates $x^2 e^{-3x}$

and $D^2 - 4D + 13$ annihilates $e^{2x}\cos(3x)$, applying $(D+3)^3$ and $D^2 - 4D + 13$ to the equation

yields $(D+3)^3\left(D^2 - 4D + 13\right)\left(D^3 - D^2 - 7D + 15\right)y(x) = 0$.

The corresponding homogeneous equation of $y'''(x) - y''(x) - 7y'(x) + 15y(x) = x^2 e^{-3x} + e^{2x}\cos(3x)$

is $y'''(x) - y''(x) - 7y'(x) + 15y(x) = 0$ which has auxiliary equation $d^3 - d^2 - 7d + 15 = 0$. In this

case, **Roots** is used to find the solutions of $d^3 - d^2 - 7d + 15 = 0$ and the auxiliary equation of

$(D+3)^3\left(D^2 - 4D + 13\right)\left(D^3 - D^2 - 7D + 15\right)y(x) = 0$, $(d+3)^3\left(d^2 - 4d + 13\right)\left(d^3 - d^2 - 7d + 15\right) = 0$.

Note, however, that **Solve** could be used just as easily.

```
In[22]:=
Roots[d^3-d^2-7d+15==0,d]
Out[22]=
d == -3 || d == 2 - I || d == 2 + I
In[23]:=
Roots[(d+3)^3(d^2-4d+13)(d^3-d^2-7d+15)==0,d]
Out[23]=
d == -3 || d == -3 || d == -3 || d == -3 ||
   d == 2 - I || d == 2 + I || d == 2 - 3 I ||
   d == 2 + 3 I
```

Therefore a general solution of the equation is of the form

$$y = \underbrace{c_1 e^{-3x} + e^{2x}\left(c_2 \cos(x) + c_3 \sin(x)\right)}_{y_h} +$$

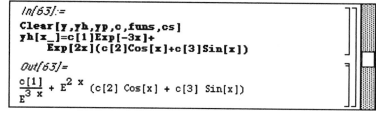

$$\underbrace{c_4 x\, e^{-3x} + c_5 x^2\, e^{-3x} + c_6 x^3 e^{-3x} + e^{2x}\left(c_7 \cos(3x) + c_8 \sin(3x)\right)}_{y_p}, \text{ where } y_p \text{ is a}$$

particular solution of $y'''(x) - y''(x) - 7y'(x) + 15y(x) = x^2 e^{-3x} + e^{2x}\cos(3x)$ and y_h is
the general solution of the corresponding homogeneous equation $y'''(x) - y''(x) - 7y'(x) + 15y(x) = 0$.
We proceed by first clearing all prior definitions of **y**, **yh**, **c**, **funs** and **cs** and then defining **yh**:

```
In[63]:=
Clear[y,yh,yp,c,funs,cs]
yh[x_]=c[1]Exp[-3x]+
   Exp[2x](c[2]Cos[x]+c[3]Sin[x])
Out[63]=
c[1]
---- + E^2 x (c[2] Cos[x] + c[3] Sin[x])
E^3 x
```

Instead of directly typing and entering the definition of **yp**, we first define a list of the functions that
appear in **yp** and name the result **funs**, then create a table of values of **c[i]** for $4 < i < 8$ and name the
result **cs**, and then finally defining **yp** by computing the dot product of **funs** and **cs**:

```
In[64]:=
funs={x Exp[-3x],x^2 Exp[-3x],
   x^3Exp[-3x],Exp[2x] Cos[3x],Exp[2x]
   Sin[3x]};
In[65]:=
cs=Table[c[i],{i,4,8}];
```

```
In[66]:=
yp[x_]=funs.cs
Out[66]=
```

$$\frac{x\ c[4]}{E^{3\ x}} + \frac{x^2\ c[5]}{E^{3\ x}} + \frac{x^3\ c[6]}{E^{3\ x}} + E^{2\ x}\ c[7]\ Cos[3\ x] +$$

$$E^{2\ x}\ c[8]\ Sin[3\ x]$$

Since we can find numbers c_4, c_5, c_6, c_7, and c_8 so that **y p[x]** satisfies

$y'''(x) - y''(x) - 7y'(x) + 15y(x) = x^2 e^{-3x} + e^{2x}\cos(3x)$, we compute and expand

yp'''-yp''-7yp'+15yp and name the result **lhseqn**:

```
In[67]:=
lhseqn=yp'''[x]-yp''[x]-7yp'[x]+
                        15yp[x]//Expand
Out[67]=
```

$$\frac{26\ c[4]}{E^{3\ x}} - \frac{20\ c[5]}{E^{3\ x}} + \frac{52\ x\ c[5]}{E^{3\ x}} + \frac{6\ c[6]}{E^{3\ x}} - \frac{60\ x\ c[6]}{E^{3\ x}} +$$

$$\frac{78\ x^2\ c[6]}{E^{3\ x}} - 40\ E^{2\ x}\ c[7]\ Cos[3\ x] -$$

$$24\ E^{2\ x}\ c[8]\ Cos[3\ x] + 24\ E^{2\ x}\ c[7]\ Sin[3\ x] -$$

$$40\ E^{2\ x}\ c[8]\ Sin[3\ x]$$

Equating the coefficients of **lhseqn** and $x^2 e^{-3x} + e^{2x}\cos(3x)$ yields the system of equations

$$\begin{cases} 26c_4 - 20c_5 + 6c_6 = 0 \\ 52c_5 - 60c_6 = 0 \\ \qquad 78c_6 = 1 \\ -40c_7 - 24c_8 = 1 \\ 24c_7 - 40c_8 = 0 \end{cases}$$

which we solve with **Solve** and then name the resulting list of

sols.

```
In[68]:=
sols=Solve[{26c[4]-20c[5]+6c[6]==0,
    52c[5]-60c[6]==0,
    78c[6]==1,
    -40c[7]-24c[8]==1,
    24c[7]-40c[8]==0}]
Out[68]=
```

$$\{\{c[5] \to \frac{5}{338}, \ c[6] \to \frac{1}{78}, \ c[4] \to \frac{37}{4394},$$
$$c[7] \to -(\frac{5}{272}), \ c[8] \to -(\frac{3}{272})\}\}$$

Then a particular solution is obtained by replacing the constants in **yp** by the values in **sols**:

```
In[69]:=
yp[x_]=yp[x]/. sols[[1]]
Out[69]=
```

$$\frac{37\ x}{4394\ E^{3\ x}} + \frac{5\ x^2}{338\ E^{3\ x}} + \frac{x^3}{78\ E^{3\ x}} - \frac{5\ E^{2\ x}\ Cos[3\ x]}{272} -$$
$$\frac{3\ E^{2\ x}\ Sin[3\ x]}{272}$$

Finally, the general solution is obtained by computing the sum of **yp** and **yh**:

```
In[70]:=
y[x_]=yh[x]+yp[x]
Out[70]=
```

$$\frac{37\ x}{4394\ E^{3\ x}} + \frac{5\ x^2}{338\ E^{3\ x}} + \frac{x^3}{78\ E^{3\ x}} + \frac{c[1]}{E^{3\ x}} -$$
$$\frac{5\ E^{2\ x}\ Cos[3\ x]}{272} + E^{2\ x}$$
$$(c[2]\ Cos[x] + c[3]\ Sin[x]) - \frac{3\ E^{2\ x}\ Sin[3\ x]}{272}$$

`100%`

§4.4 Nonhomogeneous Equations with Constant Coefficients: Variation of Parameters

Let $p(x)$, $q(x)$, and $f(x)$ be continuous on an interval I and let $y_1(x)$ and $y_2(x)$ be a fundamental set of solutions for the associated homogeneous equation of
$$y''(x) + p(x)y'(x) + q(x)y(x) = f(x), \quad y''(x) + p(x)y'(x) + q(x)y(x) = 0.$$
Then, $y_1''(x) + p(x)y_1'(x) + q(x)y_1(x) = 0$ and $y_2''(x) + p(x)y_2'(x) + q(x)y_2(x) = 0.$
This relationship is defined below in **rule**:

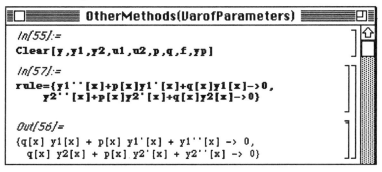

```
In[55]:=
Clear[y,y1,y2,u1,u2,p,q,f,yp]

In[57]:=
rule={y1''[x]+p[x]y1'[x]+q[x]y1[x]->0,
      y2''[x]+p[x]y2'[x]+q[x]y2[x]->0}

Out[56]=
{q[x] y1[x] + p[x] y1'[x] + y1''[x] -> 0,
  q[x] y2[x] + p[x] y2'[x] + y2''[x] -> 0}
```

Let $u_1(x)$ and $u_2(x)$ be two functions, if they exist, so that $y_p(x)=y_1(x)u_1(x)+y_2(x)u_2(x)$ is a particular solution of $y''(x) + p(x)y'(x) + q(x)y(x) = f(x)$ and $y_1u_1' + y_2u_2' = 0.$
We define **yp** and **ruletwo** below:

```
In[58]:=
yp[x_]=y1[x]u1[x]+y2[x]u2[x]

Out[58]=
u1[x] y1[x] + u2[x] y2[x]
```

```
In[59]:=
ruletwo={y1[x] u1'[x]+y2[x] u2'[x]->0}

Out[59]=
{y1[x] u1'[x] + y2[x] u2'[x] -> 0}
```

The derivative of **yp[x]** is computed below:

```
In[60]:=
D[yp[x],x]

Out[60]=
y1[x] u1'[x] + y2[x] u2'[x] + u1[x] y1'[x] +
  u2[x] y2'[x]
```

ruletwo is applied to the derivative of **yp[x]** and the result is named **first**.

```
In[61]:=
first=D[yp[x],x] /. ruletwo

Out[61]=
u1[x] y1'[x] + u2[x] y2'[x]
```

Therefore, $y_p'(x) = y_1'(x)u_1(x) + y_2'(x)u_2(x)$ $y_p''(x)$ is computed below by computing

the derivative of **first** and the result is named **second**:

```
In[62]:=
second=D[one,x]

Out[62]=
u1'[x] y1'[x] + u2'[x] y2'[x] + u1[x] y1''[x] +
  u2[x] y2''[x]
```

Then $y_p''(x) + p(x)y_p'(x) + q(x)y_p(x)$ is computed and expanded below. The result

is named **stepone**.

```
In[63]:=
stepone=second+p[x] first+q[x]yp[x]//Expand

Out[63]=
q[x] u1[x] y1[x] + q[x] u2[x] y2[x] +
  p[x] u1[x] y1'[x] + u1'[x] y1'[x] +
  p[x] u2[x] y2'[x] + u2'[x] y2'[x] +
  u1[x] y1''[x] + u2[x] y2''[x]
```

The terms containing $u_1(x)$ and $u_2(x)$ in **stepone** are then collected and the result is named **steptwo**:

```
In[64]:=
steptwo=Collect[stepone,{u1[x],u2[x]}]

Out[64]=
u1'[x] y1'[x] + u2'[x] y2'[x] +
  u1[x] (q[x] y1[x] + p[x] y1'[x] + y1''[x]) +
  u2[x] (q[x] y2[x] + p[x] y2'[x] + y2''[x])
```

And finally **rule** is applied to **steptwo** and the result is named **stepthree**.

```
In[65]:=
stepthree=steptwo /. rule

Out[65]=
u1'[x] y1'[x] + u2'[x] y2'[x]
```

100%

Thus $y_p''(x) + p(x)y_p'(x) + q(x)y_p(x) = y_1'(x)u_1'(x) + y_2'(x)u_2'(x) = f(x)$. Since

$y_1(x)u_1'(x) + y_2(x)u_2'(x) = 0$, we can solve the system $\begin{cases} y_1'(x)u_1'(x) + y_2'(x)u_2'(x) = f(x) \\ y_1(x)u_1'(x) + y_2(x)u_2'(x) = 0 \end{cases}$

for $u_1'(x)$ and $u_2'(x)$:

$$u_1'(x) = \frac{-y_2(x)f(x)}{y_1(x)y_2'(x) - y_1'(x)y_2(x)} \text{ and } u_2'(x) = \frac{y_1(x)f(x)}{y_1(x)y_2'(x) - y_1'(x)y_2(x)}.$$

It then follows that $u_1(x) = \int \dfrac{-y_2(x)f(x)}{y_1(x)y_2'(x) - y_1'(x)y_2(x)} \, dx$ and

$u_2(x) = \int \dfrac{y_1(x)f(x)}{y_1(x)y_2'(x) - y_1'(x)y_2(x)} \, dx$ so the general solution of $y''(x) + p(x)y'(x) + q(x)y(x) = f(x)$

is $y(x) = y_1(x) + y_2(x) + y_1(x)u_1(x) + y_2(x)u_2(x)$

$$= y_1(x) + y_2(x) + y_1(x)\int \frac{-y_2(x)f(x)}{y_1(x)y_2'(x) - y_1'(x)y_2(x)} \, dx + y_2(x)\int \frac{y_1(x)f(x)}{y_1(x)y_2'(x) - y_1'(x)y_2(x)} \, dx$$

More generally, if we are given the nonhomogeneous equation

$$y^{(n)}(x) + a_{n-1}(x)y^{(n-1)}(x) + \ldots + a_1(x)y'(x) + a_0(x)y(x) = \sum_{k=0}^{n} a_k(x)y^{(k)}(x) = g(x)$$

and a fundamental set of solutions $y_1(x), y_2(x), \ldots, y_n(x)$ of the associated homogeneous equation

$$y^{(n)}(x) + a_{n-1}(x)y^{(n-1)}(x) + \ldots + a_1(x)y'(x) + a_0(x)y(x) = \sum_{k=0}^{n} a_k(x)y^{(k)}(x) = 0, \text{ then}$$

we can extend the above to find $u_1(x), u_2(x), \ldots, u_n(x)$ such that $y_p(x) = \displaystyle\sum_{i=1}^{n} u_i(x)y_i(x)$ is a

particular solution of the nonhomogeneous equation.

If $\displaystyle\sum_{i=1}^{n} u_i'(x)y_i(x) = 0$, then $y_p^{(m)}(x) = \displaystyle\sum_{i=1}^{n} u_i(x)y_i^{(m)}(x)$ for $m = 0, 1, 2, \ldots, n-1$ and if

$$\sum_{i=1}^{n} u_i'(x)y_i^{(m-1)}(x) = 0 \text{ for } m = 1, 2, \ldots, n-1, \text{ then}$$

$$y_p^{(n)}(x) = \sum_{i=1}^{n} u_i(x)y_i^{(n)}(x) + \sum_{i=1}^{n} u_i'(x)y_i^{(n-1)}(x).$$

Therefore we obtain the system of n equations $\begin{cases} \displaystyle\sum_{i=1}^{n} y_i(x)u_i{}'(x) = 0 \\[2mm] \displaystyle\sum_{i=1}^{n} y_i{}'(x)u_i{}'(x) = 0 \\[1mm] \hspace{1cm}\vdots \\[1mm] \displaystyle\sum_{i=1}^{n} y_i{}^{(n-1)}(x)u_i{}'(x) = g(x) \end{cases}$ which, using Cramer's rule,

can be solved for $u_i{}'(x), i = 1, 2, \ldots, n$.

Let $W_m\big(y_1(x), y_2(x), \ldots, y_n(x)\big)$ denote the determinant obtained by replacing the mth column of

$$\begin{bmatrix} y_1(x) & y_2(x) & \cdots & y_n(x) \\ y_1{}'(x) & y_2{}'(x) & \cdots & y_n{}'(x) \\ \vdots & \vdots & \vdots & \vdots \\ y_1{}^{(n-1)}(x) & y_2{}^{(n-1)}(x) & \cdots & y_n{}^{(n-1)}(x) \end{bmatrix} \text{ by the column } \begin{bmatrix} 0 \\ 0 \\ \vdots \\ 0 \\ 1 \end{bmatrix}.$$

Then, by Cramer's rule, $u_i{}'(x) = \dfrac{g(x)W_i\big(y_1(x), y_2(x), \ldots, y_n(x)\big)}{W\big(y_1(x), y_2(x), \ldots, y_n(x)\big)}$, for $i = 1, 2, \ldots, n$, and

$$u_i(x) = \int \frac{g(x)W_i\big(y_1(x), y_2(x), \ldots, y_n(x)\big)}{W\big(y_1(x), y_2(x), \ldots, y_n(x)\big)}\, dx \text{ so}$$

$$y_p(x) = \sum_{i=1}^{n} u_i(x)y_i(x) = \sum_{i=1}^{n} y_i(x)\int \frac{g(x)W_i\big(y_1(x), y_2(x), \ldots, y_n(x)\big)}{W\big(y_1(x), y_2(x), \ldots, y_n(x)\big)}\, dx.$$

❏ EXAMPLE 4.19

Solve $y'' + \dfrac{1}{4}y = \sec\!\left(\dfrac{1}{2}x\right) + \csc\!\left(\dfrac{1}{2}x\right)$.

Solution:

A fundamental set of solutions for the associated homogeneous equations is

$$S = \left\{ \cos\!\left(\tfrac{1}{2}x\right),\ \sin\!\left(\tfrac{1}{2}x\right) \right\} \text{ and } W(S) = \begin{vmatrix} \cos\!\left(\tfrac{1}{2}x\right) & \sin\!\left(\tfrac{1}{2}x\right) \\ -\tfrac{1}{2}\sin\!\left(\tfrac{1}{2}x\right) & \tfrac{1}{2}\cos\!\left(\tfrac{1}{2}x\right) \end{vmatrix} = \frac{1}{2}.$$

Let $u_1(x) = \int \dfrac{-\sin\left(\frac{1}{2}x\right)\left(\sec\left(\frac{1}{2}x\right) + \csc\left(\frac{1}{2}x\right)\right)}{1/2} = -2\int\left(\dfrac{\sin\left(\frac{1}{2}x\right)}{\cos\left(\frac{1}{2}x\right)} + 1\right)dx = -2x + 4\mathrm{Ln}\left(\cos\left(\frac{x}{2}\right)\right)$

and $u_2(x) = \int \dfrac{\cos\left(\frac{1}{2}x\right)\left(\sec\left(\frac{1}{2}x\right) + \csc\left(\frac{1}{2}x\right)\right)}{1/2} = 2\int\left(1 + \dfrac{\cos\left(\frac{1}{2}x\right)}{\sin\left(\frac{1}{2}x\right)}\right)dx = 2x + 4\mathrm{Ln}\left(\sin\left(\frac{1}{2}x\right)\right).$

Then by Variation of Parameters,

$y_p(x) = \cos\left(\frac{1}{2}x\right)\left[-2x + 4\mathrm{Ln}\left(\cos\left(\frac{x}{2}\right)\right)\right] + \sin\left(\frac{1}{2}x\right)\left[2x + 4\mathrm{Ln}\left(\sin\left(\frac{1}{2}x\right)\right)\right]$ is a particular solution of

$y'' + \dfrac{1}{4}y = \sec\left(\frac{1}{2}x\right) + \csc\left(\frac{1}{2}x\right)$ and

$y = c_1\cos\left(\frac{1}{2}x\right) + c_2\sin\left(\frac{1}{2}x\right) + y_p(x)$

$= c_1\cos\left(\frac{1}{2}x\right) + c_2\sin\left(\frac{1}{2}x\right) + \cos\left(\frac{1}{2}x\right)\left[-2x + 4\mathrm{Ln}\left(\cos\left(\frac{x}{2}\right)\right)\right] + \sin\left(\frac{1}{2}x\right)\left[2x + 4\mathrm{Ln}\left(\sin\left(\frac{1}{2}x\right)\right)\right]$

is the general solution. ∎

■ EXAMPLE 4.20

Solve $y'' + 4y' + 13y = x\cos^2(3x)$.

Solution:

The associated homogeneous equation of the equation $y'' + 4y' + 13y = x\cos^2(3x)$ is $y'' + 4y' + 13y = 0$

which has characteristic equation $m^2 + 4m + 13 = 0$. Since the symbols **y1, y2, yc, yp, u1**, and **u2**
will be used in constructing the solution, all prior definitions are first cleared and then the characteristic
equation is solved for m:

```
≡□≡══════════ VariationOfParameters ═══════════回≡
In[4]:=
Clear[y1,y2,yc,yp,u1,u2]
Solve[m^2+4m+13==0]

Out[4]=
{{m -> -2 - 3 I}, {m -> -2 + 3 I}}
```

Since the solutions of the characteristic equation are $-2-3i$ and $-2+3i$, a fundamental set of solutions of

$y''+4y'+13y = 0$ is $\left\{e^{-2x}\cos(3x), e^{-2x}\sin(3x)\right\}$. Therefore, we define $f(x) = x\cos^2(3x)$,

$y_1(x) = e^{-2x}\cos(3x)$, and $y_2(x) = e^{-2x}\sin(3x)$.

```
In[5]:=
f[x_]=x Cos[3x]^2

Out[5]=
x Cos[3 x]^2

In[7]:=
y1[x_]=Exp[-2x]Cos[3x];
y2[x_]=Exp[-2x]Sin[3x];
```

and **wronksian** $= \begin{vmatrix} y_1(x) & y_2(x) \\ y_1'(x) & y_2'(x) \end{vmatrix} = 3e^{-4x}$.

```
In[8]:=
wronskian=Det[{{y1[x],y2[x]},
               {y1'[x],y2'[x]}}]//Simplify

Out[8]=
 3
───
E^4 x
```

To find a particular solution of $y''+4y'+13y = x\cos^2(3x)$, we first define

$$\textbf{u1prime} = \frac{-y_2(x)f(x)}{\textbf{wronskian}}:$$

```
In[9]:=
u1prime=-y2[x] f[x]/wronskian

Out[9]=
-(E^2 x x Cos[3 x]^2 Sin[3 x])
──────────────────────────────
              3
```

and then define $u_1(x) = \int \textbf{u1prime}\,dx = \int \dfrac{-y_2(x)f(x)}{\textbf{wronskian}}\,dx$.

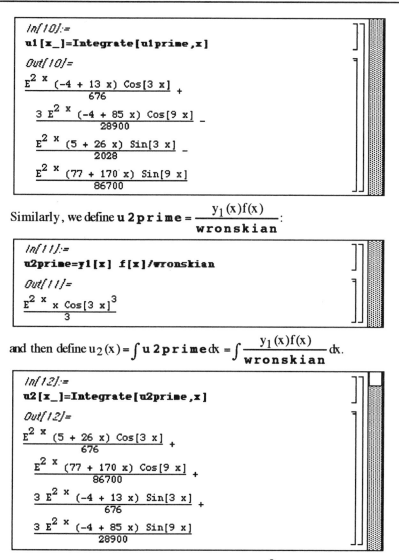

```
In[10]:=
u1[x_]=Integrate[u1prime,x]

Out[10]=
E^2 x (-4 + 13 x) Cos[3 x]
————————————————————————— +
          676

3 E^2 x (-4 + 85 x) Cos[9 x]
———————————————————————————— -
           28900

E^2 x (5 + 26 x) Sin[3 x]
———————————————————————— -
          2028

E^2 x (77 + 170 x) Sin[9 x]
———————————————————————————
          86700
```

Similarly, we define $\mathbf{u\,2\,prime} = \dfrac{y_1(x)f(x)}{\mathbf{wronskian}}$:

```
In[11]:=
u2prime=y1[x] f[x]/wronskian

Out[11]=
E^2 x x Cos[3 x]^3
——————————————————
        3
```

and then define $u_2(x) = \int \mathbf{u\,2\,prime}\,dx = \int \dfrac{y_1(x)f(x)}{\mathbf{wronskian}}\,dx$.

```
In[12]:=
u2[x_]=Integrate[u2prime,x]

Out[12]=
E^2 x (5 + 26 x) Cos[3 x]
———————————————————————— +
          676

E^2 x (77 + 170 x) Cos[9 x]
——————————————————————————— +
           86700

3 E^2 x (-4 + 13 x) Sin[3 x]
———————————————————————————— +
           676

3 E^2 x (-4 + 85 x) Sin[9 x]
————————————————————————————
           28900
```

Then a particular solution of $y''+4y'+13y = x\cos^2(3x)$ is given by
$y_p(x) = y_1(x)u_1(x) + y_2(x)u_2(x)$.

```
In[13]:=
yp[x_]=y1[x] u1[x]+y2[x] u2[x]
Out[13]=
```

$$(\text{Sin}[3\ x]\ (\frac{E^{2\ x}\ (5 + 26\ x)\ \text{Cos}[3\ x]}{676} +$$

$$\frac{E^{2\ x}\ (77 + 170\ x)\ \text{Cos}[9\ x]}{86700} +$$

$$\frac{3\ E^{2\ x}\ (-4 + 13\ x)\ \text{Sin}[3\ x]}{676} +$$

$$\frac{3\ E^{2\ x}\ (-4 + 85\ x)\ \text{Sin}[9\ x]}{28900})) \ / \ E^{2\ x} +$$

$$(\text{Cos}[3\ x]\ (\frac{E^{2\ x}\ (-4 + 13\ x)\ \text{Cos}[3\ x]}{676} +$$

$$\frac{3\ E^{2\ x}\ (-4 + 85\ x)\ \text{Cos}[9\ x]}{28900} -$$

$$\frac{E^{2\ x}\ (5 + 26\ x)\ \text{Sin}[3\ x]}{2028} -$$

$$\frac{E^{2\ x}\ (77 + 170\ x)\ \text{Sin}[9\ x]}{86700})) \ / \ E^{2\ x}$$

and the general solution of $y''+4y'+13y = 0$ is given by $y_c(x) = c_1 y_1(x) + c_2 y_2(x)$:

```
In[15]:=
yc[x_]=c[1] y1[x]+c[2] y2[x]
Out[15]=
```

$$\frac{c[1]\ \text{Cos}[3\ x]}{E^{2\ x}} + \frac{c[2]\ \text{Sin}[3\ x]}{E^{2\ x}}$$

so the general solution of $y''+4y'+13y = x \cos^2(3x)$ is given by $y(x) = y_c(x) + y_p(x)$.

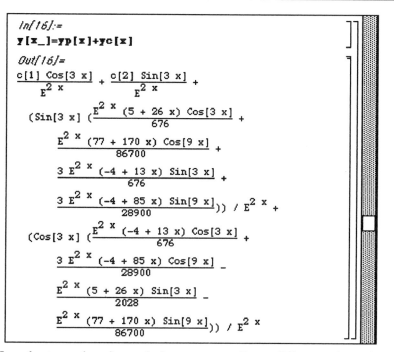

In order to graph various solutions corresponding to different values of c_1 and c_2, we first create a table of functions **tograph**:

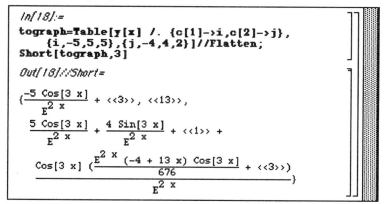

and then graph the table **tograph** on the interval $[-1,1]$:

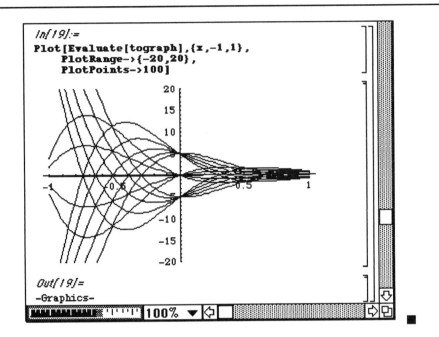

```
In[19]:=
Plot[Evaluate[tograph],{x,-1,1},
    PlotRange->{-20,20},
    PlotPoints->100]
```

```
Out[19]=
-Graphics-
```

■ EXAMPLE 4.21

Solve $y'' - 4y = \dfrac{e^{-4x}}{x^3}$.

Solution:

Proceeding as in the previous example, the characteristic equation of the associated homogeneous

equation of $y'' - 4y = \dfrac{e^{-4x}}{x^3}$ is $m^2 - 4 = 0$ which we solve for m:

```
══════════════ VariationOfParameters ══════════════
In[2]:=
Clear[y1,y2,yc,yp,u1,u2]
Solve[m^2-4==0]
Out[2]=
{{m -> 2}, {m -> -2}}
```

Then a fundamental set of solutions of $y'' - 4y = 0$ is $\left\{ e^{2x}, e^{-2x} \right\}$ so we define $f(x) = \dfrac{e^{-4x}}{x^3}$,

$y_1(x) = e^{2x}$, and $y_2(x) = e^{-2x}$:

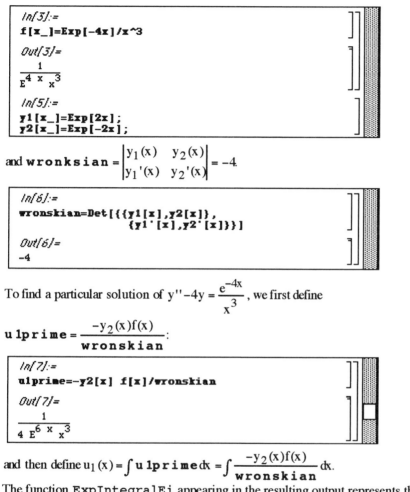

```
In[3]:=
f[x_]=Exp[-4x]/x^3
Out[3]=
   1
 -------
  4 x  3
 E    x
In[5]:=
y1[x_]=Exp[2x];
y2[x_]=Exp[-2x];
```

and $\mathbf{wronksian} = \begin{vmatrix} y_1(x) & y_2(x) \\ y_1{}'(x) & y_2{}'(x) \end{vmatrix} = -4.$

```
In[6]:=
wronskian=Det[{{y1[x],y2[x]},
               {y1'[x],y2'[x]}}]
Out[6]=
-4
```

To find a particular solution of $y'' - 4y = \dfrac{e^{-4x}}{x^3}$, we first define

$$\mathbf{u\ 1prime} = \frac{-y_2(x)f(x)}{\mathbf{wronskian}}:$$

```
In[7]:=
u1prime=-y2[x] f[x]/wronskian
Out[7]=
    1
 --------
  6 x  3
4 E    x
```

and then define $u_1(x) = \int \mathbf{u\ 1prime}\,dx = \int \dfrac{-y_2(x)f(x)}{\mathbf{wronskian}}\,dx.$

The function $\mathtt{ExpIntegralEi}$ appearing in the resulting output represents the **second exponential integral** function, $\mathrm{Ei}(z) = \int_{-z}^{\infty} \dfrac{e^{-t}}{t}\,dt.$

```
In[8]:=
u1[x_]=Integrate[u1prime,x]

Out[8]=
  -(  1   -  3 )
     2x²    x
  ───────────── +  9 ExpIntegralEi[-6 x]
     4 E⁶ˣ                  2
```

In the same manner as above we define $u2prime = \dfrac{y_1(x)f(x)}{\mathbf{wronskian}}$:

```
In[9]:=
u2prime=y1[x] f[x]/wronskian

Out[9]=
    -1
  ─────────
  4 E²ˣ x³
```

and then define $u_2(x) = \int u2prime\, dx = \int \dfrac{y_1(x)f(x)}{\mathbf{wronskian}}\, dx$.

```
In[10]:=
u2[x_]=Integrate[u2prime,x]

Out[10]=
   1   -  1
  2x²     x        ExpIntegralEi[-2 x]
  ────────    -    ──────────────────
   4 E²ˣ                    2
```

Then a particular solution of $y'' - 4y = \dfrac{e^{-4x}}{x^3}$ is $y_p(x) = y_1(x)u_1(x) + y_2(x)u_2(x)$

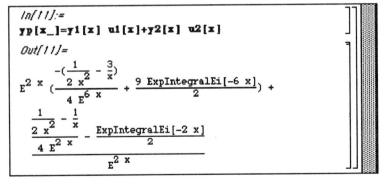

and the general solution of $y'' - 4y = 0$ is $y_c(x) = c_1\, y_1(x) + c_2\, y_2(x)$

```
In[12]:=
yc[x_]=c[1] Exp[-2x]+c[2] Exp[2x]

Out[12]=
c[1]
---- + E^2 x c[2]
E^2 x
```

Therefore the general solution of $y'' - 4y = \dfrac{e^{-4x}}{x^3}$ is $y(x) = y_c(x) + y_p(x)$.

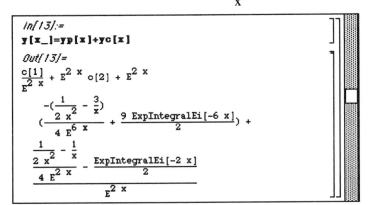

```
In[13]:=
y[x_]=yp[x]+yc[x]

Out[13]=
c[1]
---- + E^2 x c[2] + E^2 x
E^2 x
```

These solutions are graphed by creating a table of functions **tograph** corresponding to the function $y(x)$ where c_1 and c_2 are replaced by various constants.

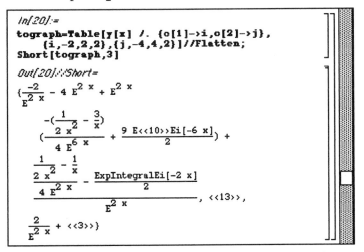

The set of functions **tograph** is then graphed on the interval [0.01,1].

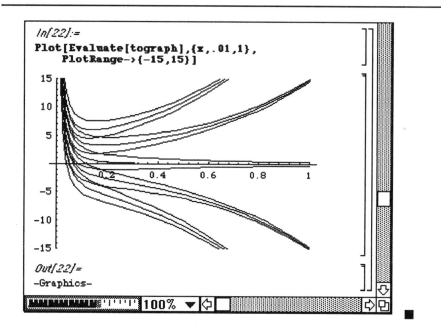

```
In[22]:=
Plot[Evaluate[tograph],{x,.01,1},
    PlotRange->{-15,15}]
```

Out[22]=
-Graphics-

■ EXAMPLE 4.22

Solve $y^{(4)} + 8y'' + 16y = \dfrac{\sin(2x)}{x}$.

Solution:

Proceeding as in the previous two examples, the associated homogeneous equation of

$y^{(4)} + 8y'' + 16y = \dfrac{\sin(2x)}{x}$ is $y^{(4)} + 8y'' + 16y = 0$ which has characteristic equation

$m^4 + 8m^2 + 16 = 0$ solved below:

```
≣□≣═══════ UariationOfParameters ═══════□≣
In[47]:=
Clear[y,yf,yc,yp,u1,u2]
Solve[x^4+8x^2+16==0]

Out[47]=
{{x -> 2 I}, {x -> -2 I}, {x -> 2 I},
   {x -> -2 I}}
```

Since the solutions $-2i$ and $2i$ have multiplicity two, a fundamental set of solutions of
$y^{(4)} + 8y'' + 16y = 0$ is $\{\cos(2x), \sin(2x), x\cos(2x), x\sin(2x)\}$. Each of these functions is
defined below as $\mathbf{yf[1]}$, $\mathbf{yf[2]}$, $\mathbf{yf[3]}$, and $\mathbf{yf[4]}$, respectively.

```
In[52]:=
yf[1][x_]=Cos[2x];
yf[2][x_]=Sin[2x];
yf[3][x_]=x Cos[2x];
yf[4][x_]=x Sin[2x];
```

Let **up[1]**, **up[2]**, **up[3]**, and **up[4]** denote u_1', u_2', u_3', and u_4', respectively, where $y_p(x)=$**yf[1]**u_1+**yf[2]**u_2+**yf[3]**u_3+**yf[4]**u_4 is a particular solution of

$y^{(4)} + 8y'' + 16y = \dfrac{\sin(2x)}{x}$. Then, by the method of variation of parameters we obtain the system

of equations $\left\{ \mathbf{y\,f[1]}^{(j)}\mathbf{u\,p[1]} + \mathbf{y\,f[2]}^{(j)}\mathbf{u\,p[2]} + \mathbf{y\,f[3]}^{(j)}\mathbf{u\,p[3]} + \mathbf{y\,f[4]}^{(j)}\mathbf{u\,p[4]} = 0 \right\}\Big|_{j=0}^{2}$,

defined below using the **Table** command; the resulting system of equations is named **equations**:

```
In[53]:=
equations=
    Table[Sum[D[yf[i][x],{x,j}] up[i],
          {i,1,4}]==0,{j,0,2}]

Out[53]=
{Cos[2 x] up[1] + Sin[2 x] up[2] +
   x Cos[2 x] up[3] + x Sin[2 x] up[4] == 0,
 -2 Sin[2 x] up[1] + 2 Cos[2 x] up[2] +
   (Cos[2 x] - 2 x Sin[2 x]) up[3] +
   (2 x Cos[2 x] + Sin[2 x]) up[4] == 0,
 -4 Cos[2 x] up[1] - 4 Sin[2 x] up[2] +
   (-4 x Cos[2 x] - 4 Sin[2 x]) up[3] +
   (4 Cos[2 x] - 4 x Sin[2 x]) up[4] == 0}
```

In addition, we obtain the equation

$\mathbf{y\,f[1]}^{(4)}\mathbf{u\,p[1]} + \mathbf{y\,f[2]}^{(4)}\mathbf{u\,p[2]} + \mathbf{y\,f[3]}^{(4)}\mathbf{u\,p[3]} + \mathbf{y\,f[4]}^{(4)}\mathbf{u\,p[4]} = g(x)$,

where $g(x) = \dfrac{\sin(2x)}{x}$, defined below as **equation4**.

```
In[54]:=
g[x_]=Sin[2x]/x

Out[54]=
Sin[2 x]
--------
   x

In[55]:=
equation4=Sum[D[yf[i][x],{x,3}] up[i],
                          {i,1,4}]==g[x]

Out[55]=
8 Sin[2 x] up[1] - 8 Cos[2 x] up[2] +
   (-12 Cos[2 x] + 8 x Sin[2 x]) up[3] +
   (-8 x Cos[2 x] - 12 Sin[2 x]) up[4] ==
   Sin[2 x]
   --------
      x
```

In order to solve the system of equations determined by the equations in **equations** and **equation4**, we use **AppendTo** to concatenate **equation4** to the end of **equations** and name the result **equations**:

```
In[56]:=
AppendTo[equations,equation4]

Out[56]=
{Cos[2 x] up[1] + Sin[2 x] up[2] +
    x Cos[2 x] up[3] + x Sin[2 x] up[4] == 0,
 -2 Sin[2 x] up[1] + 2 Cos[2 x] up[2] +
    (Cos[2 x] - 2 x Sin[2 x]) up[3] +
    (2 x Cos[2 x] + Sin[2 x]) up[4] == 0,
 -4 Cos[2 x] up[1] - 4 Sin[2 x] up[2] +
    (-4 x Cos[2 x] - 4 Sin[2 x]) up[3] +
    (4 Cos[2 x] - 4 x Sin[2 x]) up[4] == 0,
 8 Sin[2 x] up[1] - 8 Cos[2 x] up[2] +
    (-12 Cos[2 x] + 8 x Sin[2 x]) up[3] +
    (-8 x Cos[2 x] - 12 Sin[2 x]) up[4] ==
    Sin[2 x]
    --------}
       x
```

Since we must solve the system of equations **equations** for **up[1]**, **up[2]**, **up[3]**, and **up[4]**, we first define a list of the variables and name the results **ups**:

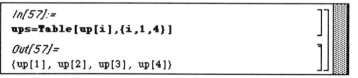

```
In[57]:=
ups=Table[up[i],{i,1,4}]

Out[57]=
{up[1], up[2], up[3], up[4]}
```

and then we solve the system of equations **equations** for the variables in the list **ups**, simplify the result, and name the resulting output **uprimetable**:

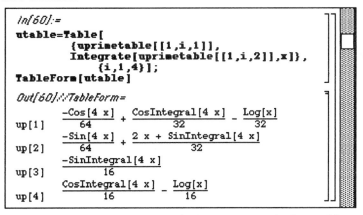

```
In[58]:=
uprimetable=Solve[equations,ups]//Simplify

Out[58]=
{{up[1] -> -1/(32 x) + Cos[4 x]/(32 x) + Sin[4 x]/16,

   up[2] -> 1/16 - Cos[4 x]/16 + Sin[4 x]/(32 x),

   up[3] -> -Sin[4 x]/(16 x), up[4] -> -Sin[2 x]^2/(8 x)}}
```

To compute u_1, u_2, u_3, and u_4 we integrate each expression in **uprimetable** and name the resulting table **utable**:

```
In[60]:=
utable=Table[
        {uprimetable[[1,i,1]],
         Integrate[uprimetable[[1,i,2]],x]},
        {i,1,4}];
TableForm[utable]

Out[60]//TableForm=

           -Cos[4 x]     CosIntegral[4 x]     Log[x]
up[1]      ---------  +  ----------------  -  ------
              64                32              32

           -Sin[4 x]     2 x + SinIntegral[4 x]
up[2]      ---------  +  ----------------------
              64                  32

           -SinIntegral[4 x]
up[3]      -----------------
                  16

           CosIntegral[4 x]     Log[x]
up[4]      ----------------  -  ------
                 16               16
```

The functions `CosIntegral` and `SinIntegral` in the resulting output represent the <u>**cosine integral function**</u>, $\mathrm{Ci}(z) = -\int_z^\infty \frac{\cos(t)}{t}\, dt = \gamma + \mathrm{Ln}(z) + \int_0^z \frac{\cos(t)-1}{t}$, and the <u>**sine integral function**</u>, $\mathrm{Si}(z) = \int_0^z \frac{\sin(t)}{t}\, dt$, respectively.

We then use the results to define $y_p(x)$ to be a particular solution of $y^{(4)} + 8y'' + 16y = \frac{\sin(2x)}{x}$.

```
In[53]:=
yp[x_]=Sum[yf[i][x] utable[[i,2]],{i,1,4}]

Out[53]=
          -Cos[4 x]     CosIntegral[4 x]
Cos[2 x] (--------- + ----------------- -
             64               32

   Log[x]          CosIntegral[4 x]     Log[x]
   ------) + x (----------------- - ------)
     32                  16              16

   Sin[2 x] - x Cos[2 x] SinIntegral[4 x] +
              ----------------------------
                          16

            -Sin[4 x]     2 x + SinIntegral[4 x]
   Sin[2 x] (--------- + ----------------------)
               64                 32
```

Similarly, $y_c(x)$ is defined to be the general solution of $y^{(4)} + 8y'' + 16y = 0$ and

$y(x) = y_c(x) + y_p(x)$ is defined to be the general solution of $y^{(4)} + 8y'' + 16y = \dfrac{\sin(2x)}{x}$.

```
In[54]:=
yc[x_]=Sum[c[i] yf[i][x],{i,1,4}]

Out[54]=
c[1] Cos[2 x] + x c[3] Cos[2 x] + c[2] Sin[2 x] +
  x c[4] Sin[2 x]

In[55]:=
y[x_]:=yc[x]+yp[x]
```

Finally, we graph $y(x)$ for various values of c_1, c_2, c_3, and c_4 by creating a table **tograph**:

```
In[70]:=
tograph=Table[y[x] /. {c[1]->i,
                       c[2]->j,
                       c[3]->k,
                       c[4]->l},
          {i,0,4,4},{j,-3,0,3},{k,0,2,2},
          {l,-2,0,2}]//Flatten;
```

and then plotting the result on the interval [.01,2]:

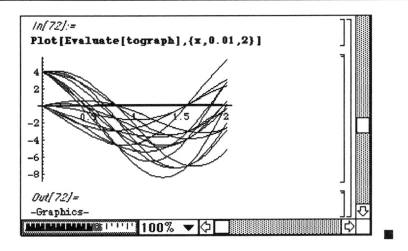

In[72]:=
Plot[Evaluate[tograph],{x,0.01,2}]

Out[72]=
-Graphics-

§4.5 Ordinary Differential Equations with Nonconstant Coefficients: Cauchy-Euler Equations

A **Cauchy-Euler** differential equation is a differential equation of the form

$$a_n x^n y^{(n)} + a_{n-1} x^{n-1} y^{(n-1)} + \ldots + a_1 x\, y' + a_0\, y = g(x),$$ where $\{a_i\}_{i=0}^n$ is a collection of constants.

Let $a x^2 y'' + b x y' + c y = 0$ be the general second order homogeneous Cauchy - Euler equation.

If $y = x^m$ for some constant m, then $y' = m x^{m-1}$ and $y'' = m(m-1) x^{m-2}$ and the general equation of order two becomes $a x^2 y'' + b x y' + c y = a x^2 m(m-1) x^{m-2} + b x m x^{m-1} + c x^m$

$$= x^m \left(a m(m-1) + b m + c \right) = 0.$$

$y = x^m$ is a solution when m is a solution of the equation $a m(m-1) + b m + c = 0$. The equation $a m(m-1) + b m + c$ is called the **auxiliary equation** of the Cauchy - Euler equation of order two.

The solutions of the auxiliary equation completely determine the general solution of the homogeneous Cauchy-Euler equation of order two.

Let m_1 and m_2 denote the two solutions of the equation $a m(m-1) + b m + c = 0$.

    (i) If $m_1 \neq m_2$ are real, then the general solution of $a x^2 y'' + b x y' + c y = 0$ is $y = c_1 x^{m_1} + c_2 x^{m_2}$;

    (ii) If $m_1 = m_2$, then the general solution of $a x^2 y'' + b x y' + c y = 0$ is $y = c_1 x^{m_1} + c_2 x^{m_1} Ln(x)$;

and

(iii) If $m_1 = \overline{m_2} = \alpha + i\beta$, $\beta \neq 0$, then the general solution of $ax^2 y'' + bx y' + cy = 0$ is

$$y = x^\alpha \left[c_1 \cos\left(\beta Ln(x)\right) + c_2 \sin\left(\beta Ln(x)\right) \right].$$

❑ EXAMPLE 4.23

Solve $3x^2 y'' - 2x y' + 2y = 0$.

Solution:

If $y = x^m$, $y' = mx^{m-1}$ and $y'' = m(m-1)x^{m-2}$ so

$3x^2 y'' - 2x y' + 2y = 3x^2 m(m-1)x^{m-2} - 2x mx^{m-1} + 2x^m = x^m \left(3m(m-1) - 2m + 2\right) = 0.$

Therefore the auxiliary equation of $3x^2 y'' - 2x y' + 2y = 0$ is $3m(m-1) - 2m + 2 = 3m^2 - 5m + 2 = 0$.

Factoring, $3m^2 - 5m + 2 = (3m-2)(m-1) = 0$ so the solutions of the auxiliary equation are $\dfrac{2}{3}$ and 1

and thus the general solution of $3x^2 y'' - 2x y' + 2y = 0$ is $y = c_1 x^{2/3} + c_2 x$. ∎

❑ EXAMPLE 4.24

Solve $x^2 y'' - x y' + y = 0$.

Solution:

Proceeding as in the previous example, the auxiliary equation of $x^2 y'' - x y' + y = 0$ is

$m(m-1) - m + 1 = m^2 - 2m + 1 = (m-1)^2 = 0$ which has the solution $m = 1$ with multiplicity two.

Therefore the general solution of $x^2 y'' - x y' + y = 0$ is $y = c_1 x + c_2 x Ln(x)$. ∎

❑ EXAMPLE 4.25

Solve $x^2 y'' - 5x y' + 10y = 0$.

Solution:

Proceeding as in the previous two examples, the auxiliary equation of $x^2 y'' - 5x y' + 10y = 0$ is

$m(m-1) - 5m + 10 = m^2 - 6m + 10 = 0$. Using the quadratic formula, we obtain that

$m = \dfrac{-4 \pm \sqrt{36 - 40}}{2} = 3 \pm i$. Therefore the general solution of $x^2 y'' - 5x y' + 10y = 0$ is

$y = x^3 \left[c_1 \cos\left(Ln(x)\right) + c_2 \sin\left(Ln(x)\right) \right]$. ∎

The auxiliary equation of higher order Cauchy-Euler equations is defined in the same way and solutions of higher order homogeneous Cauchy-Euler equations are determined in the same manner as solutions of higher order homogeneous differential equations with constant coefficients.

■ EXAMPLE 4.26

Solve (a) $x^3 y''' + 16x^2 y'' + 79x\,y' + 125y = 0$; and (b) $x^3 y''' + 9x^2 y'' + 44x\,y' + 58y = 0$.

Solution:

For (a), we use **DSolve** to directly compute the general solution of the equation,

$$y = \frac{c_1 + c_2\,x\sin\left(3\mathrm{Ln}(x)\right) + c_3\,x\cos\left(3\mathrm{Ln}(x)\right)}{x^5}.$$

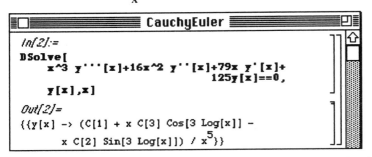

```
══════════════ CauchyEuler ══════════════
In[2]:=
DSolve[
      x^3 y'''[x]+16x^2 y''[x]+79x y'[x]+
                          125y[x]==0,
      y[x],x]
Out[2]=
{{y[x] -> (C[1] + x C[3] Cos[3 Log[x]] -
      x C[2] Sin[3 Log[x]]) / x^5}}
```

For (b), we proceed as in the previous three examples. We begin by defining **lhs** to correspond to the left - hand side of the equation $x^3 y''' + 9x^2 y'' + 44x\,y' + 58y = 0$.

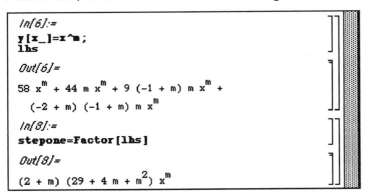

```
In[4]:=
Clear[x,y]
lhs=x^3 y'''[x]+9x^2 y''[x]+44x y'[x]+58y[x];
```

To compute the auxiliary equation of $x^3 y''' + 9x^2 y'' + 44x\,y' + 58y = 0$ we then define $y(x) = x^m$, evaluate **lhs**, and factor the result and name it **stepone**:

```
In[6]:=
y[x_]=x^m;
lhs
Out[6]=
58 x^m + 44 m x^m + 9 (-1 + m) m x^m +
   (-2 + m) (-1 + m) m x^m
In[8]:=
stepone=Factor[lhs]
Out[8]=
(2 + m) (29 + 4 m + m^2) x^m
```

The auxiliary equation of $x^3 y''' + 9x^2 y'' + 44x\,y' + 58y = 0$ is $(2 + m)\left(29 + 4m + m^2\right) = 0$. We first extract the left-hand side of the auxiliary equation and name the result **auxeq** with the **Take** command:

```
In[10]:=
auxeq=Take[stepone,2]

Out[10]=
                      2
(2 + m) (29 + 4 m + m )
```

and then solve the auxiliary equation:

```
In[12]:=
roots=Solve[auxeq==0]

Out[12]=
{{m -> -2}, {m -> -2 - 5 I}, {m -> -2 + 5 I}}
```

Since the solutions of the auxiliary equation are -2, $-2-5i$ and $-2+5i$, the general solution of $x^3y'''+9x^2y''+44x\,y'+58y = 0$ is $y(x) = c_1x^{-2} + x^{-2}\left(c_2\cos\left(5\text{Ln}\,(x)\right) + c_2\sin\left(5\text{Ln}\,(x)\right)\right)$ defined below:

```
In[15]:=
Clear[y]
y[x_]=c[1] x^(-2)+x^(-2)(c[2]Cos[5 Log[x]]+
          c[3]Sin[5 Log[x]]);
```

We then graph the general solutions for various values of c_1, c_2, and c_3 by creating a table **tograph** displayed in an abbreviated three-line form:

```
In[26]:=
tograph=Table[y[x] /. {c[1]->i,c[2]->j,
                       c[3]->k},
                 {i,-4,4,8},
                 {j,0,4,4},
                 {k,-4,0,4}]//Flatten;
Short[tograph,3]

Out[26]//Short=
 -4    4 Sin[5 Log[x]]   -4
{-- -  --------------- , -- ,
  2           2           2
 x           x           x

 -4    4 Cos[5 Log[x]] + <<1>>   -4
 -- +  --------------------- , -- + <<1>>,
  2             2               2
 x             x               x

         4             4    4 Cos[5 Log[x]]
 <<2>>, -- + <<1>>,   -- +  ---------------}
         2             2           2
        x             x           x
```

and then graphing the resulting table on the interval [.25,3]:

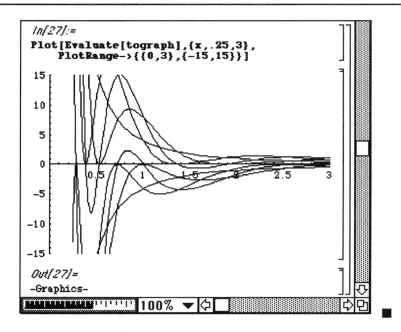

In some cases the method of variation of parameters can be used to solve nonhomogeneous Cauchy-Euler equations.

■ EXAMPLE 4.27

Solve $x^2 y'' - x y' + 5y = \dfrac{1}{x}$.

Solution:

We begin by using **DSolve** to find a fundamental set of solutions for the associated homogeneous $x^2 y'' - x y' + 5y = 0$:

```
In[3]:=
Clear[x,y,y1,y2,f]
solhom=DSolve[x^2 y''[x]-x y'[x]+5y[x]==0,y[x],x]
Out[3]=
{{y[x] -> x (C[2] Cos[2 Log[x]] - C[1] Sin[2 Log[x]])}}
```

Then a fundamental set of solutions for $x^2 y'' - x y' + 5y = 0$ is $\left\{ x \cos(2\mathrm{Ln}(x)),\, x \sin(2\mathrm{Ln}(x)) \right\}$. We

proceed by defining $f(x) = \dfrac{1}{x}$:

```
In[10]:=
f[x_]=1/x

Out[10]=
1
─
x
```

and then defining $y_1(x) = x\cos(2\text{Ln}(x))$ and $y_2(x) = x\sin(2\text{Ln}(x))$, and computing the wronskian

$$\textbf{wronskian} = \begin{vmatrix} y_1(x) & y_2(x) \\ y_1{}'(x) & y_2{}'(x) \end{vmatrix} = 2x:$$

```
In[6]:=
y1[x_]=x Cos[2Log[x]];
y2[x_]=x Sin[2Log[x]];
wronskian=Det[{{y1[x],y2[x]},
              {y1'[x],y2'[x]}}]//Simplify

Out[6]=
2 x
```

To find a particular solution, we begin by defining $\textbf{u 1p} = \dfrac{-y_2 f(x)}{\textbf{wronskian}}$ and $\textbf{u 2 p} = \dfrac{y_1 f(x)}{\textbf{wronskian}}$:

```
In[12]:=
u1p=-y2[x]f[x]/wronskian
u2p=y1[x] f[x]/wronskian

Out[11]=
-Sin[2 Log[x]]
──────────────
     2 x

Out[12]=
Cos[2 Log[x]]
─────────────
     2 x
```

and then defining $u_1(x) = \int \textbf{u 1p}\,dx = \int \dfrac{-y_2 f(x)}{\textbf{wronskian}}\,dx$ and

$u_2(x) = \int \textbf{u 2 p}\,dx = \int \dfrac{y_1 f(x)}{\textbf{wronskian}}\,dx.$

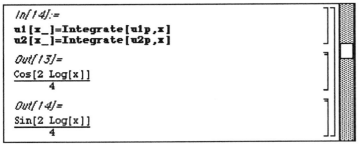

Then a particular solution of $x^2 y'' - x y' + 5y = \dfrac{1}{x}$ is $y_p(x) = y_1(x)u_1(x) + y_2(x)u_2(x)$, the general

solution of $x^2 y'' - x y' + 5y = 0$ is $y_c(x) = c_1 y_1(x) + c_2 y_2(x)$, and the general solution of

$x^2 y'' - x y' + 5y = \dfrac{1}{x}$ is $y(x) = y_c(x) + y_p(x)$, all defined and computed below:

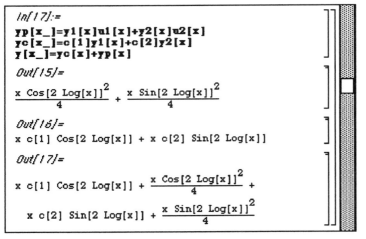

We then graph the general solution for various values of c_1 and c_2 by first creating a table of functions **tograph** and then graphing the result on the interval [.01,2]:

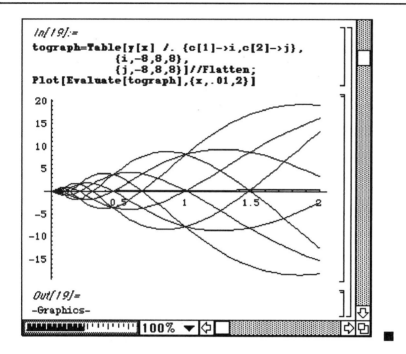

```
In[19]:=
tograph=Table[y[x] /. {c[1]->i,c[2]->j},
              {i,-8,8,8},
              {j,-8,8,8}]//Flatten;
Plot[Evaluate[tograph],{x,.01,2}]
```

```
Out[19]=
-Graphics-
```

§4.6   Ordinary Differential Equations with Nonconstant Coefficients: Exact Second-Order, Autonomous, and Equidimensional Equations

The second-order differential equation $F(x,y,y',y'')=0$ is an **exact second-order differential equation** means that there is a function ϕ such that F is the total differential of ϕ. If $F(x,y,y',y'')=0$ is exact, then $\phi=C$, C any constant, is a solution of $F(x,y,y',y'')=0$.

If $F(x,y,y',y'')=0$ is an exact second-order differential equation and ϕ is such that F is the total differential of ϕ, differentiating ϕ with respect to x yields $\dfrac{d\phi}{dx} = \dfrac{\partial\phi}{\partial x} + \dfrac{\partial\phi}{\partial y}y'+\dfrac{\partial\phi}{\partial y'}y''$. Therefore,

$$F(x,y,y',y'') = \frac{\partial\phi}{\partial x} + \frac{\partial\phi}{\partial y}y'+\frac{\partial\phi}{\partial y'}y''= f(x,y,y')y''+g(x,y,y') \text{ where}$$

$$f(x,y,y') = \frac{\partial\phi}{\partial y'} \text{ and } g(x,y,y') = \frac{\partial\phi}{\partial x} + \frac{\partial\phi}{\partial y}y'.$$

Let $p = y'$. Then, substituting into the above yields $f(x,y,p) = \dfrac{\partial\phi}{\partial p}$ and $g(x,y,p) = \dfrac{\partial\phi}{\partial x} + \dfrac{\partial\phi}{\partial y}p.$

Differentiating with respect to x, y, and p and eliminating ϕ yields the system of equations

$$\begin{cases} f_{xx} + 2pf_{xy} + p^2 f_{yy} = g_{xp} + pg_{yp} - g_y \\ \qquad f_{xp} + pf_{yp} + 2f_y = g_{pp}. \end{cases}$$ Conversely, if $f(x,y,y')y'' + g(x,y,y') = 0$ is

a differential equation that satisfies the system $\begin{cases} f_{xx} + 2pf_{xy} + p^2 f_{yy} = g_{xp} + pg_{yp} - g_y \\ \qquad f_{xp} + pf_{yp} + 2f_y = g_{pp}. \end{cases}$, then

$f(x,y,y')y'' + g(x,y,y') = 0$ is exact. In such a case, we can determine ϕ via

$\phi(x,y,p) = h(x,y) + \int f(x,y,p)\,dp$, where $p = y' = \dfrac{dy}{dx}$ as above.

❏ EXAMPLE 4.28

Solve $x\left(y'\right)^2 + y\,y' + x\,y\,y'' = 0$.

Solution:

To see that the equation is an exact equation, begin by letting $p = y'$, $f(x,y,p) = xy$, and

$g(x,y,p) = x\,p^2 + y\,p$. Then, $y\,y' - x\left(y'\right)^2 + x\,y\,y'' = 0 = f(x,y,p)y'' + g(x,y,p) = 0$ and we

must show that $f_{xx} + 2pf_{xy} + p^2 f_{yy} = g_{xp} + pg_{yp} - g_y$ and $f_{xp} + pf_{yp} + 2f_y = g_{pp}$. Computing,

$f_{xx} = 0, f_{xy} = 1, f_{yy} = 0, f_{xp} = 0, f_{yp} = 0, f_y = x, g_{xp} = 2p, g_{yp} = 1, g_y = p$, and $g_{pp} = 2x$. Then,

$f_{xx} + 2pf_{xy} + p^2 f_{yy} = 2p = 2p + p - p = g_{xp} + pg_{yp} - g_y$ and

$f_{xp} + pf_{yp} + 2f_y = 2x = g_{pp}$ so that $y\,y' - x\left(y'\right)^2 + x\,y\,y'' = 0$ is an exact equation.

Let $\phi(x,y,p) = h(x,y) + \int f(x,y,p)\,dp = h(x,y) + x\,y\,p$. The equation

$\dfrac{\partial \phi}{\partial x} + \dfrac{\partial \phi}{\partial y}p = h_x(x,y) + yp + h_y(x,y)p + x\,p^2 = g(x,y,p)$ where $g(x,y,p) = x\,p^2 + y\,p$ is satisfied

when $h(x,y) = C, C$ any constant. Therefore, $\phi(x,y,p) = \phi(x,y,y') = C + xyy'$ and

$x\,y\,y' = c_1$, c_1 any constant, is a solution to $x\left(y'\right)^2 + y\,y' + x\,y\,y'' = 0$. To compute y, we note

that the equation $xy\,y' = c_1$ is separable. Rewriting gives the equation

$y\,dy = \dfrac{c_1}{x}\,dx$ and integrating yields $\dfrac{y^2}{2} = c_1 \mathrm{Ln}(x) + c_2$, c_1 and c_2 arbitrary constants. Therefore,

the general solution of $x\left(y'\right)^2 + y\,y' + x\,y\,y'' = 0$ is $\dfrac{y^2}{2} = c_1 \mathrm{Ln}(x) + c_2$. ∎

■ **EXAMPLE 4.29**

Solve $\sec(x)\tan(x)y\,y' + \sec(x)\left(y'\right)^2 + \sec(x)y\,y'' = 0.$

Solution:

We show that the equation $\sec(x)\tan(x)y\,y' + \sec(x)\left(y'\right)^2 + \sec(x)y\,y'' = 0$ is exact by defining

$p = y'$, $f(x,y,p) = \sec(x)y$, and $g(x,y,p) = \sec(x)\tan(x)yp + \sec(x)p^2$:

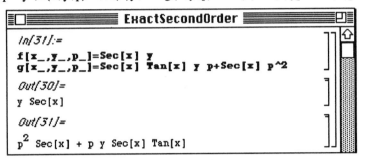

```
ExactSecondOrder
In[31]:=
f[x_,y_,p_]=Sec[x] y
g[x_,y_,p_]=Sec[x] Tan[x] y p+Sec[x] p^2

Out[30]=
y Sec[x]

Out[31]=
 2
p  Sec[x] + p y Sec[x] Tan[x]
```

All of the necessary derivatives are then computed below as are the left- and right-hand sides of the equations which must be satisfied in order that the equation be exact. Since the corresponding components are equivalent, this equation is exact.

```
In[43]:=
fxx=D[f[x,y,p],{x,2}];
fxy=D[f[x,y,p],x,y];
fyy=D[f[x,y,p],{y,2}];
fxp=D[f[x,y,p],x,p];
fyp=D[f[x,y,p],y,p];
fy=D[f[x,y,p],y];
gxp=D[g[x,y,p],x,p];
gyp=D[g[x,y,p],y,p];
gy=D[g[x,y,p],y];
gpp=D[g[x,y,p],{p,2}];

{fxx+2p fxy+p^2 fyy,gxp+p gyp-gy}
{fxp+p fyp+2 fy,gpp}

Out[42]=
                3
{y Sec[x]  + 2 p Sec[x] Tan[x] + y Sec[x] Tan[x]2,

                3
  y Sec[x]  + 2 p Sec[x] Tan[x] + y Sec[x] Tan[x]2}

Out[43]=
{2 Sec[x], 2 Sec[x]}
```

The formula used to determine the solution is defined below and appropriate derivatives determined. Note that quantities $h^{(1,0)}(x,y)$ and $h^{(0,1)}(x,y)$ below represent the derivative of h with respect to x and y, respectively.

```
In[47]:=
Clear[phi,h]
phi=h[x,y]+Integrate[f[x,y,p],p]
lhs=D[phi,x]+D[phi,y] p
lhs==g[x,y,p]

Out[45]=
h[x, y] + p y Sec[x]

Out[46]=
p y Sec[x] Tan[x] +
    p (p Sec[x] + h^(0,1)[x, y]) + h^(1,0)[x, y]

Out[47]=
p y Sec[x] Tan[x] +
    p (p Sec[x] + h^(0,1)[x, y]) + h^(1,0)[x, y] ==
   2
  p  Sec[x] + p y Sec[x] Tan[x]
```

Comparing the last two lines of output, we have:

```
In[53]:=
Clear[phi]
phi[x_,y_,p_]=p y Sec[x]+c[1]

Out[53]=
c[1] + p y Sec[x]
```

Therefore, we can use **DSolve** to obtain the solution as indicated below.

```
In[54]:=
DSolve[phi[x,y[x],y'[x]]==0,y[x],x]

Out[54]=
{{y[x] -> Sqrt[2 C[1] - 2 c[1] Sin[x]]},
  {y[x] -> -Sqrt[2 C[1] - 2 c[1] Sin[x]]}}
```

`100%` ▼

A differential equation of the form $F\left(y^{(n)}, y^{(n-1)}, \ldots, y', y\right) = 0$ is called an **autonomous**

differential equation.

Given the nth order autonomous differential equation $F\left(y^{(n)}, y^{(n-1)}, \ldots, y', y\right) = 0$, the substitution

$u(y) = y'(x)$ yields a differential equation of lower order.

We illustrate the solution of autonomous equations in the following examples.

❑ **EXAMPLE 4.30**

Solve $\dfrac{d^2 y}{dx^2} + 5\left(\dfrac{dy}{dx}\right)^2 - 4\dfrac{dy}{dx} = 0.$

Solution:

Since x does not explicitly occur in the equation $\dfrac{d^2 y}{dx^2} + 5\left(\dfrac{dy}{dx}\right)^2 - 4\dfrac{dy}{dx} = 0$, the equation is autonomous.

Let $u(y) = y'(x) = \dfrac{dy}{dx}$. By the chain rule, $\dfrac{d^2 y}{dx^2} = \dfrac{du}{dy}\dfrac{dy}{dx} = u(y)\dfrac{du}{dy}$ and substituting into the equation we

obtain the equation $u(y)\dfrac{du}{dy} + 5\left(u(y)\right)^2 - 4u(y) = 0.$ Since this equation is separable, we obtain

$\dfrac{5}{4u - 5u^2}\,du = dy$ and integrating both sides of the equation results in $\dfrac{5}{4}\mathrm{Ln}\left|\dfrac{u}{5u-4}\right| = y + c_1.$

Therefore, $\dfrac{u}{5u - 4} = ce^{4y/5}$, where c is a constant. Solving this equation for u yields

$u = \dfrac{4ce^{4y/5}}{5ce^{4y/5} - 1}.$ Since $u = y' = \dfrac{dy}{dx}$, we obtain the separable equation $\dfrac{dy}{dx} = \dfrac{4ce^{4y/5}}{5ce^{4y/5} - 1}$ and hence

$\dfrac{5ce^{4y/5} - 1}{4ce^{4y/5}}\,dy = dx.$ Integrating both sides of this equation results in $\dfrac{5y}{4} + \dfrac{5}{16ce^{4y/5}} = x + d,$

where d is a constant. ■

■ **EXAMPLE 4.31**

Solve $\dfrac{d^2 y}{dx^2} - (3y - 5)\dfrac{dy}{dx} = 0.$

Solution:

Proceeding as in the previous example, we note that the equation is autonomous since x does not explicitly occur in the equation and let u(y)=y'(x). In order to compute $y^{(n)}(x)$, we begin by defining **dy[1]=u[y[x]]** and **dy[2]=D[u[y[x]],x] /. y'[x]->u[y[x]]**. We then define **dy[n]** recursively to be **D[y[n-1],x] /. y'[x]->u[y[x]]**. The definition is of the form **dy[n_]:=dy[n]=...** so that *Mathematica* remembers the values of **dy[n]** computed. Note that **dy[n]** corresponds to the value of $y^{(n)}(x)$ using the substitution u(y)=y'(x). We then compute a table, **dys**, of **dy[1]**, **dy[2]**, and **dy[3]** and display **dys** in **ColumnForm**.

```
▤□▭▭▭▭▭▭ AutonomousEquations ▭▭▭▭▭▭▯▧
In[14]:=
Clear[dy,rule,step,lhs]
dy[1]=u[y[x]];
dy[2]=D[u[y[x]],x] /. y'[x]->u[y[x]];
dy[n_]:=dy[n]=D[dy[n-1],x] /. y'[x]->u[y[x]]

In[16]:=
dys=Table[{D[y[x],{x,n}],dy[n]},{n,1,3}];
ColumnForm[dys]

Out[16]=
{y'[x], u[y[x]]}
{y''[x], u[y[x]] u'[y[x]]}
   (3)                         2                2
{y   [x], u[y[x]] u'[y[x]]  + u[y[x]]   u''[y[x]]}
```

We then define **rule** to replace $D[y[x],\{x,n\}]=y^{(n)}(x)$ by **dy[n]**. Since **dy[n]** is defined recursively above, **dy[n]** can only be evaluated when **dy[n]** is a specific positive integer. Consequently, the command **RuleDelayed**, represented by **:>**, is used in the definition of **rule** so that **dy[n]** is only evaluated when **rule** is applied to an expression.

```
In[17]:=
rule={D[y[x],{x,n_}]:>dy[n]}
Out[17]=
{y^(n_)[x] :> dy[n]}
```

For example, in the following command we define **step** to be $y^{(3)}(x)$.

```
In[18]:=
step=D[y[x],{x,3}]
Out[18]=
 (3)
y   [x]
```

When **rule** is applied to **step**, $y^{(3)}(x)$ is replaced by **dy[3]**.

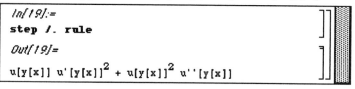

```
In[19]:=
step /. rule
Out[19]=
                   2            2
u[y[x]] u'[y[x]]  + u[y[x]]   u''[y[x]]
```

To solve the equation $\dfrac{d^2y}{dx^2} - (3y - 5)\dfrac{dy}{dx} = 0$ we define **lhs** to correspond to the left - hand side of

the equation $\dfrac{d^2y}{dx^2} - (3y - 5)\dfrac{dy}{dx} = 0$ and then apply rule to **lhs** and name the resulting output

stepone.

```
In[27]:=
lhs=y''[x]-(3y[x]-5)y'[x]

Out[27]=
-((-5 + 3 y[x]) y'[x]) + y''[x]

In[28]:=
stepone=lhs /. rule

Out[28]=
-(u[y[x]] (-5 + 3 y[x])) + u[y[x]] u'[y[x]]
```

For convenience, we then replace each occurrence of **y[x]** in **stepone** by **y** and name the resulting
output **steptwo**.

```
In[29]:=
steptwo=stepone /. y[x]->y

Out[29]=
-((-5 + 3 y) u[y]) + u[y] u'[y]
```

Therefore, we must solve the equation $u\dfrac{du}{dy} - (3y - 5)u = 0$ for u. We may proceed by recognizing that

$u\dfrac{du}{dy} - (3y - 5)u = 0$ is separable, rewriting the equation in the form $du = (3y - 5)\,dy$, and integrating
both sides of the equation. Instead, we use **DSolve** to solve the equation and name the resulting output

stepthree. The solution, $-5y + \dfrac{3y^2}{2} + c_1$, is extracted from **stepthree** with the command

stepthree[[1, 1, 2]].

```
In[30]:=
stepthree=DSolve[steptwo==0,u[y],y]

Out[30]=
{{u[y] -> -5 y + 3 y²/2 + C[1]}}
```

Since $u(y) = \dfrac{dy}{dx} = -5y + \dfrac{3y^2}{2} + c_1$ is separable, an implicit solution is obtained by integrating both

sides of the equation $\dfrac{dy}{-5y + \dfrac{3y^2}{2} + c_1} = dx$. In the following command, we compute $\int \dfrac{dy}{-5y + \dfrac{3y^2}{2} + c_1}$

and name the result **stepfour**.

```
In[31]:=
stepfour=Integrate[1/stepthree[[1,1,2]],y]

Out[31]=
       -5 + 3 y
2 ArcTan[Sqrt[-25 + 6 C[1]]]
   Sqrt[-25 + 6 C[1]]
```

Since $\int dx = x - C[2]$, where $C[2]$ represents an arbitrary constant, an implicit solution is given by

$$\frac{2\tan^{-1}\left(\dfrac{3y-5}{\sqrt{6C[1]-25}}\right)}{\sqrt{6C[1]-25}} - x + C[2] = 0.$$ To graph the solution, we begin by defining

$$\mathbf{sol} = \frac{2\tan^{-1}\left(\dfrac{3y-5}{\sqrt{6C[1]-25}}\right)}{\sqrt{6C[1]-25}} - x + C[2] = \mathbf{stepfour - x + C[2]}.$$

```
In[32]:=
sol=stepfour-x+C[2]

Out[32]=
            -5 + 3 y
     2 ArcTan[Sqrt[-25 + 6 C[1]]]
-x + ---------------------------- + C[2]
        Sqrt[-25 + 6 C[1]]
```

We will replace `C[1]` and `C[2]` by various numbers defined below in **pairs**. Then, graphs of

$$\frac{2\tan^{-1}\left(\dfrac{3y-5}{\sqrt{6C[1]-25}}\right)}{\sqrt{6C[1]-25}} - x + C[2] = 0.$$ To graph the solution, we begin by defining

$$\frac{2\tan^{-1}\left(\dfrac{3y-5}{\sqrt{6C[1]-25}}\right)}{\sqrt{6C[1]-25}} - x + C[2]$$ for particular values of $C[1]$ and $C[2]$ corresponding to the level curves

of **sol** corresponding to 0 for particular values of C[1] and C[2]. We then define **g[i]** to be the contour graph of **sol** on the rectangle $[-7,7] \times [-7,7]$ when C[1] is replaced by the first member of the ith ordered pair of **pairs** and C[2] is replaced by the second member of the ith ordered pair of **pairs**. The option **Contours->{0}** specifies that the only contour graphed correspond to the graph of **sol**=0, the option **ContourShading->False** specifies that the resulting space between contours is not to be shaded, and the option **DisplayFunction->Identity** specifies that the resulting graphics object is not displayed. Finally, we create a table of nine graphs, **graphs**, corresponding to the graph of **sol** for each of the nine members of **pairs**.

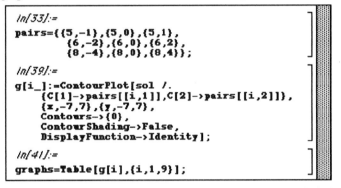

```
In[33]:=
pairs={{5,-1},{5,0},{5,1},
       {6,-2},{6,0},{6,2},
       {8,-4},{8,0},{8,4}};
In[39]:=
g[i_]:=ContourPlot[sol /.
    {C[1]->pairs[[i,1]],C[2]->pairs[[i,2]]},
    {x,-7,7},{y,-7,7},
    Contours->{0},
    ContourShading->False,
    DisplayFunction->Identity];
In[41]:=
graphs=Table[g[i],{i,1,9}];
```

The list of graphs **graphs** is then displayed simultaneously below.

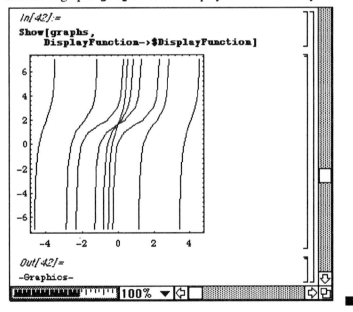

```
In[42]:=
Show[graphs,
    DisplayFunction->$DisplayFunction]
```

```
Out[42]=
-Graphics-
```

An equation in which all the y terms that appear have the same degree is called an **equidimensional (in y)** equation. An equidimensional (in y) equation can be converted to an equation of lower degree with the substitution $y(x)=e^{u(x)}$.

❑ EXAMPLE 4.32

Solve $e^{-2x}\left(y\dfrac{d^2y}{dx^2} - \left(\dfrac{dy}{dx}\right)^2\right) - 2x(1+x)y^2 = 0.$

Solution:

The equation $e^{-2x}\left(y\dfrac{d^2y}{dx^2} - \left(\dfrac{dy}{dx}\right)^2\right) - 2x(1+x)y^2 = 0$ is an equidimensional equation since each

y term, $y\dfrac{d^2y}{dx^2}, \left(\dfrac{dy}{dx}\right)^2$, and y^2 has degree 2.

Let $y(x)=e^{u(x)}$. Then, $y'=\dfrac{dy}{dx} = e^{u(x)}u'(x) = y\,u'(x)$ and

$y''=\dfrac{d^2y}{dx^2} = y'u'(x) + y\,u''(x) = y\,u'(x)u'(x) + y\,u''(x) = y\left((u'(x))^2 + u''(x)\right).$ Substituting into

the equation results in the equation $e^{-2x}\left(yy\left((u'(x))^2 + u''(x)\right) - \left(y\,u'(x)\right)^2\right) - 2x(1+x)y^2 = 0$ which

when simplified and factored yields $y^2\left(e^{-2x}u'' - 2x(1+x)\right) = 0.$ Then $e^{-2x}u'' - 2x(1+x) = 0$ so

$u'' = 2x(1+x)e^{2x}.$ Integrating twice, using integration by parts, we obtain

$u(x) = e^{2x}\left(\dfrac{1}{4} - \dfrac{1}{2}x + \dfrac{1}{2}x^2\right) + c_1x + c_2,$ where c_1 and c_2 represent arbitrary constants. Since $y(x) = e^{u(x)}$

we conclude that $y(x) = e^{e^{2x}\left(1/4 - x/2 + x^2/2\right) + c_1x + c_2}$ is the solution. ∎

◼ EXAMPLE 4.33

Solve $4x^{3/2}\left(y^2\dfrac{d^3y}{dx^3} - 3y\dfrac{dy}{dx}\dfrac{d^2y}{dx^2} + 2\left(\dfrac{dy}{dx}\right)^3\right) + \left(1 + 4x^{3/2}\cos(x)\right)y^3 = 0.$

Solution:

As in the previous example, the equation

$$4x^{3/2}\left(y^2\frac{d^3y}{dx^3}-3y\frac{dy}{dx}\frac{d^2y}{dx^2}+2\left(\frac{dy}{dx}\right)^3\right)+\left(1+4x^{3/2}\cos(x)\right)y^3=0$$ is an equidimensional equation

since each y term, $y^2\dfrac{d^3y}{dx^3}$, $y\dfrac{dy}{dx}\dfrac{d^2y}{dx^2}$, $\left(\dfrac{dy}{dx}\right)^3$, and y^3 has degree 3.

In order to compute $y^{(n)}(x)$, we begin by defining **dy[0]=Exp[u[x]]** and **dy[1]=y[x]u'[x]**. We then define **dy[n]** recursively to be **D[y[n-1],x] /. y'[x]->y[x]u'[x]**. The definition is of the form **dy[n_]:=dy[n]=...** so that *Mathematica* remembers the values of **dy[n]** computed. Note that **dy[n]** corresponds to the value of $y^{(n)}(x)$ using the substitution $y(x)=e^{u(x)}$. We then compute a table of **dy[1], dy[2]**, and **dy[3]** and display the result in **ColumnForm**.

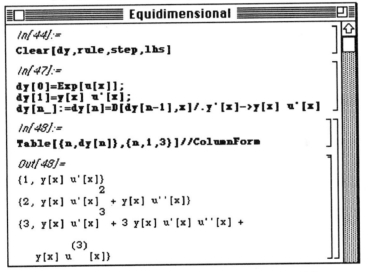

We then define **rule** to replace **D[y[x],{x,n}]**=$y^{(n)}(x)$ by **dy[n]**. Since **dy[n]** is defined recursively above, **dy[n]** can only be evaluated when **dy[n]** is a specific positive integer. Consequently, the command **RuleDelayed**, represented by **:>**, is used in the definition of **rule** so that **dy[n]** is only evaluated when **rule** is applied to an expression.

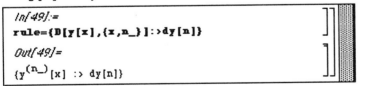

For example, in the following commands **step** is defined to be $y^{(3)}(x)$ and then when **rule** is applied to **step**, $y^{(3)}(x)$ is replaced by **dy[3]**.

```
In[50]:=
step=D[y[x],{x,3}]
Out[50]=
y(3)[x]
In[51]:=
step /. rule
Out[51]=
y[x] u'[x]³ + 3 y[x] u'[x] u''[x] + y[x] u(3)[x]
```

To solve $4x^{3/2}\left(y^2\dfrac{d^3y}{dx^3} - 3y\dfrac{dy}{dx}\dfrac{d^2y}{dx^2} + 2\left(\dfrac{dy}{dx}\right)^3\right) + \left(1 + 4x^{3/2}\cos(x)\right)y^3 = 0$ we begin by defining

lhs to be the left-hand side of the equation:

```
In[52]:=
lhs=4x^(3/2)(y[x]^2 y'''[x]-3y[x]y'[x]y''[x]+
          2y'[x]^3)+
          (1+4x^(3/2)Cos[x])y[x]^3
Out[52]=
(1 + 4 x3/2 Cos[x]) y[x]3 +
   4 x3/2 (2 y'[x]3 - 3 y[x] y'[x] y''[x] +
      y[x]2 y(3)[x])
```

and then apply **rule** to **lhs**, factor the result, and name the resulting output **stepone**.

```
In[54]:=
stepone=lhs /. rule//Factor
Out[54]=
y[x]3 (1 + 4 x3/2 Cos[x] + 4 x3/2 u(3)[x])
```

Then, we must solve the equation $1 + 4x^{3/2}\cos(x) + 4x^{3/2}u^{(3)}(x) = 0$ for u. Note that

$1 + 4x^{3/2}\cos(x) + 4x^{3/2}u^{(3)}(x)$ is extracted from **stepone** with **stepone[[1, 1, 2]]**.

Solving $1 + 4x^{3/2}\cos(x) + 4x^{3/2}u^{(3)}(x) = 0$ for $u^{(3)}(x)$ yields

$u^{(3)}(x) = \dfrac{-1\left(1 + 4x^{3/2}\cos(x)\right)}{4x^{3/2}}$. The expression $\dfrac{-1\left(1 + 4x^{3/2}\cos(x)\right)}{4x^{3/2}}$ is extracted from the list

steptwo with **steptwo[[1, 1, 2]]**.

```
In[57]:=
steptwo=Solve[stepone[[2]]==0,u'''[x]]
Out[57]=
```

$$\{\{u^{(3)}[x] \;\to\; \frac{-(1 + 4\ x^{3/2}\ Cos[x])}{4\ x^{3/2}}\}\}$$

To compute $u(x)$ we first compute $u''(x) = \int \dfrac{-1\left(1 + 4x^{3/2}\cos(x)\right)}{4x^{3/2}}\,dx = \int$ **steptwo[[1, 1, 2]]** dx

and name the result **stepthree** then compute

$u'(x) = \int\left(\dfrac{1}{2\sqrt{x}} - \sin(x) + c_1\right)dx = \int$ **stepthree** dx and name the result **stepfour**.

```
In[59]:=
stepthree=Integrate[steptwo[[1,1,2]],x]+c[1]
Out[59]=
 1
----- + c[1] - Sin[x]
2 Sqrt[x]
In[60]:=
stepfour=Integrate[stepthree,x]+c[2]
Out[60]=
Sqrt[x] (1 + Sqrt[x] c[1]) + c[2] + Cos[x]
```

Last we compute $u(x) = \int$ **stepfour** $dx = \int\left(\sqrt{x}\left(1 + c_1\sqrt{x}\right) + \cos(x) + c_2\right)dx$

$$= \frac{x\left(4\sqrt{x} + 3c_1 x + 6c_2\right)}{6} + \sin(x) + c_3 \text{, where } c_1, c_2, \text{ and}$$

c_3 represent arbitrary constants, and name the result **stepfive**. Since $y(x) = e^{u(x)}$, the general

solution is given by $y(x) = e^{\text{stepfive}}$, defined below as **sol**.

```
In[61]:=
stepfive=Integrate[stepfour,x]+c[3]

Out[61]=
x (4 Sqrt[x] + 3 x c[1] + 6 c[2])
--------------------------------- + c[3] + Sin[x]
             6

In[62]:=
sol=Exp[stepfive]

Out[62]=
Power[E,  x (4 Sqrt[x] + 3 x c[1] + 6 c[2])  +
          ---------------------------------
                       6
   c[3] + Sin[x]]
```

To graph the solution, **sol**, for various values of c_1, c_2, and c_3, we first create an array of ordered triples, **array**. Since **array** is a two-dimensional array and we wish to have a list of ordered triples, we use **Flatten** to convert **array** to a list of ordered triples, **triples**. An abbreviated two-line form of **triples** is then displayed.

We then define **tograph** to be a table of functions consisting of **sol** with **c[1]** replaced by the first member of the ith ordered triple of **triples**, **c[2]** replaced by the second member of the ith ordered triple of **triples**, and **c[3]** replaced by the third member of the ith ordered triple of **triples**, for i=1, 2, 3, ... ,9.

```
In[77]:=
array=Table[{i,j,0},{i,-1,1},{j,-1,1}];
triples=Flatten[array,1];
Short[triples,2]

Out[77]//Short=
{{-1, -1, 0}, {-1, 0, 0}, {-1, 1, 0}, {0, -1, 0},
   <<3>>, {1, 0, 0}, {1, 1, 0}}

In[78]:=
tograph=Table[sol/.{c[1]->triples[[i,1]],
     c[2]->triples[[i,2]],
     c[3]->triples[[i,3]]},{i,1,9}];
```

We then graph the table of functions **tograph** on the interval [0,1].

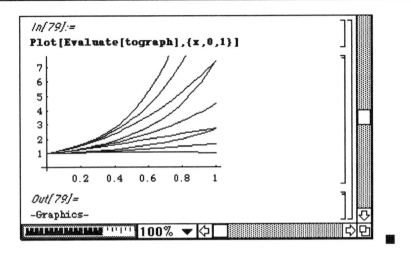

```
In[79]:=
Plot[Evaluate[tograph],{x,0,1}]
```

Out[79]=
-Graphics-

■ **EXAMPLE 4.34**

Solve $2\left(\dfrac{dy}{dx}\right)^3 + 2y\left(\dfrac{dy}{dx}\right)^2 - 3y\,\dfrac{dy}{dx}\,\dfrac{d^2y}{dx^2} + 17y^2\,\dfrac{dy}{dx} - 2y^2\,\dfrac{d^2y}{dx^2} + y^2\,\dfrac{d^3y}{dx^2} = 0.$

Solution:

As in the previous two examples, the equation

$2\left(\dfrac{dy}{dx}\right)^3 + 2y\left(\dfrac{dy}{dx}\right)^2 - 3y\,\dfrac{dy}{dx}\,\dfrac{d^2y}{dx^2} + 17y^2\,\dfrac{dy}{dx} - 2y^2\,\dfrac{d^2y}{dx^2} + y^2\,\dfrac{d^3y}{dx^2} = 0$ is equidimensional

since each y term, $\left(\dfrac{dy}{dx}\right)^3$, $y\left(\dfrac{dy}{dx}\right)^2$, $\dfrac{dy}{dx}\,\dfrac{d^2y}{dx^2}$, $y^2\,\dfrac{dy}{dx}$, $y^2\,\dfrac{d^2y}{dx^2}$ and $y^2\,\dfrac{d^3y}{dx^2}$ has degree 3.

Proceeding as in the previous example, we first define **lhs** to be the left-hand side of the equation and then apply rule to **lhs**, factor the result and name the resulting output **stepone**.

```
╔══════════════ Equidimensional ══════════════╗
 In[80]:=
 lhs=2y'[x]^3+2y[x]y'[x]^2-3y[x]y'[x]y''[x]+
           17y[x]^2y'[x]-
           2y[x]^2y''[x]+y[x]^2y'''[x]

 Out[80]=
                          2                  2               3
 17 y[x]  y'[x] + 2 y[x] y'[x]  + 2 y'[x]  -
        2
   2 y[x]  y''[x] - 3 y[x] y'[x] y''[x] +
        2  (3)
   y[x]  y   [x]

 In[81]:=
 stepone=lhs /. rule//Factor

 Out[81]=
       3                               (3)
 y[x]  (17 u'[x] - 2 u''[x] + u   [x])
╚══════════════════════════════════════════════╝
```

We must then solve the equation $u''' - 2u'' + 17u' = 0$. Note that $u''' - 2u'' + 17u'$ is extracted from **stepone** with **stepone[[2]]**. The characteristic equation of $u''' - 2u'' + 17u' = 0$ is $m^3 - 2m^2 + 17m = 0$. Factoring and completing the square yields

$$m^3 - 2m^2 + 17m = m\left(m^2 - 2m + 17\right) = m\left[(m-1)^2 + 16\right] = m\left[(m-1) - 4i\right]\left[(m-1) + 4i\right] = 0$$

so $m = 0$ or $m = 1 \pm 4i$. Therefore, the general solution of $u''' - 2u'' + 17u' = 0$ is

$u(x) = c_1 + e^x\left(c_2 \sin(4x) + c_3 \cos(4x)\right)$. Below we use **DSolve** to solve $u''' - 2u'' + 17u' = 0$ and name the resulting output **steptwo**.

```
╔══════════════════════════════════════════════╗
 In[82]:=
 steptwo=DSolve[stepone[[2]]==0,u[x],x]

 Out[82]=
                        x
 {{u[x] -> C[1] + E  C[3] Cos[4 x] -
      x
     E  C[2] Sin[4 x]}}
╚══════════════════════════════════════════════╝
```

Then the general solution of

$$2\left(\frac{dy}{dx}\right)^3 + 2y\left(\frac{dy}{dx}\right)^2 - 3y\frac{dy}{dx}\frac{d^2y}{dx^2} + 17y^2\frac{dy}{dx} - 2y^2\frac{d^2y}{dx^2} + y^2\frac{d^3y}{dx^2} = 0 \text{ is}$$

$y(x) = e^{u(x)} = e^{c_1 + e^x\left(c_2 \sin(4x) + c_3 \cos(4x)\right)}$, defined below as **sol**.

```
In[83]:=
sol=Exp[steptwo[[1,1,2]]]

Out[83]=
E C[1] + E^x C[3] Cos[4 x] - E^x C[2] Sin[4 x]
```

To graph the solution, **sol**, for various values of c_1, c_2, and c_3, we first create an array of ordered triples, **array**. Since **array** is a two-dimensional array and we wish to have a list of ordered triples, we use **Flatten** to convert **array** to a list of ordered triples, **triples**. An abbreviated two-line form of **triples** is then displayed.

We then define **tograph** to be a table of functions consisting of **sol** with **c[1]** replaced by the first member of the ith ordered triple of **triples**, **c[2]** replaced by the second member of the ith ordered triple of **triples**, and **c[3]** replaced by the third member of the ith ordered triple of **triples**, for i=1, 2, 3, ... ,9.

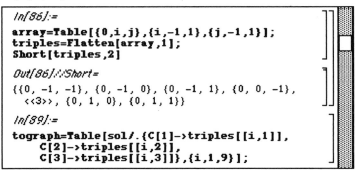

```
In[86]:=
array=Table[{0,i,j},{i,-1,1},{j,-1,1}];
triples=Flatten[array,1];
Short[triples,2]

Out[86]//Short=
{{0, -1, -1}, {0, -1, 0}, {0, -1, 1}, {0, 0, -1},
  <<3>>, {0, 1, 0}, {0, 1, 1}}

In[89]:=
tograph=Table[sol/.{C[1]->triples[[i,1]],
     C[2]->triples[[i,2]],
     C[3]->triples[[i,3]]},{i,1,9}];
```

We then graph the set of functions, **tograph**, on the interval [−2,0]:

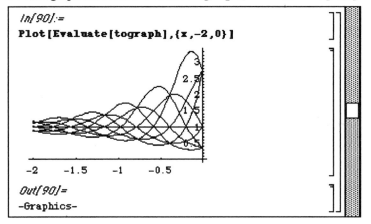

```
In[90]:=
Plot[Evaluate[tograph],{x,-2,0}]
```

```
Out[90]=
-Graphics-
```

and on the interval [0,1].

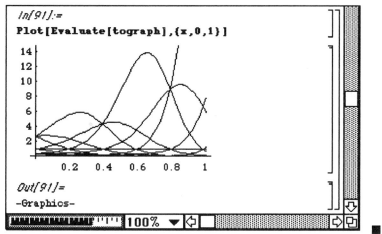

Chapter 5: Applications of Higher Order Differential Equations

Mathematica commands used in **Chapter 5** include:

AppendTo	Line	PlotStyle
AspectRatio	Module	PrependTo
Axes	NDSolve	ReplaceAll
D	NIntegrate	Show
DisplayFunction	Part	Simplify
DSolve	Partition	Solve
Evaluate	Plot	Table
Graphics	PlotRange	Ticks
GraphicsArray	Point	
GrayLevel	PointSize	

§5.1 Simple Harmonic Motion

Suppose that a mass is attached to an elastic spring which is suspended from a rigid support such as a ceiling. According to Hooke's law, the spring exerts a restoring force in the upward direction which is proportional to the displacement of the spring. Mathematically, this is stated as follows:

> **Hooke's Law:** $F = ks$ where $k>0$ is the constant of proportionality or spring constant, and s is displacement of the spring.

Using this law and assuming that x(t) represents the position of the mass, we obtain the initial value problem

$$m\frac{d^2x}{dt^2} + kx = 0$$

$$x(0) = \alpha, \quad x'(0) = \beta.$$

Note that the initial conditions give the initial position and velocity, respectively. The solution x(t) to this problem represents the position of the mass at time t. Based on the assumptions made in deriving the differential equation (the positive direction is down), positive values of x(t) indicate that the mass is beneath the equilibrium position while negative values of x(t) indicate that the mass is above the equilibrium position. We investigate solutions to this initial value problem under varying conditions below.

■ EXAMPLE 5.1

A mass weighing 60 lb. stretches a spring 6 inches. Use **DSolve** to determine the function x(t) which describes the motion of the mass if the mass is released with zero initial velocity 12 inches below the equilibrium position.

Solution:

First, the spring constant k must be determined from the given information. By Hooke's law, F = ks, so we have 60 = k(0.5). Therefore, k = 120 lb/ft. Next, the mass m must be determined using F = mg. In this case, 60 = m(32), so m = 15/8 slugs. Since k/m = 64 and 12 inches is equivalent to 1 foot, the initial value problem which should be solved is

$$\frac{d^2x}{dt^2} + 64x = 0$$

$x(0) = 1, \quad x'(0) = 0.$

This problem is solved below with **DSolve** and the resulting output is named **de1**. The solution is explicitly extracted from **de1** with **de1[[1,1,2]]** and named **sol[t]**.

```
≡□≡≡≡≡≡≡≡≡≡≡≡ HarmonicMotion ≡≡≡≡≡≡≡≡≡≡▢≡
In[2]:=
Clear[de1]
de1=DSolve[{x''[t]+64x[t]==0,x[0]==1,
        x'[0]==0},x[t],t]

Out[2]=
{{x[t] -> Cos[8 t]}}

In[3]:=
sol[t_]=de1[[1,1,2]]

Out[3]=
Cos[8 t]
```

We then graph **sol[t]** on the interval $[0,\pi/2]$.

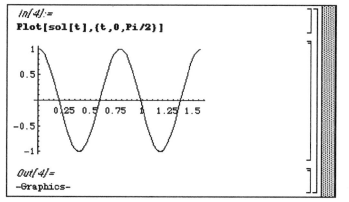

```
In[4]:=
Plot[sol[t],{t,0,Pi/2}]




Out[4]=
-Graphics-
```

In order to better understand the relationship between the formula obtained above and the motion of the mass on the spring, an alternate approach is taken below. First, the function **zigzag** is defined to produce a list of points joined by line segments to represent the graphics of a spring.

```
In[6]:=
Clear[
spring,zigzag,length,points,pairs]
zigzag[{a_,b_},{c_,d_},n_,eps_]:=
    Module[{length,points,pairs,zigzag},

    length=d-b;

    points=Table[b+i length/n,
            {i,1,n-1}];

    pairs=Table[
        {a+(-1)^i eps,points[[i]]},
                {i,1,n-1}];

    PrependTo[pairs,{a,b}];

    AppendTo[pairs,{c,d}];

    Line[pairs]
    ]
```

Below, the function **spring** is defined. This function produces the graphics of a point (the mass attached to the end of the spring) as well as that of the spring obtained with **zigzag**.

```
In[7]:=
spring[t_]:=
Show[Graphics[
        {zigzag[{0,sol[t]},{0,1},20,.025],
        PointSize[.1],
        Point[{0,sol[t]}]}],
    Axes->Automatic,
    Ticks->None,
    AspectRatio->1,
    PlotRange->{{-1/2,1/2},{-1,1}},
    DisplayFunction->Identity]
```

A list of graphics cells is produced in **graphs** below for values of t = 0 to t = π/4. The symbol ***** represents the multiplication symbol. The same results would be obtained if **(15*4)** were replaced by **(15 4)**. This list is then partitioned into groups of four in **toshow** for use within a **GraphicsArray**. The plots displayed show the position of the mass at these values of time. However, they cannot be animated since they are contained in a **GraphicsArray**. In order to achieve animation, the spring function should be used with a **Do** command. After the list of graphics cells is created with the **Do** command, the resulting graphics can be animated to view the motion of the mass.

```
In[10]:=
graphs=Table[spring[t],
    {t,0,Pi/4,Pi/(15*4)}];
toshow=Partition[graphs,4];
Show[GraphicsArray[toshow]]
```

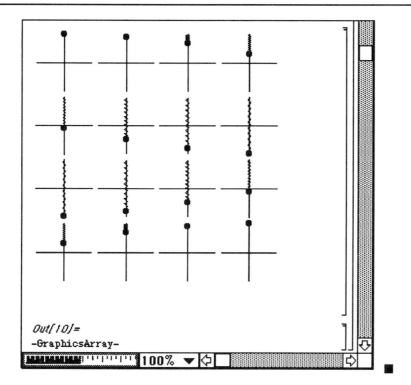

Out[10]=
-GraphicsArray-

■ EXAMPLE 5.2
Determine how varying the value of the initial position α affects the solution of the initial value problem

$$\begin{cases} \dfrac{d^2x}{dt^2} + 4x = 0 \\ x(0) = \alpha, \quad x'(0) = 0 \end{cases} \quad \text{by considering values of } \alpha = 1, \ 4, \text{ and } -2.$$

Solution:
The first step is to determine the solution to each of the three problems in **de2**, **de3**, and **de4** using **DSolve**.

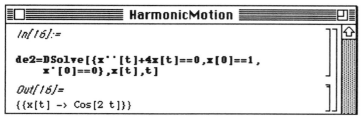

HarmonicMotion

In[16]:=

```
de2=DSolve[{x''[t]+4x[t]==0,x[0]==1,
    x'[0]==0},x[t],t]
```

Out[16]=
{{x[t] -> Cos[2 t]}}

```
In[17]:=

de3=DSolve[{x''[t]+4x[t]==0,x[0]==4,
    x'[0]==0},x[t],t]
Out[17]=
{{x[t] -> 4 Cos[2 t]}}
In[18]:=

de4=DSolve[{x''[t]+4x[t]==0,x[0]==-2,
    x'[0]==0},x[t],t]
Out[18]=
{{x[t] -> -2 Cos[2 t]}}
```

These three solutions are then plotted simultaneously in varying levels of gray to show the resulting motion of the mass on the spring.

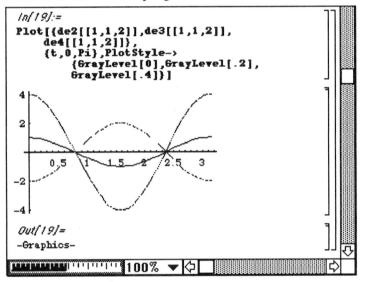

```
In[19]:=
Plot[{de2[[1,1,2]],de3[[1,1,2]],
    de4[[1,1,2]]},
    {t,0,Pi},PlotStyle->
        {GrayLevel[0],GrayLevel[.2],
        GrayLevel[.4]}]
```

```
Out[19]=
-Graphics-
```

Notice that the initial position only affects the amplitude of the function (and direction in the case of the negative initial position). The mass passes through the equilibrium (x = 0) at the same time in all three cases. ∎

∎ EXAMPLE 5.3

Determine how varying the value of the initial velocity β affects the solution of the initial value problem

$$
\begin{cases}
\dfrac{d^2x}{dt^2} + 4x = 0 \\
x(0) = 0, \quad x'(0) = \beta
\end{cases}
\quad \text{by considering values of } \beta = 1, \ 4, \text{ and } -2.
$$

Solution:

Again, the first step in the solution process is to determine the solution of each of the three problems in **de2**, **de3**, and **de4** using **DSolve**.

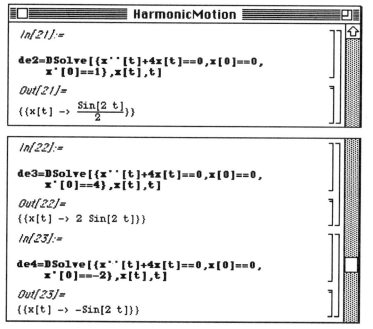

These solutions are then plotted simultaneously in varying levels of gray to demonstrate the resulting motion.

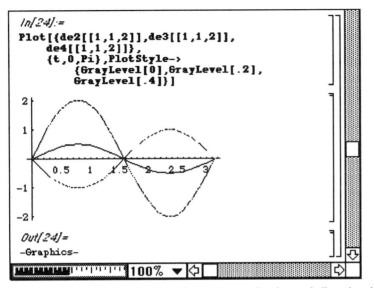

```
In[24]:=
Plot[{de2[[1,1,2]],de3[[1,1,2]],
    de4[[1,1,2]]},
    {t,0,Pi},PlotStyle->
        {GrayLevel[0],GrayLevel[.2],
        GrayLevel[.4]}]
```

```
Out[24]=
-Graphics-
```

Notice that the initial velocity affects the amplitude (and direction in the case of the negative initial velocity) of each function. The mass passes through the equilibrium ($x = 0$) at the same time in all three cases. ∎

§5.2 Damped Motion

Since the differential equation derived in **Section 3.1** disregarded all retarding forces acting on the motion of the mass, a more realistic model which takes these forces into account is needed. Studies in mechanics reveal that resistive forces due to damping are proportional to a power of the velocity of the motion. Hence, $F_R = c \dfrac{dx}{dt}$ or $F_R = c\left(\dfrac{dx}{dt}\right)^2$, where $c > 0$, are typically used to represent the damping force. Following similar procedures which lead to the differential equation in **Section 3.1** which modeled simple harmonic motion, we have the following initial value problem assuming that $F_R = c \dfrac{dx}{dt}$:

$$m\frac{d^2x}{dt^2} + c\frac{dx}{dt} + kx = 0$$

$$x(0) = \alpha, \ x'(0) = \beta$$

Clearly, from our experience with second order ordinary differential equations with constant coefficients in **Chapter 4**, the solutions to initial value problems of this type greatly depend on the values of m, k, and c.

Suppose that we assume that solutions of the differential equation have the form $x(t)=e^{rt}$. Note that m is not used in the exponent as it was in **Chapter 4** to avoid confusion with the mass m. Otherwise, this calculation is identical to those followed in **Chapter 4**. Since $x'(t)=re^{rt}$ and $x''(t)=r^2e^{rt}$, we have by substitution into the differential equation

$$mr^2e^{rt} + cre^{rt} + ke^{rt} = 0, \text{ so } e^{rt}\left(mr^2 + cr + k\right) = 0.$$

The solutions to the auxiliary (or characteristic) equation are

$$r = \frac{-c \pm \sqrt{c^2 - 4mk}}{2a}.$$

Hence, the solution depends on the value of the quantity $(c^2 - 4mk)$. In fact, problems of this type are characterized by the value of $(c^2 - 4mk)$ as follows:

Case I : $c^2 - 4mk > 0$
This situation is said to be **overdamped** since the damping coefficient c is large in comparison with the spring constant k.

Case II : $c^2 - 4mk = 0$
This situation is described as **critically damped** since the resulting motion is oscillatory with a slight decrease in the damping coefficient c.

Case III : $c^2 - 4mk < 0$
This situation is called **underdamped**, because the damping coefficient c is small in comparison with the spring constant k.

■ **EXAMPLE 5.4**

Classify the following differential equation as overdamped, underdamped, or critically damped. Also, solve the corresponding initial value problem using the given initial conditions and investigate the behavior of the resulting solutions.

$$\frac{d^2x}{dt^2} + 8\frac{dx}{dt} + 16x = 0$$

(a) $x(0)=0, \ x'(0)=1$

(b) $x(0)=-0.5, \ x'(0)=5$

Solution:

In this case, m = 1, c = 8, and k = 16. Thus, $c^2 - 4mk = 64-(4)(16)=0$, so this situation is critically damped.
(a) **DSolve** is used to solve the first initial value problem. Clearly, in this case, the solution is always positive and approaches zero as t approaches infinity as illustrated in the plot of the function.

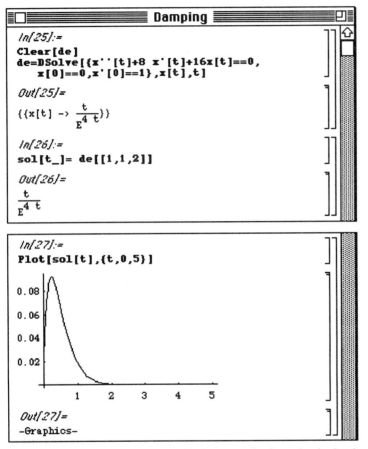

```
≣▭≣≣≣≣≣≣≣≣≣≣≣≣≣ Damping ≣≣≣≣≣≣≣≣≣≣≣≣≣▣≣
In[25]:=
Clear[de]
de=DSolve[{x''[t]+8 x'[t]+16x[t]==0,
    x[0]==0,x'[0]==1},x[t],t]

Out[25]=
```
$$\{\{x[t] \rightarrow \frac{t}{E^4\,t}\}\}$$
```
In[26]:=
sol[t_]= de[[1,1,2]]

Out[26]=
```
$$\frac{t}{E^4\,t}$$
```
In[27]:=
Plot[sol[t],{t,0,5}]
```

```
Out[27]=
-Graphics-
```

To further understand the relationship between the formula obtained and the motion of the spring, similar
commands are used as those employed in the section on harmonic motion. In this case, the function **p** is
defined below which produces the graphics of a point (the mass on the end of the spring) moving along
the vertical axis. A list of graphics cells is produced in **graphs** using **p** for values of t from t = 0 to
t = 2.5. This list is then partitioned into groups of three in array for use with a **GraphicsArray**.

```
In[45]:=
Clear[p]

In[46]:=
p[t_]:=p[t]=Show[Graphics[{PointSize[.1],
                   Point[{0,sol[t]}]}],
    Axes->Automatic,
    Ticks->{None,{0,.05,.1}},
    PlotRange->{{-1,1},{0,0.1}},
    DisplayFunction->Identity]

In[51]:=
graphs=Table[p[t],{t,0,2.5,2.5/14}];
array=Partition[graphs,3];
```

The array of graphics cells generated above is displayed below. Note that animation cannot be used with these graphics. In order to animate the graphics of the spring, a **Do** command should be used to create the list of graphics cells.

```
In[52]:=
Show[GraphicsArray[array]]
```

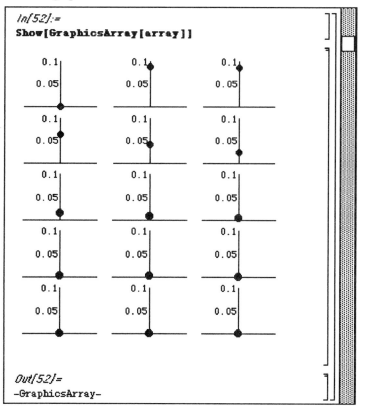

```
Out[52]=
-GraphicsArray-
```

(b) **DSolve** is used to solve the second initial value problem. These results indicate the importance of the initial conditions on the resulting motion. In this case, the position is negative initially, but the positive initial velocity causes the function to become positive before approaching zero.

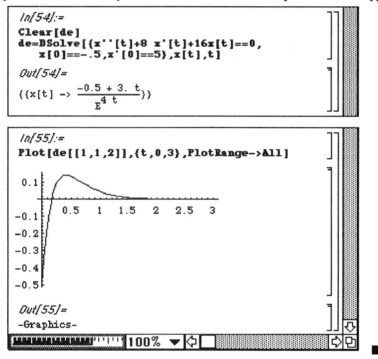

■ EXAMPLE 5.5

Classify the following differential equation as overdamped, underdamped, or critically damped. Also, solve the corresponding initial value problem using the given initial conditions and investigate the behavior of the resulting solutions.

$$\frac{d^2x}{dt^2} + 5\frac{dx}{dt} + 4x = 0$$

(a) $x(0)=1,\ x'(0)=-1$

(b) $x(0)=1,\ x'(0)=-6$

Solution:

In this case, m = 1, c = 5, and k = 4. Thus, $c^2 - 4mk = 25-(4)(4)=4$, so this situation is overdamped.

(a) **DSolve** is used to solve the first initial value problem.

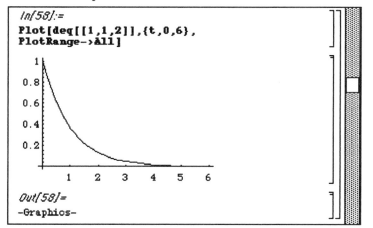

Clearly, in this case, the solution is always positive and approaches zero as t approaches infinity as illustrated in the plot of the function.

The damping is so great that even though the initial position is positive, the negative initial velocity is not large enough to force the position to go negative.

(b) **DSolve** is used to solve the second initial value problem. These results indicate the importance of the initial conditions on the resulting motion. In this case, the position is positive initially, but the larger negative initial velocity causes the function to become negative before approaching zero.

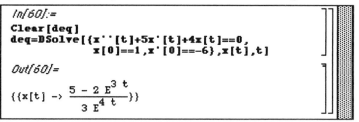

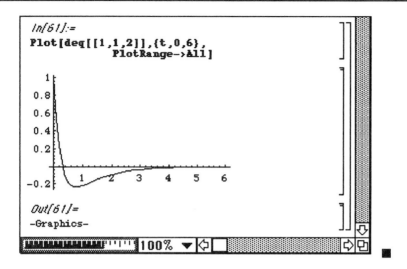

■ EXAMPLE 5.6

Classify the following differential equation as overdamped, underdamped, or critically damped. Also, solve the corresponding initial value problem using the given initial conditions and investigate the behavior of the resulting solutions.

$$\frac{d^2x}{dt^2} + \frac{dx}{dt} + 16x = 0$$

$x(0) = 0, \ x'(0) = 1.$

Solution:

In this case, $m = 1$, $c = 1$, and $k = 16$. Thus, $c^2 - 4mk = 1 - (4)(16) = -63$, so this situation is underdamped. **DSolve** is used to solve the initial value problem.

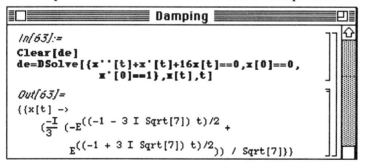

Since the values of r in the solution $x(t) = e^{rt}$ are complex conjugates, the **Trigonometry.m** package is loaded to obtain a real-valued solution. (Note, however, that *Mathematica* can plot the solution in its original form involving imaginary components.) **ComplexToTrig** applies Euler's formula to the solution in **de** to obtain the resulting real-valued function called **solution**. (In many cases, **ComplexToTrig** must be followed by **Simplify** to obtain the desired result; in other cases, the command **ComplexExpand** can be used successfully.) A function of t is then defined as **sol** for later use.

```
In[64]:=
<<Trigonometry.m

In[65]:=
solution=ComplexToTrig[de[[1,1,2]]]

Out[65]=
  2 Sin[3 Sqrt[7] t]
         2
  ─────────────────
   3 Sqrt[7] E^(t/2)

In[67]:=
Clear[sol]
sol[t_]=solution

Out[67]=
  2 Sin[3 Sqrt[7] t]
         2
  ─────────────────
   3 Sqrt[7] E^(t/2)
```

Solutions of this type have several interesting properties. First, the trigonometric component of the solution causes the motion to oscillate. Also, the exponential portion forces the solution to approach zero as t approaches infinity. These qualities are illustrated in **plot** below.

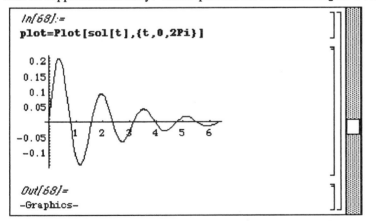

```
In[68]:=
plot=Plot[sol[t],{t,0,2Pi}]
```

```
Out[68]=
-Graphics-
```

Notice also that the solution is bounded above and below by the exponential term of the solution and its reflection through the horizontal axis. This is illustrated with the simultaneous display of **plot** and

plot1.

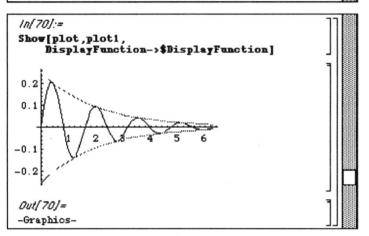

```
In[69]:=
plot1=Plot[{2/(3 Sqrt[7]) Exp[-t/2],
      -2/(3 Sqrt[7]) Exp[-t/2]},
    {t,0,2Pi},PlotStyle->{GrayLevel[.5],
            GrayLevel[.5]},
    DisplayFunction->Identity];
```

```
In[70]:=
Show[plot,plot1,
    DisplayFunction->$DisplayFunction]
```

```
Out[70]=
-Graphics-
```

Other questions of interest include: (1) When does the mass first pass through its equilibrium point? (2) What is the maximum displacement of the mass on the spring?

The time at which the mass passes through $x = 0$ can be determined in several ways. Since the solution involves the sine function, the solution equals zero at the time that $\sin(x)$ first equals zero after $t = 0$. Hence, **Solve** can be used to find out when the argument of the sine function in **sol** equals zero. On the other hand, **FindRoot** can be used with an initial guess obtained from the earlier plots to yield an identical result.

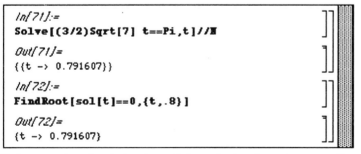

```
In[71]:=
Solve[(3/2)Sqrt[7] t==Pi,t]//N

Out[71]=
{{t -> 0.791607}}

In[72]:=
FindRoot[sol[t]==0,{t,.8}]

Out[72]=
{t -> 0.791607}
```

Unfortunately, the derivative of the solution is rather complicated due to the exponential term. Therefore, **Solve** fails to determine the time at which the first derivative of the solution is equal to zero.

However, **FindRoot** can be employed to determine this time. Then, the result found in **cp1** can be used to find the corresponding position.

```
In[73]:=
cp=Solve[D[sol[t],t]==0,t]

Out[73]=
{}

In[74]:=
cp1=FindRoot[sol'[t]==0,{t,.35}]

Out[74]=
{t -> 0.364224}
```

```
In[75]:=
cp1[[1,2]]

Out[75]=
0.364224

In[76]:=
sol[cp1[[1,2]]]

Out[76]=
0.555672 Sin[0.546336 Sqrt[7]]
-----------------------------------
           Sqrt[7]

In[77]:=
N[%]

Out[77]=
0.208377
```

Another interesting characteristic of solutions to underdamped problems is the time between successive maxima of the solution, called the **quasiperiod**. This quantity is found below by, first, determining the time at which the second maximum occurs with **FindRoot**. Then, the difference between these values of t is taken to obtain the value of 1.58321.

```
In[78]:=
cp2=FindRoot[sol'[t]==0,{t,1.9}]

Out[78]=
{t -> 1.94744}

In[79]:=
cp2[[1,2]]-cp1[[1,2]]

Out[79]=
1.58321
```

To further investigate the solution, animation can be used with commands similar to those previously used to observe the motion of the mass on the spring. Again, the function **p** is defined. The graphics cells generated in **graphs** below are then partitioned into groups of five.

```
In[114]:=
Clear[p,graphs,array]
p[t_]:=p[t]=Show[Graphics[{PointSize[.1],
        Point[{0,sol[t]}]}],
    Axes->Automatic,
    AxesOrigin->{0,0},
    Ticks->{None,{0,.1}},
    PlotRange->{{-1,1},{-.1,.15}},
    DisplayFunction->Identity]
In[116]:=
graphs=Table[p[t],{t,0,4.5,4.5/44}];
array=Partition[graphs,5];
```

This array of graphics cells is shown below. As was the case with the previous examples, a **Do** command should be used with **p** to produce a list of graphics cells which can be animated. To see the motion of the mass, graphs should be viewed across each row from left to right.

```
In[117]:=
Show[GraphicsArray[array]]
```

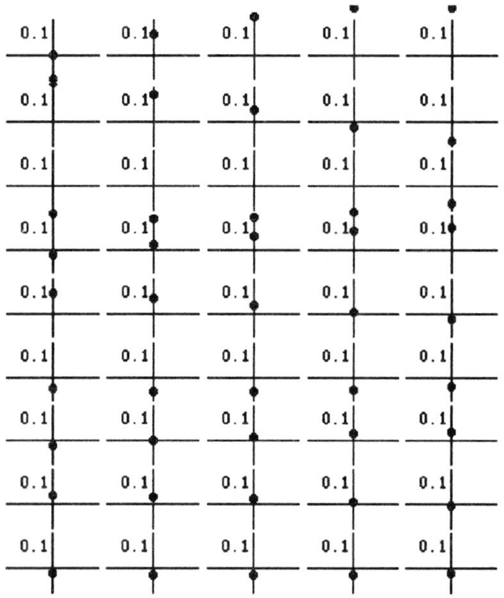

```
Out[117]=
-GraphicsArray-
```

100%

■ EXAMPLE 5.7

Suppose that we have the initial value problem

$$\frac{d^2x}{dt^2} + c\frac{dx}{dt} + 6x = 0$$

$$x(0) = 0, \quad x'(0) = 1$$

where $c = 2\sqrt{6}$, $4\sqrt{6}$, and $\sqrt{6}$. Determine how the value of c affects the solution of the initial value problem.

Solution:

DSolve is used to find the solution of the initial value problem for each value of c. Note that each case results in a different classification:

$c = 2\sqrt{6}$ Critically Damped

$c = 4\sqrt{6}$ Overdamped

$c = \sqrt{6}$ Underdamped

```
≣□≣══════════════ Damping ═══════════════⊞≣
In[13]:=
de1=DSolve[{y''[x]+2 Sqrt[6] y'[x]+6y[x]==0,
   y[0]==0,y'[0]==1},y[x],x]

Out[13]=
{{y[x] -> ───────}}
          E^Sqrt[6] x
```

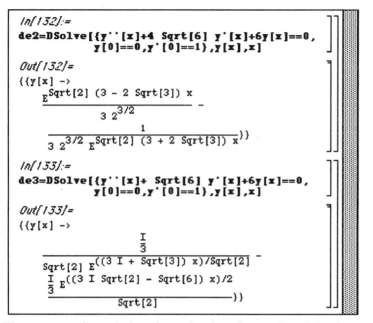

The corresponding solutions for each value of c are plotted simultaneously. Note that since the values of c vary more widely than those considered in the previous example, the behavior of the solutions differ more obviously as well.

```
In[134]:=
plot1=Plot[de1[[1,1,2]],{x,0,6},
    DisplayFunction->Identity];

In[135]:=
plot2=Plot[de2[[1,1,2]],{x,0,6},
    DisplayFunction->Identity,
    PlotStyle->GrayLevel[.2]];

In[136]:=
plot3=Plot[de3[[1,1,2]],{x,0,6},
    DisplayFunction->Identity,
    PlotStyle->GrayLevel[.4]];

    Plot::plnr:
        Comp<<8>>tion[{x}, <<2>>][x]
            is not a machine-size real number at
```

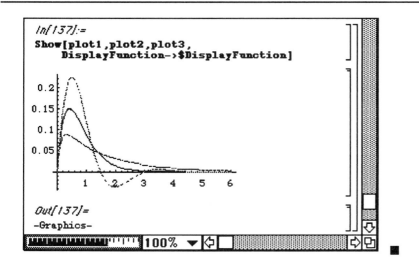

```
In[137]:=
Show[plot1,plot2,plot3,
    DisplayFunction->$DisplayFunction]
```

```
Out[137]=
-Graphics-
```

§5.3 Forced Motion

In some cases, the motion of the spring is influenced by an external driving force, f(t). Mathematically, this force is included in the differential equation which models the situation as follows:

$$m\frac{d^2x}{dt^2} = -kx - c\frac{dx}{dt} + f(t).$$

Hence, the resulting initial value problem is

$$m\frac{d^2x}{dt^2} + c\frac{dx}{dt} + kx = f(t)$$

$$x(0) = \alpha, \ x'(0) = \beta.$$

Therefore, differential equations modeling forced motion are nonhomogeneous and require the method of Undetermined Coefficients or Variation of Parameters for solution. We first consider forced motion which is undamped.

■ EXAMPLE 5.8

Investigate the effect that the forcing function:

(a) f(t) = 0,

(b) f(t) = 1,

(c) f(t) = cos(t),

(d) f(t) = sin(t)

has on the result of the initial value problem:

$$\frac{d^2x}{dt^2} + 4x = f(t)$$

$x(0) = 0, \; x'(0) = 0.$

Solution:

(a) **DSolve** is used to solve the first initial value problem. However, this problem could have been solved without any calculations whatsoever. Since f(t) = 0, there is no external driving force. Also, the initial position is zero as is the initial velocity. Therefore, the mass is never placed into motion, so it remains in its equilibrium position x = 0.

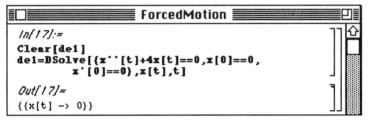

```
In[17]:=
Clear[de1]
de1=DSolve[{x''[t]+4x[t]==0,x[0]==0,
         x'[0]==0},x[t],t]

Out[17]=
{{x[t] -> 0}}
```

(b) **DSolve** is used to solve the second initial value problem.

```
In[18]:=
de2=DSolve[{x''[t]+4x[t]==1,x[0]==0,
         x'[0]==0},x[t],t]

Out[18]=
{{x[t] -> Sin[t]^2/2}}
```

The forcing function in this case is constant. The mass moves to a distance .5 units from equilibrium and then returns to the equilibrium. The external force then causes the mass to repeat this motion.

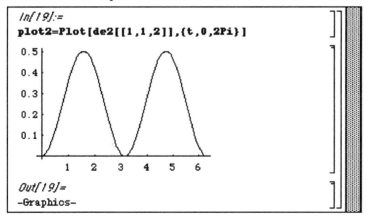

```
In[19]:=
plot2=Plot[de2[[1,1,2]],{t,0,2Pi}]
```

```
Out[19]=
-Graphics-
```

(c) As in the previous cases, **DSolve** is used to find the solution to this nonhomogeneous initial value problem.

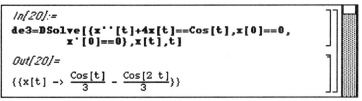

Note that in this case, however, the external force is a periodic function of time. Therefore, the external force varies with time and the solution is periodic since it is the sum of two periodic functions. The resulting motion is shown in **plot3** below.

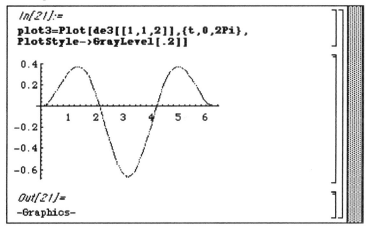

(d) This situation is similar to that in part(c).

```
In[22]:=
de4=DSolve[{x''[t]+4x[t]==Sin[t],x[0]==0,
       x'[0]==0},x[t],t]
Out[22]=
                4 Cos[t/2] Sin[t/2]^3
{{x[t] ->  ---------------------}}
                        3
```

The forcing function is, again, a periodic function of time. The motion shown in **plot4** points out that the value of the forcing function is different from the external force in (c) at different values of time. Hence, the resulting motion is not the same as that of part(c).

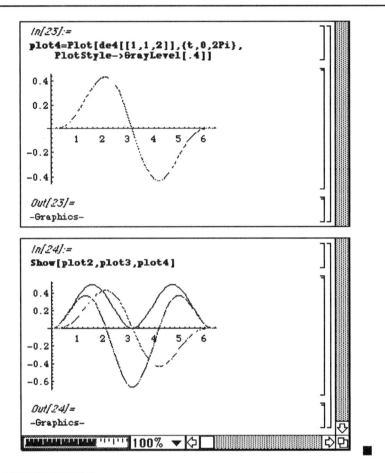

```
In[23]:=
plot4=Plot[de4[[1,1,2]],{t,0,2Pi},
    PlotStyle->GrayLevel[.4]]
```

```
Out[23]=
-Graphics-
```

```
In[24]:=
Show[plot2,plot3,plot4]
```

```
Out[24]=
-Graphics-
```

100% ▼

■ EXAMPLE 5.9

Investigate the effect that the forcing function:

(a) $f(t) = \cos(2t)$

(b) $f(t) = \sin(2t)$

has on the result of the initial value problem:

$$\frac{d^2x}{dt^2} + 4x = f(t)$$

$$x(0) = 0, \quad x'(0) = 0.$$

Solution:

(a) This situation is interesting in that the forcing function is a component of the homogeneous (or complimentary) solution, $x_c(t) = C_1\cos(2t) + C_2\sin(2t)$. Therefore, the nonperiodic function t sin(2t) appears in the solution.

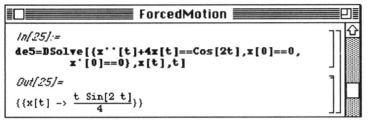

This causes the solution to approach infinity as t approaches infinity. This phenomenon is called pure resonance and is illustrated in **plot5** below.

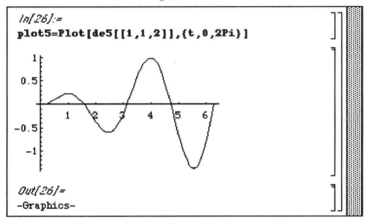

(b) In this case, the forcing function is also a solution to the corresponding homogeneous equation.

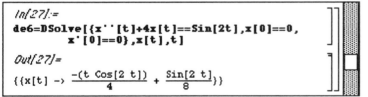

Hence, the nonperiodic function tsin(2t) causes the value of x in **plot6** to approach infinity with t as in the case above.

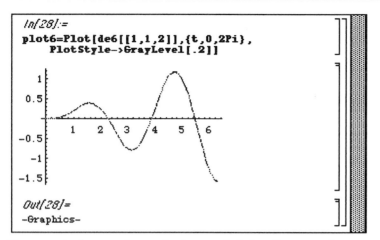

The two solutions are shown simultaneously below for comparison.

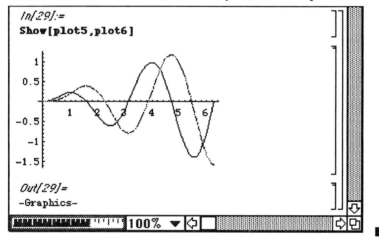

■ EXAMPLE 5.10

Investigate the effect that slightly changing the value of the argument of the forcing function f(t) = cos(2t) has on the solution of the initial value problem given in the previous example. In this case, use the functions:

(a) f(t) = cos(1.9t)

(b) f(t) =cos(2.1t)

with the initial value problem:

$$\frac{d^2x}{dt^2} + 4x = f(t)$$

$$x(0) = 0, \ x'(0) = 0.$$

Solution:

The solution of each initial value problem is determined with **DSolve**.

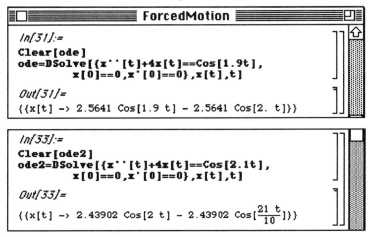

Note that each solution is periodic and bounded since the solutions involve cosine functions only. These solutions are then plotted to reveal the unusual behavior of the curves. If the solutions are plotted only over a small interval, then resonance seems to be present.

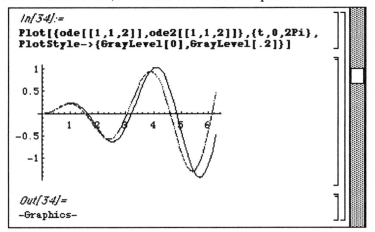

However, the functions given with **DSolve** clearly indicate that there is no resonance. This is further indicated with the second plot.

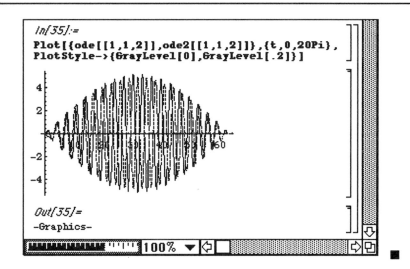

Let us further investigate initial value problems of the form

$$\frac{d^2x}{dt^2} + \omega^2 x = F\cos(\beta t), \; \omega \neq \beta$$

$$x(0) = 0, \; x'(0) = 0.$$

The homogeneous (complimentary) solution is given by $x_c(t) = C_1\cos(\omega t) + C_2\sin(\omega t)$. Using the Method of Undetermined Coefficients, the particular solution is given by $x_p(t) = A\cos(\beta t) + B\sin(\beta t)$. The corresponding derivatives of this solution are:

$$x_p'(t) = -A\beta\sin(\beta t) + B\beta\cos(\beta t) \; \text{ and } x_p''(t) = -A\beta^2\cos(\beta t) + B\beta^2\sin(\beta t).$$

Substitution into the nonhomogeneous equation and equating the corresponding coefficients yields:

$$A = \frac{F}{\omega^2 - \beta^2} \; \text{ and } B = 0.$$

Therefore, the solution is

$$x(t) = C_1\cos(\omega t) + C_2\sin(\omega t) + \frac{F}{\omega^2 - \beta^2}\cos(\beta t).$$

Application of the initial conditions yields the solution

$$x(t) = \frac{F}{\omega^2 - \beta^2}\left(\cos(\beta t) - \cos(\omega t)\right).$$

Using the trigonometric identity $\dfrac{1}{2}\left[\cos(A - B) - \cos(A + B)\right] = \sin A \sin B$ we have

$$x(t) = \frac{-2F}{\omega^2 - \beta^2} \sin\left(\frac{(\omega + \beta)t}{2}\right) \sin\left(\frac{(\omega - \beta)t}{2}\right).$$

These solutions are of interest because of their motion. Notice that the solution can be represented as

$$x(t) = A(t)\sin\left(\frac{(\omega + \beta)t}{2}\right) \text{ where } A(t) = \frac{-2F}{\omega^2 - \beta^2}\sin\left(\frac{(\omega - \beta)t}{2}\right).$$

Therefore, when the quantity $(\omega - \beta)$ is small, $(\omega + \beta)$ is relatively large in comparison. Hence, the function

$\sin\left(\dfrac{(\omega + \beta)t}{2}\right)$ oscillates quite frequently since its period is $\dfrac{\pi}{(\omega + \beta)}$. Meanwhile, the function $\sin\left(\dfrac{(\omega - \beta)t}{2}\right)$

oscillates relatively slowly since $\dfrac{\pi}{|\omega - \beta|}$ is its period. Hence, the function $\dfrac{-2F}{\omega^2 - \beta^2}\sin\left(\dfrac{(\omega - \beta)t}{2}\right)$ forms an

envelope for the solution.

We examine problems of this type in the example below:

■ EXAMPLE 5.11

Investigate the effect that the forcing function:
(a) $f(t) = \cos(3t)$
(b) $f(t) = \cos(5t)$
has on the result of the initial value problem:

$$\frac{d^2x}{dt^2} + 4x = f(t)$$

$$x(0) = 0, \ x'(0) = 0.$$

Solution:

(a) Notice that in this case, the particular solution using the Method of Undetermined Coefficients is $x_p(t) = C_1\cos(3t) + C_2\sin(3t)$. This is shown below in the solution obtained with **DSolve**.

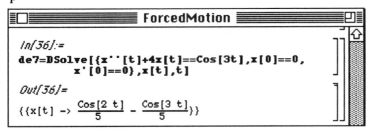

```
In[36]:=
de7=DSolve[{x''[t]+4x[t]==Cos[3t],x[0]==0,
        x'[0]==0},x[t],t]

Out[36]=
{{x[t] -> Cos[2 t]   Cos[3 t]
          --------- - --------}}
             5           5
```

This solution is plotted below in **plot7**.

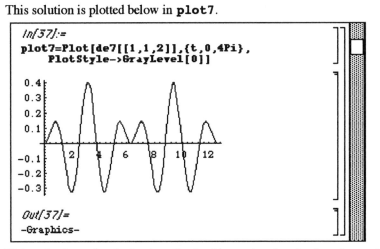

Using the formula obtained earlier in this section for the functions which "envelops" the solution, we have $(-2/5)\sin(t/2)$ and $(2/5)\sin(t/2)$ in this case.

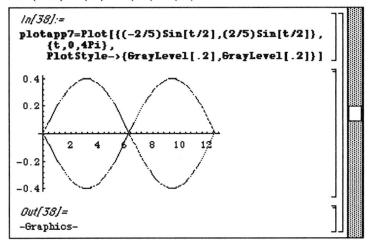

The two solutions are shown simultaneously below for comparison.

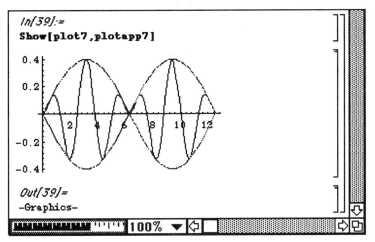

In[39]:=
Show[plot7,plotapp7]

Out[39]=
-Graphics-

(b) Finally, the solution and enveloping function are determined and plotted for the case with f(t)=cos(5t).

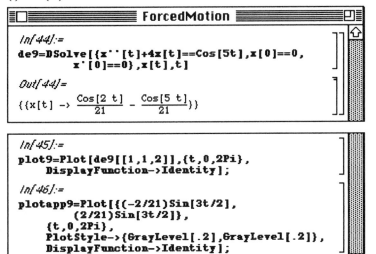

ForcedMotion

In[44]:=
de9=DSolve[{x''[t]+4x[t]==Cos[5t],x[0]==0,
     x'[0]==0},x[t],t]

Out[44]=
$$\{\{x[t] \rightarrow \frac{Cos[2\ t]}{21} - \frac{Cos[5\ t]}{21}\}\}$$

In[45]:=
plot9=Plot[de9[[1,1,2]],{t,0,2Pi},
    DisplayFunction->Identity];

In[46]:=
plotapp9=Plot[{(-2/21)Sin[3t/2],
        (2/21)Sin[3t/2]},
    {t,0,2Pi},
    PlotStyle->{GrayLevel[.2],GrayLevel[.2]},
    DisplayFunction->Identity];

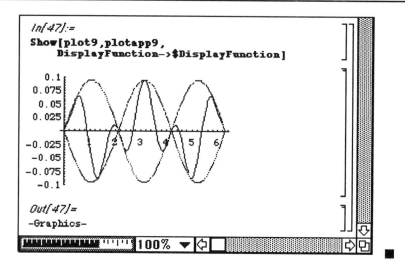

```
In[47]:=
Show[plot9,plotapp9,
    DisplayFunction->$DisplayFunction]
```

```
Out[47]=
-Graphics-
```

■ EXAMPLE 5.12

Determine the effect of the forcing function $f(t) = e^{-t}\cos(2t)$ on the result of the initial value problem:

$$\frac{d^2x}{dt^2} + 4x = f(t)$$

$x(0) = 0, \; x'(0) = 0.$

Solution:

The initial value problem is solved with **DSolve**. (Note: In a problem such as this which involves a complicated forcing function, a great deal of time is necessary for **DSolve** to determine the solution. Hence, the methods discussed earlier for solving nonhomogeneous differential equations should be considered.)

```
≣□≣                    ForcedMotion ≣
In[49]:=
Clear[de3]
de3=DSolve[{x''[t]+4x[t]==Exp[-t] Cos[2t],
        x[0]==0,x'[0]==0},
        x[t],t]
Out[49]=
{{x[t] -> (2 Cos[2 t] - 2 E  Cos[2 t] -
        8 Sin[2 t] + 9 E  Sin[2 t]) / (34 E )}}
```

Note that $\lim_{t \to \infty} f(t) = \lim_{t \to \infty}\left(e^{-t}\cos(2t)\right) = 0.$

Hence, the effect of the forcing function decreases with time, so the solution approaches the oscillatory solution to the homogeneous problem with $f(t) = 0$. This is illustrated in the plot below.

de 3[[1, 1, 2]] extracts the solution, $\dfrac{2\cos(2t) - 8\sin(2t) - e^2\left(2\cos(2t) - 9\sin(2t)\right)}{34e^t}$, from **de 3**.

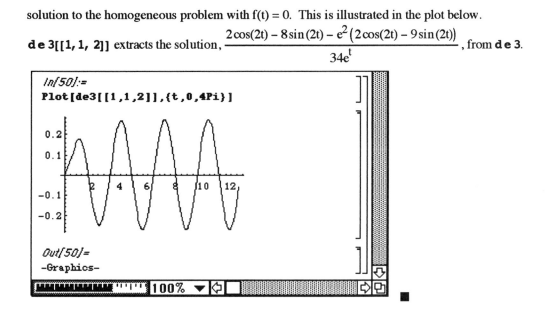

Let us now consider spring problems which involve forces due to damping as well as external forces. In particular, consider the following initial value problem:

$$m\frac{d^2x}{dt^2} + c\frac{dx}{dt} + kx = \alpha\cos(\lambda t)$$

$$x(0) = x_0, \ x'(0) = v_0.$$

Problems of this nature have solutions of the form

$$x(t) = h(t) + s(t)$$

where $\lim\limits_{t\to\infty} h(t) = 0$ and $s(t) = C_1\cos(\lambda t) + C_2\sin(\lambda t)$.

$h(t)$ is typically called the **transient** solution while $s(t)$ is known as the **steady-state** solution. Hence, as t approaches infinity, the solution $x(t)$ approaches the steady-state solution. Note that the steady-state solution simply corresponds to the particular solution obtained through the Method of Undetermined Coefficients or Variation of Parameters.

■ **EXAMPLE 5.13**

Consider the initial value problem:

$$\frac{d^2x}{dt^2} + 4\frac{dx}{dt} + 13x = \cos(t)$$

$$x(0) = 0, \ x'(0) = 1.$$

Solution:

First, **DSolve** is used to obtain the solution to this nonhomogeneous problem. (The Method of Undetermined Coefficients could be used to find this solution as well.)

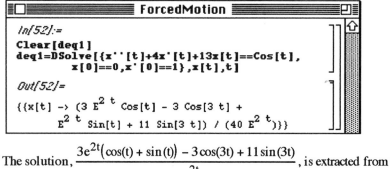

```
In[52]:=
Clear[deq1]
deq1=DSolve[{x''[t]+4x'[t]+13x[t]==Cos[t],
        x[0]==0,x'[0]==1},x[t],t]

Out[52]=
{{x[t] -> (3 E^2 t Cos[t] - 3 Cos[3 t] +
        E^2 t Sin[t] + 11 Sin[3 t]) / (40 E^2 t)}}
```

The solution, $\dfrac{3e^{2t}\big(\cos(t) + \sin(t)\big) - 3\cos(3t) + 11\sin(3t)}{40e^{2t}}$, is extracted from the list **deq1** with the

command **deq1[[1, 1, 2]]**.

The solution is then plotted over the interval $[0,5\pi]$ in **plot1** to illustrate the behavior of this solution.

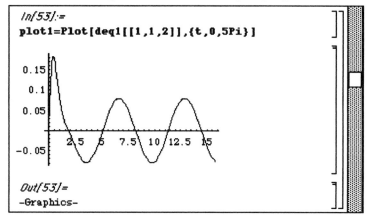

```
In[53]:=
plot1=Plot[deq1[[1,1,2]],{t,0,5Pi}]

Out[53]=
-Graphics-
```

Next, the steady-state solution is plotted over the same interval in **ssplot** so that it can be displayed simultaneously with **plot1**.

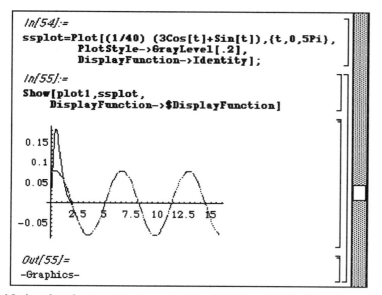

Notice that the two curves appear identical for t > 2.5. The reason for this is shown in the subsequent plot of the transient solution which becomes quite small near t = 2.5.

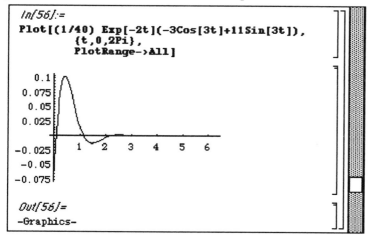

Notice also that the steady-state solution corresponds to the particular solution to the nonhomogeneous differential equation as verified below by defining $ss(t) = \dfrac{3\cos(t) + \sin(t)}{40}$:

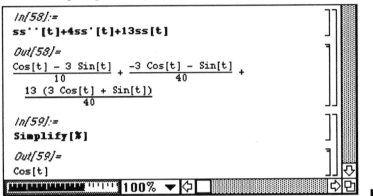

```
In[57]:=
ss[t_]=(1/40) (3Cos[t]+Sin[t])

Out[57]=
3 Cos[t] + Sin[t]
─────────────────
        40
```

and then computing and simplifying ss''(t)+4ss'(t)+13ss(t):

```
In[58]:=
ss''[t]+4ss'[t]+13ss[t]

Out[58]=
Cos[t] - 3 Sin[t]     -3 Cos[t] - Sin[t]
───────────────── +   ─────────────────── +
       10                     40

   13 (3 Cos[t] + Sin[t])
   ──────────────────────
            40

In[59]:=
Simplify[%]

Out[59]=
Cos[t]
```

100% ▼

■ EXAMPLE 5.14

Solve the initial value problem:

$$\frac{d^2x}{dt^2} + x = \begin{cases} 1, \, 0 \le t \le 1 \\ -1, \, 1 < t \le 2 \end{cases}$$

$x(0)=a, \ x'(0)=b.$

Solution:

Since the forcing function is piecewise defined, we must take special care in solving this problem. First, the homogeneous (complimentary) solution is determined with **DSolve** and the resulting list is named **hm**. The solution is extracted from **hm** with the command **hm[[1,1,2]]** and then the symbols C[1] and C[2], which represent arbitrary constants, are replaced by a and b, respectively, and the resulting expression is named **func**.

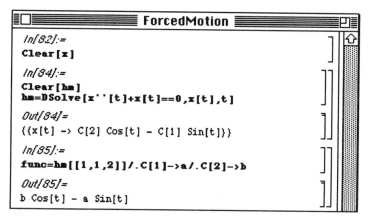

```
In[82]:=
Clear[x]
In[84]:=
Clear[hm]
hm=DSolve[x''[t]+x[t]==0,x[t],t]
Out[84]=
{{x[t] -> C[2] Cos[t] - C[1] Sin[t]}}
In[85]:=
func=hm[[1,1,2]]/.C[1]->a/.C[2]->b
Out[85]=
b Cos[t] - a Sin[t]
```

func is then defined as the function **sol**. The forcing function is then defined in **force** and plotted over the interval [0,2] to verify its definition.

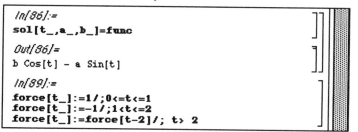

```
In[86]:=
sol[t_,a_,b_]=func
Out[86]=
b Cos[t] - a Sin[t]
In[89]:=
force[t_]:=1/;0<=t<=1
force[t_]:=-1/;1<t<=2
force[t_]:=force[t-2]/; t> 2
```

The forcing function is extended over the interval [2,6] below. This extension is plotted in **plotg** for later use.

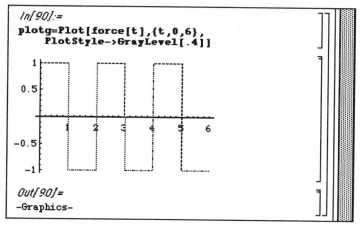

```
In[90]:=
plotg=Plot[force[t],{t,0,6},
    PlotStyle->GrayLevel[.4]]
```

```
Out[90]=
-Graphics-
```

According to the Method of Undetermined Coefficients, the particular solution is assumed to be $x_p(t) = c$ (constant). Hence, after substitution into the differential, the particular solution is easily found to be $x_p(t) = 1$ on [0,1] and $x_p(t) = -1$ on [1,2]. Therefore, the solution is defined in the form of **x** below. This solution is extended in a similar manner as g was above.

```
In[100]:=
Clear[x]
x[t_,a_,b_]:=sol[t,a,b]+1/;0<=t<=1
x[t_,a_,b_]:=sol[t,a,b]-1/;1<t<=2
x[t_,a_,b_]:=x[t-2,a,b]/;t>2
```

The solution is plotted in **plote** below. Finally, the forcing function and the solution are displayed simultaneously.

```
In[101]:=
plote=Plot[x[t,1,1],{t,0,6},
    DisplayFunction->Identity]

Out[101]=
-Graphics-
```

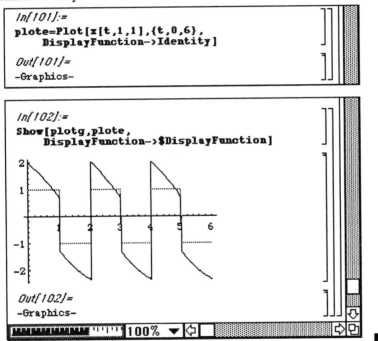

```
In[102]:=
Show[plotg,plote,
    DisplayFunction->$DisplayFunction]



Out[102]=
-Graphics-
```

§5.4 L-R-C Circuits

Second order nonhomogeneous linear ordinary differential equations arise in the study of electrical circuits after the application of Kirchhoff's law. Suppose that i(t) is the current in an L-R-C series electrical circuit where L, R, and C represent the inductance, resistance, and capacitance of the circuit, respectively. The following voltage drops have been obtained from experimental data:

$L\dfrac{di}{dt}$ = voltage drop across the inductor

iR   = voltage drop across the resistor

$\dfrac{1}{C}q$ = voltage drop across the capacitor, where $i = \dfrac{dq}{dt}$ and q(t) = the charge on the capacitor.

According to Kirchoff's law, the sum of these voltage drops is equivalent to the voltage E(t) impressed on the circuit. Therefore, we have the differential equation

$$L\dfrac{di}{dt} + Ri + \dfrac{1}{C}q = E(t).$$

Using the fact that $i = \dfrac{dq}{dt}$, we also have $\dfrac{di}{dt} = \dfrac{d^2q}{dt^2}$. Therefore, the equation becomes

$$L\dfrac{d^2q}{dt^2} + R\dfrac{dq}{dt} + \dfrac{1}{C}q = E(t)$$

which can be solved with the Method of Undetermined Coefficients or the Method of Variation of Parameters.

■ EXAMPLE 5.15

Consider the L-R-C circuit problem with L = 1 henry, R = 40 ohms, C = 4000 farads, and E(t) = 24 volts. Determine the steady-state current to this initial value problem with zero initial current and zero initial charge. Also determine the current i(t) for this circuit.

Solution:

Using the values indicated, the initial value problem is:

$$\dfrac{d^2q}{dt^2} + 40\dfrac{dq}{dt} + 4000q = 24, \quad q(0) = 0, \quad q'(0) = 0.$$

DSolve is used to obtain the solution to this nonhomogeneous problem in **cir1** below.

```
≡□▤══════════ LRCCircuits ═══════════▣▤
In[30]:=
Clear[cir1]
cir1=DSolve[{q''[t]+40q'[t]+4000q[t]==24,
q[0]==0,q'[0]==0},q[t],t]

Out[30]=
               3 E^20 t - 3 Cos[60 t] - Sin[60 t]
{{q[t] -> ─────────────────────────────────────}}
                        500 E^20 t

In[32]:=
Clear[sol]
sol[t_]=cir1[[1,1,2]]

Out[32]=
3 E^20 t - 3 Cos[60 t] - Sin[60 t]
──────────────────────────────────
          500 E^20 t
```

The solution is then extracted from the output list and defined as the function **sol[t]**. (The Method of Undetermined Coefficients could be used to find this solution as well.) Observation of the formula obtained indicates that the solution approaches 3/500 as t approaches infinity. This limit is indicated in the graph **q** as well.

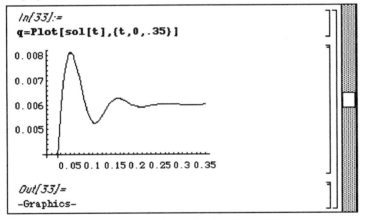

```
In[33]:=
q=Plot[sol[t],{t,0,.35}]
```

```
Out[33]=
-Graphics-
```

The current i(t) for this circuit is obtained by differentiating **sol**. This function is also plotted in **current**.

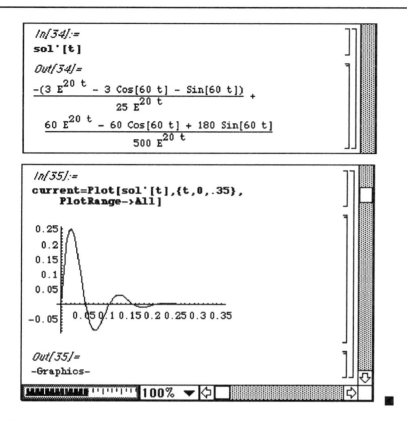

§5.5 Deflection of a Beam

An important mechanical model involves the deflection of a long beam which is supported at one or both ends. Assuming that in its undeflected form the beam is horizontal, then the deflection of the beam can be expressed as a function of x. In particular, let $y = -s(x)$ represent this deflection with x being the distance from one end of the beam and y the measurement of the vertical deflection from the equilibrium position. The initial value problem which models this situation is derived as follows:

Let $m(x)$ equal the turning moment of the force relative to the point x and $w(x)$ represent the weight distribution of the beam. These two functions are related by the equation:

$$\frac{d^2 m}{dx^2} = w.$$

Also, the turning moment is proportional to the curvature of the beam. Hence,

$$m = \frac{EI\,s''}{\left(\sqrt{1+(s')^2}\right)^3}$$

where E and I are constants related to the composition of the beam and the shape and size of a cross section of the beam, respectively. Notice that this equation is, unfortunately, nonlinear. However, this difficulty is overcome with an approximation. For small values of s, the denominator of the right-hand side of the equation can be approximated by the constant 1. Therefore, the equation is simplified to

$m = EI\,s''(x).$

This equation is linear and can be differentiated twice to obtain

$$m'' = EI\,\frac{d^4 s}{dx^4}.$$

This equation can then be used with the equation above relating m(x) and w(x) to obtain the single fourth order linear nonhomogeneous differential equation

$$EI\,\frac{d^4 s}{dx^4} = w(x).$$

Initial and boundary conditions for this problem may vary. In most cases, two conditions are given for each end of the beam. Some of these conditions include: s=0, s'=0 (fixed end); s''=0, s'''=0 (free end); s=0, s''=0 (simple support); and s'=0, s'''=0 (sliding clamped end).

Solutions to this initial-boundary value problem are investigated below for various conditions and values of w(x).

The following example investigates the effects that a nonconstant weight distribution function w(x) has on the solution to the boundary value problems discussed above.

■ EXAMPLE 5.16

Solve the beam equation given above assuming that the weight distribution w(x) is constant on the interval [0,1]. Solve the problem using the boundary conditions:

(a) s(0)=0, s'(0)=0, (fixed end at x = 0); s(1)=0, s''(1)=0 (simple support at x = 1)

(b) s(0)=0, s'(0)=0, (fixed end at x = 0); s''(1)=0, s'''(1)=0 (free end at x = 1)

(c) s(0)=0, s'(0)=0, (fixed end at x = 0); s'(1)=0, s'''(1)=0 (sliding clamped end at x = 1)

(d) s(0)=0, s'(0)=0, (fixed end at x = 0); s(1)=0, s'(1)=0 (fixed end at x = 1)

Note the differences brought about by these conditions.

Solution:

DSolve is used to obtain the solution to this nonhomogeneous problem. In **del**, the solution which depends on the parameters w, e, and i is given. This formula is then extracted and defined as the function **beam**. Therefore, the solution for various values of the parameters can easily be determined. The solution is plotted in **p1** for the particular values e = 1, i = 1, and w = 48.

```
≡≡≡≡≡≡≡≡≡≡≡≡≡ Beam ≡≡≡≡≡≡≡≡≡≡≡

In[1]:=
de1=DSolve[{e i D[s[x],{x,4}]==w,
s[0]==0,s'[0]==0,s[1]==0,s''[1]==0},s[x],x]

Out[1]=
             w x^2 (3 - 5 x + 2 x^2)
{{s[x] ->  ─────────────────────── }}
                    48 e i

In[2]:=
beam[e_,i_,w_,x_]=de1[[1,1,2]]

Out[2]=
 w x^2 (3 - 5 x + 2 x^2)
─────────────────────────
         48 e i
```

```
In[3]:=
p1=Plot[beam[1,1,48,x],{x,0,1},
    PlotRange->All,AspectRatio->1,
    DisplayFunction->Identity]

Out[3]=
-Graphics-
```

Similar steps are followed to determine the solution to each of the other three boundary value problems. The corresponding plots are named **p2**, **p3**, and **p4**.

```
In[4]:=
de2=DSolve[{e i D[s[x],{x,4}]==w,
    s[0]==0,s'[0]==0,s''[1]==0,s'''[1]==0},
         s[x],x]

Out[4]=
             w x^2 (6 - 4 x + x^2)
{{s[x] ->  ───────────────────── }}
                  24 e i

In[5]:=
beam2[e_,i_,w_,x_]=de2[[1,1,2]]

Out[5]=
 w x^2 (6 - 4 x + x^2)
───────────────────────
        24 e i

In[6]:=
p2=Plot[beam2[1,1,48,x],{x,0,1},
    PlotRange->All,AspectRatio->1,
    PlotStyle->GrayLevel[.2],
    DisplayFunction->Identity]

Out[6]=
-Graphics-
```

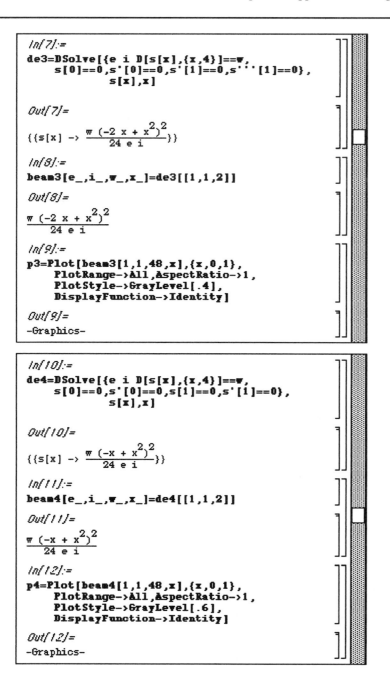

```
In[7]:=
de3=DSolve[{e i D[s[x],{x,4}]==w,
    s[0]==0,s'[0]==0,s'[1]==0,s'''[1]==0},
        s[x],x]
```

```
Out[7]=
                  w (-2 x + x )
{{s[x] -> ───────────────────}}
                    24 e i
```

```
In[8]:=
beam3[e_,i_,w_,x_]=de3[[1,1,2]]
```

```
Out[8]=
           2 2
w (-2 x + x )
─────────────────
    24 e i
```

```
In[9]:=
p3=Plot[beam3[1,1,48,x],{x,0,1},
    PlotRange->All,AspectRatio->1,
    PlotStyle->GrayLevel[.4],
    DisplayFunction->Identity]
```

```
Out[9]=
-Graphics-
```

```
In[10]:=
de4=DSolve[{e i D[s[x],{x,4}]==w,
    s[0]==0,s'[0]==0,s[1]==0,s'[1]==0},
        s[x],x]
```

```
Out[10]=
                  2 2
          w (-x + x )
{{s[x] -> ───────────────}}
              24 e i
```

```
In[11]:=
beam4[e_,i_,w_,x_]=de4[[1,1,2]]
```

```
Out[11]=
         2 2
w (-x + x )
─────────────────
    24 e i
```

```
In[12]:=
p4=Plot[beam4[1,1,48,x],{x,0,1},
    PlotRange->All,AspectRatio->1,
    PlotStyle->GrayLevel[.6],
    DisplayFunction->Identity]
```

```
Out[12]=
-Graphics-
```

In order to compare the effects that the varying boundary conditions have on the resulting solutions, all four plots are displayed together with **Show**.

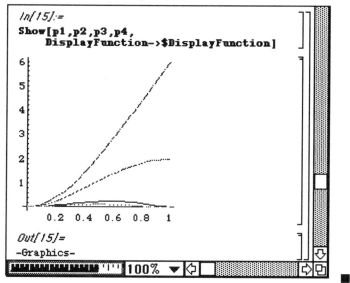

```
In[15]:=
Show[p1,p2,p3,p4,
    DisplayFunction->$DisplayFunction]
```

```
Out[15]=
-Graphics-
```

§5.6 The Simple Pendulum

Another situation which leads to a second order ordinary differential equation is that of the simple pendulum. In this case, a mass m is attached to the end of rod of length L which is suspended from a rigid support. Since the motion is best determined in terms of the angular displacement t, we let $\theta = 0$ correspond to the rod hanging vertically.

The objective is to find the motion of the mass as a function of θ, an initial position, and an initial velocity. Assuming that the pendulum is allowed to rotate without friction, the only force acting on the pendulum is that of gravity. Newton's second law and the relationship $s = L\theta$ are used to establish the following initial value problem which models this situation:

$$L\frac{d^2\theta}{dt^2} + g \sin \theta = 0, \ \theta(0) = \theta_0, \ \theta'(0) = v_0.$$

Notice that this differential equation is nonlinear. However, this nonlinear equation can be approximated by making use of the power series expansion of $\sin \theta$ given below:

$$\sin \theta = \theta - \frac{\theta^3}{3!} + \frac{\theta^5}{5!} - + \cdots$$

Hence, for small displacements (i.e., small values of θ), we have the approximation $\sin\theta \approx \theta$.

Therefore, the initial value problem becomes

$$L\frac{d^2\theta}{dt^2} + g\,\theta = 0,\ \theta(0) = \theta_0,\ \theta'(0) = v_0.$$

Notice that this problem is linear and can be easily solved.

■ EXAMPLE 5.17

Determine the solution to both versions of the pendulum problem (nonlinear and linear approximation) for values of g and L such that $g/L = 1$, an initial position of $\theta(0)=0$, and an initial velocity of $\theta'(0)=2$.

Solution:

In order to solve the nonlinear equation, we employ *Mathematica*'s numerical differential equation solving command **NDSolve**. For convenience, the function $\theta(t)$ is entered in *Mathematica* as x(t).

After defining the differential equation as **eqn**, this solution is given on the interval [0,10] in **soln1** below and plotted in **plot1**.

```
≡□≡≡≡≡≡≡ SimplePendulum ≡≡≡≡≡□≡
In[1]:=
eqn=x''[t]+Sin[x[t]]==0

Out[1]=
Sin[x[t]] + x''[t] == 0

In[2]:=
soln1=NDSolve[{eqn,x[0]==0,x'[0]==2},x[t],
    {t,0,10}]

Out[2]=
{{x[t] -> InterpolatingFunction[{0., 10.}, <>][t]}}
```

```
In[3]:=
plot1=Plot[x[t]/.soln1,{t,0,10},
    PlotRange->All,
    DisplayFunction->Identity];
```

The solution to the linear approximation is then determined with **DSolve**. For convenience and later use, this solution is found in terms of the parameters a and b. The solution to the linear case is called **pen** and is defined as a function of t, a and b.

```
In[4]:=
eq=DSolve[{x''[t]+x[t]==0,x[0]==a,x'[0]==b},
x[t],t]

Out[4]=
{{x[t] -> a Cos[t] + b Sin[t]}}

In[5]:=
pen[t_,a_,b_]=eq[[1,1,2]]

Out[5]=
a Cos[t] + b Sin[t]
```

This function is then plotted **approx1** for the values of a = 0 and b = 2 to correspond to the initial position and velocity values used in the nonlinear case. The two solutions are then displayed together for comparison. Notice that these functions vary greatly as t increases.

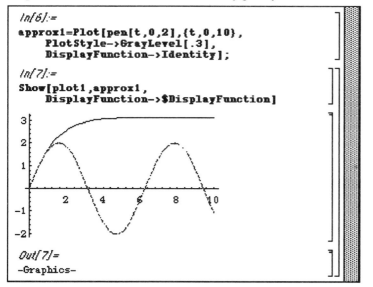

```
In[6]:=
approx1=Plot[pen[t,0,2],{t,0,10},
    PlotStyle->GrayLevel[.3],
    DisplayFunction->Identity];

In[7]:=
Show[plot1,approx1,
    DisplayFunction->$DisplayFunction]
```

```
Out[7]=
-Graphics-
```

However, for small displacements, the two solutions yield similar results as illustrated below in **plot12** and **approx12** when these functions are plotted on the interval [0,1.5].

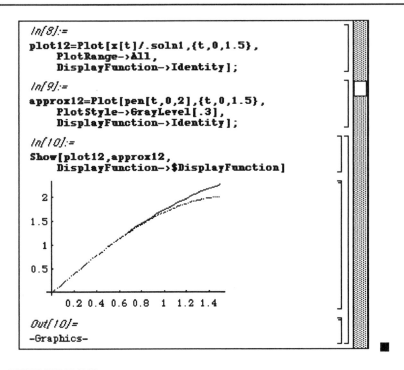

```
In[8]:=
plot12=Plot[x[t]/.soln1,{t,0,1.5},
    PlotRange->All,
    DisplayFunction->Identity];
In[9]:=
approx12=Plot[pen[t,0,2],{t,0,1.5},
    PlotStyle->GrayLevel[.3],
    DisplayFunction->Identity];
In[10]:=
Show[plot12,approx12,
    DisplayFunction->$DisplayFunction]
```

```
Out[10]=
-Graphics-
```

■ EXAMPLE 5.18

Using values of g and L such that $g/L = 1$, determine the family of solutions to the nonlinear pendulum equation with **NDSolve** for the following initial conditions:

(a) $\theta(0) = 0, \theta'(0) = 2$

(b) $\theta(0) = 2, \theta'(0) = 0$

(c) $\theta(0) = -2, \theta'(0) = 0$

(d) $\theta(0) = 0, \theta'(0) = -1$

(e) $\theta(0) = 0, \theta'(0) = -2$

(f) $\theta(0) = 1, \theta'(0) = -1$

(g) $\theta(0) = -1, \theta'(0) = 1$

Also, determine the approximate solution to the corresponding linear differential equation with **DSolve**. Plot the solutions in both cases.

Solution:

Since the first two sets of initial conditions were considered in the previous two examples, we begin with the problem corresponding to the initial conditions in (c). This problem is solved with **NDSolve** in **soln3** below over the interval [0,10].

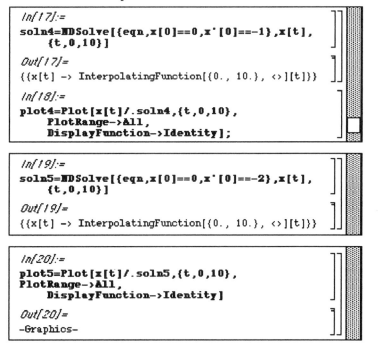

The result is then plotted in **plot3** where the plot is suppressed initially.

Solutions to the other problems are determined in a similar manner in the commands which follow.

```
In[21]:=
soln6=NDSolve[{eqn,x[0]==1,x'[0]==-1},x[t],
{t,0,10}]

Out[21]=
{{x[t] -> InterpolatingFunction[{0., 10.}, <>][t]}}

In[22]:=
plot6=Plot[x[t]/.soln6,{t,0,10},
     PlotRange->All,
     DisplayFunction->Identity];
```

```
In[23]:=
soln7=NDSolve[{eqn,x[0]==-1,x'[0]==1},x[t],
     {t,0,10}]

Out[23]=
{{x[t] -> InterpolatingFunction[{0., 10.}, <>][t]}}

In[24]:=
plot7=Plot[x[t]/.soln7,{t,0,10},
     PlotRange->All,
     DisplayFunction->Identity];
```

Upon completion of these commands, the plots are displayed simultaneously in **nonlin**.

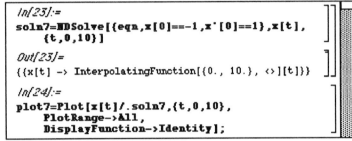

```
In[25]:=
nonlin=Show[plot1,plot2,plot3,
     plot4,plot5,plot6,plot7,
     DisplayFunction->$DisplayFunction]
```

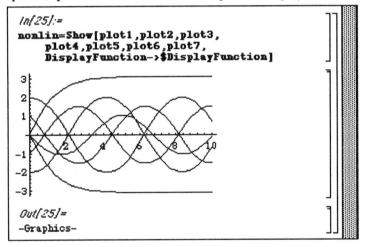

```
Out[25]=
-Graphics-
```

Since the solution to the linear approximation was determined in **pen** in a prior example, we again use this calculation to determine and plot the solutions to the corresponding initial value problems.

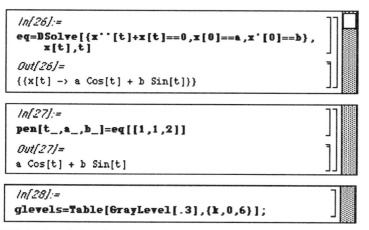

In[26]:=
```
eq=DSolve[{x''[t]+x[t]==0,x[0]==a,x'[0]==b},
    x[t],t]
```
Out[26]=
```
{{x[t] -> a Cos[t] + b Sin[t]}}
```

In[27]:=
```
pen[t_,a_,b_]=eq[[1,1,2]]
```
Out[27]=
```
a Cos[t] + b Sin[t]
```

In[28]:=
```
glevels=Table[GrayLevel[.3],{k,0,6}];
```

This is done below in **linear**. Note that these curves are plotted using the option **GrayLevel** which is defined as the list **glevels**.

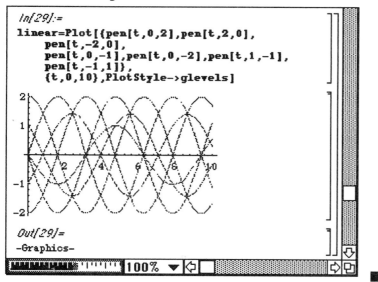

In[29]:=
```
linear=Plot[{pen[t,0,2],pen[t,2,0],
    pen[t,-2,0],
    pen[t,0,-1],pen[t,0,-2],pen[t,1,-1],
    pen[t,-1,1]},
    {t,0,10},PlotStyle->glevels]
```

Out[29]=
-Graphics-

Suppose that the pendulum undergoes a damping force which is proportional to the instantaneous velocity. Hence, the force due to damping is given as

$$F_D = -b\frac{d\theta}{dt}.$$

Incorporating this force into the sum of the forces acting on the pendulum, we have

$$L\frac{d^2\theta}{dt^2} + b\frac{d\theta}{dt} + g\sin\theta = 0, \quad \theta(0) = \theta_0, \quad \theta'(0) = v_0.$$

We now investigate the properties of this nonlinear differential equation.

■ **EXAMPLE 5.19**

Use **NDSolve** to investigate the solutions to the damped pendulum equation

$$\frac{d^2\theta}{dt^2} + 0.5\frac{d\theta}{dt} + \sin\theta = 0, \quad \theta(0) = \theta_0, \quad \theta'(0) = v_0.$$

using the following initial conditions:

(a) $\theta(0) = 1$, $\theta'(0) = 0$; (b) $\theta(0) = -1, \theta'(0) = 0$; (c) $\theta(0) = 0.5$, $\theta'(0) = 0$; (d) $\theta(0) = -0.5$, $\theta'(0) = -1$;
(e) $\theta(0) = 0$, $\theta'(0) = 1$; (f) $\theta(0) = 0$, $\theta'(0) = -1$; (g) $\theta(0) = 0, \theta'(0) = 2$; (h) $\theta(0) = 0, \theta'(0) = -2$;
(i) $\theta(0) = 1, \theta'(0) = 1$; (j) $\theta(0) = 1, \theta'(0) = -1$; (k) $\theta(0) = -1, \theta'(0) = 1$; (l) $\theta(0) = -1, \theta'(0) = -1$;
(m) $\theta(0) = 1$, $\theta'(0) = 2$; (n) $\theta(0) = 1$, $\theta'(0) = 3$; (o) $\theta(0) = -1, \theta'(0) = 4$; (p) $\theta(0) = -1, \theta'(0) = 5$;
(q) $\theta(0) = -1, \theta'(0) = 2$; (r) $\theta(0) = -1, \theta'(0) = 3$; (s) $\theta(0) = 1, \theta'(0) = -4$; (t) $\theta(0) = 1, \theta'(0) = -5$.

Solution:

Notice that, in this case, the damping coefficient is relatively small compared to the other coefficients. If this nonlinear differential equation is approximated (as was the illustrated with the simple pendulum), then we can observe the value of ($b^2 - 4 Lg$). Here, that value is $-15/4$. (Thus, the linearized equation is underdamped.) In order to observe the properties of solutions to this problem, the solution to the initial value problem given in (a) is found with **NDSolve**. To facilitate the calculations in this problem, the function $s[i,j]$ which uses **NDSolve** to solve this initial value problem with initial position i and initial velocity j is defined below.

```
≣▭▬▬▬▬▬▬ DampedPendulum ▬▬▬▬▬▭≣

In[2]:=
Clear[eq,s]
eq=x''[t]+.5 x'[t]+Sin[x[t]]==0

Out[2]=
Sin[x[t]] + 0.5 x'[t] + x''[t] == 0

In[3]:=
s[i_,j_]:=
    NDSolve[{eq,x[0]==i,x'[0]==j},x[t],
        {t,0,15}];
```

The function **s** is used with the values of initial position given in **table1** below and **j** = 0 to generate a list of four interpolating functions.

```
In[5]:=
table1={-1,-.5,.5,1};
sols1=Table[s[table1[[i]],0][[1,1,2]],{i,1,4}]

Out[5]=
{InterpolatingFunction[{0., 15.}, <>][t],
  InterpolatingFunction[{0., 15.}, <>][t],
  InterpolatingFunction[{0., 15.}, <>][t],
  InterpolatingFunction[{0., 15.}, <>][t]}
```

These four solutions are plotted below in **one**.

```
In[6]:=
one=Plot[Evaluate[sols1],{t,0,15}]
```

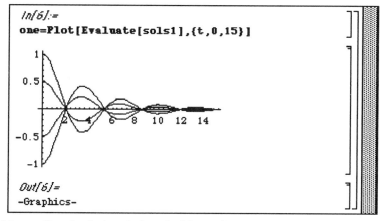

```
Out[6]=
-Graphics-
```

Next, a list of solutions is found with **s** using **i** = 0 and values of initial velocity given in **table2**.

```
In[8]:=
table2={-2,-1,1,2};
sols2=Table[s[0,table2[[i]]][[1,1,2]],{i,1,4}]

Out[8]=
{InterpolatingFunction[{0., 15.}, <>][t],
  InterpolatingFunction[{0., 15.}, <>][t],
  InterpolatingFunction[{0., 15.}, <>][t],
  InterpolatingFunction[{0., 15.}, <>][t]}
```

These functions are plotted below in **two**.

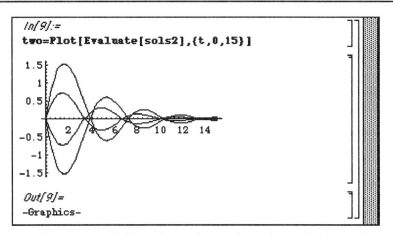

```
In[9]:=
two=Plot[Evaluate[sols2],{t,0,15}]
```

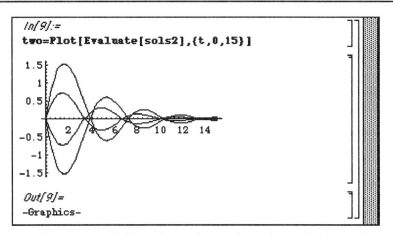

```
Out[9]=
-Graphics-
```

The solutions to the initial value problem using values of initial position and initial velocity listed in the ordered pairs in **table3** are determined in **sol3** below.

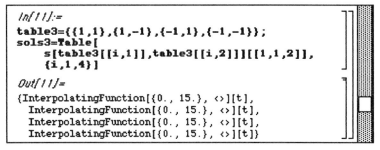

```
In[11]:=
table3={{1,1},{1,-1},{-1,1},{-1,-1}};
sols3=Table[
    s[table3[[i,1]],table3[[i,2]]][[1,1,2]],
    {i,1,4}]
```

```
Out[11]=
{InterpolatingFunction[{0., 15.}, <>][t],
 InterpolatingFunction[{0., 15.}, <>][t],
 InterpolatingFunction[{0., 15.}, <>][t],
 InterpolatingFunction[{0., 15.}, <>][t]}
```

These functions are plotted below in **three**.

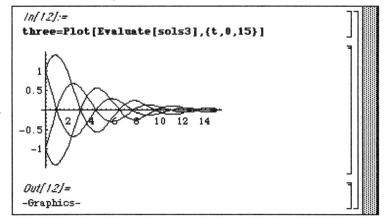

```
In[12]:=
three=Plot[Evaluate[sols3],{t,0,15}]
```

```
Out[12]=
-Graphics-
```

Similar calculations are performed with **s** for the initial position and velocity values given in **table4** below.

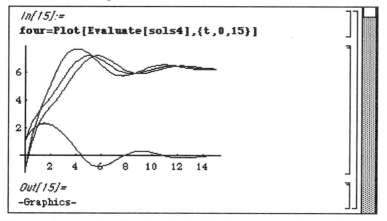

```
In[14]:=
table4={{1,2},{1,3},{-1,4},{-1,5}};
sols4=Table[
    s[table4[[i,1]],table4[[i,2]]][[1,1,2]],
    {i,1,4}]
Out[14]=
{InterpolatingFunction[{0., 15.}, <>][t],
  InterpolatingFunction[{0., 15.}, <>][t],
  InterpolatingFunction[{0., 15.}, <>][t],
  InterpolatingFunction[{0., 15.}, <>][t]}
```

These solutions are plotted below in **four**.

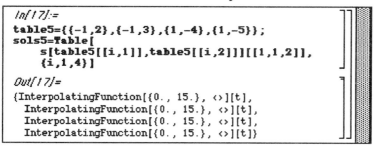

```
In[15]:=
four=Plot[Evaluate[sols4],{t,0,15}]
```

```
Out[15]=
-Graphics-
```

The solutions are then found with the initial positions and velocities in **table5** below.

```
In[17]:=
table5={{-1,2},{-1,3},{1,-4},{1,-5}};
sols5=Table[
    s[table5[[i,1]],table5[[i,2]]][[1,1,2]],
    {i,1,4}]
Out[17]=
{InterpolatingFunction[{0., 15.}, <>][t],
  InterpolatingFunction[{0., 15.}, <>][t],
  InterpolatingFunction[{0., 15.}, <>][t],
  InterpolatingFunction[{0., 15.}, <>][t]}
```

The solutions are plotted in **five** below.

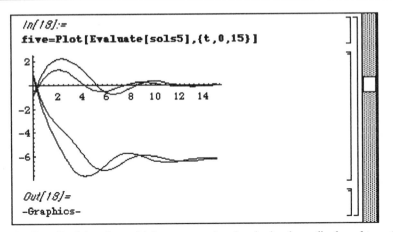

Finally, all of the plots which were previously obtained are displayed together.

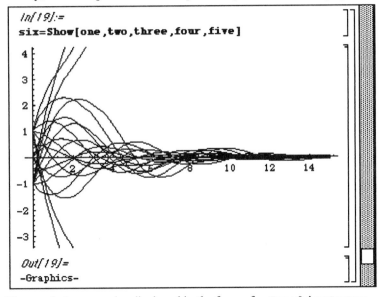

These solutions are also displayed in the form of a **GraphicsArray**.

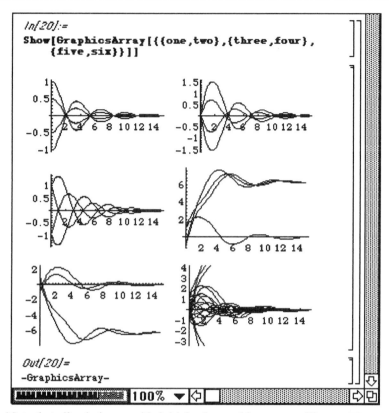

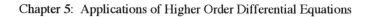

Note that all solutions to this initial value problem are oscillatory. ■

Chapter 6: Power Series Solutions of Ordinary Differential Equations

Mathematica commands used in **Chapter 6** include:

Abs	If	PlotPoints
BesselJ	Integer	PlotRange
ColumnForm	Integrate	PlotStyle
D	LaguerreL	PowerExpand
Dashing	LegendreP	ReplaceAll
DisplayFunction	Logical Expand	Series
DSolve	Map	Show
Evaluate	Module	Simplify
Expand	NDSolve	Solve
ExpandAll	Normal	Sum
GraphicsArray	Part	Table
Head	Partition	TableForm
HermiteH	Plot	

§6.1 Power Series Review

This section quickly reviews basic properties of power series that will be used in later sections. These properties and proofs of the major theorems can be found in most elementary calculus books.

Definitions:

Let x_0 be a number. A **power series** in $(x - x_0)$ is a series of the form $\sum_{n=0}^{\infty} a_n (x - x_0)^n$, where a_n is a number for all values of n.

The power series $\sum_{n=0}^{\infty} a_n (x - x_0)^n$ always converges for $x = x_0$. If there is a positive number h so that the power series $\sum_{n=0}^{\infty} a_n (x - x_0)^n$ converges absolutely for all values of x in the interval $(x_0 - h, x_0 + h)$ and diverges for all values of x in the interval $(-\infty, x_0 - h) \cup (x_0 + h, +\infty)$ the power series has **radius of convergence** h. In this case, the power series may or may not converge for $x = x_0 - h$ and may or may not converge for $x = x_0 + h$.

If the power series converges absolutely for all values of x, the power series has infinite radius of convergence.

A power series may be differentiated and integrated term-by-term on its interval of convergence. Precisely, if the power series $\sum_{n=0}^{\infty} a_n (x - x_0)^n$ has radius of convergence $h > 0$ (h may be $+ \infty$), then the

function $f(x) = \sum_{n=0}^{\infty} a_n (x - x_0)^n$ has derivatives of all orders on its interval of convergence and

$$f'(x) = \sum_{n=0}^{\infty} n a_n (x - x_0)^{n-1}, f''(x) = \sum_{n=0}^{\infty} n(n - 1) a_n (x - x_0)^{n-2}, \text{ and so on.}$$

If the power series $\sum_{n=0}^{\infty} a_n (x - x_0)^n$ has radius of convergence $h > 0$ (h may be $+ \infty$), then the

series $\sum_{n=0}^{\infty} \frac{a_n}{n+1} (x - x_0)^{n+1}$ has radius of convergence h. In fact, if $f(x) = \sum_{n=0}^{\infty} a_n (x - x_0)^n$ then

$$\int f(x) \, dx = \sum_{n=0}^{\infty} \frac{a_n}{n+1} (x - x_0)^{n+1} + c.$$

Definition:

Let f be a function with derivatives of all orders on an interval (a,b) and let $a < x_0 < b$. The **Taylor series**

for f about $x = x_0$ is $\sum_{k=0}^{\infty} \frac{f^{(k)}(x_0)}{k!} (x - x_0)^k = f(x_0) + f'(x_0)(x - x_0) + \ldots + \frac{f^{(n)}(x_0)}{n!} (x - x_0)^n + \ldots$

The **Maclaurin series for f** is the Taylor series for f about $x = 0$: $\sum_{k=0}^{\infty} \frac{f^{(k)}(0)}{k!} x^k$.

The **nth degree Taylor polynomial for f about $x = x_0$ is** $\sum_{k=0}^{n} \frac{f^{(k)}(x_0)}{k!} (x - x_0)^k$; the

nth degree Maclaurin polynomial for f is the nth degree Taylor polynomial for f about $x=0$:

$$\sum_{k=0}^{n} \frac{f^{(k)}(0)}{k!} x^k.$$

Definition:

A function f is **analytic at x = x_0** means that $f(x) = \displaystyle\sum_{n=0}^{\infty} \frac{f^{(n)}(x_0)}{n!}(x - x_0)^n$ has radius of

convergence h > 0.

Taylor's Theorem:

Let f be a function with derivatives of all orders on an interval (a,b) and let a<x_0<b.

If $x \in (a, b)$ and $x \neq x_0$, there is a number z between x and x_0 so that

$$f(x) = \sum_{k=0}^{n} \frac{f^{(k)}(x_0)}{k!}(x - x_0)^k + R_n(x), \text{ where } R_n(x) = \frac{f^{(n+1)}(z)}{(n+1)!}(x - x_0)^{n+1}. \text{ Moreover, if}$$

$\displaystyle\lim_{n \to \infty} R_n(x) = 0$ for every x in the interval (a,b), then $f(x) = \displaystyle\sum_{k=0}^{\infty} \frac{f^{(k)}(x_0)}{k!}(x - x_0)^k$.

❏ **EXAMPLE 6.1**

Let $f(x) = e^x$. (a) Find the Maclaurin series for f(x); (b) find the nth Maclaurin polynomial for f(x);

and (c) show that $e^x = \displaystyle\sum_{k=0}^{\infty} \frac{1}{k!}x^k$ for every value of x.

Solution:

Since $f^{(k)}(x) = e^x$ for every k, $f^{(k)}(0) = 1$ and thus the Maclaurin series for f(x) is

$\displaystyle\sum_{k=0}^{\infty} \frac{f^{(k)}(0)}{k!}x^k = \sum_{k=0}^{\infty} \frac{1}{k!}x^k$. The nth Maclaurin polynomial is $\displaystyle\sum_{k=0}^{n} \frac{1}{k!}x^k$ and, by Taylor's Theorem,

$R_n(x) = \dfrac{f^{(n+1)}(z)}{(n+1)!}x^{n+1} = \dfrac{e^z}{(n+1)!}x^{n+1}$. If x < 0, then since z is between x and 0, $e^z \leq 1$ and

$\left| R_n(x) \right| = \left| \dfrac{e^z}{(n+1)!}x^{n+1} \right| \leq \dfrac{|x|^{n+1}}{(n+1)!}$ so that $\displaystyle\lim_{n \to \infty}\left| R_n(x) \right| \leq \lim_{n \to \infty} \dfrac{|x|^{n+1}}{(n+1)!} = 0$. If x > 0, then $e^z \leq e^x$

and $\displaystyle\lim_{n \to \infty}\left| R_n(x) \right| \leq e^x \lim_{n \to \infty} \dfrac{x^{n+1}}{(n+1)!} = 0$. In any case, $\displaystyle\lim_{n \to \infty} R_n(x) = 0$ so $e^x = \displaystyle\sum_{k=0}^{\infty} \dfrac{1}{k!}x^k$ for every

value of x. ∎

❑ EXAMPLE 6.2

A **geometric series** is a series of the form $\displaystyle\sum_{n=0}^{\infty} a\,r^n$. If $|r| < 1$, the geometric series $\displaystyle\sum_{n=0}^{\infty} a\,r^n$

converges and $\displaystyle\sum_{n=0}^{\infty} a\,r^n = \frac{a}{1-r}$. If $|r| \geq 1$, the geometric series $\displaystyle\sum_{n=0}^{\infty} a\,r^n$ diverges.

Find the Maclaurin series for $f(x) = \dfrac{x}{2 - 3x^2}$ and find the Taylor series for $g(x) = \dfrac{1}{3x - 5}$ about $x = 1$.

Solution:

In this case, we rewrite

$$f(x) = \frac{x}{2 - 3x^2} = x\frac{1}{2 - 3x^2} = \frac{x}{2}\frac{1}{1 - \frac{3}{2}x^2} = \frac{x}{2}\sum_{n=0}^{\infty}\left(\frac{3}{2}x^2\right)^n = \sum_{n=0}^{\infty}\frac{3^n}{2^{n+1}}x^{2n+1} \text{ and}$$

$$g(x) = \frac{1}{3x - 5} = \frac{1}{3(x-1) + 3 - 5} = \frac{1}{3(x-1) - 2} = \frac{1}{-2}\frac{1}{1 - \frac{3}{2}(x-1)} = \frac{-1}{2}\sum_{n=0}^{\infty}\left(\frac{3}{2}(x-1)\right)^n$$

$$= \sum_{n=0}^{\infty}\frac{-3^n}{2^{n+1}}(x-1)^n. \quad\blacksquare$$

Frequently used Maclaurin series are listed in the table below.

Table 6.1	
Frequently used Maclaurin Series	
$\dfrac{1}{1-x} = \displaystyle\sum_{n=0}^{\infty} x^n,\ x \in (-1,1)$	$e^x = \displaystyle\sum_{n=0}^{\infty}\frac{1}{n!}x^n,\ x \in (-\infty,\infty)$
$\sin(x) = \displaystyle\sum_{n=0}^{\infty}\frac{(-1)^n x^{2n+1}}{(2n+1)!},\ x \in (-\infty,\infty)$	$\cos(x) = \displaystyle\sum_{n=0}^{\infty}\frac{(-1)^n x^{2n}}{(2n)!},\ x \in (-\infty,\infty)$
$\text{Ln}(1+x) = \displaystyle\sum_{n=1}^{\infty}\frac{(-1)^{n-1} x^n}{n},\ x \in (-1,1]$	$\tan^{-1}(x) = \displaystyle\sum_{n=0}^{\infty}\frac{(-1)^n x^{2n+1}}{(2n+1)!},\ x \in [-1,1]$
$\sinh(x) = \displaystyle\sum_{n=0}^{\infty}\frac{x^{2n+1}}{(2n+1)!},\ x \in (-\infty,\infty)$	$\cosh(x) = \displaystyle\sum_{n=0}^{\infty}\frac{x^{2n}}{(2n)!},\ x \in (-\infty,\infty)$

■ **EXAMPLE 6.3**

Let $f(x) = (x-1) \sin(x) + \cos\left(\dfrac{x}{3}\right) \sin\left(\dfrac{x}{2}\right)$. Find and graph the Maclaurin polynomial for $f(x)$ of

degree one, four, seven, and ten.

Solution:

We begin by defining and graphing f on the interval $[-2\pi, 2\pi]$.

```
▐█▌▬▬▬▬▬▬▬▬▬▬▬ SeriesExample ▬▬▬▬▬▬▬▬▬▐▣▐

 In[63]:=
 Clear[f]
 f[x_]=(x-1)Sin[x]+Cos[x/3]Sin[x/2];
 plotf=Plot[f[x],{x,-2Pi,2Pi},
     PlotStyle->Dashing[{.01,.02}],
     DisplayFunction->Identity]

 Out[63]=
 -Graphics-
```

The command `Series[f[x],{x,a,n}]` computes the power series for f(x) about x=a to order

$(x-a)^n$. The symbol $O[x, a]^{n+1}$, or, if a = 0, $O[x]^{n+1}$, represents the higher order terms that are

omitted from the series. In this case, `Series` is used to compute the power series for f(x) about x=0 to

order one, four, seven, and ten. The resulting list of series is named `approxes` and displayed in

column form with `ColumnForm`.

```
 In[65]:=
 approxes=Table[Series[f[x],{x,0,n}],
             {n,1,10,3}];
 ColumnForm[approxes]

 Out[65]=
```

$$\frac{-x}{2} + O[x]^2$$

$$\frac{-x}{2} + x^2 + \frac{17 x^3}{144} - \frac{x^4}{6} + O[x]^5$$

$$\frac{-x}{2} + x^2 + \frac{17 x^3}{144} - \frac{x^4}{6} - \frac{2071 x^5}{311040} + \frac{x^6}{120} +$$

$$\frac{80291 x^7}{470292480} + O[x]^8$$

$$\frac{-x}{2} + x^2 + \frac{17 x^3}{144} - \frac{x^4}{6} - \frac{2071 x^5}{311040} + \frac{x^6}{120} +$$

$$\frac{80291 x^7}{470292480} - \frac{x^8}{5040} - \frac{337079 x^9}{135444234240} + \frac{x^{10}}{362880} +$$

$$O[x]^{11}$$

The command `Normal[expression]` is used to convert the expression `expression` to normal

form. If `expression` is a power series, then this is accomplished by truncating the series by removing the higher order terms. In the following command, **Normal** is applied to each element of **approxes**, and the resulting list is named **tograph**. Consequently, the Maclaurin polynomial for $f(x)$ of degree one is $\dfrac{-x}{2}$, degree 4 is $\dfrac{-x}{2} + x^2 + \dfrac{17}{144}x^3 - \dfrac{1}{6}x^4$, of degree seven is

$$\dfrac{-x}{2} + x^2 + \dfrac{17}{144}x^3 - \dfrac{1}{6}x^4 - \dfrac{2071}{311040}x^5 + \dfrac{1}{120}x^6 + \dfrac{80291}{470292480}x^7, \text{ and of degree ten is}$$

$$\dfrac{-x}{2} + x^2 + \dfrac{17}{144}x^3 - \dfrac{1}{6}x^4 - \dfrac{2071}{311040}x^5 + \dfrac{1}{120}x^6 + \dfrac{80291}{470292480}x^7 - \dfrac{1}{5040}x^8 -$$

$$\dfrac{337079}{135444234240}x^9 + \dfrac{1}{362880}x^{10}.$$

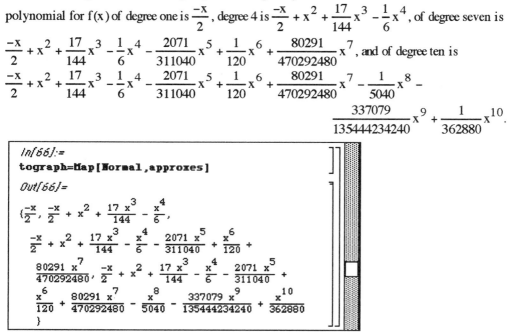

To see that the Maclaurin polynomial approximates f, we define the function **p[j]** which
1. declares the variable **plotj** local to the function **p**;
2. graphs the jth element of **tograph** on the interval $[-2\pi, 2\pi]$ and names the resulting graph, which is not displayed, **plotj**; and
3. shows the graphics objects **plotf** and **plotj** simultaneously. Note that **p[j]** is not actually displayed; the graph is displayed with the command
Show[p[j],DisplayFunction->$DisplayFunction].

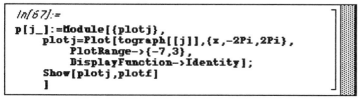

We then compute **p[j]** for j=1, 2, 3, and 4. The resulting list of four graphics objects is named **graphs**, then partitioned into two element subsets and named **array**. The resulting two-dimensional array is displayed as a graphics array.

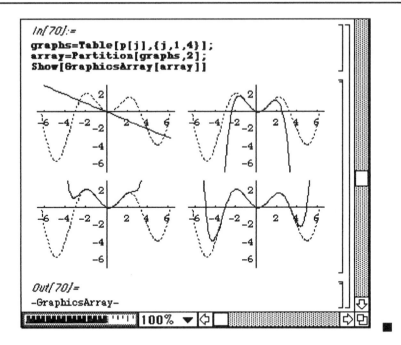

```
In[70]:=
graphs=Table[p[j],{j,1,4}];
array=Partition[graphs,2];
Show[GraphicsArray[array]]
```

```
Out[70]=
-GraphicsArray-
```

100%

■ EXAMPLE 6.4

Let $f(x) = \sin\left(\dfrac{x}{2}\right) + \cos(2x)$. Compute and graph the Taylor polynomial of degree five for

$f(x)$ about $x = \dfrac{\pi}{2}$.

Solution:

Proceeding as in the previous example, we begin by defining and graphing f on the interval $[0,\pi]$. Note that the resulting graph, which is not displayed since the option **DisplayFunction->Identity** is included, is named **plotf** for future use.

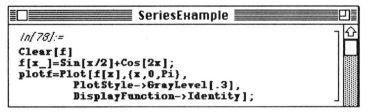

```
SeriesExample
In[78]:=
Clear[f]
f[x_]=Sin[x/2]+Cos[2x];
plotf=Plot[f[x],{x,0,Pi},
        PlotStyle->GrayLevel[.3],
        DisplayFunction->Identity];
```

Then **Series** is used to compute the power series for f(x) about $x=\pi/2$ to order $(x-\pi/2)^5$. The resulting series is named **serf**.

```
In[79]:=
serf=Series[f[x],{x,Pi/2,5}]

Out[79]=
```

$$(-1 + \frac{1}{\text{Sqrt}[2]}) + \frac{\frac{-Pi}{2} + x}{2^{3/2}} +$$

$$(2 - 2^{-(7/2)}) \; (\frac{-Pi}{2} + x)^2 - \frac{(\frac{-Pi}{2} + x)^3}{3 \; 2^{9/2}} +$$

$$(-(\frac{2}{3}) + \frac{1}{3 \; 2^{15/2}}) \; (\frac{-Pi}{2} + x)^4 + \frac{(\frac{-Pi}{2} + x)^5}{15 \; 2^{17/2}} +$$

$$O[\frac{-Pi}{2} + x]^6$$

Then **Normal** is used to convert **serf** to a normal expression; the resulting output is named **poly**. Therefore the Taylor polynomial of five for f(x) about x=π/2 is

$$\left(\frac{1}{\sqrt{2}} - 1\right) + \frac{1}{2\sqrt{2}}\left(x - \frac{\pi}{2}\right) + \left(2 - \frac{1}{2^{7/2}}\right)\left(x - \frac{\pi}{2}\right)^2 - \frac{1}{3 \cdot 2^{9/2}}\left(x - \frac{\pi}{2}\right)^3 +$$

$$\left(\frac{1}{3 \cdot 2^{15/2}} - \frac{2}{3}\right)\left(x - \frac{\pi}{2}\right)^4 + \frac{1}{15 \cdot 2^{17/2}}\left(x - \frac{\pi}{2}\right)^5.$$

```
In[80]:=
poly=Normal[serf]

Out[80]=
```

$$-1 + \frac{1}{\text{Sqrt}[2]} + \frac{\frac{-Pi}{2} + x}{2^{3/2}} +$$

$$(2 - 2^{-(7/2)}) \; (\frac{-Pi}{2} + x)^2 - \frac{(\frac{-Pi}{2} + x)^3}{3 \; 2^{9/2}} +$$

$$(-(\frac{2}{3}) + \frac{1}{3 \; 2^{15/2}}) \; (\frac{-Pi}{2} + x)^4 + \frac{(\frac{-Pi}{2} + x)^5}{15 \; 2^{17/2}}$$

Then **poly** is graphed on the interval [0,π]. The resulting graphics object is named **plotpoly** then displayed simultaneously with **plotf**. From the resulting graph we see that **poly** appears to approximate f well on the interval [1,2].

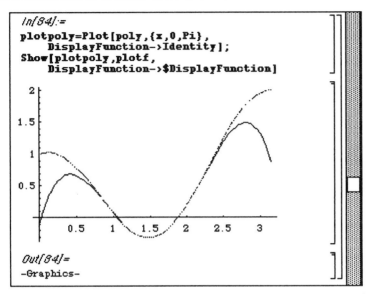

```
In[84]:=
plotpoly=Plot[poly,{x,0,Pi},
    DisplayFunction->Identity];
Show[plotpoly,plotf,
    DisplayFunction->$DisplayFunction]
```

```
Out[84]=
-Graphics-
```

The difference between **poly** and f(x) is then graphed on the interval [1,2].

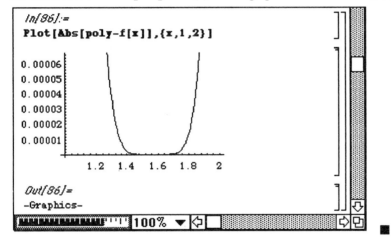

```
In[86]:=
Plot[Abs[poly-f[x]],{x,1,2}]
```

```
Out[86]=
-Graphics-
```

§6.2 Power Series Solutions about Ordinary Points

Definition:

$a_2(x)y''(x) + a_1(x)y'(x) + a_0(x)y(x) = 0$ and let $p(x) = \dfrac{a_1(x)}{a_2(x)}$ and $q(x) = \dfrac{a_0(x)}{a_2(x)}$. Then,

$a_2(x)y''(x) + a_1(x)y'(x) + a_0(x)y(x) = 0$ is equivalent to $y''(x) + p(x)(x)y'(x) + q(x)y(x) = 0$.
A number x_0 is an **ordinary point** means that both $p(x)$ and $q(x)$ are analytic at x_0. If x_0 is not an ordinary
point, then x_0 is called a **singular point**.

❑ **EXAMPLE 6.5**

Find the general solution of $\left(4 - x^2\right)\dfrac{dy}{dx} + y(x) = 0$ subject to the condition $y(0) = 1$.

Solution:

Let $y(x) = \displaystyle\sum_{n=0}^{\infty} a_n x^n$. Then, $y'(x) = \dfrac{dy}{dx} = \displaystyle\sum_{n=1}^{\infty} n a_n x^{n-1}$ and

$$\left(4 - x^2\right)\dfrac{dy}{dx} + y(x) = \left(4 - x^2\right)\sum_{n=1}^{\infty} n a_n x^{n-1} + \sum_{n=0}^{\infty} a_n x^n = \sum_{n=1}^{\infty}\left[4n a_n x^{n-1} - n a_n x^{n+1}\right] + \sum_{n=0}^{\infty} a_n x^n$$

$$= \left(4a_1 + a_0\right) + \left(8a_2 + a_1\right)x + \sum_{n=2}^{\infty}\left(a_n + 4(n+1)a_{n+1} - (n-1)a_{n-1}\right)x^n = 0.$$

Equating coefficients yields

$a_1 = \dfrac{-a_0}{4}$, $a_2 = \dfrac{-a_1}{8} = \dfrac{a_0}{32}$, and $a_n + 4(n+1)a_{n+1} - (n-1)a_{n-1} = 0$ means

$a_{n+1} = \dfrac{(n-1)a_{n-1} - a_n}{4(n+1)}$.

When $y(0)=1$, $a_0=1$. Then the first eleven coefficients of the series for y are:

n	a_n	n	a_n	n	a_n
0	1	4	$\dfrac{11}{2048}$	8	$\dfrac{1843}{8388608}$
1	$\dfrac{-1}{4}$	5	$\dfrac{-31}{8192}$	9	$\dfrac{-4859}{33554432}$
2	$\dfrac{1}{32}$	6	$\dfrac{69}{65536}$	10	$\dfrac{12767}{268435456}$
3	$\dfrac{-3}{28}$	7	$\dfrac{-187}{262144}$		

In this case the equation $\left(4 - x^2\right)\dfrac{dy}{dx} + y(x) = 0$ is separable and we can compute a closed form for y

by rewriting the equation first as $\left(4 - x^2\right)dy = -y\,dx$ and then as $\dfrac{-dy}{y} = \dfrac{dx}{4 - x^2}$. Integrating yields

$$\text{Ln}\,(y) = \frac{\text{Ln}|x - 2| - \text{Ln}|x + 2|}{4} = \text{Ln}\left|\frac{x - 2}{x + 2}\right|^{1/4} + c.\ \text{Since } y(0) = 1, \text{ we obtain } y = \left|\frac{x - 2}{x + 2}\right|^{1/4}.$$

The recursive relationship obtained above for the coefficients is used with *Mathematica* below to obtain the first ten coefficients in **table**. These coefficients are then used to generate the Maclaurin polynomial, **poly**, of order ten.

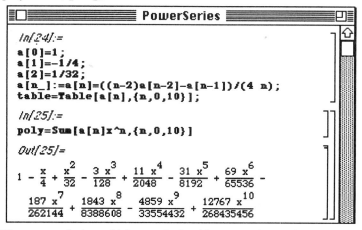

```
In[24]:=
a[0]=1;
a[1]=-1/4;
a[2]=1/32;
a[n_]:=a[n]=((n-2)a[n-2]-a[n-1])/(4 n);
table=Table[a[n],{n,0,10}];

In[25]:=
poly=Sum[a[n]x^n,{n,0,10}]

Out[25]=
```

$$1 - \frac{x}{4} + \frac{x^2}{32} - \frac{3\,x^3}{128} + \frac{11\,x^4}{2048} - \frac{31\,x^5}{8192} + \frac{69\,x^6}{65536} -$$
$$\frac{187\,x^7}{262144} + \frac{1843\,x^8}{8388608} - \frac{4859\,x^9}{33554432} + \frac{12767\,x^{10}}{268435456}$$

The exact solution which was obtained by separating variables is plotted below in **ploty**.

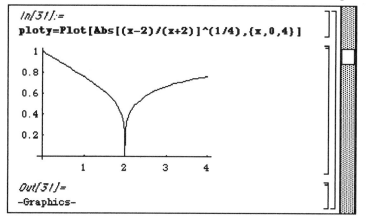

```
In[31]:=
ploty=Plot[Abs[(x-2)/(x+2)]^(1/4),{x,0,4}]
```

```
Out[31]=
-Graphics-
```

The approximating polynomial (of order ten) is also plotted for comparison.

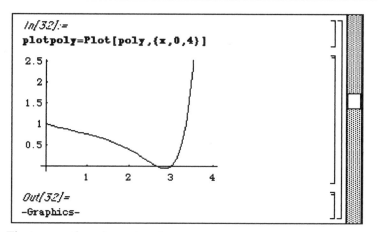

The two are then shown together.

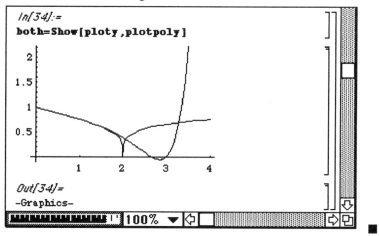

■ EXAMPLE 6.6

Find a power series solution of $y'' + \dfrac{\sin(x)}{x} y' + y = \cos(x)$ subject to $y(0) = 1$ and $y'(0) = -1$.

Solution:

We begin by trying unsuccessfully to use **DSolve** to solve the equation. Since **DSolve** does not solve the equation, we then define **lhs** to be the left-hand side of the equation,

$y'' + \dfrac{\sin(x)}{x} y' + y$, and **rhs** to be the right - hand side, $\cos(x)$.

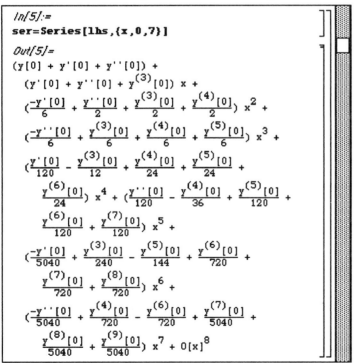

```
╔══════════════════════ PowerSeries ══════════════════════╗
In[1]:=
DSolve[y''[x]+Sin[x]/x y'[x]+y[x]==Cos[x],
     y[x],x]

Out[1]=
DSolve[y[x] + Sin[x] y'[x] + y''[x] == Cos[x],
                 ─────────
                     x

   y[x], x]

In[3]:=
lhs=y''[x]+Sin[x]/x y'[x]+y[x];
rhs=Cos[x];
```

We then use **Series** to compute the power series expansion of **lhs** about x=0 and name the resulting output **ser**.

Note that $\dfrac{\sin(x)}{x}$ is undefined when x = 0. Nevertheless, since $\underset{x \to 0}{\text{Lim}} \dfrac{\sin(x)}{x} = 1$

Mathematica is able to compute the correct series expansion.

```
In[5]:=
ser=Series[lhs,{x,0,7}]

Out[5]=
(y[0] + y'[0] + y''[0]) +

   (y'[0] + y''[0] + y^(3)[0]) x +

   (-y'[0]  + y''[0] + y^(3)[0] + y^(4)[0]) x^2 +
    ──────    ─────    ────────   ────────
      6         2         2          2

   (-y''[0] + y^(3)[0] + y^(4)[0] + y^(5)[0]) x^3 +
    ───────   ────────   ────────   ────────
       6         6          6          6

   (y'[0]  - y^(3)[0] + y^(4)[0] + y^(5)[0] +
    ─────    ────────   ────────   ────────
     120        12         24         24

    y^(6)[0]) x^4 + (y''[0] - y^(4)[0] + y^(5)[0] +
    ────────         ──────   ────────   ────────
       24              120       36         120

    y^(6)[0] + y^(7)[0]) x^5 +
    ────────   ────────
      120         120

   (-y'[0]  + y^(3)[0] - y^(5)[0] + y^(6)[0] +
    ──────    ────────   ────────   ────────
     5040       240        144        720

    y^(7)[0] + y^(8)[0]) x^6 +
    ────────   ────────
      720         720

   (-y''[0] + y^(4)[0] - y^(6)[0] + y^(7)[0] +
    ───────   ────────   ────────   ────────
      5040      720        720        5040

    y^(8)[0] + y^(9)[0]) x^7 + O[x]^8
    ────────   ────────
      5040       5040
```

The initial conditions specify that y(0)=1 and y'(0)=−1 so the symbols **y[0]** and **y'[0]** in **ser** are replaced by these values and the resulting series is named **serone**.

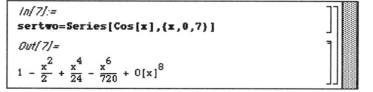

```
In[6]:=
serone=ser /.{y[0]->1,y'[0]->-1}

Out[6]=
```

$$y''[0] + (-1 + y''[0] + y^{(3)}[0]) \; x \; +$$

$$(\frac{1}{6} + \frac{y''[0]}{2} + \frac{y^{(3)}[0]}{2} + \frac{y^{(4)}[0]}{2}) \; x^2 \; +$$

$$(\frac{-y''[0]}{6} + \frac{y^{(3)}[0]}{6} + \frac{y^{(4)}[0]}{6} + \frac{y^{(5)}[0]}{6}) \; x^3 \; +$$

$$(-(\frac{1}{120}) - \frac{y^{(3)}[0]}{12} + \frac{y^{(4)}[0]}{24} + \frac{y^{(5)}[0]}{24} \; +$$

$$\frac{y^{(6)}[0]}{24}) \; x^4 + (\frac{y''[0]}{120} - \frac{y^{(4)}[0]}{36} + \frac{y^{(5)}[0]}{120} \; +$$

$$\frac{y^{(6)}[0]}{120} + \frac{y^{(7)}[0]}{120}) \; x^5 \; +$$

$$(\frac{1}{5040} + \frac{y^{(3)}[0]}{240} - \frac{y^{(5)}[0]}{144} + \frac{y^{(6)}[0]}{720} + \frac{y^{(7)}[0]}{720} \; +$$

$$\frac{y^{(8)}[0]}{720}) \; x^6 + (\frac{-y''[0]}{5040} + \frac{y^{(4)}[0]}{720} - \frac{y^{(6)}[0]}{720} \; +$$

$$\frac{y^{(7)}[0]}{5040} + \frac{y^{(8)}[0]}{5040} + \frac{y^{(9)}[0]}{5040}) \; x^7 + O[x]^8$$

In the same manner as above, we define **sertwo** to be the power series of cos(x) about x=0.

```
In[7]:=
sertwo=Series[Cos[x],{x,0,7}]

Out[7]=
```

$$1 - \frac{x^2}{2} + \frac{x^4}{24} - \frac{x^6}{720} + O[x]^8$$

Since two power series are equal means that their corresponding coefficients are equal, **LogicalExpand** is used to equate the coefficients of the series **serone** and **sertwo**. The resulting system of equations is named **equations**:

In[8]:=
equations=LogicalExpand[serone==sertwo]

Out[8]=

$-1 + y''[0] == 0$ && $-1 + y''[0] + y^{(3)}[0] == 0$ &&

$\frac{2}{3} + \frac{y''[0]}{2} + \frac{y^{(3)}[0]}{2} + \frac{y^{(4)}[0]}{2} == 0$ &&

$\frac{-y''[0]}{6} + \frac{y^{(3)}[0]}{6} + \frac{y^{(4)}[0]}{6} + \frac{y^{(5)}[0]}{6} == 0$ &&

$-(\frac{1}{20}) - \frac{y^{(3)}[0]}{12} + \frac{y^{(4)}[0]}{24} + \frac{y^{(5)}[0]}{24} +$

$\frac{y^{(6)}[0]}{24} == 0$ && $\frac{y''[0]}{120} - \frac{y^{(4)}[0]}{36} + \frac{y^{(5)}[0]}{120} +$

$\frac{y^{(6)}[0]}{120} + \frac{y^{(7)}[0]}{120} == 0$ &&

$\frac{1}{630} + \frac{y^{(3)}[0]}{240} - \frac{y^{(5)}[0]}{144} + \frac{y^{(6)}[0]}{720} + \frac{y^{(7)}[0]}{720} +$

$\frac{y^{(8)}[0]}{720} == 0$ && $\frac{-y''[0]}{5040} + \frac{y^{(4)}[0]}{720} - \frac{y^{(6)}[0]}{720} +$

$\frac{y^{(7)}[0]}{5040} + \frac{y^{(8)}[0]}{5040} + \frac{y^{(9)}[0]}{5040} == 0$

then **equations** is solved for the unknowns and the resulting solution list is named **roots**.

In[9]:=
roots=Solve[equations]

Out[9]=

$\{\{y''[0] \to 1,\ y^{(3)}[0] \to 0,\ y^{(4)}[0] \to -(\frac{7}{3}),$

$y^{(5)}[0] \to \frac{10}{3},\ y^{(6)}[0] \to \frac{1}{5},$

$y^{(7)}[0] \to -(\frac{554}{45}),\ y^{(8)}[0] \to \frac{1741}{63},$

$y^{(9)}[0] \to \frac{358}{105}\}\}$

To display the first few terms of the series, we first compute the power series for y(x) about x=0 and name the resulting series **sery**:

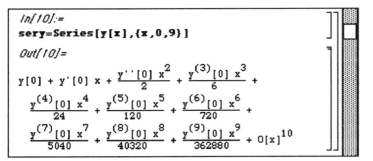

and then replace y(0) by 1, y'(0) by –1, and the remaining unknowns by the values specified in **roots**. The resulting series is converted to a normal expression with **Normal** and the resulting output is named **solapprox**.

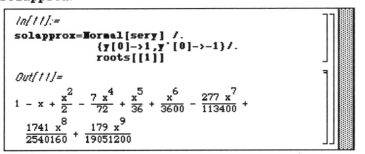

Finally, we graph **solapprox** on the interval [0,3]:

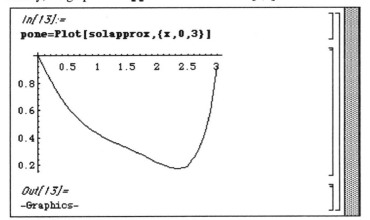

and compare the result to one obtained with **NDSolve**.

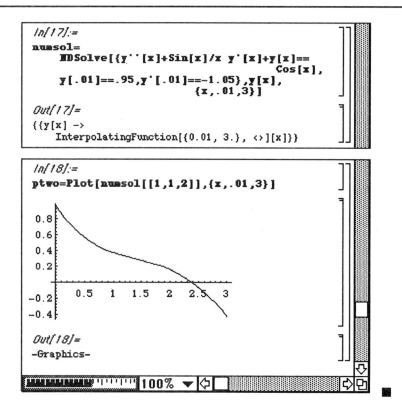

EXAMPLE 6.7

Solve the equations $y''-xy'+3y = 0$, $y''-2xy'+2 \cdot 3y = 0$, $y''-xy'+4y = 0$ and $y''-2xy'+2 \cdot 4y = 0$.

Solution:

Proceeding as in the previous example, we use **DSolve** to solve the first two equations. The symbol **Erfi[x]** represents the imaginary error function: $\operatorname{erfi}(x) = -i\,\operatorname{erf}(ix) = -i\dfrac{2}{\sqrt{\pi}}\int_0^{ix} e^{-t^2}\,dt$.

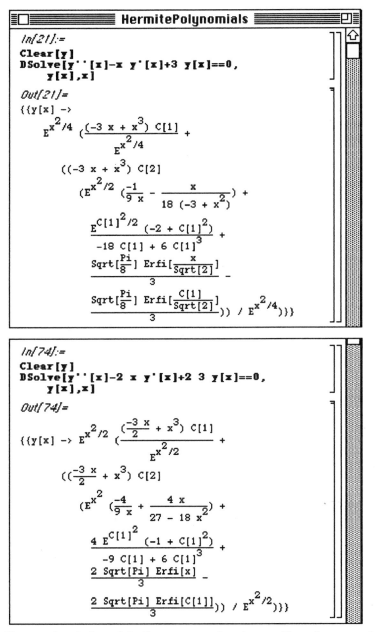

Since each solution of the first two equations is complicated, we search for a simpler solution with power series for the second two equations. Note that we could also use power series for the first two equations. We begin by defining **lhs** to be the left - hand side of the equation $y'' - x\,y' + 4y = 0$:

```
In[51]:=
Clear[x,y,lhs,serleft,eqs,roots,sol]
In[52]:=
lhs=y''[x]-x y'[x]+4y[x];
```

and then defining **serleft** to be the power series for **lhs** about x=0.

```
In[53]:=
serleft=Series[lhs,{x,0,6}]
Out[53]=
```

$$(4\ y[0] + y''[0]) + y^{(3)}[0]\ x +$$
$$(y''[0] + \frac{y^{(4)}[0]}{2})\ x^2 +$$
$$(\frac{y^{(3)}[0]}{6} + \frac{y^{(5)}[0]}{6})\ x^3 + \frac{y^{(6)}[0]\ x^4}{24} +$$
$$(\frac{-y^{(5)}[0]}{120} + \frac{y^{(7)}[0]}{120})\ x^5 +$$
$$(\frac{-y^{(6)}[0]}{360} + \frac{y^{(8)}[0]}{720})\ x^6 + O[x]^7$$

Since the coefficient of x^i must be zero for all values of i, we use **LogicalExpand** to equate the coefficients of **serleft** and 0, the right-hand side of the equation, and name the resulting system of equations **eqs**.

```
In[54]:=
eqs=LogicalExpand[serleft==0]
Out[54]=
```

$$4\ y[0] + y''[0] == 0\ \&\&\ y^{(3)}[0] == 0\ \&\&$$
$$y''[0] + \frac{y^{(4)}[0]}{2} == 0\ \&\&$$
$$\frac{y^{(3)}[0]}{6} + \frac{y^{(5)}[0]}{6} == 0\ \&\&\ \frac{y^{(6)}[0]}{24} == 0\ \&\&$$
$$\frac{-y^{(5)}[0]}{120} + \frac{y^{(7)}[0]}{120} == 0\ \&\&$$
$$\frac{-y^{(6)}[0]}{360} + \frac{y^{(8)}[0]}{720} == 0$$

We then solve **eqs** in terms of y(0) and y'(0) and name the resulting list of solutions **roots**.

```
In[55]:=
roots=Solve[eqs,
         Evaluate[Table[D[y[x],{x,i}],
            {i,2,8}] /. x->0]]
Out[55]=
{{y''[0] -> -4 y[0], y^(3)[0] -> 0,
   y^(4)[0] -> 8 y[0], y^(5)[0] -> 0,
   y^(6)[0] -> 0, y^(7)[0] -> 0, y^(8)[0] -> 0}}
```

Then the solution is obtained by computing the power series for y(x) about x=0 and then replacing each unknown by the values obtained in **roots**. The resulting series is named **sol**.

```
In[56]:=
sol=Series[y[x],{x,0,5}] /. roots[[1]]
Out[56]=
```
$$y[0] - 2\ y[0]\ x^2 + \frac{y[0]\ x^4}{3} + O[x]^6$$

We solve $y''-2x\,y'+2 \bullet 4y = 0$ in the exact same manner as above.

```
Clear[x,y,lhs,serleft,eqs,roots,sol]
In[59]:=
lhs=y''[x]-2 x y'[x]+2 4y[x];
```

```
In[60]:=
serleft=Series[lhs,{x,0,6}]
Out[60]=
```

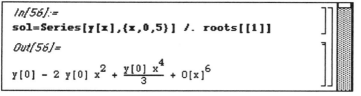

$$(8\ y[0] + y''[0]) + y^{(3)}[0]\ x +$$
$$\left(2\ y''[0] + \frac{y^{(4)}[0]}{2}\right)\ x^2 +$$
$$\left(\frac{y^{(3)}[0]}{3} + \frac{y^{(5)}[0]}{6}\right)\ x^3 + \frac{y^{(6)}[0]\ x^4}{24} +$$
$$\left(\frac{-y^{(5)}[0]}{60} + \frac{y^{(7)}[0]}{120}\right)\ x^5 +$$
$$\left(\frac{-y^{(6)}[0]}{180} + \frac{y^{(8)}[0]}{720}\right)\ x^6 + O[x]^7$$

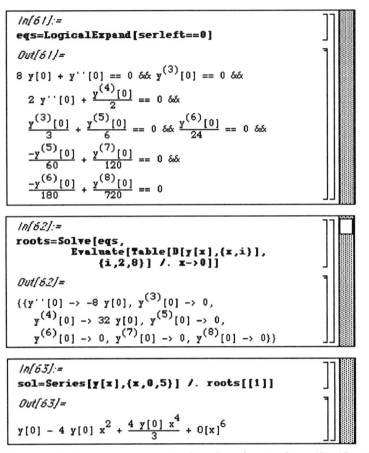

```
In[61]:=
eqs=LogicalExpand[serleft==0]

Out[61]=

8 y[0] + y''[0] == 0 && y^(3)[0] == 0 &&

   2 y''[0] + y^(4)[0]/2 == 0 &&

   y^(3)[0]/3 + y^(5)[0]/6 == 0 && y^(6)[0]/24 == 0 &&

   -y^(5)[0]/60 + y^(7)[0]/120 == 0 &&

   -y^(6)[0]/180 + y^(8)[0]/720 == 0
```

```
In[62]:=
roots=Solve[eqs,
        Evaluate[Table[D[y[x],{x,i}],
             {i,2,8}] /. x->0]]

Out[62]=

{{y''[0] -> -8 y[0], y^(3)[0] -> 0,
    y^(4)[0] -> 32 y[0], y^(5)[0] -> 0,
    y^(6)[0] -> 0, y^(7)[0] -> 0, y^(8)[0] -> 0}}
```

```
In[63]:=
sol=Series[y[x],{x,0,5}] /. roots[[1]]

Out[63]=

y[0] - 4 y[0] x^2 + 4 y[0] x^4/3 + O[x]^6
```

However, in this case, we compute the value of **sol** when y(0)=12 and y'(0)=0.

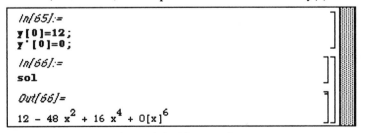

```
In[65]:=
y[0]=12;
y'[0]=0;

In[66]:=
sol

Out[66]=

12 - 48 x^2 + 16 x^4 + O[x]^6
```

We obtain the same polynomial with **HermiteH[4,x]**.

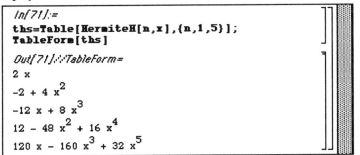

```
In[67]:=
HermiteH[4,x]
Out[67]=
         2       4
12 - 48 x  + 16 x
```

In fact, the **Hermite polynomials**, $H_n(x)$, are solutions of the ordinary differential equation

$\dfrac{d^2y}{dx^2} - 2x\dfrac{dy}{dx} + 2ny = 0$ and satisfy the **orthogonality relation** $\int_{-\infty}^{\infty} H_n(x)H_m(x)e^{-x^2}\,dx = 0$

for $m \neq n$.

The *Mathematica* command **HermiteH[n,x]** yields the Hermite polynomial $H_n(x)$. The following commands compute a table of the first five Hermite polynomials, name the resulting table **ths**, and then display **ths** in **TableForm**.

```
In[71]:=
ths=Table[HermiteH[n,x],{n,1,5}];
TableForm[ths]
Out[71]//TableForm=
2 x
         2
-2 + 4 x
            3
-12 x + 8 x
         2       4
12 - 48 x  + 16 x
              3       5
120 x - 160 x  + 32 x
```

Note that the ith element of **ths** is extracted from **ths** with **ths[[i]]** and represents the Hermite polynomial $H_i(x)$.

We then illustrate that $\int_{-\infty}^{\infty} H_n(x)H_m(x)e^{-x^2}\,dx = 0$ for $m \neq n$ by computing

$\int_{-\infty}^{\infty}$ **ths[[i]] ths[[j]]** $e^{-x^2}\,dx$ for $i = 1,2,3,4,5$ and $j = 1,2,3,4,5$ and displaying the result in **TableForm**.

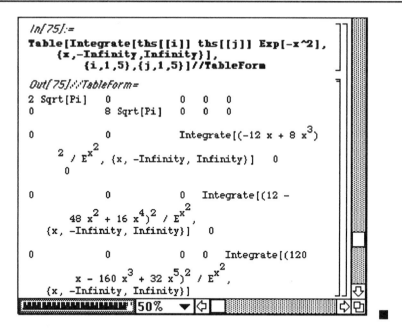

```
In[75]:=
Table[Integrate[ths[[i]] ths[[j]] Exp[-x^2],
    {x,-Infinity,Infinity}],
        {i,1,5},{j,1,5}]//TableForm

Out[75]//TableForm=
2 Sqrt[Pi]    0                  0    0    0
0             8 Sqrt[Pi]         0    0    0

0             0                  Integrate[(-12 x + 8 x³)
      2   x²
       / E   , {x, -Infinity, Infinity}]    0
    0

0             0                  0    Integrate[(12 -
      48 x² + 16 x⁴)² / E^x²,
    {x, -Infinity, Infinity}]    0

0             0                  0    0    Integrate[(120
      x - 160 x³ + 32 x⁵)² / E^x²,
    {x, -Infinity, Infinity}]
```

■ EXAMPLE 6.8

Let n and m be fixed positive integers. **Legendre's equation** is the equation

$$\left(1 - x^2\right)y'' - 2x\,y' + \left[n\,(n + 1) - \frac{m^2}{1 - x^2}\right]y = 0. \text{ If } m = 0, \text{find the general solution of}$$

Legendre's equation.

Solution:

When m = 0, Legendre's equation becomes $\left(1 - x^2\right)y'' - 2x\,y' + n\,(n + 1)y = 0.$

Let $y(x) = \displaystyle\sum_{k=0}^{\infty} a_k x^k.$ Then, $y'(x) = \dfrac{dy}{dx} = \displaystyle\sum_{k=1}^{\infty} k a_k x^{k-1}$, $y''(x) = \displaystyle\sum_{k=2}^{\infty} k(k-1) a_k x^{k-2}$ and

$$\left(1 - x^2\right)y'' - 2x\,y' + n\,(n+1)y = \left(1 - x^2\right)\sum_{k=2}^{\infty} k(k-1)a_k x^{k-2} - 2x \sum_{k=1}^{\infty} k a_k x^{k-1} + n(n+1) \sum_{k=0}^{\infty} a_k x^k$$

$$= \sum_{k=2}^{\infty} k(k-1)a_k x^{k-2} - \sum_{k=2}^{\infty} k(k-1)a_k x^k - \sum_{k=1}^{\infty} 2k a_k x^k + \sum_{k=0}^{\infty} n(n+1)a_k x^k$$

$$= \left(2a_2 + n(n+1)a_0\right) + \left(6a_3 - 2a_1 + n(n+1)a_1\right)x +$$

$$\sum_{k=2}^{\infty}\left[(k+2)(k+1)a_{k+2} - k(k-1)a_k - 2ka_k + n(n+1)a_k\right]x^k$$

$$= \left(2a_2 + n(n+1)a_0\right) + \left(6a_3 - 2a_1 + n(n+1)a_1\right)x +$$

$$\sum_{k=2}^{\infty}\left[(n(n+1) - k(k+1))a_k + \left(k^2 + 3k + 1\right)a_{k+2}\right]x^k.$$

Equating coefficients yields $2a_2 + n(n+1)a_0 = 0$, $6a_3 - 2a_1 + n(n+1)a_1 = 0$, and

$$(n(n+1) - k(k+1))a_k + \left(k^2 + 3k + 1\right)a_{k+2} = 0.$$

Then, $a_2 = \dfrac{-n(n+1)}{2}a_0$, $a_3 = \dfrac{-\left(n^2 + n - 2\right)}{6}a_1$, and $a_{k+2} = \dfrac{k(k+1) - n(n+1)}{k^2 + 3k + 1}a_k$ which means

$$a_k = \dfrac{(k-2)(k-1) - n(n+1)}{(k-2)^2 + 3(k-2) + 1}a_{k-2}.$$

Note that when n is odd, $a_{n+2i}=0$ for i=1, 2, 3, ... and when n is even, $a_{n+2i}=0$ for i=1, 2, 3, ...

Therefore, if n is odd and $a_0=0$, then the solution of Legendre's equation

$$\left(1 - x^2\right)y'' - 2xy' + n(n+1)y = 0 \text{ is a polynomial of degree n; if n is even and } a_1 = 0 \text{ the solution}$$

of Legendre's equation is a polynomial of degree n.

In the following, we define the function **coeffs[i0,i1,n,m]** which yields the first m+1 coefficients

of the power series solution $\sum_{k=0}^{\infty}a_k x^k$ of Legendre's equation $\left(1 - x^2\right)y'' - 2xy' + n(n+1)y = 0$

which satisfy $a_0 = $ **i0** and $a_1 = $ **i1**.

coeffs[i0,i1,n,m] works by

1. Declaring the variable **a** local to the function **coeffs**;
2. Defining **a[0]=i0** and **a[1]=i1**;

3. Defines **a[2]** to be $\dfrac{-n(n+1)}{2}$**a[0]** and **a[3]** to be $\dfrac{-\left(n^2 + n - 2\right)}{6}$**a[1]**;

4. Defines **a[k]** to be $\dfrac{(k-1)(k-1) - n(n+1)}{(k-2)^2 + 3(k-2) + 1}$**a[k - 2]**; and

5. Produces a table of ordered pairs **{j,a[j]}** for j=0, ... , **m**

The definition of **a[k]** is of the form **a[k_]:=a[k]=...** so that *Mathematica* "remembers" the values of **a[k]** computed since **a[k]** is defined recursively in terms of **a[k-2]**.

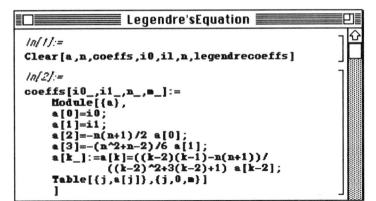

```
In[1]:=
Clear[a,n,coeffs,i0,i1,n,legendrecoeffs]

In[2]:=
coeffs[i0_,i1_,n_,m_]:=
    Module[{a},
    a[0]=i0;
    a[1]=i1;
    a[2]=-n(n+1)/2 a[0];
    a[3]=-(n^2+n-2)/6 a[1];
    a[k_]:=a[k]=((k-2)(k-1)-n(n+1))/
            ((k-2)^2+3(k-2)+1) a[k-2];
    Table[{j,a[j]},{j,0,m}]
    ]
```

For example, the first eight coefficients of the power series solution for

$$\left(1 - x^2\right)y'' - 2xy' + 5(5+1)y = 0$$ which satisfy $a_0 = 1$ and $a_1 = 1$ are computed below with

coeffs[[1,1,5,7]]:

```
In[3]:=
coeffs[1,1,5,7]

Out[3]=
{{0, 1}, {1, 1}, {2, -15}, {3, -(14/3)}, {4, 360/11},

 {5, 84/19}, {6, -(3600/319)}, {7, 0}}
```

Summarizing the above, the first eight coefficients are listed in the following table:

n	a_n	n	a_n
0	1	4	$\dfrac{360}{11}$
1	1	5	$\dfrac{84}{19}$
2	-15	6	$\dfrac{-3600}{319}$
3	$\dfrac{-14}{3}$	7	0

Note that since $a_7 = 0$, $a_n = 0$ for all odd values of n greater than 7. Therefore one solution of

$$\left(1 - x^2\right)y'' - 2xy' + 5(5+1)y = 0$$ is the polynomial $x - \dfrac{14}{3}x^3 + \dfrac{84}{19}x^5$.

Similarly, we compute the first nine coefficients of the power series solution for

$\left(1 - x^2\right)y'' - 2xy' + 6(6 + 1)y = 0$ with $a_0 = 1$ and $a_1 = 1$ with **coeffs[[1, 1, 6, 8]]**.

```
In[4]:=
coeffs[1,1,6,8]

Out[4]=
{{0, 1}, {1, 1}, {2, -21}, {3, -(20/3)}, {4, 756/11},
 {5, 200/19}, {6, -(1512/29)}, {7, -(2400/779)}, {8, 0}}
```

We summarize the results in the following table:

n	a_n	n	a_n	n	a_n
0	1	3	$\dfrac{-20}{3}$	6	$\dfrac{-1512}{29}$
1	1	4	$\dfrac{756}{11}$	7	$\dfrac{-2400}{779}$
2	-21	5	$\dfrac{200}{19}$	8	0

Since $a_8 = 0$, it follows that $a_n = 0$ for all even integers greater than 8. Therefore, one solution of

$\left(1 - x^2\right)y'' - 2xy' + 6(6 + 1)y = 0$ is the polynomial $1 - 21x^2 + \dfrac{756}{11}x^4 - \dfrac{1512}{29}x^6$.

With the following command, we define the function **legendrecoeffs**. **legendrecoeffs[n]**

1. Declares the variable **a** local to the command **legendrecoeffs**;

2. If **n** is even, defines **a[0]** to be $\dfrac{(-1)^{n/2}(n - 1)!!}{n!!}$, **a[1]** to be 0, and **m** to be 0;

   if **n** is odd, defines **a[0]** to be 0, **a[1]** to be $\dfrac{(-1)^{(n-1)/2}n!!}{(n - 1)!!}$, and **m** to be 1;

3. Defines **a[2]** to be $\dfrac{-n(n + 1)}{2}$ **a[0]** and **a[3]** to be $\dfrac{-\left(n^2 + n - 2\right)}{6}$ **a[1]**;

4. Defines **a[k]** to be $\dfrac{(k - 2)(k - 1) - n(n + 1)}{(k - 2)^2 + 3(k - 2) + 1}$ **a[k - 2]**; and

5. Produces a table of ordered pairs **{j,a[j]}** for **j=m, m+2, m+4, ... , n**.

Therefore, if n is odd, $a_0 = 0$, and $a_1 = \dfrac{(-1)^{(n-1)/2}n!!}{(n - 1)!!}$, **legendrecoffs[n]** yields the

coefficients of the polynomial solution of $\left(1 - x^2\right)y'' - 2xy' + n(n + 1)y = 0$; and if n is even,

$a_0 = \dfrac{(-1)^{n/2}(n-1)!!}{n!!}$, and $a_1 = 0$, **legendrecoeffs[n]** yields the coefficients of the

polynomial solution of $\left(1 - x^2\right)y'' - 2xy' + n(n+1)y = 0$.

The symbol **!!** represents the **Factorial2** function:

$n!! = \begin{cases} n(n-2) \cdot \cdots \cdot 2, \text{if } n \text{ is even} \\ n(n-2) \cdot \cdots \cdot 1, \text{if } n \text{ is odd} \end{cases}$.

```
In[5]:=
legendrecoeffs[n_]:=
    Module[{a},
    {a[0],a[1],m}=If[Head[n/2]===Integer,
        {(-1)^(n/2) (n-1)!!/n!!,0,0},
        {0,(-1)^((n-1)/2) n!!/(n-1)!!,1}];
    a[2]=-n(n+1)/2 a[0];
    a[3]=-(n^2+n-2)/6 a[1];
    a[k_]:=a[k]=((k-2)(k-1)-n(n+1))/
            ((k-2)^2+3(k-2)+1) a[k-2];
    Table[{j,a[j]},{j,m,n,2}]
    ]
```

For example, the result of entering the command **legendrecoeffs[5]**, shown below:

```
In[6]:=
legendrecoeffs[5]

Out[6]=
{{1, 15/8}, {3, -(35/4)}, {5, 315/38}}
```

means that the solution of $\left(1 - x^2\right)y'' - 2xy' + 5(5+1)y = 0$ with $a_0 = 0$ and $a_1 = \dfrac{15}{8}$ is

$\dfrac{15}{8}x - \dfrac{35}{4}x^3 + \dfrac{315}{38}x^5$ and the result of entering the command **legendrecoeffs[7]**, shown

below:

```
In[7]:=
legendrecoeffs[7]

Out[7]=
{{1, -(35/16)}, {3, 315/16}, {5, -(3465/76)}, {7, 45045/1558}}
```

means that the solution of $\left(1 - x^2\right)y'' - 2xy' + 7(7+1)y = 0$ with $a_0 = 0$ and $a_1 = \dfrac{-35}{16}$ is

$\dfrac{-35}{16}x + \dfrac{315}{16}x^3 - \dfrac{3465}{76}x^5 + \dfrac{45045}{1558}x^7$.

More generally, the **Legendre polynomials**, $P_n(x)$, are the polynomial solutions of the equation $\left(1 - x^2\right)y'' - 2x\,y' + n\,(n + 1)y = 0$ that satisfy the **orthogonality relation** $\int_{-1}^{1} P_n(x)P_m(x)\,dx = 0$ for $n \neq m$ and are obtained with the built-in *Mathematica* command **LegendreP[n, x]**.

With the following commands, we compute a table of the first eight Legendre polynomials, name the resulting table **legendres**, and then display **legendres** in **TableForm**.

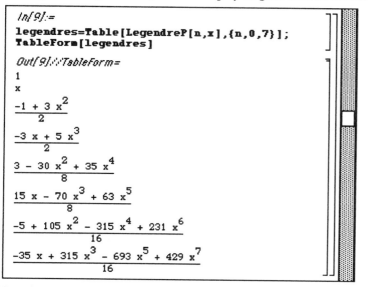

Note that **legendres[[i]]** represents the Legendre polynomial $P_i(x)$.

We then graph the first eight Legendre polynomials, **legendres**, on the interval $[-1, 1]$:

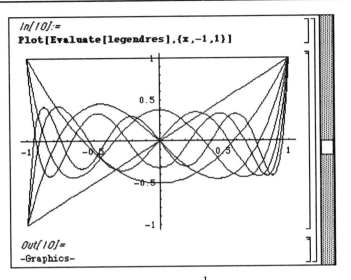

```
In[10]:=
Plot[Evaluate[legendres],{x,-1,1}]
```

```
Out[10]=
-Graphics-
```

and verify the orthogonality relation $\int_{-1}^{1} P_n(x) P_m(x) dx = 0$ for $n \neq m$ by evaluating

$\int_{-1}^{1}$ **legendres[[i]] legendres[[j]]** dx for $i = 1, 2, \ldots, 8$ and $j = 1, 2, \ldots, 8$

```
In[12]:=
ortho=Table[
    Integrate[
        legendres[[i]]legendres[[j]],
        {x,-1,1}],
    {i,1,8},{j,1,8}];
TableForm[ortho]
```

Out[12]//TableForm=

2	0	0	0	0	0	0	0
0	$\frac{2}{3}$	0	0	0	0	0	0
0	0	$\frac{2}{5}$	0	0	0	0	0
0	0	0	$\frac{2}{7}$	0	0	0	0
0	0	0	0	$\frac{2}{9}$	0	0	0
0	0	0	0	0	$\frac{2}{11}$	0	0
0	0	0	0	0	0	$\frac{2}{13}$	0
0	0	0	0	0	0	0	$\frac{2}{15}$

100% ▼

§6.3 Power Series Solutions about Regular Singular Points

Definition:
Let x_0 be a singular point of $y''(x) + p(x)(x)y'(x) + q(x)y(x) = 0$. x_0 is a **regular singular point** means that both $(x - x_0)\,p(x)$ and $(x - x_0)^2\,q(x)$ are analytic at $x = x_0$. If x_0 is not a regular singular point, x_0 is called an **irregular singular point.**

❑ **EXAMPLE 6.9**

Classify the singular points of each of the equations $y'' + \dfrac{y'}{x} + \left(1 - \dfrac{\mu^2}{x^2}\right)y = 0$,

$(x - \alpha)(x - \beta)y'' + \dfrac{1}{2}\left[2x - (\alpha + \beta)\right]y' - \left[p^2 x + q^2\right]y = 0$ (the **Baer equation**), and

$x^2 y'' + x\,y' + \left(x^2 - \mu^2\right)y = 0$ (the **Bessel equation**).

Solution:

The singular point $x = 0$ of $y'' + \dfrac{y'}{x} + \left(1 - \dfrac{\mu^2}{x^2}\right)y = 0$ is a regular singular point;

the singular points $x = \alpha$ and $x = \beta$ of the Baer equation

$(x - \alpha)(x - \beta)y'' + \dfrac{1}{2}\left[2x - (\alpha + \beta)\right]y' - \left[p^2 x + q^2\right]y = 0$ are regular singular points; and

the singular point $x = 0$ of the Bessel equation $x^2 y'' + x\,y' + \left(x^2 - \mu^2\right)y = 0$ is a regular singular

point. ■

Let $x = 0$ be a regular singular point of the equation $y''(x) + p(x)y'(x) + q(x)y(x) = 0$. Define **l h s** to be the expression (corresponding to the left - hand side of the equation)
$y''(x) + p(x)y'(x) + q(x)y(x)$:

```
┌──────────────────── SingularPoints ─────────────────────┐
│ In[86]:=                                                 │
│ Clear[equation,serp,serq,sery]                           │
│ In[87]:=                                                 │
│ lhs=y''[x]+p[x] y'[x]+q[x] y[x]                          │
│ Out[87]=                                                 │
│ q[x] y[x] + p[x] y'[x] + y''[x]                          │
└──────────────────────────────────────────────────────────┘
```

Since x=0 is a regular singular point, xp(x) is analytic at x=0 so

$$p(x) = \sum_{n=0}^{\infty} p_n x^{n-1} = \frac{p_0}{x} + p_1 + p_2 x + p_3 x^2 + p_4 x^3 + \dots$$ Define **serp** to be the first five terms

of the series for p(x) as follows where **cp[n]** denotes p_n:

```
In[88]:=
serp=Sum[cp[n] x^(n-1),{n,0,4}]

Out[88]=
cp[0]
----- + cp[1] + x cp[2] + x^2 cp[3] + x^3 cp[4]
  x
```

Similarly, since x=0 is a regular singular point, $x^2 q(x)$ is analytic at x=0 so

$$q(x) = \sum_{n=0}^{\infty} p_n x^{n-2} = \frac{q_0}{x^2} + \frac{q_1}{x} + q_2 + q_3 x + q_4 x^2 + q_5 x^3 + \dots$$ Define **serq** to be the first six

terms of the series for q(x) as follows where **cq[n]** denotes q_n:

```
In[89]:=
serq=Sum[cq[n] x^(n-2),{n,0,5}]

Out[89]=
cq[0]    cq[1]
-----  + ----- + cq[2] + x cq[3] + x^2 cq[4] +
 x^2       x

 x^3 cq[5]
```

Assume there is a constant r so that $y(x) = x^r \sum_{n=0}^{\infty} a_n x^n$ is a solution of $y''+py'+qy = 0$.

Define **sery** to be the first four terms of $x^r \sum_{n=0}^{\infty} a_n x^n$ as follows:

```
In[90]:=
sery=Expand[x^r Sum[a[n] x^n,{n,0,3}]]

Out[90]=
x^r a[0] + x^(1 + r) a[1] + x^(2 + r) a[2] + x^(3 + r) a[3]
```

Then, $\dfrac{d}{dx}\left(x^r \sum_{n=0}^{\infty} a_n x^n\right) = \dfrac{d}{dx}\left(\sum_{n=0}^{\infty} a_n x^{n+r}\right) = \sum_{n=0}^{\infty} \dfrac{d}{dx}\left(a_n x^{n+r}\right) = \sum_{n=0}^{\infty} a_n (n+r) x^{n+r-1}$.

We compute and expand the derivative of **sery** and name the result **seryprime**.

```
In[91]:=
seryprime=D[sery,x]//Expand
Out[91]=
```

$$r \, x^{-1+r} \, a[0] + x^r \, a[1] + r \, x^r \, a[1] +$$
$$2 \, x^{1+r} \, a[2] + r \, x^{1+r} \, a[2] + 3 \, x^{2+r} \, a[3] +$$
$$r \, x^{2+r} \, a[3]$$

Similarly, $\dfrac{d^2}{dx^2}\left(x^r \sum_{n=0}^{\infty} a_n x^n\right) = \dfrac{d}{dx}\left(\sum_{n=0}^{\infty} a_n \, (n+r) x^{n+r-1}\right) = \sum_{n=0}^{\infty} \dfrac{d}{dx}\left(a_n \, (n+r) x^{n+r-1}\right)$

$$= \sum_{n=0}^{\infty} a_n \, (n+r)(n+r-1) x^{n+r-2}.$$

We also compute and expand the second derivative of **sery** and name the result **seryprimetwo**:

```
In[92]:=
seryprimetwo=D[sery,{x,2}]//Expand
Out[92]=
```

$$-(r \, x^{-2+r} \, a[0]) + r^2 \, x^{-2+r} \, a[0] +$$
$$r \, x^{-1+r} \, a[1] + r^2 \, x^{-1+r} \, a[1] + 2 \, x^r \, a[2] +$$
$$3 \, r \, x^r \, a[2] + r^2 \, x^r \, a[2] + 6 \, x^{1+r} \, a[3] +$$
$$5 \, r \, x^{1+r} \, a[3] + r^2 \, x^{1+r} \, a[3]$$

Since $y''+py'+qy = 0$, substituting each series into the equation yields

$$\left(\sum_{n=0}^{\infty} a_n \, (n+r)(n+r-1) x^{n+r-2}\right) + \left(\sum_{n=0}^{\infty} p_n x^{n-1}\right)\left(\sum_{n=0}^{\infty} a_n \, (n+r) x^{n+r-1}\right) +$$

$$\left(\sum_{n=0}^{\infty} q_n x^{n-2}\right)\left(x^r \sum_{n=0}^{\infty} a_n x^n\right) = 0.$$

At this point, we can take advantage of *Mathematica*'s symbolic manipulation abilities by replacing **y''[x]** by **seryprimetwo**, **p[x]** by **serp**, **y'[x]** by **seryprime**, **q[x]** by **serq**, and **y[x]** by **sery** in **lhs** and simplifying the result; we name the result **lhstwo**.

In[93]:=
```
lhstwo=Expand[lhs /. {y[x]->sery,
    y'[x]->seryprime,y''[x]->seryprimetwo,
    p[x]->serp,q[x]->serq}]
```

Out[93]=

$-(r\ x^{-2+r}\ a[0]) + r^2\ x^{-2+r}\ a[0] +$

$\quad r\ x^{-1+r}\ a[1] + r^2\ x^{-1+r}\ a[1] + 2\ x^r\ a[2] +$

$\quad 3\ r\ x^r\ a[2] + r^2\ x^r\ a[2] + 6\ x^{1+r}\ a[3] +$

$\quad 5\ r\ x^{1+r}\ a[3] + r^2\ x^{1+r}\ a[3] +$

$\quad r\ x^{-2+r}\ a[0]\ cp[0] + x^{-1+r}\ a[1]\ cp[0] +$

$\quad r\ x^{-1+r}\ a[1]\ cp[0] + 2\ x^r\ a[2]\ cp[0] +$

$\quad r\ x^r\ a[2]\ cp[0] + 3\ x^{1+r}\ a[3]\ cp[0] +$

$\quad r\ x^{1+r}\ a[3]\ cp[0] + r\ x^{-1+r}\ a[0]\ cp[1] +$

$\quad x^r\ a[1]\ cp[1] + r\ x^r\ a[1]\ cp[1] +$

$\quad 2\ x^{1+r}\ a[2]\ cp[1] + r\ x^{1+r}\ a[2]\ cp[1] +$

$\quad 3\ x^{2+r}\ a[3]\ cp[1] + r\ x^{2+r}\ a[3]\ cp[1] +$

$\quad r\ x^r\ a[0]\ cp[2] + x^{1+r}\ a[1]\ cp[2] +$

$\quad r\ x^{1+r}\ a[1]\ cp[2] + 2\ x^{2+r}\ a[2]\ cp[2] +$

$\quad r\ x^{2+r}\ a[2]\ cp[2] + 3\ x^{3+r}\ a[3]\ cp[2] +$

$\quad r\ x^{3+r}\ a[3]\ cp[2] + r\ x^{1+r}\ a[0]\ cp[3] +$

$\quad x^{2+r}\ a[1]\ cp[3] + r\ x^{2+r}\ a[1]\ cp[3] +$

$\quad 2\ x^{3+r}\ a[2]\ cp[3] + r\ x^{3+r}\ a[2]\ cp[3] +$

$\quad 3\ x^{4+r}\ a[3]\ cp[3] + r\ x^{4+r}\ a[3]\ cp[3] +$

$\quad r\ x^{2+r}\ a[0]\ cp[4] + x^{3+r}\ a[1]\ cp[4] +$

$\quad r\ x^{3+r}\ a[1]\ cp[4] + 2\ x^{4+r}\ a[2]\ cp[4] +$

$\quad r\ x^{4+r}\ a[2]\ cp[4] + 3\ x^{5+r}\ a[3]\ cp[4] +$

$\quad r\ x^{5+r}\ a[3]\ cp[4] + x^{-2+r}\ a[0]\ cq[0] +$

```
 -1 + r
x        a[1] cq[0] + x  a[2] cq[0] +
 1 + r                     -1 + r
x        a[3] cq[0] + x          a[0] cq[1] +
 r                 1 + r
x  a[1] cq[1] + x        a[2] cq[1] +
 2 + r                 r
x        a[3] cq[1] + x  a[0] cq[2] +
 1 + r                 2 + r
x        a[1] cq[2] + x        a[2] cq[2] +
 3 + r                 1 + r
x        a[3] cq[2] + x        a[0] cq[3] +
 2 + r                 3 + r
x        a[1] cq[3] + x        a[2] cq[3] +
 4 + r                 2 + r
x        a[3] cq[3] + x        a[0] cq[4] +
 3 + r                 4 + r
x        a[1] cq[4] + x        a[2] cq[4] +
 5 + r                 3 + r
x        a[3] cq[4] + x        a[0] cq[5] +
 4 + r                 5 + r
x        a[1] cq[5] + x        a[2] cq[5] +
 6 + r
x        a[3] cq[5]
```

Using **lhstwo**, we see that the coefficient of x^{r-2} on the left - hand side of

$$\left(\sum_{n=0}^{\infty} a_n (n+r)(n+r-1)x^{n+r-2}\right) + \left(\sum_{n=0}^{\infty} p_n x^{n-1}\right)\left(\sum_{n=0}^{\infty} a_n (n+r)x^{n+r-1}\right) +$$

$$\left(\sum_{n=0}^{\infty} q_n x^{n-2}\right)\left(x^r \sum_{n=0}^{\infty} a_n x^n\right) = 0 \text{ is}$$

$$-r a_0 + r^2 a_0 + r a_0 p_0 + a_0 q_0 = a_0\left(r^2 + (p_0 - 1)r + q_0\right) = a_0\left(r(r-1) + p_0 r + q_0\right).$$

```
In[95]:=
eqn=Coefficient[lhstwo,x^(-2+r)]
Out[95]=
            2
-(r a[0]) + r  a[0] + r a[0] cp[0] + a[0] cq[0]
```

Equating coefficients we must have that $\left(r^2 + (p_0 - 1)r + q_0\right) = \left(r(r-1) + p_0 r + q_0\right) = 0.$

The equation $r^2 + (p_0 - 1)r + q_0 = 0$ is called the **indicial equation** of the problem.
We can then solve this equation for r:

```
In[96]:=
Solve[eqn==0,r]
Out[96]=
{{r -> (1 - cp[0] + Sqrt[1 - 2 cp[0] + cp[0]^2 -
          4 cq[0]]) / 2},
  {r -> (1 - cp[0] -
          Sqrt[1 - 2 cp[0] + cp[0]^2 - 4 cq[0]]) / 2}}
```
`100%`

Therefore the two roots of the indicial equation $r^2 + (p_0 - 1)r + q_0 = 0$ are

$$r_1 = \frac{1 - p_0 + \sqrt{1 - 2p_0 + p_0^2 - 4q_0}}{2} \text{ and } r_2 = \frac{1 - p_0 - \sqrt{1 - 2p_0 + p_0^2 - 4q_0}}{2};$$

$$r_1 - r_2 = \sqrt{1 - 2p_0 + p_0^2 - 4q_0}.$$

If $r_1 - r_2 = \sqrt{1 - 2p_0 + p_0^2 - 4q_0}$ is NOT an integer, then there are two linearly independent

solutions of the problem of the form $y_1(x) = x^{r_1} \sum_{n=0}^{\infty} a_{1n}x^n$ and $y_2(x) = x^{r_2} \sum_{n=0}^{\infty} a_{2n}x^n$.

If $r_1 - r_2 = \sqrt{1 - 2p_0 + p_0^2 - 4q_0}$ is a non - zero integer, then there are two linearly independent

solutions of the problem of the form $y_1(x) = x^{r_1} \sum_{n=0}^{\infty} a_{1n}x^n$ and

$$y_2(x) = cy_1(x)\,Ln(x) + x^{r_2} \sum_{n=0}^{\infty} a_{2n}x^n.$$

If $r_1 - r_2 = \sqrt{1 - 2p_0 + p_0^2 - 4q_0} = 0$, then there are two linearly independent

solutions of the problem of the form $y_1(x) = x^{r_1} \sum_{n=0}^{\infty} a_{1n}x^n$ and $y_2(x) = y_1(x)\,Ln(x) + x^{r_1} \sum_{n=0}^{\infty} a_{2n}x^n$.

In any case, if $y_1(x)$ is a solution of the problem, a second linearly independent solution is given

by $y_2(x) = y_1(x)\int \frac{e^{-\int p(x)\,dx}}{[y_1(x)]^2}\,dx.$

❑ **EXAMPLE 6.10**

Find the general solution of $xy'' + (1 + x)y' - \frac{1}{16x}y = 0.$

Solution:

In this case, $p(x) = \dfrac{1}{x} + 1$ and $q(x) = \dfrac{-1}{16x^2}$ so the indicial equation is $r^2 + (1-1)r - \dfrac{1}{16} = 0$

which has roots $r_1 = \dfrac{1}{4}$ and $r_2 = \dfrac{-1}{4}$.

Since r_1 and r_2 do not differ by an integer, there are two linearly independent solutions of the form

$$y_1(x) = x^{1/4} \sum_{n=0}^{\infty} a_n x^n = \sum_{n=0}^{\infty} a_n x^{n+1/4} \text{ and } y_2(x) = x^{-1/4} \sum_{n=0}^{\infty} b_n x^n = \sum_{n=0}^{\infty} b_n x^{n-1/4}.$$

Replacing y in the equation $xy'' + (1+x)y' - \dfrac{1}{16x} y = 0$ by $y_1(x)$ yields

$$x \dfrac{d^2}{dx^2}\left(\sum_{n=0}^{\infty} a_n x^{n+1/4} \right) + (1+x)\dfrac{d}{dx}\left(\sum_{n=0}^{\infty} a_n x^{n+1/4} \right) - \dfrac{1}{16x} \sum_{n=0}^{\infty} a_n x^{n+1/4} =$$

$$\sum_{n=0}^{\infty}\left(n+\dfrac{1}{4}\right)\left(n-\dfrac{3}{4}\right)a_n x^{n-3/4} + \sum_{n=0}^{\infty}\left(n+\dfrac{1}{4}\right)a_n x^{n-3/4} + \sum_{n=0}^{\infty}\left(n+\dfrac{1}{4}\right)a_n x^{n+1/4} - \sum_{n=0}^{\infty}\dfrac{a_n}{16} x^{n-3/4} =$$

$$\left[\left(\dfrac{1}{4}\right)\left(\dfrac{-3}{4}\right)a_0 + \left(\dfrac{1}{4}\right)a_0 - \dfrac{a_0}{16}\right]x^{-3/4} +$$

$$\sum_{n=1}^{\infty}\left[\left(n+\dfrac{1}{4}\right)\left(n-\dfrac{3}{4}\right)a_n + \left(n+\dfrac{1}{4}\right)a_n + \left(n-\dfrac{3}{4}\right)a_{n-1} - \dfrac{a_n}{16}\right]x^{n-3/4} =$$

$$\sum_{n=1}^{\infty}\left[\dfrac{a_n\, n(2n+1)}{2} + \dfrac{a_{n-1}(4n-3)}{4}\right]x^{n-3/4} = 0.$$

Equating coefficients yields $a_n = a_{n-1}\dfrac{(3-4n)}{2n(2n+1)}$. Then, for $a_0 \ne 0$, $a_1 = \dfrac{-a_0}{6}$, $a_2 = \dfrac{a_0}{24}$, $a_3 = \dfrac{-a_0}{112}$,

$a_4 = \dfrac{13a_0}{8064}$, $a_5 = \dfrac{-221a_0}{887040}$,.... and

$$y_1(x) = a_0\left(x^{1/4} - \dfrac{1}{6}x^{5/4} + \dfrac{1}{24}x^{9/4} - \dfrac{1}{112}x^{13/4} + \dfrac{13}{8064}x^{17/4} - \dfrac{221}{887040}x^{21/4} + ... \right).$$

Similarly, replacing y in the equation $xy'' + (1+x)y' - \dfrac{1}{16x} y = 0$ by $y_2(x)$ yields

$$\sum_{n=0}^{\infty}\left(n-\dfrac{1}{4}\right)\left(n-\dfrac{5}{4}\right)b_n x^{n-5/4} + \sum_{n=0}^{\infty}\left(n-\dfrac{1}{4}\right)b_n x^{n-5/4} + \sum_{n=0}^{\infty}\left(n-\dfrac{1}{4}\right)b_n x^{n-1/4} - \sum_{n=0}^{\infty}\dfrac{b_n}{16} x^{n-5/4} =$$

$$\sum_{n=1}^{\infty}\left[\left(n-\dfrac{1}{4}\right)\left(n-\dfrac{5}{4}\right)b_n + \left(n-\dfrac{1}{4}\right)b_n + \left(n-\dfrac{5}{4}\right)b_{n-1} - \dfrac{b_n}{16}\right]x^{n-5/4} =$$

$$\sum_{n=1}^{\infty}\left[\frac{b_n\, n\,(2n-1)}{2}+\frac{b_{n-1}\,(4n-5)}{4}\right]x^{n-5/4}=0 \text{ and equating coefficients yields}$$

$$b_n=b_{n-1}\frac{(5-4n)}{2n\,(2n-1)}.$$

Then, for $b_0 \neq 0$, $b_1=\dfrac{b_0}{2}$, $b_2=\dfrac{-b_0}{8}$, $b_3=\dfrac{7b_0}{240}$, $b_4=\dfrac{-11b_0}{1920}$, $b_5=\dfrac{11b_0}{11520}$,... and

$$y_2(x)=b_0\left(x^{-1/4}+\frac{1}{2}x^{3/4}-\frac{1}{8}x^{7/4}+\frac{7}{240}x^{11/4}-\frac{11}{1920}x^{15/4}+\frac{11}{11520}x^{19/4}+...\right).$$

Therefore, the general solution of $xy''+(1+x)y'-\dfrac{1}{16x}y=0$ is $y(x)=c_1y_1(x)+c_2y_2(x)$, where c_1 and c_2 are arbitrary constants. ∎

❑ EXAMPLE 6.11

Find the general solution of $xy''+(2-x)y'+\dfrac{1}{4x}y=0.$

Solution:

In this case, $p(x)=\dfrac{2}{x}-1$ and $q(x)=\dfrac{1}{4x^2}$ so the indicial equation is $r^2+(2-1)r+\dfrac{1}{4}=0$

which has equal roots $r_1=\dfrac{-1}{2}$ and $r_2=\dfrac{-1}{2}$. Then, there is a solution of the form

$$y_1(x)=x^{-1/2}\sum_{n=0}^{\infty}a_n x^n=\sum_{n=0}^{\infty}a_n x^{n-1/2}.\text{ Replacing y in the equation by }y_1\text{ yields}$$

$$x\frac{d^2}{dx^2}\left(\sum_{n=0}^{\infty}a_n x^{n-1/2}\right)+(2-x)\frac{d}{dx}\left(\sum_{n=0}^{\infty}a_n x^{n-1/2}\right)+\frac{1}{4x}\left(\sum_{n=0}^{\infty}a_n x^{n-1/2}\right)=$$

$$x\sum_{n=0}^{\infty}\left(n-\frac{1}{2}\right)\left(n-\frac{3}{2}\right)a_n x^{n-5/2}+(2-x)\sum_{n=0}^{\infty}\left(n-\frac{1}{2}\right)a_n x^{n-3/2}+\frac{1}{4x}\sum_{n=0}^{\infty}a_n x^{n-1/2}=$$

$$\sum_{n=0}^{\infty}\left(n-\frac{1}{2}\right)\left(n-\frac{3}{2}\right)a_n x^{n-3/2}+\sum_{n=0}^{\infty}2\left(n-\frac{1}{2}\right)a_n x^{n-3/2}-\sum_{n=0}^{\infty}\left(n-\frac{1}{2}\right)a_n x^{n-1/2}+\sum_{n=0}^{\infty}\frac{a_n}{4}x^{n-3/2}=$$

$$\sum_{n=1}^{\infty}\left[\left(n-\frac{1}{2}\right)\left(n-\frac{3}{2}\right)a_n+2\left(n-\frac{1}{2}\right)a_n-\left(n-\frac{3}{2}\right)a_{n-1}+\frac{a_n}{4}\right]x^{n-3/2}=$$

$$\sum_{n=1}^{\infty}\left[a_n n^2-\frac{2n-3}{2}a_{n-1}\right]x^{n-3/2}=0.\text{ Then, equating coefficients we obtain }a_n=a_{n-1}\frac{2n-3}{2n^2}.$$

Therefore, for $a_0 \neq 0$, $a_1 = \dfrac{-a_0}{2}$, $a_2 = \dfrac{-a_0}{16}$, $a_3 = \dfrac{-a_0}{96}$, $a_4 = \dfrac{-5a_0}{3072}$, and $a_5 = \dfrac{-7a_0}{30720}$ so

$$y_1(x) = a_0\left(x^{-1/2} - \frac{1}{2}x^{1/2} - \frac{1}{16}x^{3/2} - \frac{1}{96}x^{5/2} - \frac{5}{3072}x^{7/2} - \frac{7}{30720}x^{9/2} - \cdots\right).$$

Since the roots of the indicial equation are equal, there is a second linearly independent solution of the form $y_2(x) = y_1(x)\mathrm{Ln}(x) + \sum_{n=1}^{\infty} b_n x^{n-1/2}$. Substituting y_2 in the equation yields

$$x\frac{d^2}{dx^2}\left(y_1(x)\mathrm{Ln}(x) + \sum_{n=1}^{\infty} b_n x^{n-1/2}\right) + (2-x)\frac{d}{dx}\left(y_1(x)\mathrm{Ln}(x) + \sum_{n=1}^{\infty} b_n x^{n-1/2}\right) +$$

$$\frac{1}{4x}\left(y_1(x)\mathrm{Ln}(x) + \sum_{n=1}^{\infty} b_n x^{n-1/2}\right) =$$

$$x\left[\frac{-y_1(x)}{x^2} + \frac{2y_1'(x)}{x} + y_1''(x)\mathrm{Ln}(x) + \sum_{n=1}^{\infty}\left(n-\frac{1}{2}\right)\left(n-\frac{3}{2}\right)b_n x^{n-5/2}\right] +$$

$$(2-x)\left[\frac{y_1(x)}{x} + y_1'(x)\mathrm{Ln}(x) + \sum_{n=1}^{\infty}\left(n-\frac{1}{2}\right)b_n x^{n-3/2}\right] +$$

$$\frac{y_1(x)\mathrm{Ln}(x)}{4x} + \frac{1}{4x}\sum_{n=1}^{\infty} b_n x^{n-1/2} =$$

$$\left(xy_1''(x) + (2-x)y_1'(x) + \frac{1}{4x}y_1(x)\right)\mathrm{Ln}(x) +$$

$$x\left[\frac{-y_1(x)}{x^2} + \frac{2y_1'(x)}{x} + \sum_{n=1}^{\infty}\left(n-\frac{1}{2}\right)\left(n-\frac{3}{2}\right)b_n x^{n-5/2}\right] +$$

$$(2-x)\left[\frac{y_1(x)}{x} + \sum_{n=1}^{\infty}\left(n-\frac{1}{2}\right)b_n x^{n-3/2}\right] + \frac{1}{4x}\sum_{n=1}^{\infty} b_n x^{n-1/2} =$$

$$x\left[\frac{-y_1(x)}{x^2} + \frac{2y_1'(x)}{x} + \sum_{n=1}^{\infty}\left(n-\frac{1}{2}\right)\left(n-\frac{3}{2}\right)b_n x^{n-5/2}\right] + (2-x)\left[\frac{y_1(x)}{x} + \sum_{n=1}^{\infty}\left(n-\frac{1}{2}\right)b_n x^{n-3/2}\right] +$$

$$\frac{1}{4x}\sum_{n=1}^{\infty} b_n x^{n-1/2} =$$

$$\frac{y_1(x)}{x} + 2y_1'(x) - y_1(x) + \sum_{n=1}^{\infty}\left(n - \frac{1}{2}\right)\left(n - \frac{3}{2}\right)b_n x^{n-3/2} + 2\sum_{n=1}^{\infty}\left(n - \frac{1}{2}\right)b_n x^{n-3/2} -$$

$$\sum_{n=1}^{\infty}\left(n - \frac{1}{2}\right)b_n x^{n-1/2} + \sum_{n=1}^{\infty}\frac{b_n}{4}x^{n-3/2} =$$

$$\frac{y_1(x)}{x} + 2y_1'(x) - y_1(x) + \sum_{n=1}^{\infty}n^2 b_n x^{n-3/2} - \sum_{n=1}^{\infty}\left(n - \frac{1}{2}\right)b_n x^{n-1/2} =$$

$$\frac{y_1(x)}{x} + 2y_1'(x) - y_1(x) + b_1 x^{-1/2} + \sum_{n=2}^{\infty}\left[n^2 b_n - \left(n - \frac{3}{2}\right)b_{n-1}\right]x^{n-3/2} = 0 \text{ so}$$

$$b_1 x^{-1/2} + \sum_{n=2}^{\infty}\left[n^2 b_n - \left(n - \frac{3}{2}\right)b_{n-1}\right]x^{n-3/2} = y_1(x) - \frac{y_1(x)}{x} - 2y_1'(x).$$

Since $y_1(x) = a_0\left(x^{-1/2} - \frac{1}{2}x^{1/2} - \frac{1}{16}x^{3/2} - \frac{1}{96}x^{5/2} - \frac{5}{3072}x^{7/2} - \frac{7}{30720}x^{9/2} + \ldots\right)$, we obtain

$$b_1 x^{-1/2} + \sum_{n=2}^{\infty}\left[n^2 b_n - \left(n - \frac{3}{2}\right)b_{n-1}\right]x^{n-3/2} = a_0\left(2x^{-1/2} - \frac{1}{4}x^{1/2} + \frac{1}{384}x^{5/2} + \frac{1}{1536}x^{7/2} + \ldots\right)$$

and equating coefficients yields $b_1 = 2a_0$, $\frac{-b_1}{2} + 4b_2 = \frac{-a_0}{4}$, $\frac{-3b_2}{2} + 9b_3 = 0$, $\frac{-5b_3}{2} + 16b_4 = \frac{a_0}{384}$,

$\frac{-7b_4}{2} + 25b_5 = \frac{a_0}{1536}$,... and thus $b_1 = 2a_0$, $b_2 = \frac{3a_0}{16}$, $b_3 = \frac{a_0}{32}$, $b_4 = \frac{31a_0}{6144}$, $b_5 = \frac{3a_0}{4096}$,... so

$$y_2(x) = y_1(x)Ln(x) + \sum_{n=1}^{\infty}b_n x^{n-1/2} =$$

$$y_1(x)Ln(x) + a_0\left(2x^{1/2} + \frac{3}{16}x^{3/2} + \frac{1}{32}x^{5/2} + \frac{31}{6144}x^{7/2} + \frac{3}{4096}x^{9/2} + \ldots\right); \text{ the}$$

general solution is $y(x) = c_1 y_1(x) + c_2 y_2(x)$, where c_1 and c_2 denote arbitrary constants. ∎

■ EXAMPLE 6.12

Find the general solution of $y'' - \left(\frac{2}{3x} - 1\right)y' + \left(\frac{4}{9x^2} + x\right)y = 0.$

Solution:

Proceeding as in the previous two examples, we identify $p(x) = \dfrac{-2}{3x} + 1$ and $q(x) = \dfrac{4}{9x^2} + x$ so that the

indicial equation is $r^2 + \left(\dfrac{-2}{3} - 1\right)r + \dfrac{4}{9} = 0$. In this case, factoring yields the equation

$r^2 - \dfrac{5}{3}r + \dfrac{4}{9} = \left(r - \dfrac{4}{3}\right)\left(r - \dfrac{1}{3}\right) = 0$ so that the solutions of the indicial equation are $r_1 = \dfrac{4}{3}$ and $r_2 = \dfrac{1}{3}$

and $r_1 - r_2 = 1$. In this case the roots of the indicial equation differ by an integer so we search for one

solution of the form $y_1(x) = \displaystyle\sum_{n=0}^{\infty} a_n x^{n+4/3}$ and another linearly independent solution of the form

$y_2(x) = cy_1(x)\text{Ln}(x) + \displaystyle\sum_{n=0}^{\infty} b_n x^{n+1/3}$, where c is a constant that may be zero.

Let $y_1(x) = \displaystyle\sum_{n=0}^{\infty} a_n x^{n+4/3}$. Then, $y_1'(x) = \displaystyle\sum_{n=0}^{\infty} a_n\left(n + \dfrac{4}{3}\right)x^{n+1/3}$ and

$y_1''(x) = \displaystyle\sum_{n=0}^{\infty} a_n\left(n + \dfrac{4}{3}\right)\left(n + \dfrac{1}{3}\right)x^{n-2/3}$. If $y_1(x)$ is a solution of $y'' - \left(\dfrac{2}{3x} - 1\right)y' + \left(\dfrac{4}{9x^2} + x\right)y = 0$,

substituting $y_1(x)$ into the equation results in the equation

$$\displaystyle\sum_{n=0}^{\infty} a_n\left(n + \dfrac{4}{3}\right)\left(n + \dfrac{1}{3}\right)x^{n-2/3} - \left(\dfrac{2}{3x} - 1\right)\displaystyle\sum_{n=0}^{\infty} a_n\left(n + \dfrac{4}{3}\right)x^{n+1/3} + \left(\dfrac{4}{9x^2} + x\right)\displaystyle\sum_{n=0}^{\infty} a_n x^{n+4/3} = 0$$

and simplifying yields

$$\displaystyle\sum_{n=0}^{\infty} a_n\left(n + \dfrac{4}{3}\right)\left(n + \dfrac{1}{3}\right)x^{n-2/3} + \displaystyle\sum_{n=0}^{\infty} \dfrac{-2a_n}{3}\left(n + \dfrac{4}{3}\right)x^{n-2/3} + \displaystyle\sum_{n=0}^{\infty} a_n\left(n + \dfrac{4}{3}\right)x^{n+1/3} +$$

$$\displaystyle\sum_{n=0}^{\infty} \dfrac{4a_n}{9}x^{n-2/3} + \displaystyle\sum_{n=0}^{\infty} a_n x^{n+7/3} = 0$$

and then

$$\displaystyle\sum_{n=0}^{\infty} a_n n(n+1)x^{n-2/3} + \displaystyle\sum_{n=0}^{\infty} a_n\left(n + \dfrac{4}{3}\right)x^{n+1/3} + \displaystyle\sum_{n=0}^{\infty} a_n x^{n+7/3} = 0. \text{ Expanding and reindexing}$$

produces $\left(2a_1 + \dfrac{4}{3}a_0\right)x^{1/3} + \left(6a_2 + \dfrac{7}{3}a_1\right)x^{4/3} +$

$$\displaystyle\sum_{n=0}^{\infty}\left[(n+4)(n+3)a_{n+3} + \left(n + \dfrac{10}{3}\right)a_{n+2} + a_n\right]x^{n+7/3} = 0.$$

Since the coefficient of $x^{i-2/3}$ must be 0 for all values of i,

$$\begin{cases} 2a_1 + \dfrac{4}{3}a_0 = 0 \\[2mm] 6a_2 + \dfrac{7}{3}a_1 = 0 \\[2mm] (n+4)(n+3)a_{n+3} + \left(n + \dfrac{10}{3}\right)a_{n+2} + a_n = 0 \text{ for } n = 0,\ldots \end{cases}$$

We then use *Mathematica* to find the values of a_n. With the following command, we solve for a_1 in terms of a_0 and name the resulting output **a1**:

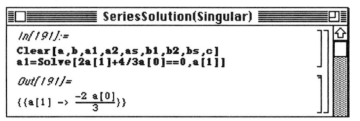

```
In[191]:=
Clear[a,b,a1,a2,as,b1,b2,bs,c]
a1=Solve[2a[1]+4/3a[0]==0,a[1]]
Out[191]=
{{a[1] -> -2 a[0]/3}}
```

Similarly, we solve for a_2 in terms of a_1 and then replace a_1 by the value obtained in **a1**:

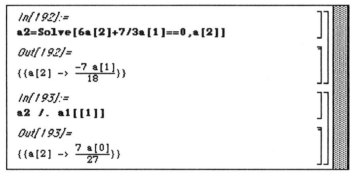

```
In[192]:=
a2=Solve[6a[2]+7/3a[1]==0,a[2]]
Out[192]=
{{a[2] -> -7 a[1]/18}}
In[193]:=
a2 /. a1[[1]]
Out[193]=
{{a[2] -> 7 a[0]/27}}
```

Finally, we solve $(n+4)(n+3)a_{n+3} + \left(n + \dfrac{10}{3}\right)a_{n+2} + a_n = 0$ for a_{n+3}, name the resulting

output **a n 3**:

```
In[194]:=
an3=Solve[(n+4)(n+3)a[n+3]+
    (n+10/3)a[n+2]+a[n]==0,a[n+3]]
Out[194]=
{{a[3 + n] ->
    -(3 a[n] + 10 a[2 + n] + 3 n a[2 + n]))
    ──────────────────────────────────
         3 (12 + 7 n + n²)              }}
```

and then replace each occurrence of n in **an3** by n–3:

```
In[195]:=
an3 /. n->n-3//ExpandAll
Out[195]=
{{a[n] ->
    -3 a[-3 + n]     a[-1 + n]     3 n a[-1 + n]
    ────────────  -  ─────────  -  ─────────────}}
     3 n + 3 n²      3 n + 3 n²     3 n + 3 n²
```

The result means that for $n \geq 3$, $a_n = \dfrac{-3a_{n-3} - a_{n-1} - 3n\,a_{n-1}}{3n^2 + 3n}$.

Let $y_2(x) = cy_1(x)\text{Ln}(x) + \sum\limits_{n=0}^{\infty} b_n x^{n+1/3} = c\text{Ln}(x)\sum\limits_{n=0}^{\infty} a_n x^{n+4/3} + \sum\limits_{n=0}^{\infty} b_n x^{n+1/3}$ be a second

linearly independent solution of $y'' - \left(\dfrac{2}{3x} - 1\right)y' + \left(\dfrac{4}{9x^2} + x\right)y = 0$. Then,

$$y_2'(x) = \sum_{n=0}^{\infty} ca_n x^{n+1/3} + c\text{Ln}(x)y'(x) + \sum_{n=0}^{\infty} b_n\left(n + \frac{1}{3}\right)x^{n-2/3} \text{ and}$$

$$y_2''(x) = \sum_{n=0}^{\infty} ca_n\left(n + \frac{1}{3}\right)x^{n-2/3} + c\text{Ln}(x)y''(x) + \sum_{n=0}^{\infty} b_n\left(n + \frac{1}{3}\right)\left(n - \frac{2}{3}\right)x^{n-5/3}. \text{ Substituting}$$

into the equation yields

$$\sum_{n=0}^{\infty} ca_n\left(n + \frac{1}{3}\right)x^{n-2/3} + c\text{Ln}(x)y''(x) + \sum_{n=0}^{\infty} b_n\left(n + \frac{1}{3}\right)\left(n - \frac{2}{3}\right)x^{n-5/3} -$$

$$\left(\frac{2}{3x} - 1\right)\left[\sum_{n=0}^{\infty} ca_n x^{n+1/3} + c\text{Ln}(x)y'(x) + \sum_{n=0}^{\infty} b_n\left(n + \frac{1}{3}\right)x^{n-2/3}\right] +$$

$$\left(\frac{4}{9x^2} + x\right)\left[cy_1(x)\text{Ln}(x) + \sum_{n=0}^{\infty} b_n x^{n+1/3}\right] = 0$$

and simplifying results in

$$\sum_{n=0}^{\infty} n(n-1)b_n x^{n-5/3} + \sum_{n=0}^{\infty} \left[\frac{c(3n-1)}{3}a_n + \frac{3n+1}{3}b_n \right] x^{n-2/3} +$$

$$\sum_{n=0}^{\infty} ca_n x^{n+1/3} + \sum_{n=0}^{\infty} b_n x^{n+4/3} = 0.$$

Expanding and reindexing results in

$$\left(\frac{-c}{3}a_0 + \frac{1}{3}b_0 \right)x^{-2/3} + \left(2b_2 + \frac{2c}{3}a_1 + \frac{4}{3}b_1 + ca_0 \right)x^{1/3} +$$

$$\sum_{n=0}^{\infty} \left[(n+3)(n+2)b_{n+3} + \frac{3(3n+5)}{3}a_{n+2} + \frac{3n+7}{3}b_{n+2} + ca_{n+1} + b_n \right] x^{n+4/3} = 0.$$

Since the coefficient of $x^{i-2/3}$ is 0 for all values of i,

$$\begin{cases} \dfrac{-c}{3}a_0 + \dfrac{1}{3}b_0 = 0 \\[2mm] 2b_2 + \dfrac{2c}{3}a_1 + \dfrac{4}{3}b_1 + ca_0 = 0 \\[2mm] (n+3)(n+2)b_{n+3} + \dfrac{3(3n+5)}{3}a_{n+2} + \dfrac{3n+7}{3}b_{n+2} + ca_{n+1} + b_n = 0 \text{ for } n = 0,\dots \end{cases}$$

In the same manner as above, we solve for b_0 in terms of a_0 and name the resulting output **b0**:

```
In[197]:=
Clear[a,b]
b0=Solve[-c/3a[0]+1/3b[0]==0,b[0]]

Out[197]=
{{b[0] -> c a[0]}}
```

and then we solve for b_2 in terms of a_1, a_0, and b_1. b_1 is arbitrary.

```
In[198]:=
b2=Solve[2b[2]+2c/3a[1]+4/3b[1]+c a[0]==0,
             b[2]]

Out[198]=
                -(3 c a[0] + 2 c a[1] + 4 b[1])
{{b[2] -> ───────────────────────────────}}
                              6
```

Finally, we solve $(n+3)(n+2)b_{n+3} + \dfrac{3(3n+5)}{3}a_{n+2} + \dfrac{3n+7}{3}b_{n+2} + ca_{n+1} + b_n = 0$ for b_{n+3},

name the resulting output **b n 3**:

```
In[199]:=
bn3=Solve[(n+3)(n+2)b[n+3]+3(3n+5)/3a[n+2]+
        (3n+7)/3b[n+2]+c a[n+1]+b[n]==0,
        b[n+3]]

Out[199]=
{{b[3 + n] ->
    -(3 c a[1 + n] + 15 a[2 + n] +
        9 n a[2 + n] + 3 b[n] + 7 b[2 + n] +
                                          2
        3 n b[2 + n]) / (3 (6 + 5 n + n ))}}
```

and then replace each occurrence of n in **bn3** by n–3.

```
In[200]:=
bn=bn3 /. n->n-3//ExpandAll

Out[200]=
{{b[n] ->
    -3 c a[-2 + n]     12 a[-1 + n]
    ──────────────  +  ──────────── -
         2                    2
    -3 n + 3 n         -3 n + 3 n
    9 n a[-1 + n]      3 b[-3 + n]
    ────────────── -  ──────────── +
         2                    2
    -3 n + 3 n         -3 n + 3 n
    2 b[-1 + n]       3 n b[-1 + n]
    ──────────── -    ─────────────}}
         2                    2
    -3 n + 3 n         -3 n + 3 n
```

The result means that for $n \geq 3$, $b_n = \dfrac{-3ca_{n-2} + 12a_{n-1} - 9na_{n-1} - 3b_{n-3} + 2b_{n-1} - 3nb_{n-1}}{3n^2 - 3n}$.

Since we have computed recurrence relations that yield the value of a_n and b_n for all values of n, we may construct our solutions $y_1(x)$ and $y_2(x)$.

In the following commands, we define $a_0=1$, $a_1=-2/3$, $a_2=7/27$, and then a_n as above.

```
In[205]:=
Clear[a,b]
a[0]=1;
a[1]=-2/3;
a[2]=7/27;
a[n_]:=a[n]=(-3a[n-3]-a[n-1]-3n a[n-1])/
        (3n+3n^2)
```

then compute a table of values of a_n for n=0, 1, 2, ... , 10, name the resulting table **as**, and display **as** in

TableForm.

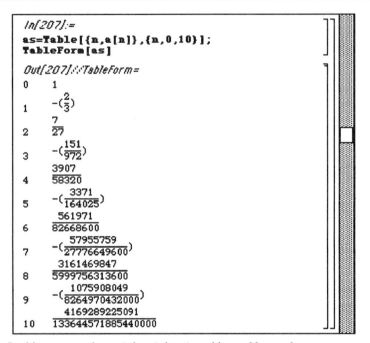

In this case, we let c=1, b_0=1, b_1=1, and b_2 and b_n as above:

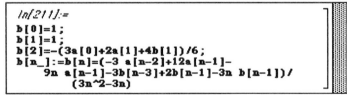

and then compute a table of values of b_n for n=0, 1, 2, ... , 10, name the resulting table **bs**, and display **bs** in **TableForm**.

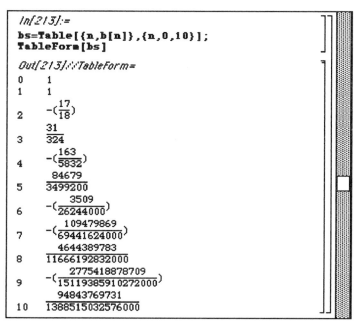

```
In[213]:=
bs=Table[{n,b[n]},{n,0,10}];
TableForm[bs]
```

Out[213]//TableForm=

0	1
1	1
2	$-(\frac{17}{18})$
3	$\frac{31}{324}$
4	$-(\frac{163}{5832})$
5	$\frac{84679}{3499200}$
6	$-(\frac{3509}{26244000})$
7	$-(\frac{109479869}{69441624000})$
8	$\frac{4644389783}{11666192832000}$
9	$-(\frac{2775418878709}{15119385910272000})$
10	$\frac{94843769731}{1388515032576000}$

We then compute the first eleven terms of the series for $y_1(x)$ and name the resulting function **y1approx**:

```
In[214]:=
y1approx[x_]=Sum[a[n]x^(n+4/3),{n,0,10}]
```

Out[214]=

$$x^{4/3} - \frac{2\,x^{7/3}}{3} + \frac{7\,x^{10/3}}{27} - \frac{151\,x^{13/3}}{972} +$$

$$\frac{3907\,x^{16/3}}{58320} - \frac{3371\,x^{19/3}}{164025} + \frac{561971\,x^{22/3}}{82668600} -$$

$$\frac{57955759\,x^{25/3}}{27776649600} + \frac{3161469847\,x^{28/3}}{5999756313600} -$$

$$\frac{1075908049\,x^{31/3}}{8264970432000} + \frac{4169289225091\,x^{34/3}}{133644571885440000}$$

and compute the first eleven terms of the series for $y_2(x)$ and name the result **y2approx**.

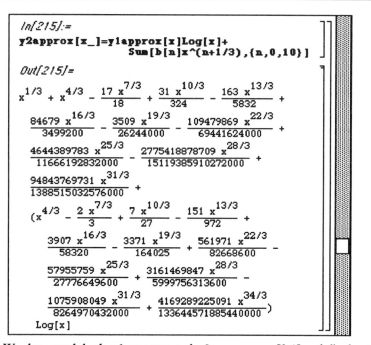

In[215]:=
```
y2approx[x_]=y1approx[x]Log[x]+
              Sum[b[n]x^(n+1/3),{n,0,10}]
```

Out[215]=

$$x^{1/3} + x^{4/3} - \frac{17\,x^{7/3}}{18} + \frac{31\,x^{10/3}}{324} - \frac{163\,x^{13/3}}{5832} +$$

$$\frac{84679\,x^{16/3}}{3499200} - \frac{3509\,x^{19/3}}{26244000} - \frac{109479869\,x^{22/3}}{69441624000} +$$

$$\frac{4644389783\,x^{25/3}}{11666192832000} - \frac{2775418878709\,x^{28/3}}{15119385910272000} +$$

$$\frac{94843769731\,x^{31/3}}{1388515032576000} +$$

$$(x^{4/3} - \frac{2\,x^{7/3}}{3} + \frac{7\,x^{10/3}}{27} - \frac{151\,x^{13/3}}{972} +$$

$$\frac{3907\,x^{16/3}}{58320} - \frac{3371\,x^{19/3}}{164025} + \frac{561971\,x^{22/3}}{82668600} -$$

$$\frac{57955759\,x^{25/3}}{27776649600} + \frac{3161469847\,x^{28/3}}{5999756313600} -$$

$$\frac{1075908049\,x^{31/3}}{8264970432000} + \frac{4169289225091\,x^{34/3}}{133644571885440000})$$

```
Log[x]
```

We then graph both **y1approx** and **y2approx** on [0,1] and display the two graphs as a graphics array. Note that several error messages are generated when *Mathematica* graphs **y2approx**, due to the logarithm term, but the resulting graphs are displayed correctly.

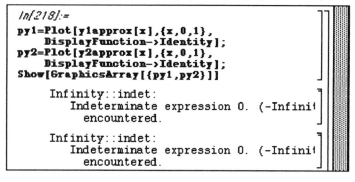

In[218]:=
```
py1=Plot[y1approx[x],{x,0,1},
    DisplayFunction->Identity];
py2=Plot[y2approx[x],{x,0,1},
    DisplayFunction->Identity];
Show[GraphicsArray[{py1,py2}]]

    Infinity::indet:
        Indeterminate expression 0. (-Infini1
            encountered.

    Infinity::indet:
        Indeterminate expression 0. (-Infini1
            encountered.
```

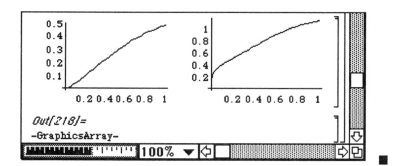

-GraphicsArray-

■ EXAMPLE 6.13

The **Laguerre equation** is the equation $xy''+(\alpha+1-x)y'+\lambda y=0$.

In this case, $p(x)=\dfrac{\alpha+1-x}{x}=\dfrac{\alpha+1}{x}-1$ and $q(x)=\dfrac{\lambda}{x}=\dfrac{0}{x^2}+\dfrac{\lambda}{x}$ so the indicial equation

is $r^2+\big((\alpha+1)-1\big)r=r^2+\alpha r=0$ which has roots $r=0$ and $r=-\alpha$.

If $\alpha=0$ and $\lambda=2$, then the Laguerre equation becomes $xy''+(1-x)y'+2y=0$ and there is a series

solution of the form $y_1(x)=\displaystyle\sum_{n=0}^{\infty}a_n x^n$.

In the following, we define **lhs** to be the left-hand side of the Laguerre equation.

```
▤□▤▤▤▤▤▤▤▤▤▤▤▤▤ LaguerreEquation ▤▤▤▤▤▤▤▤▤
In[18]:=
Clear[y,a,lhs,sery,seryprime,lhstwo]
In[19]:=
lhs=x y''[x]+(1-x)y'[x]+2 y[x]
Out[19]=
2 y[x] + (1 - x) y'[x] + x y''[x]
```

To aid in calculations, we define **s e r y** to be the first five terms of $y_1(x)=\displaystyle\sum_{n=0}^{\infty}a_n x^n$.

```
In[20]:=
sery=Sum[a[n] x^n,{n,0,4}]+O[x]^5
Out[20]=
                        2          3          4
a[0] + a[1] x + a[2] x  + a[3] x  + a[4] x  +
        5
   O[x]
```

Then, $\dfrac{d}{dx} y_1(x) = \dfrac{d}{dx} \left(\displaystyle\sum_{n=0}^{\infty} a_n x^n \right) = \displaystyle\sum_{n=1}^{\infty} n\, a_n x^{n-1}$ and $\dfrac{d^2}{dx^2} y_1(x) = \dfrac{d^2}{dx^2} \left(\displaystyle\sum_{n=0}^{\infty} a_n x^n \right)$

$$= \sum_{n=2}^{\infty} n(n-1) a_n x^{n-2}.$$

The following commands compute the first and second derivatives of **sery** and name the results **seryprime** and **seryprimetwo**, respectively.

```
In[21]:=
seryprime=D[sery,x]

Out[21]=
a[1] + 2 a[2] x + 3 a[3] x  + 4 a[4] x  + O[x]
                          2            3        4

In[22]:=
seryprimetwo=D[sery,{x,2}]

Out[22]=
2 a[2] + 6 a[3] x + 12 a[4] x  + O[x]
                             2        3
```

Replacing y in the equation $xy'' + (1-x) y' + 2y = 0$ by y_1 yields

$$x\left(\sum_{n=2}^{\infty} n(n-1) a_n x^{n-2} \right) + (1-x)\left(\sum_{n=1}^{\infty} n\, a_n x^{n-1} \right) + 2\left(\sum_{n=0}^{\infty} a_n x^n \right) = 0. \ \text{Then,}$$

$$\sum_{n=2}^{\infty} n(n-1) a_n x^{n-1} + \sum_{n=1}^{\infty} n\, a_n x^{n-1} - \sum_{n=1}^{\infty} n\, a_n x^n + \sum_{n=0}^{\infty} 2 a_n x^n = 0,$$

$$\sum_{n=1}^{\infty} (n+1)n\, a_{n+1} x^n + \sum_{n=0}^{\infty} (n+1) a_{n+1} x^n - \sum_{n=1}^{\infty} n\, a_n x^n + \sum_{n=0}^{\infty} 2 a_n x^n = 0,$$

$$(a_1 + 2a_0) + \sum_{n=1}^{\infty} \left[(n+1)n a_{n+1} + (n+1) a_{n+1} - n a_n + 2 a_n \right] x^n = 0 \text{ and finally}$$

$$(a_1 + 2a_0) + \sum_{n=1}^{\infty} \left[(n+1)^2 a_{n+1} + (2-n) a_n \right] x^n = 0.$$

Equating coefficients, we obtain that $a_1 = -2a_0$, $4a_2 + a_1 = 0$ so $a_2 = \dfrac{-a_1}{4} = \dfrac{a_0}{2}$,

$9a_3 = 0$ so $a_3 = 0$ and thus for $n \geq 3$, $a_n = 0$.

Therefore, $y_1(x) = a_0 - 2a_0 x + \dfrac{a_0}{2} x^2$.

The following shows the first few terms of this series obtained by replacing **y''[x]** by

seryprimetwo, **y'[x]** by **seryprime**, and **y[x]** by **sery** in **lhs** and expanding the result. The result is named **lhstwo**.

```
In[23]:=
lhstwo=lhs /. {y''[x]->seryprimetwo,
               y'[x]->seryprime,
               y[x]->sery}//Expand
Out[23]=
(2 a[0] + a[1]) + (a[1] + 4 a[2]) x +
   9 a[3] x  + (-a[3] + 16 a[4]) x  + O[x]
```

The following calculations show the result in the special case when a_0=1. First, we define **a[0]=1** and then we use **LogicalExpand** to equate the coefficients in the equation **lhstwo**=0. The resulting system of linear equations is named **eqs**.

```
In[24]:=
a[0]=1
Out[24]=
1
In[25]:=
eqs=LogicalExpand[lhstwo==0]
Out[25]=
2 + a[1] == 0 && a[1] + 4 a[2] == 0 &&
   9 a[3] == 0 && -a[3] + 16 a[4] == 0
```

The system of linear equations **eqs** is solved with the **Solve** command and the resulting values are named **roots**.

```
In[26]:=
roots=Solve[eqs]
Out[26]=
{{a[1] -> -2, a[2] -> 1/2, a[3] -> 0, a[4] -> 0}}
```

The resulting values are resubstituted into **sery** and the result is named **sersol**. Note, however, that **sersol** contains the term $O[x]^5$ which represents the omitted higher order terms.

```
In[31]:=
sersol=sery /. roots[[1]]
Out[31]=
                2
1 - 2 x + x /2 + O[x]
```

Since we know that the value of all higher order terms is identically zero, we use the command **Normal** to remove $O[x]^5$ and name the resulting function $y1[x]$ which represents the desired solution.

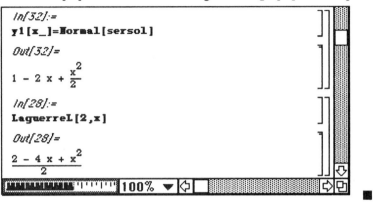

```
In[32]:=
y1[x_]=Normal[sersol]
Out[32]=
          2
         x
1 - 2 x + --
          2
In[28]:=
LaguerreL[2,x]
Out[28]=
          2
2 - 4 x + x
-----------
     2
```

Bessel's equation is the ordinary differential equation $x^2 y'' + x y' + \left(x^2 - \mu^2\right) y = 0.$

■ EXAMPLE 6.14

Find a power series solution of $x^2 y'' + x y' + \left(x^2 - 4\right) y = 0.$

Solution:

The built-in *Mathematica* functions **BesselJ[n,x]** and **BesselY[n,x]** represent the Bessel functions of the first kind, $J_n(x)$, and second kind, $Y_n(x)$, respectively, which form a linearly independent solution of Bessel's equation. *Mathematica* can compute the general solution of Bessel's equation in terms of **BesselJ** and **BesselY** as shown below:

```
═════════════════ Bessel'sEquation ═════════════════
In[39]:=
gensol=DSolve[
    x^2 y''[x]+x y'[x]+(x^2-mu^2)y[x]==0,
                                    y[x],x]
Out[39]=
{{y[x] ->
              2
    BesselY[Sqrt[mu ], x] C[1] +
              2
    BesselJ[Sqrt[mu ], x] C[2]}}
In[40]:=
PowerExpand[gensol[[1,1,2]]]
Out[40]=
BesselY[mu, x] C[1] + BesselJ[mu, x] C[2]
```

Consequently, we could use **DSolve** to find the general solution of the differential equation $x^2 y'' + x y' + \left(x^2 - 4 \right) y = 0.$ Instead, we first compute the solutions of the indicial equation

for Bessel's equation.

The indicial equation of Bessel's equation is $r^2 + (1 - 1) - \mu^2 = 0$, solved below.

```
In[6]:=
ieq=r^2+(1-1)-mu^2==0;
Solve[ieq,r]

Out[6]=
{{r -> mu}, {r -> -mu}}
```

In this case, $\mu = 2$ so we will find a power series solution of $x^2 y'' + x y' + \left(x^2 - 4 \right) y = 0$ of the form

$$y(x) = \sum_{n=0}^{\infty} a_n x^{n+2}.$$ We begin by defining **lhs** to be $x^2 y'' + x y' + \left(x^2 - 4 \right) y$ and **sery** to be the

first nine terms of the series $\sum_{n=0}^{\infty} a_n x^{n+2}$.

```
In[15]:=
lhs=x^2 y''[x]+x y'[x]+(x^2-4)y[x];

In[16]:=
sery=Sum[a[i] x^(i+2),{i,0,8}]+O[x]^11

Out[16]=
a[0] x^2 + a[1] x^3 + a[2] x^4 + a[3] x^5 +
    a[4] x^6 + a[5] x^7 + a[6] x^8 + a[7] x^9 +
    a[8] x^10 + O[x]^11
```

We then compute the derivative of **sery**, corresponding to $\dfrac{d}{dx}\left(\sum_{n=0}^{\infty} a_n x^{n+2} \right) = \sum_{n=0}^{\infty} a_n (n + 2) x^{n+1}$,

name the resulting output **seryprime**:

```
In[17]:=
seryprime=D[sery,x]

Out[17]=

2 a[0] x + 3 a[1] x  + 4 a[2] x  + 5 a[3] x  +
                    2            3            4

   6 a[4] x  + 7 a[5] x  + 8 a[6] x  +
          5            6            7

   9 a[7] x  + 10 a[8] x  + 0[x]
          8             9       10
```

and compute the second derivative of **sery**, corresponding to

$$\frac{d^2}{dx^2}\left(\sum_{n=0}^{\infty} a_n x^{n+2}\right) = \sum_{n=0}^{\infty} a_n (n+2)(n+1)x^n,$$ and the resulting output **seryprimetwo**.

```
In[18]:=
seryprimetwo=D[sery,{x,2}]

Out[18]=

2 a[0] + 6 a[1] x + 12 a[2] x  + 20 a[3] x  +
                            2            3

   30 a[4] x  + 42 a[5] x  + 56 a[6] x  +
           4            5            6

   72 a[7] x  + 90 a[8] x  + 0[x]
           7            8       9
```

If $y(x) = \displaystyle\sum_{n=0}^{\infty} a_n x^{n+2}$ is a solution of $x^2 y'' + x y' + \left(x^2 - 4\right) y = 0$, substitution into the equation

results in $x^2 \displaystyle\sum_{n=0}^{\infty} a_n (n+2)(n+1)x^n + x \sum_{n=0}^{\infty} a_n (n+2)x^{n+1} + \left(x^2 - 4\right)\sum_{n=0}^{\infty} a_n x^{n+2} = 0.$

We compute the first few terms of this series by replacing **y[x]**, **y'[x]**, and **y''[x]** in **lhs** by **sery**, **seryprime**, and **seryprimetwo**, respectively, and name the resulting output **steptwo**.

```
In[19]:=
steptwo=lhs /. {y[x]->sery,y'[x]->seryprime,
              y''[x]->seryprimetwo}

Out[19]=

5 a[1] x  + (a[0] + 12 a[2]) x  +
        3                     4

   (a[1] + 21 a[3]) x  + (a[2] + 32 a[4]) x  +
                     5                     6

   (a[3] + 45 a[5]) x  + (a[4] + 60 a[6]) x  +
                     7                     8

   (a[5] + 77 a[7]) x  + (a[6] + 96 a[8]) x   +
                     9                     10

   0[x]
      11
```

Simplifying yields $\displaystyle\sum_{n=0}^{\infty} a_n n(n+4)x^{n+2} + \sum_{n=0}^{\infty} a_n x^{n+4} = 0$ and reindexing we obtain

$5a_1 x^3 + \displaystyle\sum_{n=0}^{\infty}\left[a_n + a_{n+2}(n+2)(n+6)\right]x^{n+4} = 0.$ Therefore, $a_1 = 0$ and $a_{n+2} = \dfrac{-a_n}{(n+2)(n+6)}$.

We compute the first several values of a_n by using **LogicalExpand** to form the system of equations obtained from equating the coefficients of the equation **steptwo**=0:

```
In[21]:=
eqs=LogicalExpand[steptwo==0]

Out[21]=
5 a[1] == 0 && a[0] + 12 a[2] == 0 &&
  a[1] + 21 a[3] == 0 && a[2] + 32 a[4] == 0 &&
  a[3] + 45 a[5] == 0 && a[4] + 60 a[6] == 0 &&
  a[5] + 77 a[7] == 0 && a[6] + 96 a[8] == 0
```

and then using **Solve** to find the values of a_n for n=2, 4, 5, 8. Note that since a_1=0, a_n=0 for all odd values of n.

```
In[22]:=
Solve[eqs,{a[2],a[4],a[6],a[8]}]

Out[22]=
{{a[8] -> a[0]/2211840, a[6] -> -a[0]/23040, a[4] -> a[0]/384,
  a[2] -> -a[0]/12}}
```

Below, we compute a table of the first ten Bessel functions of the first kind, $J_n(x)$, name the resulting list of functions **besseltable**, and then graph the list of functions **besseltable** on [0,15].

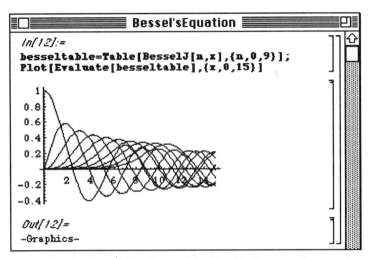

In[12]:=
```
besseltable=Table[BesselJ[n,x],{n,0,9}];
Plot[Evaluate[besseltable],{x,0,15}]
```

Out[12]=
```
-Graphics-
```

To see that the Bessel functions of the first kind are regular, we then compute the first several terms of the power series for each member of **besseltable**, name the resulting table **besser**, and display **besser** in **TableForm**.

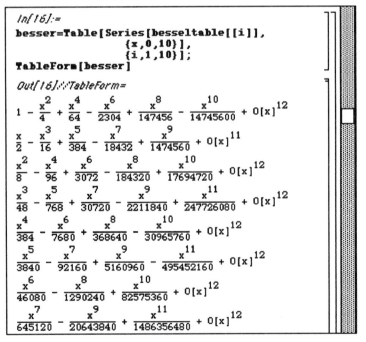

In[16]:=
```
besser=Table[Series[besseltable[[i]],
                  {x,0,10}],
                {i,1,10}];
TableForm[besser]
```

Out[16]//TableForm=

$$1 - \frac{x^2}{4} + \frac{x^4}{64} - \frac{x^6}{2304} + \frac{x^8}{147456} - \frac{x^{10}}{14745600} + O[x]^{12}$$

$$\frac{x}{2} - \frac{x^3}{16} + \frac{x^5}{384} - \frac{x^7}{18432} + \frac{x^9}{1474560} + O[x]^{11}$$

$$\frac{x^2}{8} - \frac{x^4}{96} + \frac{x^6}{3072} - \frac{x^8}{184320} + \frac{x^{10}}{17694720} + O[x]^{12}$$

$$\frac{x^3}{48} - \frac{x^5}{768} + \frac{x^7}{30720} - \frac{x^9}{2211840} + \frac{x^{11}}{247726080} + O[x]^{12}$$

$$\frac{x^4}{384} - \frac{x^6}{7680} + \frac{x^8}{368640} - \frac{x^{10}}{30965760} + O[x]^{12}$$

$$\frac{x^5}{3840} - \frac{x^7}{92160} + \frac{x^9}{5160960} - \frac{x^{11}}{495452160} + O[x]^{12}$$

$$\frac{x^6}{46080} - \frac{x^8}{1290240} + \frac{x^{10}}{82575360} + O[x]^{12}$$

$$\frac{x^7}{645120} - \frac{x^9}{20643840} + \frac{x^{11}}{1486356480} + O[x]^{12}$$

$$\frac{x^8}{10321920} - \frac{x^{10}}{371589120} + O[x]^{12}$$

$$\frac{x^9}{185794560} - \frac{x^{11}}{7431782400} + O[x]^{12}$$

100% ▼

Chapter 7: Applications of Power Series

Mathematica commands used in **Chapter 7** include:

`Apart`	`Evaluate`	`PlotPoints`
`AspectRatio`	`FindRoot`	`PlotRange`
`AxesOrigin`	`GrayLevel`	`PlotStyle`
`Cancel`	`Integrate`	`PlotVectorField`
`ContourPlot`	`Limit`	`ReplaceAll`
`Contours`	`ListPlot`	`Show`
`ContourShading`	`Log`	`Solve`
`D`	`N`	`Table`
`DisplayFunction`	`Part`	
`DSolve`	`Plot`	

§7.1 Applications of Power Series Solutions to Cauchy-Euler Equations

Because of their form, power series solutions can be used to solve Cauchy-Euler equations. We illustrate this method of solution and compare the results with those obtained in earlier sections through the following examples. We begin by showing that, in most cases, the result of the power series method only approximates an exact solution.

❑ **EXAMPLE 7.1**

Solve the Cauchy-Euler equation $x^2y'' - 2xy' - 10y = 0$, $y(1)=5$, $y'(1)=4$ using a power series solution about the point $x = 0$.

Solution:

We first place the ordinary differential equation in the form $y'' - \dfrac{2}{x}y' - \dfrac{10}{x^2}y = 0$. Then $p(x) = -\dfrac{2}{x}$ and

$q(x) = -\dfrac{10}{x^2}$, so we see that $xp(x)$ and $x^2 q(x)$ are both analytic functions. Therefore, $x = 0$ is a regular singular point of the differential equation.

Therefore, we assume a series solution of the form $y = x^r \displaystyle\sum_{n=0}^{\infty} a_n x^n = \displaystyle\sum_{n=0}^{\infty} a_n x^{n+r}$.

Differentiating, we obtain the following for substitution into the Cauchy-Euler equation to find r.

$y' = \displaystyle\sum_{n=0}^{\infty} a_n (n + r) x^{n+r-1}$ and $y'' = \displaystyle\sum_{n=0}^{\infty} a_n (n + r - 1)(n + r) x^{n+r-2}$.

Substituting into the differential equation and simplifying, we have:

$$x^2 \sum_{n=0}^{\infty} a_n (n+r-1)(n+r) x^{n+r-2} - 2x \sum_{n=0}^{\infty} a_n (n+r) x^{n+r-1} - 10 \sum_{n=0}^{\infty} a_n x^{n+r} = 0$$

$$\sum_{n=0}^{\infty} a_n (n+r-1)(n+r) x^{n+r} - \sum_{n=0}^{\infty} 2a_n (n+r) x^{n+r} - \sum_{n=0}^{\infty} 10a_n x^{n+r} = 0$$

$$\sum_{n=0}^{\infty} a_n \left[(n+r-1)(n+r) - 2(n+r) - 10 \right] x^{n+r} = 0.$$

Therefore, for n=0, we have:

$$a_0 \left[(r-1)(r) - 2r - 10 \right] x^r = 0$$

so the indicial equation is $r^2 - 3r - 10 = 0$ with roots $r_1 = -2$ and $r_2 = 5$.
In general with $r = -2$, we have

$$\sum_{n=0}^{\infty} a_n \left[(n-3)(n-2) - 2(n-2) - 10 \right] x^{n-2} = 0$$

which leads to the following equation when all coefficients are equated to zero:

$$a_n \left[(n-3)(n-2) - 2(n-2) - 10 \right] = 0.$$

Hence, $a_0 = 0$ for all n except for those values which satisfy $\left[(n-3)(n-2) - 2(n-2) - 10 \right] = 0.$

Simplifying this expression leads to $n^2 - 7n = 0$, so the values of n such that $a_n \ne 0$ are $n = 0$ and $n = 7$.

The power series, thus, becomes $y = c_0 x^{0-2} + c_7 x^{7-2} = c_0 x^{-2} + c_7 x^5$. Application of the initial conditions yields the following system of equations which must be solved for c_0 and c_7:

$$\begin{cases} c_0 + c_7 = 5 \\ -2c_0 + 5c_7 = 4 \end{cases}.$$

The solution of this system is $c_0 = 3$ and $c_7 = 2$, so the solution is $y = 3x^{-2} + 2x^5$.

Solving this problem using the usual method for Cauchy-Euler equations, we assume that $y = x^m$ and find m. Differentiating, we have $y' = mx^{m-1}$ and $y'' = m(m-1)x^{m-2}$. Substituting into the differential equation, we have $x^2 m(m-1)x^{m-2} - 2xmx^{m-1} - 10x^m = (m^2 - m - 2m - 10)x^m = (m^2 - 3m - 10)x^m = 0$. Therefore, $m = -2$ and $m = 5$, so the general solution of the Cauchy-Euler is the same as that obtained by the power series solution. ∎

■ EXAMPLE 7.2

Approximate the solution of the Cauchy-Euler equation $x^2 y'' - 2xy' - 10y = 0$, with initial conditions $y(1) = 5$, $y'(1) = 4$ using a power series solution about the point $x = 1$.

Solution:

We place the Cauchy - Euler equation in the form $y'' - \dfrac{2}{x} y' - \dfrac{10}{x^2} y = 0.$ Then $p(x) = -\dfrac{2}{x}$ and

$q(x) = -\dfrac{10}{x^2}$, so we see that $p(x)$ and $q(x)$ are both analytic at $x = 1$. Therefore, $x = 1$ is an ordinary point of the differential equation.

Therefore, we assume a series solution of the form $y = \displaystyle\sum_{n=0}^{\infty} a_n (x - 1)^n$.

Differentiating, we obtain the following for substitution into the Cauchy-Euler equation to find r.

$$y' = \sum_{n=0}^{\infty} a_n n (x - 1)^{n-1} = \sum_{n=1}^{\infty} a_n n (x - 1)^{n-1} \text{ and } y'' = \sum_{n=1}^{\infty} a_n (n - 1) n (x - 1)^{n-2}$$

$$= \sum_{n=2}^{\infty} a_n (n - 1) n (x - 1)^{n-2}.$$

Substituting into the differential equation and expanding, we have:

$$x^2 \sum_{n=2}^{\infty} a_n (n - 1) n (x - 1)^{n-2} - 2x \sum_{n=1}^{\infty} a_n n (x - 1)^{n-1} - 10 \sum_{n=0}^{\infty} a_n (x - 1)^n = 0$$

$$x^2 \left[a_2 + 2a_3 (x - 1) + 3a_4 (x - 1)^2 + \cdots \right] - 2x \left[a_1 + 2a_2 (x - 1) + 3a_3 (x - 1)^2 + \cdots \right]$$

$$- 10 \left[a_0 + a_1 (x - 1) + a_2 (x - 1)^2 + \cdots \right] = 0.$$

Hence, this power series must be expanded and all coefficients equated to zero. We use *Mathematica* to perform this task. First, the Taylor polynomial of order 5 with substitution of the appropriate initial conditions is determined in **ser** below. This approximation of the solution is then substituted into the differential equation in **equation**.

```
================ CauchyEuler(Series) =================

In[20]:=
ser=Series[y[x],{x,1,5}]/.
            {y[1]->5,y'[1]->4};

In[21]:=
equation=x^2 D[ser,{x,2}]-
            2x D[ser,x]-10ser==0

Out[21]=
```

$(-58 + y''[1]) + (-48 + y^{(3)}[1]) (-1 + x) +$

$(-6 y''[1] + y^{(3)}[1] + \frac{y^{(4)}[1]}{2}) (-1 + x)^2 +$

$(\frac{-5 y^{(3)}[1]}{3} + \frac{2 y^{(4)}[1]}{3} + \frac{y^{(5)}[1]}{6})$

$(-1 + x)^3 + O[-1 + x]^4 == 0$

Corresponding powers of $(-1+x)$ are equated with **LogicalExpand** in **lineqs**.

```
In[22]:=
lineqs=LogicalExpand[equation]

Out[22]=
```

$-58 + y''[1] == 0 \&\& -48 + y^{(3)}[1] == 0 \&\&$

$-6 y''[1] + y^{(3)}[1] + \frac{y^{(4)}[1]}{2} == 0 \&\&$

$\frac{-5 y^{(3)}[1]}{3} + \frac{2 y^{(4)}[1]}{3} + \frac{y^{(5)}[1]}{6} == 0$

The system of equations in **lineqs** is then solved

```
In[23]:=
values=Solve[lineqs]

Out[23]=
```

$\{\{y''[1] \to 58, y^{(3)}[1] \to 48, y^{(4)}[1] \to 600,$

$y^{(5)}[1] \to -1920\}\}$

and the values of the derivatives are substituted into the expression in **ser**. The resulting expression is called **sol**. **Normal** is then applied to remove the higher order term from **sol** so that the approximate solution **y1** can be defined.

```
In[24]:=
sol=ser/.values[[1]]
Out[24]=
5 + 4 (-1 + x) + 29 (-1 + x)^2 + 8 (-1 + x)^3 +
   25 (-1 + x)^4 - 16 (-1 + x)^5 + O[-1 + x]^6
In[25]:=
y1[x_]=Normal[sol]
Out[25]=
5 + 4 (-1 + x) + 29 (-1 + x)^2 + 8 (-1 + x)^3 +
   25 (-1 + x)^4 - 16 (-1 + x)^5
```

We now plot the approximate solution and the exact solution to the equation $y=3x^{-1}+2x^5$ found by using the standard method for solving a Cauchy-Euler equation which was discussed in **Chapter 4**. Below, **plotapp1** plots the approximate solution using dashing and **exact** plots the actual solution.

```
In[26]:=
plotapp1=Plot[y1[x],{x,.1,1.9},
    PlotStyle->Dashing[{.02}],
    DisplayFunction->Identity];
In[28]:=
yexact[x_]:=3x^(-1)+2 x^5
exact=Plot[yexact[x],{x,.1,1.9},
    DisplayFunction->Identity];
```

Notice that the approximation improves near x = 1 which is the center of the radius of convergence of the series solution.

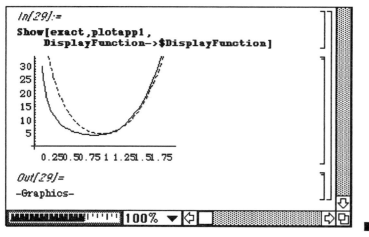

```
In[29]:=
Show[exact,plotapp1,
    DisplayFunction->$DisplayFunction]
```

```
Out[29]=
-Graphics-
```

We now investigate a Cauchy-Euler problem which involves a logarithmic function.

■ EXAMPLE 7.3

Solve the Cauchy-Euler equation $4x^2y'' + y=0$ using a power series solution about the point $x = 0$.

Solution:

We place the differential equation in the form $y'' + \dfrac{1}{4x^2}y = 0$. Then $p(x)=0$ and $q(x)=\dfrac{1}{4x^2}$,

so we see that $xp(x)$ and $x^2q(x)$ are both analytic functions. Therefore, $x = 0$ is a regular singular point of the differential equation.

Therefore, we assume a series solution of the form $y = x^r \displaystyle\sum_{n=0}^{\infty} a_n x^n = \sum_{n=0}^{\infty} a_n x^{n+r}$.

The left-hand side of the equation is entered as **lhs** below and appropriate form of the series solution is entered in **sery**.

```
▤▢▤▤▤▤▤▤ CauchyEuler(Series) ▤▤▤▤▢▤
In[38]:=
lhs=4 x^2 y''[x]+y[x];
In[39]:=
sery=Expand[x^r Sum[a[n] x^n,{n,0,5}]];
```

Differentiating, we obtain the following for substitution into the Cauchy-Euler equation to find r.

$$y' = \sum_{n=0}^{\infty} a_n (n+r) x^{n+r-1} \quad \text{and} \quad y'' = \sum_{n=0}^{\infty} a_n (n+r-1)(n+r) x^{n+r-2}.$$

The first derivative is taken with *Mathematica* in **seryprime** below

```
In[41]:=
seryprime=D[sery,x]//Expand;
Short[seryprime]
Out[41]//Short=
r x^{-1 + r} a[0] + <<10>>
```

and, similarly, the second derivative in **seryprimetwo**.

```
In[43]:=
seryprimetwo=D[sery,{x,2}]//Expand;
Short[seryprimetwo]

Out[43]//Short=
         -2
-(r  x    + <<1>>  a[0]) + <<15>>
```

Substituting into the differential equation and simplifying, we have:

$$4x^2 \sum_{n=0}^{\infty} a_n (n+r-1)(n+r) x^{n+r-2} + \sum_{n=0}^{\infty} a_n x^{n+r} = 0$$

$$\sum_{n=0}^{\infty} 4a_n (n+r-1)(n+r) x^{n+r} + \sum_{n=0}^{\infty} a_n x^{n+r} = 0$$

$$\sum_{n=0}^{\infty} a_n \left[4(n+r-1)(n+r) + 1 \right] x^{n+r} = 0.$$

The first and second derivatives are substituted into the left-hand side of the differential equation and expanded in **lhstwo** below.

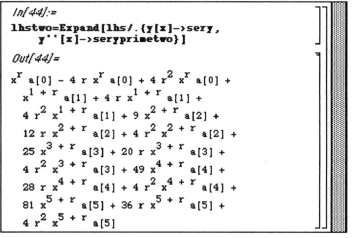

```
In[44]:=
lhstwo=Expand[lhs/.{y[x]->sery,
    y''[x]->seryprimetwo}]

Out[44]=
 r            r          2  r
x  a[0] - 4 r x  a[0] + 4 r  x  a[0] +
    1 + r              1 + r
   x      a[1] + 4 r x      a[1] +
      2  1 + r             2 + r
   4 r  x      a[1] + 9 x      a[2] +
          2 + r         2  2 + r
   12 r x      a[2] + 4 r  x      a[2] +
       3 + r             3 + r
   25 x      a[3] + 20 r x      a[3] +
      2  3 + r              4 + r
   4 r  x      a[3] + 49 x      a[4] +
          4 + r         2  4 + r
   28 r x      a[4] + 4 r  x      a[4] +
       5 + r             5 + r
   81 x      a[5] + 36 r x      a[5] +
      2  5 + r
   4 r  x      a[5]
```

Therefore, for n=0, we have:

$$a_0 \left[4(r-1)r+1 \right] x^r = 0$$

so the indicial equation is $4r^2 - 4r + 1 = 0$ with roots $r_1 = r_2 = \dfrac{1}{2}$.

We find and solve the indicial equation with *Mathematica*.

```
In[45]:=
eqn=Coefficient[lhstwo,x^r]
Out[45]=
a[0] - 4 r a[0] + 4 r² a[0]
```

```
In[46]:=
Solve[eqn==0,r]
Out[46]=
{{r -> 1/2}, {r -> 1/2}}
```

In general with r = –2, we have

$$\sum_{n=0}^{\infty} a_n \left[4\left(n - \frac{1}{2}\right)\left(n + \frac{1}{2}\right) + 1 \right] x^{n+1/2} = \sum_{n=0}^{\infty} a_n \left[4\left(n^2 - \frac{1}{4}\right) + 1 \right] x^{n+1/2} = \sum_{n=0}^{\infty} a_n \, 4n^2 x^{n+1/2} = 0.$$

This implies that $a_n = 0$ for all n except n = 0. We see this below by replacing r with 1/2 in **lhsthree**.

Since the left-hand side of the equation must equate to zero, each coefficient of the term given must be zero as indicated above.

```
In[47]:=
lhsthree=lhstwo/.r->1/2
Out[47]=
4 x^{3/2} a[1] + 16 x^{5/2} a[2] + 36 x^{7/2} a[3] +
   64 x^{9/2} a[4] + 100 x^{11/2} a[5]
```
`100% ▼`

Therefore, according to the power series solution to this problem, we have the one solution $y_1 = x^{1/2}$, and we can obtain a second linearly independent solution

y_2 using $y_2(x) = y_1(x) \int \dfrac{e^{-\int p(x)\,dx}}{|y_1(x)|^2} dx$.

Recall that the equation is assumed to have the form:

$y''(x) + p(x)y'(x) + q(x)y(x) = 0$, so we have that $p(x) = 0$ and $q(x) = \dfrac{1}{4x^2}$. Hence, the second

linearly independent solution is given by $x^{1/2} \int \dfrac{1}{\left[x^{1/2}\right]^2} dx = x^{1/2} \int \dfrac{1}{x} dx = x^{1/2} \text{Log}(x)$.

Therefore, the general solution is $y = C_1 x^{1/2} + C_2 x^{1/2} \text{Log}(x)$.

```
In[47]:=
lhsthree=lhstwo/.r->1/2

Out[47]=
    3/2            5/2            7/2
4 x    a[1] + 16 x    a[2] + 36 x    a[3] +
      9/2             11/2
  64 x    a[4] + 100 x     a[5]
```

Solving this problem using the usual method for Cauchy-Euler equations, we assume that $y = x^m$ and find m. Differentiating, we have $y' = mx^{m-1}$ and $y'' = m(m-1)x^{m-2}$. Substituting into the differential equation, we have $4x^2 m(m-1)x^{m-2} + x^m = (4m^2 - 4m + 1)x^m = 0$. Therefore, $m_1 = m_2 = 1/2$, so the general solution of the Cauchy-Euler equation is the same as that obtained by the power series solution. ■

§7.2 The Hypergeometric Equation

An equation of particular interest in applied mathematics is the second order linear differential equation with non-constant coefficients,

$$x(1-x)y'' + [c - (a+b+1)x]y' - aby = 0$$

where a, b, and c are constants. This differential equation is commonly known as the **hypergeometric equation**.

■ **EXAMPLE 7.4**

Solve the hypergeometric equation using a power series solution about the point $x = 0$.

Solution:

We place the differential equation in the form $y'' + \dfrac{[c-(a+b+1)x]}{x(1-x)} y' - \dfrac{ab}{x(1-x)} y = 0$.

Then $p(x) = \dfrac{[c-(a+b+1)x]}{x(1-x)}$ and $q(x) = -\dfrac{ab}{x(1-x)} y$,

so we see that $xp(x)$ and $x^2 q(x)$ are both analytic functions at $x = 0$. Therefore, $x = 0$ is a regular singular point of the differential equation.

We enter the left-hand side of this equation in **hyplhs** and the first five terms of the approximate solution using a power series solution about a regular singular point in **sery**.

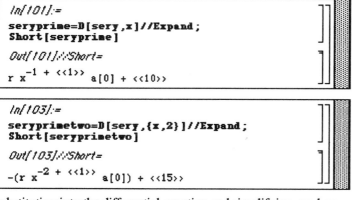

```
================ Hypergeometric ================
In[98]:=
Clear[a,b,c]
hyplhs=x(1-x)y''[x]+(c-(a+b+1)x)y'[x]-
                                a b y[x];
In[99]:=
sery=Expand[x^r Sum[a[n] x^n,{n,0,5}]]
Out[99]=
 r          1 + r          2 + r
x  a[0] + x      a[1] + x      a[2] +
    3 + r          4 + r          5 + r
   x      a[3] + x      a[4] + x      a[5]
```

Differentiating, we obtain the following for substitution into the hypergeometric equation to find r.

$$y' = \sum_{n=0}^{\infty} a_n (n+r) x^{n+r-1} \quad \text{and} \quad y'' = \sum_{n=0}^{\infty} a_n (n+r-1)(n+r) x^{n+r-2}.$$

We obtain the first and second derivatives of the approximate solution in **seryprime** and **seryprimetwo** below.

```
In[101]:=
seryprime=D[sery,x]//Expand;
Short[seryprime]
Out[101]//Short=
   -1
r x   + <<1>> a[0] + <<10>>
```

```
In[103]:=
seryprimetwo=D[sery,{x,2}]//Expand;
Short[seryprimetwo]
Out[103]//Short=
     -2
-(r x   + <<1>> a[0]) + <<15>>
```

Substituting into the differential equation and simplifying, we have:

$$x(1-x) \sum_{n=0}^{\infty} a_n (n+r-1)(n+r) x^{n+r-2} + \left[c - (a+b+1)x\right] \sum_{n=0}^{\infty} a_n (n+r) x^{n+r-1} -$$

$$ab \sum_{n=0}^{\infty} a_n x^{n+r} = 0$$

$$\sum_{n=0}^{\infty} a_n (n+r-1)(n+r) x^{n+r-1} - \sum_{n=0}^{\infty} a_n (n+r-1)(n+r) x^{n+r} + \sum_{n=0}^{\infty} c a_n (n+r) x^{n+r-1} -$$

$$\sum_{n=0}^{\infty} (a+b+1) a_n (n+r) x^{n+r} - \sum_{n=0}^{\infty} ab\, a_n x^{n+r} = 0$$

$$a_0\, r(r-1) x^{r-1} + c a_0 r\, x^{r-1} + \sum_{n=1}^{\infty} a_n \left[(n+r-1)(n+r) + c(n+r) \right] x^{n+r-1} -$$

$$\sum_{n=0}^{\infty} a_n \left[(n+r-1)(n+r) + (a+b+1)(n+r) + ab \right] x^{n+r} = 0.$$

Hence the indicial equation is $r(r-1) + cr = 0$, so the roots are $r=0$ and $r=1-c$.

These calculations are performed with *Mathematica* below.

```
In[105]:=
hyplhstwo=Expand[hyplhs/.{y[x]->sery,
    y'[x]->seryprime,y''[x]->seryprimetwo}];
Short[hyplhstwo]

Out[105]//Short=
-(r x^-1 + <<1>> a[0]) + <<70>>
```

```
In[107]:=
Clear[eqn]
eqn=Coefficient[hyplhstwo,x^(r-1)]

Out[107]=
-(r a[0]) + c r a[0] + r^2 a[0]

In[108]:=
Solve[eqn==0,r]

Out[108]=
{{r -> 1 - c}, {r -> 0}}
```

For $r=0$, we have

$$\sum_{n=1}^{\infty} a_n \left[(n-1)n + cn \right] x^{n-1} - \sum_{n=0}^{\infty} a_n \left[(n-1)n + (a+b+1)n + ab \right] x^n = 0,$$

so re-indexing leads to

$$\sum_{n=1}^{\infty} \left\{ a_n \left[n(n-1+c) \right] - a_{n-1} \left[(n-2)(n-1) + (a+b+1)(n-1) + ab \right] \right\} x^{n-1} = 0.$$

The expression which results from substituting $r = 0$ into the power series expansion is given next.

Hence, the first few coefficients can be determined from this result with **LogicalExpand**. We choose, however, in this case, to simply enter the recursive formula.

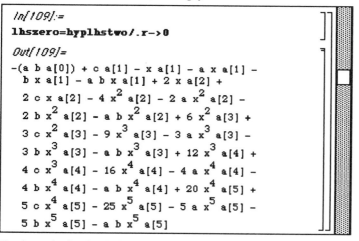

We then obtain the following recursion relationship from the earlier calculations for determining the coefficients:

$$a_n = \frac{a_{n-1}\left[(n-2)(n-1)+(a+b+1)(n-1)+ab\right]}{\left[n(n-1+c)\right]}, \quad n \geq 1.$$

This formula is entered below as **cf** with **cf[0]** representing a_0. Note that we assume that $a_0=1$. We then compute the first three coefficients a_1, a_2, and a_3.

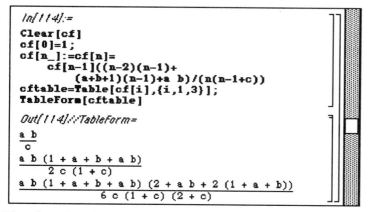

Therefore, the solution given when r=0 and $a_0=1$ is

$$y_1(x) = \left(1 + \frac{ab}{c}x + \frac{ab(a+1)(b+1)}{2(1+c)c}x^2 + \frac{ab(a+1)(b+1)(a+2)(b+2)}{3\cdot 2(2+c)(1+c)c}x^3 + \cdots \right).$$

The approximate solution is entered as **y1**.

```
In[116]:=
Clear[y1,a,b,c]
y1[x_,a_,b_,c_]=Sum[cf[i] x^i,{i,0,3}]
Out[116]=
          a b x     a b (1 + a + b + a b) x^2
1 +  ───── + ───────────────────────────  +
            c              2 c (1 + c)
   (a b (1 + a + b + a b)

        (2 + a b + 2 (1 + a + b)) x^3) /
      (6 c (1 + c) (2 + c))
```

The function **p[i]** plots the solution for values of a=i, b=1, and c=1.

```
In[117]:=
p[i_]:=Plot[y1[x,i,1,1],{x,0,.99},
     PlotStyle->GrayLevel[i/10],
     DisplayFunction->Identity];
```

This function is evaluated for i=1 to i=6, and the plots which result are displayed simultaneously.

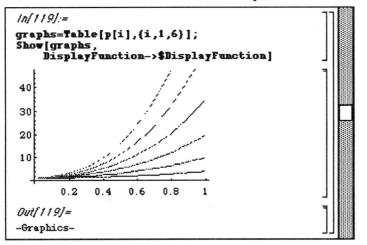

```
In[119]:=
graphs=Table[p[i],{i,1,6}];
Show[graphs,
     DisplayFunction->$DisplayFunction]
```

```
Out[119]=
-Graphics-
```

We define a similar function **q[i]** which plots the solution using a=1, b=i, and c=5. Again, the plots which result are viewed simultaneously.

```
In[121]:=
Clear[q]
q[i_]:=Plot[y1[x,1,i,5],{x,0,.99},
    PlotStyle->GrayLevel[i/10],
    DisplayFunction->Identity];
```

```
In[124]:=
Clear[graphs]
graphs=Table[q[i],{i,1,6}];
Show[graphs,
    DisplayFunction->$DisplayFunction]
```

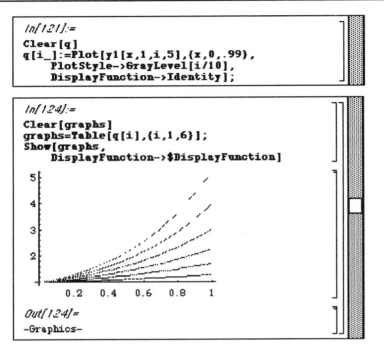

```
Out[124]=
-Graphics-
```

We also define the function r[i] to plot the solution for a=2, b=1, and c=i for values of i from i=1 to i=6.

```
In[126]:=
Clear[r]
r[i_]:=Plot[y1[x,2,1,i],{x,0,.99},
    PlotStyle->GrayLevel[i/10],
    DisplayFunction->Identity];
```

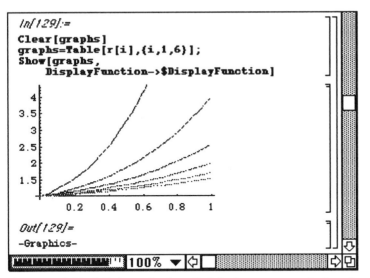

```
In[129]:=
Clear[graphs]
graphs=Table[r[i],{i,1,6}];
Show[graphs,
     DisplayFunction->$DisplayFunction]
```

```
Out[129]=
-Graphics-
```

The sum of this infinite series is called the hypergeometric function which is denoted F(a,b,c;x). The built-in *Mathematica* function **Hypergeometric2F1** represents this sum. Notice that c cannot equal a nonpositive integer.

We now determine the solution obtained with r=1−c. Substitution into the power series and differential equation yields the following expression: (Note that as we did with r=0, we are using a_n to represent the coefficients in this series solution. However, they should not be confused with those found with r=0.)

$$\sum_{n=1}^{\infty} a_n \left[(n-c)(n+1-c) + c(n+1-c) \right] x^{n-c} -$$

$$\sum_{n=0}^{\infty} a_n \left[(n-c)(n+1-c) + (a+b+1)(n+1-c) + ab \right] x^{n+1-c} = 0.$$

When we re-index, we obtain

$$\sum_{n=1}^{\infty} a_n \left[(n-c)(n+1-c) + c(n+1-c) \right] x^{n-c} -$$

$$\sum_{n=1}^{\infty} a_{n-1} \left[(n-c)(n-1-c) + (a+b+1)(n-c) + ab \right] x^{n-c} = 0.$$

Hence, we have the following recursion relationship used for finding the coefficients

$$a_n = \frac{a_{n-1} \left[(n-c)(n-1-c) + (a+b+1)(n-c) + ab \right]}{\left[(n-c)(n+1-c) + c(n+1-c) \right]}, \ n \geq 1.$$

Assuming that $a_0=1$, the second solution is then

$$y_2(x)=x^{1-c}\left(1+\frac{(a-c+1)(b-c+1)}{(2-c)}x+\frac{(a-c+1)(b-c+1)(a-c+2)(b-c+2)}{2(2-c)(3-c)}x^2+\cdots\right)$$

(Notice that c cannot equal an integer greater than or equal to 2 in the above formula.) Substituting $a=a-c+1$, $b=b-c+1$, and $c=2-c$ into this formula for $y_2(x)$, we see that $y_2(x)$ can be represented in terms of the hypergeometric function as $y_2(x)=x^{1-c}F(a-c+1,b-c+1,2-c;x)$. Therefore, the general solution to the hypergeometric equation is $y=C_1y_1(x)+C_2y_2(x)=C_1\ F(a,b,c;x)+C_2x^{1-c}F(a-c+1,b-c+1,2-c;x)$ where C_1 and C_2 are arbitrary. ∎

■ EXAMPLE 7.5

Find one solution of $x(1-x)y''+[b-(b+2)x]y'-by=0$ using the hypergeometric equation.

Solution:

We see that in this case, $a=1$, and $c=b$. Hence, one solution of this equation is $F(1,b,b;x)$. When these parameters are substituted into the formula obtained for $F(a,b,c;x)$ we find that $F(1,b,b;x)=1+x+x^2+x^3+....$, the geometric series which converges for $|x|<1$. Therefore, one solution to this differential equation is

$$F(1,b,b;x)=1+x+x^2+x^3+\cdots=\frac{1}{1-x}.$$

We use *Mathematica* to verify these results. First, the appropriate constants are substituted into the formula cf to obtain the coefficients. The result below shows that all coefficients are 1.

```
≡□≡══════════ Hypergeometric ══════════⊡≡
In[133]:=
a=1;
b=1;
c=1;
xtable=Table[cf[i],{i,1,6}]

Out[133]=
{1, 1, 1, 1, 1, 1}
```

Substituting these coefficients into the power series yields the desired solution.

```
In[135]:=

Clear[x,sol]
sol[x_]=Sum[cf[i] x^i,{i,0,6}]

Out[135]=

1 + x + x  + x  + x  + x  + x
         2    3    4    5    6
```
100%

Some equations can be transformed into the hypergeometric equation so that the formula derived above may be used to find the solution. We illustrate this helpful transformation in the following example.

❏ EXAMPLE 7.6

Use the hypergeometric equation to solve $3t(1+t)\dfrac{d^2y}{dt^2}+t\dfrac{dy}{dt}-y=0.$

Solution:

Notice that the form of this equation is similar to that of the hypergeometric equation. Hence, this equation as well as any equation of the form

$$\left(t^2+At+B\right)\frac{d^2y}{dt^2}+\left(Ct+D\right)\frac{dy}{dt}+Ey=0$$

can be transformed into the hypergeometric equation by making use of the change of variable

$$x=\frac{t-t_0}{t_1-t_0}$$

where t_0 and t_1 are distinct roots of the polynomial (t^2+At+B). In this case, the roots are easily observed to be $t_0=-1$ and $t_1=0$. Therefore, we let

$$x=\frac{t+1}{1}=t+1.$$

Note that in order to change variables, we must compute the following derivatives using the chain rule:

$$\frac{dy}{dt}=\frac{dy}{dx}\frac{dx}{dt}=\frac{dy}{dx}\ \text{ since }\ \frac{dx}{dt}=1.$$

Similarly, $\dfrac{d^2y}{dt^2}=\dfrac{d^2y}{dx^2}$, so the transformed equation becomes

$$3(x-1)\left[1+(x-1)\right]\frac{d^2y}{dx^2}+(x-1)\frac{dy}{dx}-y=0$$

$$3x(x-1)\frac{d^2y}{dx^2}+(x-1)\frac{dy}{dx}-y=0$$

$$3x(1-x)\frac{d^2y}{dx^2}+(1-x)\frac{dy}{dx}+y=0.$$

Now, dividing by 3, we place the equation in the form of the hypergeometric equation:

$$x(1-x)\frac{d^2y}{dx^2}+\left(\frac{1}{3}-\frac{x}{3}\right)\frac{dy}{dx}+\frac{1}{3}y=0$$

where $c=\dfrac{1}{3}$, $(a+b+1)=\dfrac{1}{3}$, and $ab=-\dfrac{1}{3}$. Solving this system of equations for a and b, we find that

$(a,b)=\left(\dfrac{1}{3},\ -1\right)$ or $(a,b)=\left(-1,\ \dfrac{1}{3}\right).$ Notice, however, that the hypergeometric function is symmetric

in a and b, so these two ordered pairs lead to the same solution.
Therefore, the solution can be written as

$$y(x)=C_1F\left(-1,\ \frac{1}{3},\ \frac{1}{3}\ ;\ x\right)+C_2x^{1-\frac{1}{3}}F\left(-\frac{1}{3},\ 1,\ \frac{5}{3}\ ;\ x\right),\ \text{so}$$

$$y(t)=C_1F\left(-1,\ \frac{1}{3},\ \frac{1}{3}\ ;\ t+1\right)+C_2(t+1)^{\frac{2}{3}}F\left(-\frac{1}{3},\ 1,\ \frac{5}{3}\ ;\ t+1\right).$$

We use the built-in function **Hypergeometric2F1** to plot solutions to the equation for various initial conditions below by, first, defining the function **draw[i]** which plots the solution using C_1=i and C_2=1.

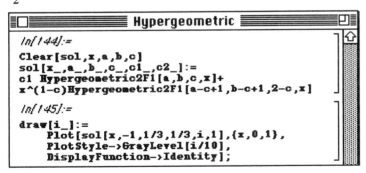

A table of plots is then determined and displayed simultaneously.

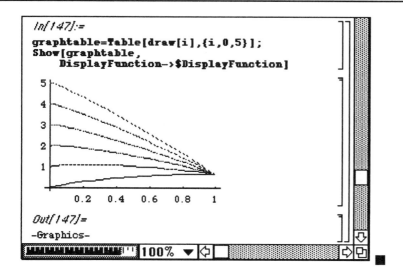

```
In[147]:=
graphtable=Table[draw[i],{i,0,5}];
Show[graphtable,
    DisplayFunction->$DisplayFunction]

Out[147]=
-Graphics-
```

§7.3 The Vibrating Cable

Suppose that we have a flexible cable of length b which is suspended vertically from a fixed point, x=0. Suppose, also, that this cable is allowed to make small displacements in the vertical plane. Our goal is to determine the position of the cable at time t. We accomplish this objective by modeling the situation with a partial differential equation, one that involves finding a solution which depends on more than one variable. (This problem is discussed more as the Wave Equation in **Chapter 12**.) The differential equation which models this particular situation is

$$\frac{\partial^2 u}{\partial t^2} = g\frac{\partial}{\partial x}\left[(b-x)\frac{\partial u}{\partial x}\right]$$

where g represents the gravitational constant. A general method of solution to this problem will be discussed in **Chapter 12**. For now, we simply assume a solution of the form u(x,t)=y(x) cos(ωt+c) where y(x) represents the amplitude of the solution curve. Differentiation of this function yields

$$\frac{\partial u}{\partial x} = y'(x)\cos(\omega t+c) \quad \text{and} \quad \frac{\partial^2 u}{\partial t^2} = -\omega^2 y(x)\cos(\omega t+c).$$

Substituting into the differential equation, we have

$$-\omega^2 y(x)\cos(\omega t+c) = g\frac{\partial}{\partial x}\left[(b-x)y'(x)\cos(\omega t+c)\right]$$

$$= g\left[(b-x)y''(x)\cos(\omega t+c) - y'(x)\cos(\omega t+c)\right].$$

Simplifying and canceling the cosine terms yields the following ordinary differential equation

$$\left(b-x\right)y''(x)-y'+\frac{\omega^2}{g}y(x)=0.$$

Therefore, if we let $\lambda^2=\dfrac{\omega^2}{g}$ and change variables with $z=\left(b-x\right)$, we transform to a differential equation

depending on z In order to do this, we compute $\dfrac{dy}{dx}=\dfrac{dy}{dz}\dfrac{dz}{dx}=-\dfrac{dy}{dz}$ and $\dfrac{d^2y}{dx^2}=\dfrac{d}{dx}\left(\dfrac{dy}{dx}\right)=\dfrac{d^2y}{dz^2}.$

Substitution then yields the equation $z\dfrac{d^2y}{dz^2}+\dfrac{dy}{dz}+\lambda^2y=0$ which can be transformed into Bessel's

equation with the change of variable $s=2\lambda z^{1/2}$.

Therefore, we must use the Chain Rule to compute

$$\frac{dy}{dz}=\frac{dy}{ds}\frac{ds}{dz}=\lambda z^{-1/2}\frac{dy}{ds}\text{ and }\frac{d^2y}{dz^2}=-\frac{1}{2}\lambda z^{-3/2}\frac{dy}{ds}+\lambda^2z^{-1}\frac{d^2y}{ds^2}.$$

Hence, we have

$$z\left[-\frac{1}{2}\lambda z^{-3/2}\frac{dy}{ds}+\lambda^2z^{-1}\frac{d^2y}{ds^2}\right]+\lambda z^{-1/2}\frac{dy}{ds}+\lambda^2y=0\text{ or }$$

$$\frac{d^2y}{ds^2}+\frac{1}{s}\frac{dy}{ds}+y=0,\text{ the Bessel equation of order zero.}$$

A solution to this equation is $y(s)=J_0(s)$, so $y(z)=J_0(2\lambda z^{1/2})$. Hence in terms of x, the solution is

$y(x)=J_0(2\lambda(b-x)^{1/2})$. The fixed-end condition $y(0)=0$ must also be satisfied, so $y(0)=J_0(2\lambda b^{1/2})=0$.

Therefore, $2\lambda b^{1/2}$ must equal a zero of the $J_0(x)$. If we let α_n represent the nth zero of $J_0(x)$, then the

following relationship must hold

$$2\lambda b^{1/2}=\alpha_n\text{ , so }\lambda=\frac{\alpha_n}{2b^{1/2}}\text{ and }\lambda^2=\frac{\alpha_n^2}{4b}.\text{ Hence, }\omega^2=\frac{\alpha_n^2g}{4b}.$$

This yields the solution

$$y(x)=J_0\left(2\omega\sqrt{\frac{b-x}{g}}\right)=J_0\left(\alpha_n\sqrt{\frac{g}{b}}\sqrt{\frac{b-x}{g}}\right)=J_0\left(\alpha_n\sqrt{\frac{b-x}{b}}\right),\ n=1,2,3,\dots\ .$$

For each value of n, the solution is called a normal mode. The first sixteen normal modes are plotted below with the aid of *Mathematica*.

■ EXAMPLE 7.7

Determine the first sixteen normal modes of the cable problem with b =1 and g = 32.

Solution:

We first plot the Bessel function of the first kind and approximate the first sixteen zeros.

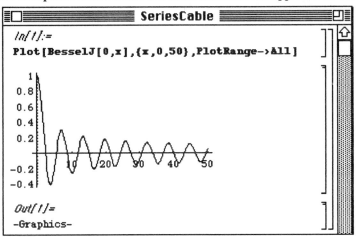

```
In[1]:=
Plot[BesselJ[0,x],{x,0,50},PlotRange->All]

Out[1]=
-Graphics-
```

We then use **FindRoot** using the initial guesses from **guess**=2.5 to **guess**=47.5 using increments of 3.

```
In[3]:=
zeros=Table[FindRoot[BesselJ[0,x]==0,
        {x,guess}],
        {guess,2.5,47.5,3}];
Short[zeros,3]

Out[3]//Short=
{{x -> 2.40483}, {x -> 5.52008}, {x -> 8.65373},
  {x -> 11.7915}, {<<1>>}, <<9>>, {x -> 40.0584},
  {x -> 46.3412}}
```

We then define the function **height** which represents the distance of the cable of the vertical axis

```
In[5]:=
Clear[height]
height[x_,i_]:=BesselJ[0,zeros[[i,1,2]]]*
                Sqrt[1-x]]//Chop
```

and a function **cabeldraw** which plots the cable. This is done by generating a list of ordered pairs called **pairs** using **height** and the quantity 1 –x. (Note that 1 –x is used in order that the top point of the cable appear at 1 and the bottom point at 0.) The points in **pairs** are then plotted and joined with **ListPlot**.

```
In[12]:=
Clear[cabledraw]
cabledraw[s_]:=Module[{pairs,x},
    pairs=Table[{height[x,s],1-x},
                {x,0,1,1/40}];
    ListPlot[pairs,PlotJoined->True,
                AspectRatio->1,
                Ticks->{None,Automatic},
                DisplayFunction->Identity]
    ];
```

The graph of the cable is generated using the zeros of the Bessel function found in **zeros**.

```
In[15]:=
ctable=Table[cabledraw[j],
        {j,1,Length[zeros]}];
graphs=Partition[ctable,4];
Show[GraphicsArray[graphs],
    DisplayFunction->$DisplayFunction]
```

These plots are displayed below. The behavior of these curves as the index of the zero α_n of the Bessel function increases is clear. The frequency also increases.

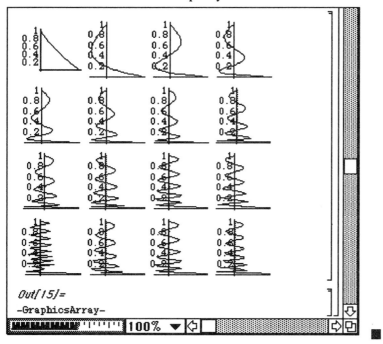

```
Out[15]=
-GraphicsArray-
```

Chapter 8: Introduction to the Laplace Transform

Mathematica commands used in **Chapter 8** include:

Apart	Integrate	Series
Cancel	InverseLaplaceTrans	Show
Clear	form	Simplify
Collect	LaplaceTransform	Solve
DisplayFunction	Normal	Sum
Drop	Numerator	Table
Evaluate	Part	TableForm
Expand	Partition	Ticks
Flatten	Plot	Together
GraphicsArray	PlotRange	
If	ReplaceAll	

§8.1 The Laplace Transform: Preliminary Definitions and Notation

Definition:

Let f(t) be a function defined on the interval $[0,+\infty)$. The **Laplace transform** of f(t) is the function (of s)

$$L\{f\}(s) = \int_0^\infty e^{-st} f(t)\, dt.$$

❏ EXAMPLE 8.1

Compute $L\{f\}(s)$ if f(t) = t.

Solution:

To compute $L\{f\}(s) = \int_0^\infty e^{-st}\, t\, dt$ use Integration by Parts with $u = t$ and $dv = e^{-st}$.

Then, $du = dt$ and $v = \dfrac{-1}{s} e^{-st}$ so $L\{f\}(s) = \int_0^\infty e^{-st}\, t\, dt = \lim_{M\to\infty}\left(\dfrac{-te^{-st}}{s}\bigg|_{t=0}^{t=M}\right) + \dfrac{1}{s}\int_0^\infty e^{-st}\, dt$

$$= 0 - \dfrac{1}{s^2}\lim_{M\to\infty}\left(e^{-st}\bigg|_{t=0}^{t=M}\right) = \dfrac{1}{s^2}. \quad \blacksquare$$

■ EXAMPLE 8.2

Compute $L\{f\}(s)$ if (a) f(t) = t^3 ; (b) f(t) = sin (at); and (c) f(t) = cos (at).

Solution:

For (b) and (c), we will use the package **LaplaceTransform** contained in the **Calculus** folder (or directory) so we begin by loading the package. For (a), we use the definition of the Laplace transform and compute $\int_0^A t^3 e^{-st}\,dt$ and name the resulting output **stepone**.

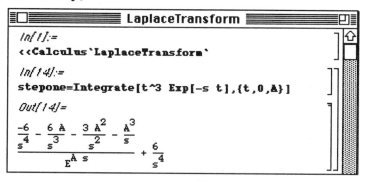

Then, the Laplace transform of $f(t) = t^3$ is

$$\text{Lim}_{A\to\infty} \int_0^A t^3 e^{-st}\,dt = \text{Lim}_{A\to\infty}\ \textbf{stepone} = \text{Lim}_{A\to\infty}\left(\frac{-6s^{-4} - 6As^{-3} - 3A^2 s^{-2} - A^3 s^{-1}}{e^{As}} + \frac{6}{s^4}\right) = \frac{6}{s^4}.$$

For (b) and (c) we use the command **LaplaceTransform[f[t],t,s]** which attempts to compute $L\{f\}(s)$.

```
In[15]:=
LaplaceTransform[Sin[a t],t,s]

Out[15]=
    a
  ------
   2    2
  a  + s

In[16]:=
LaplaceTransform[Cos[a t],t,s]

Out[16]=
    s
  ------
   2    2
  a  + s
```
`100% ▼`

In this case, **LaplaceTransform** is successful and we see that the Laplace transform of $f(t) = \sin(a\,t)$ is $\dfrac{a}{a^2 + s^2}$ and the Laplace transform of $f(t) = \cos(a\,t)$ is $\dfrac{s}{a^2 + s^2}$. ∎

Laplace transforms of frequently used functions are shown in the following table.

Table 8.1
Frequently Used Elementary Laplace Transforms

$f(t)$	$L\{f\}(s)$
1. 1	$\dfrac{1}{s}, s > 0$
2. e^{at}	$\dfrac{1}{s-a}, s > a$
3. t^n, n a positive integer	$\dfrac{n!}{s^{n+1}}$
4. $\sin(at)$	$\dfrac{a}{s^2 + a^2}$
5. $\cos(at)$	$\dfrac{s}{s^2 + a^2}$
6. $\sinh(at)$	$\dfrac{a}{s^2 - a^2}$
7. $\cosh(at)$	$\dfrac{s}{s^2 - a^2}$
8. $e^{at}\sin(bt)$	$\dfrac{b}{(s-a)^2 + b^2}$
9. $e^{at}\cos(bt)$	$\dfrac{s-a}{(s-a)^2 + b^2}$
10. $t\sin(at)$	$\dfrac{2as}{\left(s^2 + a^2\right)^2}$
11. $t\cos(at)$	$\dfrac{s^2 - a^2}{\left(s^2 + a^2\right)^2}$
12. $e^{at}t^n$, n a positive integer	$\dfrac{n!}{(s-a)^{n+1}}$
13. $\dfrac{1}{\sqrt{t}}$	$\sqrt{\dfrac{\pi}{s}}$
14. $\sqrt{t}$	$\dfrac{\sqrt{\pi}}{2s^{3/2}}$

15. $\dfrac{e^{bt} - e^{at}}{t}$	$Ln\left(\dfrac{s-a}{s+b}\right)$
16. $-Ln(t) - \gamma$	$\dfrac{Ln(s)}{s}$
17. $\dfrac{2(1 - \cos(at))}{t}$	$Ln\left(\dfrac{s^2 + a^2}{s^2}\right)$

The Laplace transform has many operational properties. For example, if both $L\{f\}(s)$ and $L\{g\}(s)$ exist and a and b are numbers, then

$$L\{af + bg\}(s) = \int_0^\infty e^{-st}\left(af(t) + bg(t)\right)dt = a\int_0^\infty e^{-st}f(t)dt + b\int_0^\infty e^{-st}g(t)dt$$
$$= aL\{f\}(s) + bL\{g\}(s).$$

The following theorem and corollary show that the Laplace transform of the derivative of a given function can be obtained from the Laplace transform of the function.

Theorem:

Let $f(t)$ be a continuous function on $[0,\infty)$ and suppose that $f'(t)$ is piecewise-continuous on $[0,\infty)$. Moreover, suppose that there are constants $C, b,$ and N so that $|f(t)| \le Ce^{bt}$ for $t \ge N$. Then,

using Integration by Parts with $u = e^{-st}$ and $dv = f'(t)$ yields

$$L\{f'\}(s) = \int_0^\infty e^{-st}f'(t) = \lim_{M \to \infty}\left(e^{-st}f(t)\Big|_{t=0}^{t=M}\right) + s\int_0^\infty e^{-st}f(t)dt = -f(0) + sL\{f\}(s)$$
$$= sL\{f\}(s) - f(0). \quad \square$$

Note that a function f which has the property that $|f(t)| \le Ce^{bt}$ for $t \ge N$ for some constants $C, b,$ and N is said to be of **exponential order**.

Corollary:

In fact, if $f^{(i)}(t)$ is a continuous function on $[0,\infty)$ for $i=0, 1, \dots, n-1$ and $f^{(n)}(t)$ is piecewise-continuous on $[0,\infty)$, then

$$L\{f^{(n)}\}(s) = -f^{(n-1)}(0) + sL\{f^{(n-1)}\}(s) = s^n L\{f\}(s) - s^{n-1}f(0) - \dots - sf^{(n-2)}(0) - f^{(n-1)}(0). \quad \square$$

❑ EXAMPLE 8.3

Let $f(t) = (3t-1)^3$. Compute the Laplace transform of $f(t), f'(t)$, and $f''(t)$.

Solution:

Since $f(t) = (3t-1)^3 = 27t^3 - 27t^2 + 9t - 1$ and the Laplace transform of t^n is $\dfrac{n!}{s^{n+1}}$, the Laplace

transform of $f(t)$ is $27\dfrac{3!}{s^4} - 27\dfrac{2!}{s^3} + 9\dfrac{1}{s^2} - \dfrac{1}{s} = \dfrac{162 - 54s + 9s^2 - s^3}{s^4}$. By the above, the Laplace

transform of $f'(t) = 9(3t-1)^2$ is

$$s\dfrac{162 - 54s + 9s^2 - s^3}{s^4} - f(0) = \dfrac{162 - 54s + 9s^2 - s^3}{s^3} + 1 = \dfrac{9\left(18 - 6s + s^2\right)}{s^3}$$ and the Laplace transform

of $f''(t) = 54(3t-1)$ is $s^2\dfrac{162 - 54s + 9s^2 - s^3}{s^4} - sf(0) - f'(0) = \dfrac{54(3-s)}{s^2}$. ∎

Similarly, the Laplace transform of the integral of a given function can also be obtained from the Laplace transform of the function as stated in the following theorem.

Theorem:

Let $f(t)$ be a piecewise-continuous function on $[0,\infty)$. Moreover, suppose there are constants C, b>0 and N so that $|f(t)| \le Ce^{bt}$ for $t \ge 0$. Then, $L\left\{\int_0^t f(\alpha)\, d\alpha\right\}(s) = \dfrac{L\{f\}(s)}{s}$.

❑ EXAMPLE 8.4

Compute $f(t)$ if $L\{f\}(s) = \dfrac{1}{s(s+2)}$.

Solution:

Since the Laplace transform of e^{-2t} is $\dfrac{1}{s+2}$, the Laplace transform of $\int_0^t e^{-2\alpha}\, d\alpha = \dfrac{1 - e^{-2t}}{2}$ is

$\dfrac{1}{s(s+2)}$. The same result is obtained by computing the partial fraction decomposition

$\dfrac{1}{s(s+2)} = \dfrac{1}{2s} - \dfrac{1}{2(2+s)}$ and noting that the Laplace transform of $\dfrac{1}{2}$ is $\dfrac{1}{2s}$ and the Laplace transform

of $\dfrac{-e^{-2t}}{2}$ is $\dfrac{-1}{2(2+s)}$. ∎

In addition to the above operational properties, additional frequently used properties of the Laplace transform are summarized in the table below.

Table 8.2
Frequently Used Elementary Operational Properties

$f(t)$	$L\{f\}(s)=F(s)$
1. $e^{at}f(t)$	$F(s-a)$
2. $t^n f(t)$, n any integer	$(-1)^n \dfrac{d^n}{ds^n} F(s)$
3. $f^{(n)}(t)$	$s^n F(s) - \sum_{i=0}^{n-1} s^{n-1-i} f^{(i)}(0)$
4. $\int_0^t f(\alpha)\, d\alpha$	$\dfrac{F(s)}{s}$
5. $\int_0^t f(\alpha) g(t-\alpha)\, d\alpha$	$F(s)G(s)$
6. $\dfrac{f(t)}{t}$	$\int_s^\infty F(\alpha)\, d\alpha$
7. $f(t) = f(t+P)$	$\dfrac{1}{1-e^{-Ps}} \int_0^P e^{-st} f(t)\, dt$

Let $u_a(t) = u(t-a) = \begin{cases} 0 \text{ if } t < a \\ 1 \text{ if } t \ge a \end{cases}$. Then, in addition to the above we have:

8. $u_a(t)$	$\dfrac{e^{-as}}{s}, s > 0$
9. $f(t-a)u_a(t)$	$e^{-as}F(s)$

■ **EXAMPLE 8.5**
Find the Laplace transform of $f(t) = 1 - t$ if $0 \le t \le 1$ and $f(t) = f(t-1)$ if $t > 1$.

Solution:
We begin by defining and graphing f. Note that f is the periodic extension of the function $1-x$ on the interval $[0,1]$.

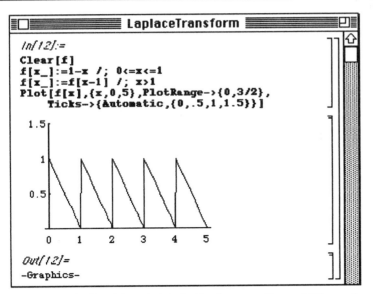

Since f is periodic with period 1, the Laplace transform of f(t) may be computed using (7) in the table

above and is given by $\dfrac{1}{1-e^{-s}}\int_0^1 e^{-st}f(t)\,dt$. With the following command we compute

$\int_0^1 (1-t)e^{-st}\,dt$ and name the result **stepone**:

```
In[16]:=
stepone=Integrate[(1-t)Exp[-s t],{t,0,1}]

Out[16]=
  1 - s    1
  -----  + ---
   s^2      s      1 - s
  -------  -  -----
    E^s        s^2
```

we then simplify **stepone**, name the result **steptwo**, compute $\dfrac{1}{1-e^{-s}}$ **steptwo**, and

name the result **1f**. Therefore the Laplace transform of f(t) is $\dfrac{1-e^{s}+se^{s}}{s^{2}\left(e^{s}-1\right)}$.

```
In[18]:=
steptwo=Together[stepone]
Out[18]=

1 - E^s + E^s s
───────────────
    E^s s^2

In[19]:=
1f=1/(1-Exp[-s]) steptwo//Simplify
Out[19]=

1 - E^s + E^s s
───────────────
 (-1 + E^s) s^2
```

Definition:

f(t) is the **inverse Laplace transform** of F(s) means that L{f}(s)=F(s) and we write $L^{-1}\{F(s)\}(t)=f(t)$.

❑ EXAMPLE 8.6

Find the inverse Laplace transform of $\dfrac{4}{s^2 + 16}$.

Solution:

The inverse Laplace transform of $\dfrac{4}{s^2 + 16}$ is $\sin(4t)$ since $L\{\sin(4t)\} = \dfrac{4}{s^2 + 16}$. ∎

Note that since the Laplace transform is linear, so is the inverse Laplace transform:
if a and b are numbers, then $L^{-1}\{af(t) + bg(t)\} = aL^{-1}\{f(t)\} + bL^{-1}\{g(t)\}$.

❑ EXAMPLE 8.7

Find the inverse Laplace transform of $\dfrac{s^5 + 6s^4 + 73s^2 - 36s + 52}{s^2(s-2)^2\left(s^2 + 4s + 13\right)}$.

Solution:

Computing the partial fraction decomposition of $\dfrac{s^5 + 6s^4 + 73s^2 - 36s + 52}{s^2(s-2)^2\left(s^2 + 4s + 13\right)}$ yields

$$\frac{s^5 + 6s^4 + 73s^2 - 36s + 52}{s^2(s-2)^2\left(s^2 + 4s + 13\right)} = \frac{1}{s^2} + \frac{4}{(s-2)^2} + \frac{s+5}{s^2 + 4s + 13}$$ and completing the square yields

$$\frac{s^5 + 6s^4 + 73s^2 - 36s + 52}{s^2(s-2)^2\left(s^2 + 4s + 13\right)} = \frac{1}{s^2} + \frac{4}{(s-2)^2} + \frac{s+5}{(s+2)^2 + 9} = \frac{1}{s^2} + \frac{4}{(s-2)^2} + \frac{s+2}{(s+2)^2 + 9} + \frac{3}{(s+2)^2 + 9}.$$

Since $L^{-1}\left\{\dfrac{1}{s^2}\right\}(t) = t, L^{-1}\left\{\dfrac{4}{(s-2)^2}\right\}(t) = 4te^{2t}, L^{-1}\left\{\dfrac{s+2}{(s+2)^2 + 9}\right\}(t) = e^{-2t}\cos(3t)$, and

$$L^{-1}\left\{\frac{3}{(s+2)^2 + 9}\right\}(t) = e^{-2t}\sin(3t),$$

$$L^{-1}\left\{\frac{s^5 + 6s^4 + 73s^2 - 36s + 52}{s^2(s-2)^2\left(s^2 + 4s + 13\right)}\right\}(t) = t + 4te^{2t} + e^{-2t}\cos(3t) + e^{-2t}\sin(3t). \blacksquare$$

■ EXAMPLE 8.8

Find the inverse Laplace transform of $\dfrac{1}{s^n\left(s^2 + 16\right)}$ for n = 0, 1, 2, 3, 4, 5, and 6.

Solution:

Let $il_n(t)$ denote the inverse Laplace transform of $\dfrac{1}{s^n\left(s^2 + 16\right)}$. Since the inverse Laplace transform

of $\dfrac{1}{s^2 + 16}$ is $\dfrac{1}{4}\sin(4t)$, $il_0(t) = \dfrac{1}{4}\sin(4t)$. Moreover, for $n \geq 1$,

$$L\left\{\int_0^t il_{n-1}(\alpha)\,d\alpha\right\} = \frac{1}{s}\frac{1}{s^{n-1}\left(s^2 + 16\right)} = \frac{1}{s^n\left(s^2 + 16\right)}$$ so the inverse Laplace transform of $\dfrac{1}{s^n\left(s^2 + 16\right)}$

can be computed via $\int_0^t il_{n-1}(\alpha)\,d\alpha$. $il_n(t) = \int_0^t il_{n-1}(\alpha)\,d\alpha$.

We define this recursive relationship below. The definition of `il` is of the form `il[n_]:=il[n]=...` so that *Mathematica* "remembers" the values of `il` computed. Hence, when *Mathematica* computes `il[n]`, *Mathematica* does not need to recompute `il[n-1]`.

```
▤□▤▤▤▤     InverseLaplaceTransform     ▤▣▤
In[11]:=
Clear[il]
il[0]=1/4 Sin[4t];
il[n_]:=il[n]=
    Integrate[il[n-1] /. t->a,{a,0,t}]
```

and then we compute `il[n]` for n=0, 1, 2, 3, 4, 5, and 6 and display the result in table form.

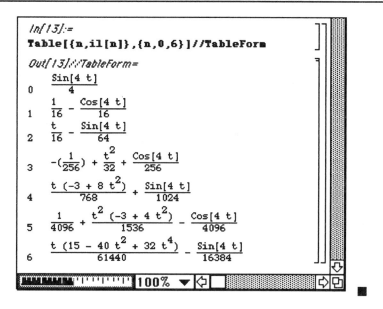

```
In[13]:=
Table[{n,il[n]},{n,0,6}]//TableForm
```

$Out[13]//TableForm=$

0	$\dfrac{Sin[4\ t]}{4}$
1	$\dfrac{1}{16} - \dfrac{Cos[4\ t]}{16}$
2	$\dfrac{t}{16} - \dfrac{Sin[4\ t]}{64}$
3	$-(\dfrac{1}{256}) + \dfrac{t^2}{32} + \dfrac{Cos[4\ t]}{256}$
4	$\dfrac{t\ (-3 + 8\ t^2)}{768} + \dfrac{Sin[4\ t]}{1024}$
5	$\dfrac{1}{4096} + \dfrac{t^2\ (-3 + 4\ t^2)}{1536} - \dfrac{Cos[4\ t]}{4096}$
6	$\dfrac{t\ (15 - 40\ t^2 + 32\ t^4)}{61440} - \dfrac{Sin[4\ t]}{16384}$

100%

■ EXAMPLE 8.9

Find the inverse Laplace transform of $\dfrac{-1 + se^{3s} + 8s^2e^{3s} + 4s^3e^{3s} + s^4e^{3s}}{s^2\left(se^{3s} - 1\right)\left(s^2 + 4s + 8\right)}$.

Solution:

In the solution of this example, we will use the command **InverseLaplaceTransform** contained in the package **LaplaceTransform** which is contained in the **Calculus** folder (or directory) so we begin by loading the package **LaplaceTransform** and defining **expression** to be

$$\dfrac{-1 + se^{3s} + 8s^2e^{3s} + 4s^3e^{3s} + s^4e^{3s}}{s^2\left(se^{3s} - 1\right)\left(s^2 + 4s + 8\right)}.$$

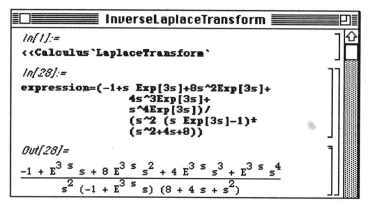

```
InverseLaplaceTransform
In[1]:=
<<Calculus`LaplaceTransform`
In[28]:=
expression=(-1+s Exp[3s]+8s^2Exp[3s]+
            4s^3Exp[3s]+
            s^4Exp[3s])/
            (s^2 (s Exp[3s]-1)*
            (s^2+4s+8))
Out[28]=
```
$$\frac{-1 + E^{3\,s}\,s + 8\,E^{3\,s}\,s^2 + 4\,E^{3\,s}\,s^3 + E^{3\,s}\,s^4}{s^2\,(-1 + E^{3\,s}\,s)\,(8 + 4\,s + s^2)}$$

We then compute the partial fraction decomposition of **expression** and name the resulting output **decomp**.

```
In[29]:=
decomp=Apart[expression]
Out[29]=
```
$$\frac{1}{s\,(-1 + E^{3\,s}\,s)} + \frac{1 + 8\,s + 4\,s^2 + s^3}{s^2\,(8 + 4\,s + s^2)}$$

Then to compute the inverse Laplace transform of $\dfrac{-1 + se^{3s} + 8s^2e^{3s} + 4s^3e^{3s} + s^4e^{3s}}{s^2\left(se^{3s} - 1\right)\left(s^2 + 4s + 8\right)}$, we must

compute the inverse Laplace transform of $\dfrac{1 + 8s + 4s^2 + s^3}{s^2\left(8 + 4s + s^2\right)}$ and the inverse Laplace transform of

$\dfrac{1}{s\left(-1 + se^{3s}\right)}$.

Completing the square yields $s^2 + 4s + 8 = (s+2)^2 + 4$. Then,

$\dfrac{1 + 8s + 4s^2 + s^3}{s^2\left(8 + 4s + s^2\right)} = \dfrac{1}{s^2\left[(s+2)^2 + 4\right]} + \dfrac{8}{s\left[(s+2)^2 + 4\right]} + \dfrac{2}{(s+2)^2 + 4} + \dfrac{s+2}{(s+2)^2 + 4}$. Using

Table 8. 1 we obtain that the inverse Laplace transform of $\dfrac{2}{(s+2)^2+4}$ is $e^{-2t}\sin(2t)$ and the inverse

Laplace transform of $\dfrac{s+2}{(s+2)^2+4}$ is $e^{-2t}\cos(2t)$.

Using **Table 8. 2**, the inverse Laplace transform of $\dfrac{8}{s\left[(s+2)^2+4\right]}=\dfrac{4}{s}-\dfrac{2}{(s+2)^2+4}$ is

$4\int_0^t e^{-2\alpha}\sin(2\alpha)\,d\alpha=1-e^{-2t}\left(\cos(2t)+\sin(2t)\right)$ and the inverse Laplace transform of

$\dfrac{1}{s^2\left[(s+2)^2+4\right]}=\dfrac{1}{8s}\dfrac{8}{s\left[(s+2)^2+4\right]}$ is

$\dfrac{1}{8}\int_0^t\left[1-e^{-2\alpha}\left(\cos(2\alpha)+\sin(2\alpha)\right)\right]d\alpha=\dfrac{-1}{16}+\dfrac{t}{8}+\dfrac{1}{16}e^{-2t}\cos(2t)$.

```
In[30]:=
InverseLaplaceTransform[1/(8+4s+s^2),s,t]

Out[30]=
-Sin[2 t]
─────────
  2 E^2 t
```

Therefore the inverse Laplace transform of

$$\dfrac{1+8s+4s^2+s^3}{s^2\left(8+4s+s^2\right)}=\dfrac{1}{s^2\left[(s+2)^2+4\right]}+\dfrac{8}{s\left[(s+2)^2+4\right]}+\dfrac{2}{(s+2)^2+4}+\dfrac{s+2}{(s+2)^2+4}$$ is

$\dfrac{-1}{16}+\dfrac{t}{8}+\dfrac{1}{16}e^{-2t}\cos(2t)+1-e^{-2t}\left(\cos(2t)+\sin(2t)\right)+e^{-2t}\sin(2t)+e^{2t}\cos(2t)$ which when

simplified is $\dfrac{15}{16}+\dfrac{t}{8}+\dfrac{1}{16}e^{-2t}\cos(2t)$ and defined below as $y_1(t)$.

```
In[31]:=
y1[t_]=15/16+t/8+1/16Cos[2t]Exp[-2t]

Out[31]=
15   t   Cos[2 t]
── + ─ + ────────
16   8    16 E^2 t
```

To compute the inverse Laplace transform of $\dfrac{1}{s\left(-1 + s\,e^{3s}\right)}$ we rewrite

$$\frac{1}{s\left(-1 + s\,e^{3s}\right)} = \frac{1}{s^2 e^{3s}}\frac{1}{1 - s^{-1}e^{-3s}} = \frac{1}{s^2 e^{3s}}\sum_{n=0}^{\infty}\left(s^{-1}e^{-3s}\right)^n = \sum_{n=0}^{\infty}\frac{e^{-3(n+1)s}}{s^{n+2}}.$$

We obtain the same results below by first defining **term** to be the expression $s^{-1}e^{-3s}$ and

geoser to be the power series for $\dfrac{1}{1 - x}$ about $x = 0$. In this case, we display the first five

terms of the series:

```
In[32]:=
term=1/(s Exp[3s])

Out[32]=
 1
-----
E^{3 s} s

In[33]:=
geoser=Series[1/(1-x),{x,0,5}]

Out[33]=
             2    3    4    5        6
1 + x + x  + x  + x  + x  + O[x]
```

and then we replace each **x** in **geoser** by term and name the resulting output **steptwo**:

```
In[34]:=
steptwo=geoser /. x->term

Out[34]=
        1           1   2       1   3       1   4
1 + --------- + (---------)  + (---------)  + (---------)  +
     E^{3 s} s      E^{3 s} s       E^{3 s} s       E^{3 s} s

       1   5           1   6
   (---------)  + O[---------]
     E^{3 s} s         E^{3 s} s
```

Using **Table 8.1**, the inverse Laplace transform of $\dfrac{1}{s^n}$ (n a positive integer) is $\dfrac{t^{n-1}}{(n-1)!}$ and using

Table 8.2 we obtain that the inverse Laplace transform of $\dfrac{e^{-as}}{s^n}$ is $\dfrac{(t-a)^{n-1}}{(n-1)!}u(t-a)$, where

$u(t - a) = \begin{cases} 1 \text{ if } t \geq a \\ 0 \text{ if } t < a \end{cases}$. Consequently, the inverse Laplace transform of $\dfrac{e^{-3(n+1)s}}{s^{n+2}}$ is

$\dfrac{(t - 3(n+1))^{n+1}}{(n+1)!} u(t - 3(n+1))$.

This relationship is defined below in **rule**. When applied to an object, **rule** replaces each expression

in the object of the form $e^{sd}s^n$ by $\dfrac{(t+d)^{1-n}}{(1-n)!} u(t+d)$. Thus, when n is negative, **rule** returns the

inverse Laplace transform of $e^{sd}s^n$.

```
In[35]:=
rule={Exp[s d_] s^n_->u[t+d]*
                (t+d)^(1-n)/(1-n)!}

Out[35]=
       s (d_)  n_        (d + t)^1 - n  u[d + t]
    {E        s    -->   ---------------------- }
                              (1 - n)!
```

Therefore the inverse Laplace transform of $\dfrac{1}{s\left(-1 + se^{3s}\right)} = \displaystyle\sum_{n=0}^{\infty} \dfrac{e^{-3(n+1)s}}{s^{n+2}}$ is

$\displaystyle\sum_{n=0}^{\infty} \dfrac{(t - 3(n+1))^{n+1}}{(n+1)!} u(t - 3(n+1))$ and the inverse Laplace transform of

$\dfrac{-1 + se^{3s} + 8s^2 e^{3s} + 4s^3 e^{3s} + s^4 e^{3s}}{s^2\left(se^{3s} - 1\right)\left(s^2 + 4s + 8\right)} = \dfrac{1 + 8s + 4s^2 + s^3}{s^2\left(8 + 4s + s^2\right)} + \dfrac{1}{s\left(-1 + se^{3s}\right)}$ is

$\dfrac{15}{16} + \dfrac{t}{8} + \dfrac{1}{16} e^{-2t} \cos(2t) + \displaystyle\sum_{n=0}^{\infty} \dfrac{(t - 3(n+1))^{n+1}}{(n+1)!} u(t - 3(n+1))$.

Note that for any value of t, the resulting series contains only finitely many terms.

Before applying **rule** to **steptwo**, we first remove the 0-term which represents the higher order omitted terms of the series and then name the resulting output **stepthree**.

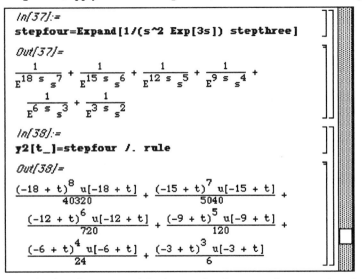

```
In[36]:=
stepthree=Normal[steptwo]
Out[36]=
1 +  1/(E^(15 s) s^5)  +  1/(E^(12 s) s^4)  +  1/(E^(9 s) s^3)  +  1/(E^(6 s) s^2)  +
     1/(E^(3 s) s)
```

To compute the first few terms of the series, we compute $\dfrac{1}{s^2 e^{3s}}$ **stepthree**, name the result

stepfour, apply **rule** to **stepfour**, and then define the resulting function to be **y2**.

```
In[37]:=
stepfour=Expand[1/(s^2 Exp[3s]) stepthree]
Out[37]=
 1/(E^(18 s) s^7)  +  1/(E^(15 s) s^6)  +  1/(E^(12 s) s^5)  +  1/(E^(9 s) s^4)  +
 1/(E^(6 s) s^3)  +  1/(E^(3 s) s^2)

In[38]:=
y2[t_]=stepfour /. rule
Out[38]=
(-18 + t)^8 u[-18 + t]/40320  +  (-15 + t)^7 u[-15 + t]/5040  +
(-12 + t)^6 u[-12 + t]/720  +  (-9 + t)^5 u[-9 + t]/120  +
(-6 + t)^4 u[-6 + t]/24  +  (-3 + t)^3 u[-3 + t]/6
```

Then the first few terms of the inverse Laplace transform are given by **y[t]=y1[t]+y2[t]** which is
defined below along with **u** and then the result is graphed on the interval [0,15].

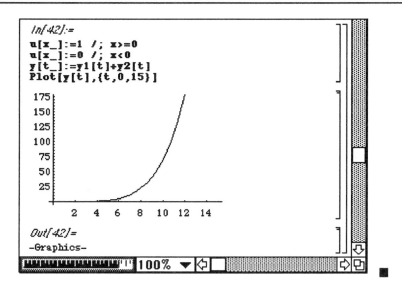

```
In[42]:=
u[x_]:=1 /; x>=0
u[x_]:=0 /; x<0
y[t_]:=y1[t]+y2[t]
Plot[y[t],{t,0,15}]
```

```
Out[42]=
-Graphics-
```

§8.2 Solving Ordinary Differential Equations with the Laplace Transform

Laplace transforms can be used to solve certain differential equations. Typically, when we use Laplace transforms to solve a differential equation for a function y, we will compute the Laplace transform of each term in the equation, solve the resulting equation for L{y}, and finally determine y by computing the inverse Laplace transform of L{y}.

❑ EXAMPLE 8.10

Use Laplace transforms to solve $y''+2y'+4y = t - e^{-t}$ subject to $y(0) = 1$ and $y'(0) = -1$.

Solution:

Let $Y(s) = L\{y\}(s)$. Then,

$$L\{y''+2y'+4y\}(s) = L\{y''\}(s) + 2L\{y'\} + 4L\{y\}(s) = s^2Y(s) - sy(0) - y'(0) + 2\lfloor sY(s) - y(0)\rfloor + 4Y(s)$$

$$= L\{t - e^{-t}\}(s) = L\{t\}(s) - L\{e^{-t}\}(s) = \frac{1}{s^2} - \frac{1}{s+1}.$$

Replacing $y(0)$ by 1 and $y'(0)$ by -1 and simplifying yields

$$Y(s)\left(s^2 + 2s + 4\right) = s + 1 + \frac{1}{s^2} - \frac{1}{s+1} = \frac{s^4 + 2s^3 + s + 1}{s^2(s+1)} \text{ so that } Y(s) = \frac{s^4 + 2s^3 + s + 1}{s^2(s+1)\left(s^2 + 2s + 4\right)}.$$

Computing the partial fraction decomposition of $\dfrac{s^4 + 2s^3 + s + 1}{s^2(s+1)\left(s^2 + 2s + 4\right)}$ yields

$Y(s) = \dfrac{1}{4s^2} - \dfrac{1}{8s} - \dfrac{1}{3(s+1)} + \dfrac{35s + 32}{24\left(s^2 + 2s + 4\right)}$ and completing the square yields

$Y(s) = \dfrac{1}{4s^2} - \dfrac{1}{8s} - \dfrac{1}{3(s+1)} + \dfrac{35s + 32}{24\left((s+1)^2 + 3\right)}$. Note that $L^{-1}\{Y\}(t) = y(t), L^{-1}\left\{\dfrac{1}{4s^2}\right\}(t) = \dfrac{t}{4}$,

$L^{-1}\left\{-\dfrac{1}{8s}\right\}(t) = \dfrac{-1}{8}, L^{-1}\left\{-\dfrac{1}{3(s+1)}\right\}(t) = \dfrac{-e^{-t}}{3}$, and

$L^{-1}\left\{\dfrac{35s + 32}{24\left((s+1)^2 + 3\right)}\right\}(t) = L^{-1}\left\{\dfrac{35(s+1)}{24\left((s+1)^2 + 3\right)} - \dfrac{3}{24\left((s+1)^2 + 3\right)}\right\}(t)$

$= \dfrac{35e^{-t}\cos\left(\sqrt{3}\,t\right)}{24} - \dfrac{\sqrt{3}\,e^{-t}\sin\left(\sqrt{3}\,t\right)}{24}$ so that

$y(t) = \dfrac{t}{4} - \dfrac{1}{8} - \dfrac{e^{-t}}{3} + \dfrac{35e^{-t}\cos\left(\sqrt{3}\,t\right)}{24} - \dfrac{\sqrt{3}\,e^{-t}\sin\left(\sqrt{3}\,t\right)}{24}$. ∎

❑ **EXAMPLE 8.11**

Solve $y' + 4y = f(t)$ subject to $y(0) = 0$ where $f(t) = \begin{cases} -1, \text{if } 0 \le t \le 1 \\ 0, \text{if } t > 1 \end{cases}$.

Solution:

As in the preceding problem, we begin by computing the Laplace transform of each term of the equation. Using the definition of the Laplace transform, we compute the Laplace transform of $f(t)$:

$\int_0^\infty e^{-st} f(t)\, dt = -\int_0^1 e^{-st}\, dt = \dfrac{-1}{s} + \dfrac{1}{se^s} = \dfrac{1 - e^s}{se^s}$. Let $Y(s)$ denote the Laplace transform of $y(t)$. Then

transforming the equation results in the equation $sY(s) - y(0) + 4Y(s) = \dfrac{1 - e^s}{se^s}$. Replacing $y(0)$ by

0 and solving for $Y(s)$ yields $Y(s) = \dfrac{1 - e^s}{e^s\left(s^2 + 4s\right)}$. To compute $y(t)$, we need to find the inverse

Laplace transform of $Y(s)$. We begin by computing the partial fraction decomposition of

$$\frac{1 - e^s}{e^s\left(s^2 + 4s\right)} : \frac{1 - e^s}{e^s\left(s^2 + 4s\right)} = \frac{1}{s(s + 4)e^s} - \frac{1}{s(s + 4)} = \frac{1}{e^s}\left(\frac{1}{4s} - \frac{1}{4(s + 4)}\right) - \frac{1}{4s} + \frac{1}{4(s + 4)}. \text{ Using}$$

Tables 8. 1 and **8. 2**, we obtain that inverse Laplace transform of $\dfrac{1}{4s}$ is $\dfrac{1}{4}$ and the inverse Laplace

transform of $\dfrac{1}{4(s + 4)}$ is $\dfrac{1}{4}e^{-4t}$ so that the inverse Laplace transform of $\dfrac{1}{e^s}\left(\dfrac{1}{4s} - \dfrac{1}{4(s + 4)}\right)$ is

$\left(\dfrac{1}{4} - \dfrac{1}{4}e^{-4(t-1)}\right)u(t - 1)$. Therefore the inverse Laplace transform of $Y(s)$, and the solution of the

problem, is $y(t) = \left(\dfrac{1}{4} - \dfrac{1}{4}e^{-4(t-1)}\right)u(t - 1) - \dfrac{1}{4} + \dfrac{1}{4}e^{-4t}$. ∎

■ EXAMPLE 8.12

Let $f(x) = \begin{cases} \sin(\pi x) \text{ if } 0 \le x \le 2 \\ \quad 0 \text{ if } x > 2 \end{cases}$. Solve $y'' + y' + y = f(x)$ subject to $y(0) = 1$ and $y'(0) = 1$.

Solution:

The solution will be constructed by computing the Laplace transform of the equation
$y'' + y' + y = f(x)$, solving the resulting equation for the Laplace transform of $y(t)$, and then finally
computing the inverse Laplace transform of the Laplace transform of y(t) to explicitly compute y(t). In
this case, we will see that *Mathematica* is particularly useful as the solution is complicated.
We begin by defining f and graphing f. Notice that f could have been defined by entering the commands

```
f[x_]:=Sin[Pi x] /; 0<=x<=2
f[x_]:=0 /; x>2
```

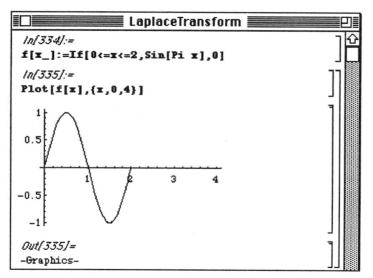

```
LaplaceTransform

In[334]:=
f[x_]:=If[0<=x<=2,Sin[Pi x],0]

In[335]:=
Plot[f[x],{x,0,4}]
```

```
Out[335]=
-Graphics-
```

We then define **lhsone** to be the Laplace transform of the left-hand side of the equation
y''+y'+y = f(x):

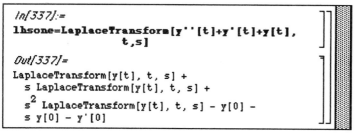

```
In[337]:=
lhsone=LaplaceTransform[y''[t]+y'[t]+y[t],
                t,s]

Out[337]=
LaplaceTransform[y[t], t, s] +
  s LaplaceTransform[y[t], t, s] +
   2
  s  LaplaceTransform[y[t], t, s] - y[0] -
  s y[0] - y'[0]
```

We then replace y[·0] by 1, y'[0] by 1, and, for convenience, we replace the object
LaplaceTransform[y[t],t,s] by **capy** and name the resulting output **lhstwo**:

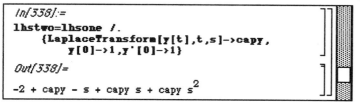

```
In[338]:=
lhstwo=lhsone /.
    {LaplaceTransform[y[t],t,s]->capy,
      y[0]->1,y'[0]->1}

Out[338]=
                                 2
-2 + capy - s + capy s + capy s
```

We then explicitly compute the Laplace transform of f(t):

$$L\{f\}(s) = \int_0^\infty f(t)\,e^{-st}\,dt = \int_0^2 \sin(\pi t)\,e^{-st}\,dt$$ and name the resulting output **r h s**.

```
In[340]:=
rhs=Integrate[Sin[Pi t] Exp[-s t],{t,0,2}]

Out[340]=
   Pi            Pi
--------- - -----------
  2    2     2 s   2    2
Pi  + s     E   (Pi  + s )
```

Next we solve the equation **lhstwo=rhs** for **capy** and name the resulting output **stepthree**. Recall that **capy** represents the Laplace transform of y(t).

```
In[342]:=
stepthree=
    Solve[lhstwo==rhs,capy]

Out[342]=
{{capy ->
     -((Pi - E^2 s Pi - 2 E^2 s Pi^2 - E^2 s Pi^2 s -
        2 E^2 s s^2 - E^2 s s^3) /
       (E^2 s Pi^2 + E^2 s Pi^2 s + E^2 s s^2 +
        E^2 s Pi^2 s^2 + E^2 s s^3 + E^2 s s^4))}}
```

Since **stepthree** is a list, we explicitly extract the Laplace transform of y(t) with the command **stepthree[[1,1,2]]** and name the result **stepfour**.

```
In[343]:=
stepfour=stepthree[[1,1,2]]

Out[343]=
-((Pi - E^2 s Pi - 2 E^2 s Pi^2 - E^2 s Pi^2 s -
   2 E^2 s s^2 - E^2 s s^3) /
  (E^2 s Pi^2 + E^2 s Pi^2 s + E^2 s s^2 +
   E^2 s Pi^2 s^2 + E^2 s s^3 + E^2 s s^4))
```

The solution is obtained with the command **InverseLaplaceTransform**.

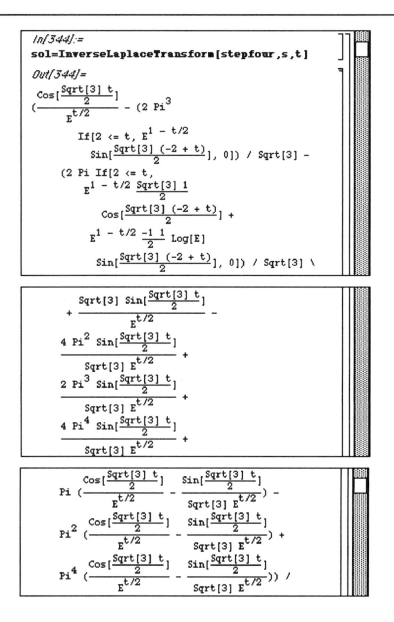

```
In[344]:=
sol=InverseLaplaceTransform[stepfour,s,t]

Out[344]=
    Cos[Sqrt[3] t]
         2
(---------------- - (2 Pi^3
       E^t/2

        If[2 <= t, E^1 - t/2
            Sin[Sqrt[3] (-2 + t)], 0]) / Sqrt[3] -
                      2
        (2 Pi If[2 <= t,
            E^1 - t/2 Sqrt[3] 1
                         2
              Cos[Sqrt[3] (-2 + t)] +
                      2
            E^1 - t/2 -1 1 Log[E]
                      2
              Sin[Sqrt[3] (-2 + t)], 0]) / Sqrt[3] \
                      2
```

```
         Sqrt[3] Sin[Sqrt[3] t]
    +                     2       -
                  E^t/2

    4 Pi^2 Sin[Sqrt[3] t]
                      2
    --------------------- +
      Sqrt[3] E^t/2
    2 Pi^3 Sin[Sqrt[3] t]
                      2
    --------------------- +
      Sqrt[3] E^t/2
    4 Pi^4 Sin[Sqrt[3] t]
                      2
    --------------------- +
      Sqrt[3] E^t/2
```

```
       Cos[Sqrt[3] t]   Sin[Sqrt[3] t]
            2                2
Pi (---------------- - ------------------) -
         E^t/2          Sqrt[3] E^t/2
       Cos[Sqrt[3] t]   Sin[Sqrt[3] t]
            2                2
Pi^2 (---------------- - ------------------) +
         E^t/2          Sqrt[3] E^t/2
       Cos[Sqrt[3] t]   Sin[Sqrt[3] t]
            2                2
Pi^4 (---------------- - ------------------)) /
         E^t/2          Sqrt[3] E^t/2
```

```
(1 - Pi² + Pi⁴) +
(-(Pi Cos[Pi t]) +
    If[2 <= t, Pi 1 Cos[Pi (-2 + t)], 0] -
    If[2 <= t, Sin[Pi (-2 + t)], 0] +
Pi² If[2 <= t, Sin[Pi (-2 + t)], 0] +
    Sin[Pi t] - Pi² Sin[Pi t]) / (1 - Pi² + Pi⁴)
```

and then finally graphed on the interval [−1,6].

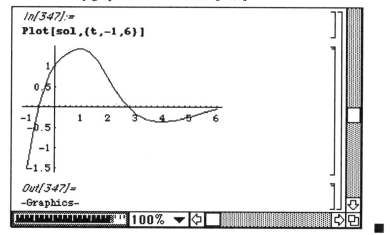

```
In[347]:=
Plot[sol,{t,-1,6}]
```

```
Out[347]=
-Graphics-
```

100%

■ EXAMPLE 8.13

Let $f(x)$ be defined recursively by $f(x) = \begin{cases} 1 \text{ if } 0 \le x < 1 \\ -1 \text{ if } 1 \le x \le 2 \end{cases}$ and $f(x) = f(x-2)$ if $x > 2$.

Solve $y''(t) + 4y'(t) + 20y(t) = f(t)$.

Solution:

We begin by first defining and graphing f and u and then displaying the resulting graphs as a graphics array. Note that the definitions of f and u could have been entered using the **If** command as in the previous example.

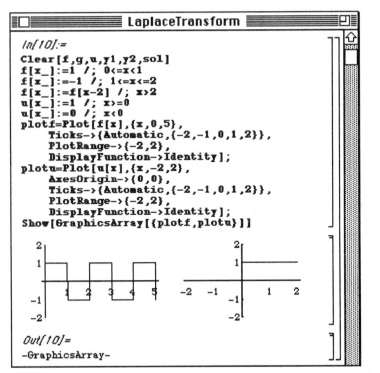

```
≡□≡≡≡≡≡≡≡≡ LaplaceTransform ≡≡≡≡≡□≡
In[10]:=
Clear[f,g,u,y1,y2,sol]
f[x_]:=1 /; 0<=x<1
f[x_]:=-1 /; 1<=x<=2
f[x_]:=f[x-2] /; x>2
u[x_]:=1 /; x>=0
u[x_]:=0 /; x<0
plotf=Plot[f[x],{x,0,5},
     Ticks->{Automatic,{-2,-1,0,1,2}},
     PlotRange->{-2,2},
     DisplayFunction->Identity];
plotu=Plot[u[x],{x,-2,2},
     AxesOrigin->{0,0},
     Ticks->{Automatic,{-2,-1,0,1,2}},
     PlotRange->{-2,2},
     DisplayFunction->Identity];
Show[GraphicsArray[{plotf,plotu}]]
```

Out[10]=
-GraphicsArray-

We then define **l h s** to be the left - hand side of the equation $y''(t) + 4y'(t) + 20y(t) = f(t)$.

```
In[11]:=
lhs=y''[x]+4y'[x]+20y[x]
Out[11]=
20 y[x] + 4 y'[x] + y''[x]
```

Let **ly** denote the Laplace transform of y. Then the Laplace transform of y' is s **ly**− y(0) and the Laplace transform of y'' is s^2**ly**−s y(0)−y'(0). These relationships are defined below in **laplacerule**. In the second command, **laplacerule** is applied to **lhs** and the simplified result is named **stepone**.

```
In[12]:=
laplacerule={y[x]->ly,
            y'[x]->s ly-y[0],
            y''[x]->s^2 ly-s y[0]-y'[0]};
In[13]:=
stepone=lhs /. laplacerule//Simplify
Out[13]=
20 ly + 4 ly s + ly s  - 4 y[0] - s y[0] - y'[0]
```

Let **lr** denote the Laplace transform of the right-hand side of the equation, f(x). We then solve the
equation $20\,\mathbf{l}\,y + 4\,\mathbf{l}\,y\,s + \mathbf{l}\,y\,s^2 - 4y(0) - sy(0) - y'(0) = \mathbf{l}\,\mathbf{r}$ for $\mathbf{l}\,y$ and name the resulting
output **steptwo**.

```
In[14]:=
steptwo=Solve[stepone==lr,ly]
Out[14]=
             -lr - 4 y[0] - s y[0] - y'[0]
{{ly -> -(----------------------------------)}}
                        2
                 20 + 4 s + s
```

To compute y, we must compute the inverse Laplace transform of **ly** which is explicitly obtained from
steptwo with **steptwo[[1,1,2]]**. We begin by collecting those terms which contain **lr** and
name the resulting output **stepthree**.

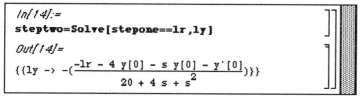

```
In[15]:=
stepthree=Collect[steptwo[[1,1,2]],lr]
Out[15]=
      lr              4 y[0]            s y[0]
-------------  +  -------------  +  -------------  +
           2                 2                 2
20 + 4 s + s      20 + 4 s + s      20 + 4 s + s
      y'[0]
-------------
           2
20 + 4 s + s
```

Since the first term of **stepthree** is the only term containing **lr**, we drop the first term from
stepthree and name the result **stepfour**.

```
In[16]:=
stepfour=Drop[stepthree,1]
Out[16]=
    4 y[0]            s y[0]            y'[0]
-------------  +  -------------  +  -------------
           2                 2                 2
20 + 4 s + s      20 + 4 s + s      20 + 4 s + s
```

Completing the square yields $20 + 4s + s^2 = (s+2)^2 + 16$. Then, the inverse Laplace transform of

$$\frac{4y(0) + sy(0) + y'(0)}{20 + 4s + s^2} = \frac{(s+2)y(0)}{(s+2)^2 + 16} + \frac{y'(0) + 2y(0)}{(s+2)^2 + 16} \text{ is}$$

$y(0)e^{-2t}\cos(4t) + \dfrac{y'(0) + 2y(0)}{4}e^{-2t}\sin(4t)$ which is defined below as $y_1(t)$.

```
In[36]:=
y1[t_]=(4 Cos[4t] y[0]+2 Sin[4t] y[0]+
         Sin[4t] y'[0])/(4 Exp[2t])

Out[36]=
(4 Cos[4 t] y[0] + 2 Sin[4 t] y[0] +
    Sin[4 t] y'[0]) / (4 E^(2 t))
```

To compute the inverse Laplace transform of $\dfrac{1\,r}{20 + 4s + s^2}$, we begin by computing $1\,r$.

The periodic function $f(x) = \begin{cases} 1 \text{ if } 0 \le x < 1 \\ -1 \text{ if } 1 \le x \le 2 \end{cases}$ and $f(x+2) = f(x)$ can be written in terms of step

functions as $f(x) = u_0(t) - 2u_1(t) + 2u_2(t) - 2u_3(t) + 2u_4(t) - \ldots$

$$= u(t) - 2u(t-1) + 2u(t-2) - 2u(t-3) + 2u(t-4) - \ldots$$

$$= u(t) + 2\sum_{n=1}^{\infty}(-1)^n u(t-n). \text{ Then}$$

$$1\,r = L\{f\}(s) = \frac{1}{s} - 2\frac{e^{-s}}{s} + 2\frac{e^{-2s}}{s} - 2\frac{e^{-3s}}{s} + \ldots = \frac{1}{s}\left[1 - 2e^{-s} + 2e^{-2s} - 2e^{-3s} + 2e^{-4s} - \ldots\right] \text{ and}$$

$$\frac{1\,r}{20 + 4s + s^2} = \frac{1}{s(20 + 4s + s^2)}\left[1 - 2e^{-s} + 2e^{-2s} - 2e^{-3s} + 2e^{-4s} - \ldots\right]$$

$$= \frac{1}{s(20 + 4s + s^2)} + 2\sum_{n=1}^{\infty}(-1)^n \frac{e^{-ns}}{s(20 + 4s + s^2)}.$$

In the following command we first extract $\dfrac{1\,r}{20 + 4s + s^2}$ from **stepthree** and name the result

stepsix:

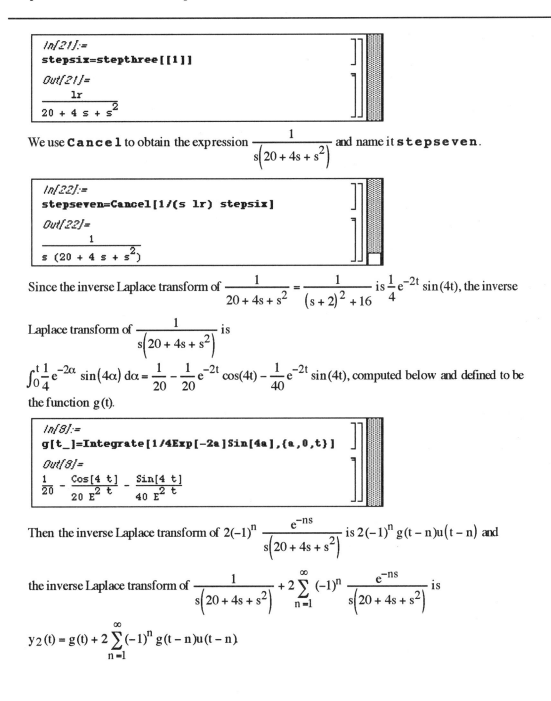

```
In[21]:=
stepsix=stepthree[[1]]
Out[21]=
     1r
 ───────────
           2
 20 + 4 s + s
```

We use **Cancel** to obtain the expression $\dfrac{1}{s\left(20+4s+s^2\right)}$ and name it **stepseven**.

```
In[22]:=
stepseven=Cancel[1/(s 1r) stepsix]
Out[22]=
        1
 ───────────────
            2
 s (20 + 4 s + s )
```

Since the inverse Laplace transform of $\dfrac{1}{20+4s+s^2}=\dfrac{1}{(s+2)^2+16}$ is $\dfrac{1}{4}e^{-2t}\sin(4t)$, the inverse

Laplace transform of $\dfrac{1}{s\left(20+4s+s^2\right)}$ is

$\int_0^t \dfrac{1}{4}e^{-2\alpha}\sin(4\alpha)\,d\alpha = \dfrac{1}{20}-\dfrac{1}{20}e^{-2t}\cos(4t)-\dfrac{1}{40}e^{-2t}\sin(4t)$, computed below and defined to be
the function $g(t)$.

```
In[8]:=
g[t_]=Integrate[1/4Exp[-2a]Sin[4a],{a,0,t}]
Out[8]=
  1    Cos[4 t]    Sin[4 t]
 ── - ───────── - ─────────
 20     2 t          2 t
      20 E         40 E
```

Then the inverse Laplace transform of $2(-1)^n\dfrac{e^{-ns}}{s\left(20+4s+s^2\right)}$ is $2(-1)^n g(t-n)u(t-n)$ and

the inverse Laplace transform of $\dfrac{1}{s\left(20+4s+s^2\right)}+2\sum_{n=1}^{\infty}(-1)^n\dfrac{e^{-ns}}{s\left(20+4s+s^2\right)}$ is

$y_2(t) = g(t)+2\sum_{n=1}^{\infty}(-1)^n g(t-n)u(t-n)$

It then follows that

$y(t) = y_1(t) + y_2(t)$

$$= y(0)e^{-2t}\cos(4t) + \frac{y'(0) + 2y(0)}{4}e^{-2t}\sin(4t) + g(t) + 2\sum_{n=1}^{\infty}(-1)^n g(t-n)u(t-n), \text{ where}$$

$$g(t) = \frac{1}{20} - \frac{1}{20}e^{-2t}\cos(4t) - \frac{1}{40}e^{-2t}\sin(4t).$$

To graph the solution for various initial conditions, we first create a table of $2(-1)n g(t-n)u(t-n)$ for n=1, 2, 3, 4, and 5:

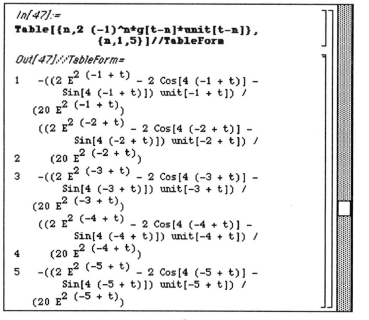

```
In[47]:=
Table[{n,2 (-1)^n*g[t-n]+unit[t-n]},
            {n,1,5}]//TableForm

Out[47]//TableForm=

 1    -((2 E^2 (-1 + t) - 2 Cos[4 (-1 + t)] -
            Sin[4 (-1 + t)]) unit[-1 + t]) /
      (20 E^2 (-1 + t))
      ((2 E^2 (-2 + t) - 2 Cos[4 (-2 + t)] -
            Sin[4 (-2 + t)]) unit[-2 + t]) /
 2      (20 E^2 (-2 + t))
 3    -((2 E^2 (-3 + t) - 2 Cos[4 (-3 + t)] -
            Sin[4 (-3 + t)]) unit[-3 + t]) /
      (20 E^2 (-3 + t))
      ((2 E^2 (-4 + t) - 2 Cos[4 (-4 + t)] -
            Sin[4 (-4 + t)]) unit[-4 + t]) /
 4      (20 E^2 (-4 + t))
 5    -((2 E^2 (-5 + t) - 2 Cos[4 (-5 + t)] -
            Sin[4 (-5 + t)]) unit[-5 + t]) /
      (20 E^2 (-5 + t))
```

and then we define $y_2(t) = g(t) + 2\sum_{n=1}^{5}(-1)^n g(t-n)u(t-n)$, sol$(t) = y_1(t) + y_2(t)$, and

inits to be the table consisting of the numbers $-1/2$ and $1/2$.

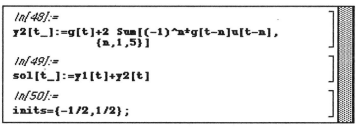

```
In[48]:=
y2[t_]:=g[t]+2 Sum[(-1)^n*g[t-n]u[t-n],
          {n,1,5}]
In[49]:=
sol[t_]:=y1[t]+y2[t]
In[50]:=
inits={-1/2,1/2};
```

We then create a table of graphs of **sol[t]** on the interval [0,5] corresponding to replacing y(0) and y'(0) by both 1/2 and –1/2 and then displaying the resulting graphics array.

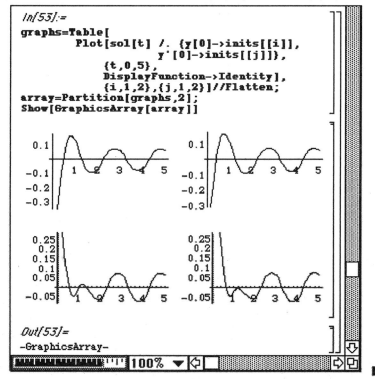

```
In[53]:=
graphs=Table[
        Plot[sol[t] /. {y[0]->inits[[i]],
                        y'[0]->inits[[j]]},
             {t,0,5},
             DisplayFunction->Identity],
             {i,1,2},{j,1,2}]//Flatten;
array=Partition[graphs,2];
Show[GraphicsArray[array]]
```

```
Out[53]=
-GraphicsArray-
```

§8.3 Some Special Equations: Delay Equations, Equations with Nonconstant Coefficients

Laplace transforms can also be used to solve certain equations with nonconstant coefficients, although Laplace transforms do not provide a general method for solving such equations, and delay equations.

■ **EXAMPLE 8.14**

Solve $y'(t) - 2y(t-1) = t$ subject to the condition $y(t) = y(0)$ for $-1 \le t \le 0$.

Solution:

Multiplying both sides of the equation by e^{-st} and integrating yields

$\int_0^\infty e^{-st} y'(t) dt - 2\int_0^\infty e^{-st} y(t-1) dt = \int_0^\infty t e^{-st} dt$. Note that $\int_0^\infty e^{-st} y'(t) dt = L\{y'\}(s) = sL\{y\}(s) - y(0)$

and $\int_0^\infty t e^{-st} dt = \dfrac{1}{s^2}$. To evaluate $\int_0^\infty e^{-st} y(t-1) dt$, let $u = t - 1$. Then,

$\int_0^\infty e^{-st} y(t-1) dt = \int_{-1}^\infty e^{-s(u+1)} y(u) du = e^{-s} \int_{-1}^\infty e^{-su} y(u) du = e^{-s} \left(\int_{-1}^0 e^{-su} y(u) du + \int_0^\infty e^{-su} y(u) du \right)$.

Since $\int_0^\infty e^{-su} y(u) du = L\{y\}(s)$ and $y(t) = y(0)$ for $-1 \le t \le 0$, we obtain

$\int_0^\infty e^{-st} y(t-1) dt = e^{-s} y(0) \int_{-1}^0 e^{-su} du + e^{-s} L\{y\}(s)$. Below we compute $\int_{-1}^0 e^{-su} du = \dfrac{e^s - 1}{s}$ and

consequently $\int_0^\infty e^{-st} y(t-1) dt = y(0) \dfrac{1 - e^{-s}}{s} + e^{-s} L\{y\}(s)$.

In addition, we clear all symbols to be used in the solution of the problem and define $\mathbf{lf} = \dfrac{1}{s^2}$.

```
▤▢▭▭▭▭▭▭ DelayEquation ▭▭▭▭▭▭▭◱
  In[1]:=
  <<Calculus`LaplaceTransform`
  In[2]:=
  Integrate[Exp[-s u],{u,-1,0}]//Together
  Out[2]=
  -1 + E^s
  ────────
     s
  In[4]:=
  Clear[s,sol,y,y1,y2,u,ly,a,c,lf]
  lf=1/s^2;
```

Then $\int_0^\infty e^{-st} y'(t) dt - 2\int_0^\infty e^{-st} y(t-1) dt = \int_0^\infty t e^{-st} dt$ is equivalent to the equation

$sL\{y\}(s) - y(0) - 2\left(e^{-s} y(0) \int_{-1}^0 e^{-su} du + e^{-s} L\{y\}(s) \right) = \dfrac{1}{s^2}$. Let $\mathbf{ly}$ denote $L\{y\}(s)$.

Then below we solve this equation for $\mathbf{ly}$ and name the resulting output **solution**. To solve the problem we must calculate the inverse Laplace transform of $\mathbf{ly}$.

```
In[5]:=
solution=Solve[s ly-y[0]-2Exp[-s] ly-
        2y[0] (1-Exp[-s])/s==lf,ly]
Out[5]=
{{ly -> -(
        -Eˢ + 2 s y[0] - 2 Eˢ s y[0] - Eˢ s² y[0]
        ────────────────────────────────────────  )}}
                -2 s² + Eˢ s³
    }
```

We extract the Laplace transform of y, $\dfrac{-e^s + 2s\,y(0) - 2e^s s\,y(0) - e^s s^2 y(0)}{-2s^2 + e^s s^3}$, from **solution**

with the command **solution[[1,1,2]]** and then compute the partial fraction decomposition

$$\frac{-e^s + 2s\,y(0) - 2e^s s\,y(0) - e^s s^2 y(0)}{-2s^2 + e^s s^3} = \frac{2(1 + 2s\,y(0))}{s^3\left(-2 + e^s s\right)} + \frac{1 + 2s\,y(0) + s^2 y(0)}{s^3} \text{ using } \mathbf{Apart}$$

and name the resulting output **stepthree**.

```
In[6]:=
stepthree=Apart[solution[[1,1,2]]]
Out[6]=
2 (1 + 2 s y[0])     1 + 2 s y[0] + s² y[0]
────────────────  +  ──────────────────────
s³ (-2 + Eˢ s)                s³
```

We first compute the inverse Laplace transform of $\dfrac{1 + 2s\,y(0) + s^2 y(0)}{s^3}$ which is extracted from

stepthree with **stepthree[[2]]**. We define the resulting function of t to be y1(t). In this case, the command **InverseLaplaceTransform** is used although using **Table 8.1** we see that the inverse

Laplace transform of $\dfrac{1 + 2s\,y(0) + s^2 y(0)}{s^3} = \dfrac{1}{s^3} + \dfrac{2y(0)}{s^2} + \dfrac{y(0)}{s}$ is $\dfrac{t^2}{2} + 2y(0)t + y(0)$.

```
In[7]:=
stepfour=stepthree[[2]]
Out[7]=
1 + 2 s y[0] + s² y[0]
───────────────────────
          3
         s
In[8]:=
y1[t_]=InverseLaplaceTransform[stepfour,s,t]
Out[8]=
 t²
 ── + y[0] + 2 t y[0]
 2
```

To compute the inverse Laplace transform of $\dfrac{2(1 + 2 s y(0))}{s^3\left(-2 + e^s\, s\right)}$, extracted from **stepthree** with

stepthree[[1]], we rewrite $\dfrac{2(1 + 2 s y(0))}{s^3\left(-2 + e^s\, s\right)} = \dfrac{2(1 + 2 s y(0))}{s^4 e^s}\,\dfrac{1}{1 - 2 s^{-1} e^{-s}}$

$$= \frac{2(1 + 2 s y(0))}{s^4 e^s}\sum_{n=0}^{\infty}\left(2 s^{-1} e^{-s}\right)^n$$

$$= 2(1 + 2 s y(0))\sum_{n=0}^{\infty} 2^n\,\frac{e^{-ns}}{s^{n+4}}$$

$$= \sum_{n=0}^{\infty}\left[2^{n+1}\,\frac{e^{-ns}}{s^{n+4}} + 2^{n+2} y(0)\,\frac{e^{-ns}}{s^{n+3}}\right].$$

We compute the first few terms of this series by defining **term** $= \dfrac{2}{s e^s}$ and then **geoseries**

to be the power series for $\dfrac{1}{1-x}$ about x = 0. The first four terms are displayed.

```
In[9]:=
term=2/(s Exp[s])
Out[9]=
 2
-----
 E^s s
In[10]:=
geoseries=Series[1/(1-x),{x,0,4}]
Out[10]=
1 + x + x^2 + x^3 + x^4 + O[x]^5
```

We then replace each **x** in **geoseries** by **term**, name the resulting series **stepfive**, use **Normal** to remove the O-term from **stepfive**:

```
In[11]:=
stepfive=geoseries /. x->term
Out[11]=
      2        2           2           2
1 + ----- + (-----)^2 + (-----)^3 + (-----)^4 + O[-----]^5
    E^s s    E^s s        E^s s       E^s s        E^s s
In[12]:=
stepsix=stepfive//Normal
Out[12]=
      16           8           4          2
1 + ------- + ------- + ------- + -----
    E^4 s^4    E^3 s^3    E^2 s^2    E^s s
```

define **s t e p s e v e n** to be the numerator of $\dfrac{2\left(1 + 2\,s\,y\,(0)\right)}{s^3\left(-2 + e^s\,s\right)}$:

```
In[13]:=
stepseven=Numerator[stepthree[[1]]]
Out[13]=
2 (1 + 2 s y[0])
```

and then expand $\dfrac{1}{s^4\,e^s}$ **s t e p s i x   s t e p s e v e n** and name the result **s t e p e i g h t**.

```
In[14]:=
stepeight=Expand[
    1/(s^4 Exp[s]) stepsix stepseven]
```

$$Out[14]=$$

$$\frac{32}{E^5 s\, s^8} + \frac{16}{E^4 s\, s^7} + \frac{8}{E^3 s\, s^6} + \frac{4}{E^2 s\, s^5} + \frac{2}{E^s\, s^4} +$$

$$\frac{64\ y[0]}{E^5 s\, s^7} + \frac{32\ y[0]}{E^4 s\, s^6} + \frac{16\ y[0]}{E^3 s\, s^5} + \frac{8\ y[0]}{E^2 s\, s^4} + \frac{4\ y[0]}{E^s\, s^3}$$

To compute the inverse Laplace transform of $\dfrac{2(1+2s\,y(0))}{s^3\left(-2+e^s\, s\right)} = \displaystyle\sum_{n=0}^{\infty}\left[2^{n+1}\dfrac{e^{-ns}}{s^{n+4}} + 2^{n+2}y(0)\dfrac{e^{-ns}}{s^{n+3}}\right]$

we note that the inverse Laplace transform of $\dfrac{e^{-at}}{s^n}$ (n a positive integer) is $\dfrac{(t-a)^{n-1}}{(n-1)!}u(t-a)$.

Then the inverse Laplace transform of $2^{n+1}\dfrac{e^{-ns}}{s^{n+4}} + 2^{n+2}y(0)\dfrac{e^{-ns}}{s^{n+3}}$ is

$$2^{n+1}\frac{(t-n)^{n+3}}{(n+3)!}u(t-n) + 2^{n+2}y(0)\frac{(t-n)^{n+2}}{(n+2)!}u(t-n)$$

$$= \frac{2^{n+1}(t-n)^{n+2}}{(n+3)!}u(t-n)\big[t+6y(0)+(2y(0)-1)\,n\big] \text{ and the inverse Laplace transform of}$$

$$\frac{2(1+2s\,y(0))}{s^3\left(-2+e^s\, s\right)} = \sum_{n=0}^{\infty}\left[2^{n+1}\frac{e^{-ns}}{s^{n+4}} + 2^{n+2}y(0)\frac{e^{-ns}}{s^{n+3}}\right] \text{ is}$$

$$y_2(t) = \sum_{n=0}^{\infty}\left[\frac{2^{n+1}(t-n)^{n+2}}{(n+3)!}u(t-n)\big[t+6y(0)+(2y(0)-1)\,n\big]\right].$$

To compute the inverse Laplace transform of **stepeight** we define **rule** below. When applied to an object, **rule** replaces each expression

in the object of the form $e^{s\,d}s^n$ by $\dfrac{(t+d)^{1-n}}{(1-n)!}u(t+d)$. Thus, when n is negative, **rule** returns the

inverse Laplace transform of $e^{s\,d}s^n$.

In[15]:=
```
rule={Exp[s d_] s^n_->u[t+d]*
                      (t+d)^(1-n)/(1-n)!}
```
Out[15]=

$$\{E^{s\ (d\_)}\ {}_s n\_\ ->\ \frac{(d\ +\ t)^{1\ -\ n}\ u[d\ +\ t]}{(1\ -\ n)!}\}$$

We then apply **rule** to **stepeight** and name the resulting function of t, y2(t).

In[16]:=
```
y2[t_]=stepeight /. rule
```
Out[16]=

$$\frac{(-5\ +\ t)^9\ u[-5\ +\ t]}{11340}\ +\ \frac{(-4\ +\ t)^8\ u[-4\ +\ t]}{2520}\ +$$

$$\frac{(-3\ +\ t)^7\ u[-3\ +\ t]}{630}\ +\ \frac{(-2\ +\ t)^6\ u[-2\ +\ t]}{180}\ +$$

$$\frac{(-1\ +\ t)^5\ u[-1\ +\ t]}{60}\ +$$

$$\frac{(-5\ +\ t)^8\ u[-5\ +\ t]\ y[0]}{630}\ +$$

$$\frac{2\ (-4\ +\ t)^7\ u[-4\ +\ t]\ y[0]}{315}\ +$$

$$\frac{(-3\ +\ t)^6\ u[-3\ +\ t]\ y[0]}{45}\ +$$

$$\frac{(-2\ +\ t)^5\ u[-2\ +\ t]\ y[0]}{15}\ +$$

$$\frac{(-1\ +\ t)^4\ u[-1\ +\ t]\ y[0]}{6}$$

We then define $u(t) = \begin{cases} 1 \text{ if } t \geq 0 \\ 0 \text{ if } t < 0 \end{cases}$, sol $(t) = y_1(t) + y_2(t)$, and create a table, **tograph**,

which consists of **sol[t]** when **y[0]** is replaced by i for i=−2, −1, 0, 1, and 2.

In[18]:=
```
u[x_]:=1 /; x>=0
u[x_]:=0 /; x<0
```
In[19]:=
```
sol[t_]:=y1[t]+y2[t]
```
In[20]:=
```
tograph=Table[sol[t] /. y[0]->i,{i,-2,2}];
```

Finally, we graph the table of functions **tograph** on the interval [0,4].

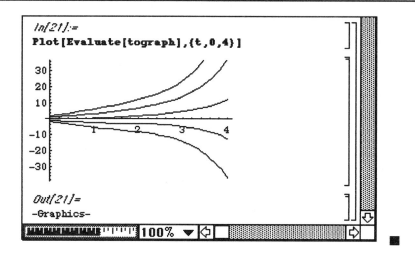

■ EXAMPLE 8.15

Let $a \neq 0$ and $b > 0$.

Solve $y'(t) + ay(t - b) = 0$ subject to the condition $y(t) = y(0)$ for $-b \leq t \leq 0$.

Solution:

Proceeding as in the previous example, we first multiply the equation $y'(t) + ay(t - b) = 0$ by

e^{-st} and integrate to obtain $e^{-st}y'(t) + ae^{-st}y(t-b) = 0$ and then

$\int_0^\infty e^{-st}y'(t)\,dt + a\int_0^\infty e^{-st}y(t-b)\,dt = 0$. Since $\int_0^\infty e^{-st}y'(t)\,dt = L\{y'\}(s)$,

$\int_0^\infty e^{-st}y'(t)\,dt = sL\{y\}(s) - y(0)$. To evaluate $a\int_0^\infty e^{-st}y(t-b)\,dt$, let $u = t - b$.

Then $du = dt$ and $a\int_0^\infty e^{-st}y(t-b)\,dt = a\int_{-b}^\infty e^{-s(u+b)}y(u)\,du$

$$= a\int_0^\infty e^{-s(u+b)}y(u)\,du + a\int_{-b}^0 e^{-s(u+b)}y(u)\,du$$

$$= ae^{-sb}\int_0^\infty e^{-su}y(u)\,du + ae^{-sb}\int_{-b}^0 e^{-su}y(u)\,du.$$

Computing $\int_{-b}^0 e^{-su}y(u)\,du$ below yields

$$a\int_0^\infty e^{-st}y(t-b)\,dt = ae^{-sb}L\{y\}(s) + ae^{-sb}\,y(0)\frac{e^{bs}-1}{s} = ae^{-sb}L\{y\}(s) + a\,y(0)\frac{1-e^{-bs}}{s}$$

and consequently $\int_0^\infty e^{-st} y'(t)\,dt + a\int_0^\infty e^{-st} y(t-b)\,dt = 0$ is equivalent to

$$sL\{y\}(s) - y(0) + ae^{-sb}L\{y\}(s) + a\,y(0)\frac{1-e^{-bs}}{s} = 0.$$

```
━━━━━━━━━━━ DelayEquation ━━━━━━━━━━━
In[1]:=
<<Calculus`LaplaceTransform`

In[2]:=
Integrate[Exp[-s u],{u,-b,0}]//Together

Out[2]=
 -1 + E^b s
 ───────────
     s
```

Let **l y** denote $L\{y\}(s)$. Then, below we solve the equation

$$s\,\mathbf{l\,y} - y(0) + ae^{-sb}\,\mathbf{l\,y} + a\,y(0)\frac{1-e^{-bs}}{s} = 0 \text{ for } \mathbf{l\,y} \text{ and name the result } \mathbf{solution}.$$

```
In[4]:=
Clear[solution,s,y,ly,a,b]
solution=Solve[s ly-y[0]+a Exp[-s b]ly+
             a y[0](1-Exp[-b s])/s==0,ly]

Out[4]=
                -(a y[0]) + a E^b s y[0] - E^b s s y[0]
{{ly -> -(────────────────────────────────────────────)}}
                        a s + E^b s s^2
```

To calculate $y(t)$ we must find the inverse Laplace transform of $-\dfrac{a\,y(0) + ae^{bs}y(0) - e^{bs}s\,y(0)}{as + e^{bs}s^2}$,

extracted from **solution** with the command **solution[[1,1,2]]**. We begin by using **Apart**

to calculate the partial fraction decomposition of $-\dfrac{a\,y(0) + ae^{bs}y(0) - e^{bs}s\,y(0)}{as + e^{bs}s^2}$ and naming

the resulting output **steptwo**.

```
In[5]:=
steptwo=Apart[solution[[1,1,2]]]

Out[5]=
 (-a + s) y[0]        a^2 y[0]
 ─────────────  +  ─────────────────
      s^2            s^2 (a + E^b s s)
```

Then we must compute the inverse Laplace transform of both $\dfrac{(-a+s)y(0)}{s^2}$, extracted from

steptwo with **steptwo[[1]]** , and $\dfrac{a^2y(0)}{s^2\left(a+e^{bs}s\right)}$, extracted from **steptwo** with

steptwo[[2]]. We begin by using **InverseLaplaceTransform** to compute the inverse

Laplace transform of $\dfrac{(-a+s)y(0)}{s^2}$ and defining the resulting function to be $y_1(t)$.

```
In[6]:=
stepthree=steptwo[[1]]
Out[6]=
(-a + s) y[0]
─────────────
      2
     s
In[7]:=
y1[t_]=InverseLaplaceTransform[stepthree,s,t]
Out[7]=
y[0] - a t y[0]
```

To compute the inverse Laplace transform of $\dfrac{a^2y(0)}{s^2\left(a+e^{bs}s\right)}$, we write

$$\frac{a^2y(0)}{s^2\left(a+e^{bs}s\right)}=\frac{a^2y(0)}{s^3e^{bs}}\frac{1}{1-\left(-a\,s^{-1}e^{-bs}\right)}=\frac{a^2y(0)}{s^3e^{bs}}\sum_{n=0}^{\infty}\left(-a\,s^{-1}e^{-bs}\right)^n$$

$$=\sum_{n=0}^{\infty}(-1)^n\,a^{n+2}y(0)\frac{e^{-(n+1)bs}}{s^{n+3}}.$$

Since the inverse Laplace transform of $\dfrac{e^{-at}}{s^n}$ (n a positive integer) is $\dfrac{(t-a)^{n-1}}{(n-1)!}u(t-a)$, the

inverse Laplace transform of $(-1)^n\,a^{n+2}y(0)\dfrac{e^{-(n+1)bs}}{s^{n+3}}$ is

$(-1)^n\,a^{n+2}y(0)\dfrac{(t-(n+1)bs)^{n+2}}{(n+2)!}u\left(t-(n+1)bs\right)$ and the inverse Laplace transform of

$$\frac{a^2y(0)}{s^2\left(a+e^{bs}s\right)}=\sum_{n=0}^{\infty}(-1)^n\,a^{n+2}y(0)\frac{e^{-(n+1)bs}}{s^{n+3}}\ \text{is}$$

$$y_2(t) = \sum_{n=0}^{\infty} (-1)^n \, a^{n+2} y(0) \frac{(t - (n+1)bs)^{n+2}}{(n+2)!} \, u(t - (n+1)bs).$$

Then $y(t) = y_1(t) + y_2(t) = y(0) - a\,y(0)t + \displaystyle\sum_{n=0}^{\infty} (-1)^n \, a^{n+2} y(0) \frac{(t - (n+1)bs)^{n+2}}{(n+2)!} \, u(t - (n+1)bs).$

We compute the first few terms of this series by defining $\mathbf{term} = \dfrac{-a}{s\,e^{bs}}$ and then $\mathbf{geoseries}$

to be the power series for $\dfrac{1}{1-x}$ about $x = 0$. The first five terms are displayed.

```
In[8]:=
term=-a/(s Exp[b s])
Out[8]=
   a
-(----)
  E^b s  s
In[9]:=
geoseries=Series[1/(1-x),{x,0,5}]
Out[9]=
1 + x + x^2 + x^3 + x^4 + x^5 + O[x]^6
```

We then replace each **x** in **geoseries** by **term**, name the result **stepfour**, and then remove the O-term from **stepfour** which represents the omitted higher order terms of the series and name the resulting output **stepfive**.

```
In[10]:=
stepfour=geoseries /. x->term
Out[10]=
       a             a              a
1 - ------ + (-(------))^2 + (-(------))^3 +
    E^b s  s     E^b s  s        E^b s  s

       a               a                 a
  (-(------))^4 + (-(------))^5 + O[-(------)]^6
    E^b s  s         E^b s  s          E^b s  s
In[11]:=
stepfive=stepfour//Normal
Out[11]=
           a^5              a^4              a^3
1 - --------------- + --------------- - --------------- +
    E^5 b s^5 s^5      E^4 b s^4 s^4     E^3 b s^3 s^3

         a^2              a
    --------------- - ------
    E^2 b s^2 s^2      E^b s  s
```

We then define **s t e p s i x** to be the numerator of $\dfrac{a^2 y(0)}{s^2\left(a + e^{bs}s\right)}$:

```
In[12]:=
stepsix=Numerator[steptwo[[2]]]

Out[12]=
 2
a  y[0]
```

and **stepseven** to be $\dfrac{1}{s^3 e^{bs}}$ **stepfive stepsix**

```
In[13]:=
stepseven=Expand[
    1/(s^3 Exp[b s]) stepfive stepsix]

Out[13]=
    7                6            5
   a  y[0]          a  y[0]      a  y[0]
-(-------) + --------- - --------- +
  E^6 b s  8   E^5 b s  7   E^4 b s  6
         s            s            s

    4                3            2
   a  y[0]          a  y[0]      a  y[0]
 --------- - --------- + ---------
  E^3 b s  5   E^2 b s  4   E^b s  3
         s            s           s
```

As in the previous example, we define **rule** so then when **rule** is applied to an object, each element of the object of the form $e^{s d} s^n$ is replaced by $\dfrac{(t+d)^{1-n}}{(1-n)!} u(t+d)$. Thus, when n is negative, **rule** returns the inverse Laplace transform of $e^{s d} s^n$.

```
In[14]:=
rule={Exp[s d_] s^n_->u[t+d]*
                    (t+d)^(1-n)/(1-n)!}

Out[14]=
         n      (d + t)^1 - n
{E^s (d_) s_ -> ------------- u[d + t]}
                  (1 - n)!
```

We then apply rule to **stepseven** and name the resulting function $y_2(t)$:

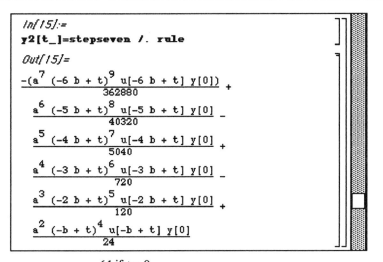

In[15]:=
```
y2[t_]=stepseven /. rule
```
Out[15]=

$$-\frac{(a^7 \ (-6\ b + t)^9 \ u[-6\ b + t]\ y[0])}{362880} +$$

$$\frac{a^6 \ (-5\ b + t)^8 \ u[-5\ b + t]\ y[0]}{40320} -$$

$$\frac{a^5 \ (-4\ b + t)^7 \ u[-4\ b + t]\ y[0]}{5040} +$$

$$\frac{a^4 \ (-3\ b + t)^6 \ u[-3\ b + t]\ y[0]}{720} -$$

$$\frac{a^3 \ (-2\ b + t)^5 \ u[-2\ b + t]\ y[0]}{120} +$$

$$\frac{a^2 \ (-b + t)^4 \ u[-b + t]\ y[0]}{24}$$

then define $u(t) = \begin{cases} 1 \text{ if } t \ge 0 \\ 0 \text{ if } t < 0 \end{cases}$ and $\mathrm{sol}(t) = y_1(t) + y_2(t)$, which corresponds to the first few terms of

the series for y(t).

In[17]:=
```
u[x_]:=1 /; x>=0
u[x_]:=0 /; x<0
```
In[18]:=
```
sol[t_]:=y1[t]+y2[t]
```

We graph various solutions corresponding to different values of a, b, and y(0). In the first case, we let a=2, b=1/2, and then graph sol(t) for y=-2, -1, 0, 1, and 2. We name the resulting graph **plotone** and do not display the result.

In[38]:=
```
a=2;
b=1/2;
tograph1=Table[sol[t] /. y[0]->i,
          {i,-2,2}];
plotone=Plot[Evaluate[tograph1],{t,0,4},
          DisplayFunction->Identity];
```

Similarly, we create a graph **plottwo** when a=-2 and b=1/2:

```
In[42]:=
a=-2;
b=1/2;
tograph2=Table[sol[t] /. y[0]->i,
            {i,-2,2}];
plottwo=Plot[Evaluate[tograph2],{t,0,4},
            DisplayFunction->Identity];
```

a graph **plotthree** when a=–4 and b=3/2:

```
In[55]:=
a=-4;
b=3/2;
tograph3=Table[sol[t] /. y[0]->i,
            {i,-2,2}];
plotthree=Plot[Evaluate[tograph3],{t,0,4},
            DisplayFunction->Identity];
```

and a graph **plotfour** when a=4 and b=3/2.

```
In[59]:=
a=4;
b=3/2;
tograph4=Table[sol[t] /. y[0]->i,
            {i,-2,2}];
plotfour=Plot[Evaluate[tograph4],{t,0,4},
            DisplayFunction->Identity];
```

All four graphs are finally displayed as a graphics array.

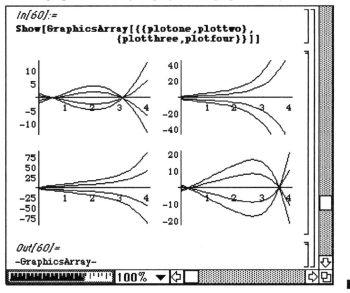

```
In[60]:=
Show[GraphicsArray[{{plotone,plottwo},
                {plotthree,plotfour}}]]
```

```
Out[60]=
-GraphicsArray-
```

In some cases, Laplace transforms can be used to solve equations with non-constant coefficients.

☐ **EXAMPLE 8.16**

Solve $tf''(t) + (8t + 1)f'(t) + (15t + 1)f(t) = 0$ subject to $f(0) = 1$ and $f'(0) = 0$.

Solution:

Let F(s) denote the Laplace transform of f(t). Then, using **Table 8.2** the Laplace transform of

$tf(t)$ is $(-1)F'(s)$, the Laplace transform of $tf'(t)$ is $(-1)\dfrac{d}{ds}(sF(s) - f(0)) = -(sF'(s) + F(s))$, and the

Laplace transform of $tf''(t)$ is $(-1)\dfrac{d}{ds}(s^2F(s) - sf(0) - f'(0)) = -(s^2F'(s) + 2sF(s) + f(0))$. This is

computed below with *Mathematica* letting **capf[s]** represent F(s).

```
                    LaplaceTransform

In[19]:=
Clear[capf,s,f]

In[20]:=
(-1)D[s capf[s]-f[0],s]

Out[20]=
-capf[s] - s capf'[s]

In[21]:=
(-1)D[s^2 capf[s]-s f[0]-f'[0],s]

Out[21]=
                       2
-2 s capf[s] + f[0] - s  capf'[s]
```
                                        100%

Computing the Laplace transform of each member of the differential equation

$tf''(t) + (8t + 1)f'(t) + (15t + 1)f(t) = 0$ and replacing f(0) by 1 and f'(0) by 0 yields the first - order

differential equation $\left(-s^2F'(s) - 2sF(s) + 1\right) - 8(sF'(s) + F(s)) + (sF(s) - 1) + 15F'(s) + F(s) = 0$.

To solve the problem, we must compute F(s) and then compute the inverse Laplace transform of F(s) to
obtain f(t).

Regrouping terms results in $\left(-15 - 8s - s^2\right)F'(s) - (s + 7)F(s) = 0$ which we see is separable. Separating

variables transforms this equation into $\dfrac{dF}{F} = \dfrac{(s + 7)\,ds}{-15 - 8s - s^2}$ which, computing the partial fraction decomposition of $\dfrac{s+7}{-15-8s-s^2}$, is equivalent to the equation $\dfrac{dF}{F} = \left(\dfrac{1}{s+5} - \dfrac{2}{s+3} \right) ds$. Integrating both

sides of the equation results in $\text{Ln}|F(s)| = \text{Ln}|s+5| - 2\text{Ln}|s+3| + c_1 = c_1 + \text{Ln}\left|\dfrac{s+5}{(s+3)^2}\right|$ so

$F(s) = c\dfrac{s+5}{(s+3)^2}$, where c_1 represents an arbitrary constant and $c = e^{c_1}$. Computing the partial

fraction decomposition of $\dfrac{s+5}{(s+3)^2}$ yields $F(s) = c\left(\dfrac{1}{s+3} + \dfrac{2}{(s+3)^2} \right)$. The inverse Laplace transform

of $\dfrac{1}{s+3}$ is e^{-3t} and the inverse Laplace transform of $\dfrac{2}{(s+3)^2}$ is $2te^{-3t}$ so the inverse Laplace transform

of $F(s)$ is $f(t) = c(1+2t)e^{-3t}$. Since $f(0) = 1$ and $f(0) = c$, we obtain that $c = 1$. Therefore, the general

solution is $f(t) = (1+2t)e^{-3t}$. ∎

■ EXAMPLE 8.17

Solve $tf''(t) + (11t+1)f'(t) + (30t+2)f(t) = 0$ subject to $f(0) = 1$ and $f'(0) = -2$.

Solution:

We begin by entering the initial conditions and the left-hand side of the differential equation, called **lhs**.

```
≡□≡≡≡≡≡≡≡≡ LaplaceTransform ≡≡≡≡≡≡≡⊡≡
In[5]:=
f[0]=1;
f'[0]=-2;
In[6]:=
lhs=t f''[t]+(11t+1)f'[t]+(30t+2)f[t]
Out[6]=
(2 + 30 t) f[t] + (1 + 11 t) f'[t] + t f''[t]
```

The appropriate rules needed to take the Laplace transform of **lhs** are given in **laplacerule** below.

```
In[7]:=
laplacerule={
t f''[t]->-(2s lf[s]-f[0]+s^2 lf'[s]),
    f''[t]->s^2 lf[s]-s f[0]-f'[0],
        t f'[t]->-(lf[s]+s lf'[s]),
            f'[t]->s lf[s]-f[0],
                t f[t]->-lf'[s],
                    f[t]->lf[s]};
```

In **stepone**, the left-hand side of the equation is expanded and **laplacerule** applied. Then, the terms which contain **lf'[s]** and **lf[s]** are collected.

```
In[8]:=
stepone=Expand[lhs]/. laplacerule

Out[8]=
2 lf[s] - s lf[s] - 30 lf'[s] - s  lf'[s] +
   11 (-lf[s] - s lf'[s])

In[9]:=
Collect[stepone,{lf'[s],lf[s]}]

Out[9]=
(-9 - s) lf[s] + (-30 - 11 s - s ) lf'[s]
```

Since this is a separable equation, we have

$$\frac{\text{lf}'[s]}{\text{lf}[s]} = \frac{9+s}{-30-11s-s^2}$$

the right-hand side of which is integrated below in **steptwo**. The result in **steptwo** is then simplified in **stepthree** to yield **lf[s]**. (Recall that the integral of the left-hand side of the above equation is **Log[lf[s]]**.)

```
In[10]:=
steptwo=Integrate[(9+s)/(-30-11s-s^2),s]

Out[10]=
-4 Log[5 + s] + 3 Log[6 + s]

In[14]:=
stepthree=Exp[steptwo]//Expand

Out[14]=
       3
(6 + s)
--------
       4
(5 + s)
```

In an attempt to compute the inverse Laplace transform of the result in **stepthree**, we produce the partial fraction decomposition of this result below with **Apart** and name the output **stepfour**.

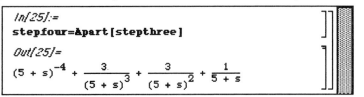

```
In[25]:=
stepfour=Apart[stepthree]

Out[25]=
          -4      3.          3         1
(5 + s)     + -------- + -------- + -----
                     3          2    5 + s
              (5 + s)    (5 + s)
```

The inverse Laplace transform is then computed with the appropriate command found in the

LaplaceTransform package and called **sol**.

```
In[25]:=
<<Calculus`LaplaceTransform`

In[27]:=
sol[t_]=
     InverseLaplaceTransform[stepfour,s,t]

Out[27]=
```
$$E^{-5\,t} + \frac{3\,t}{E^{5\,t}} + \frac{3\,t^2}{2\,E^{5\,t}} + \frac{t^3}{6\,E^{5\,t}}$$

Finally, the solution is plotted on the interval [−1,1].

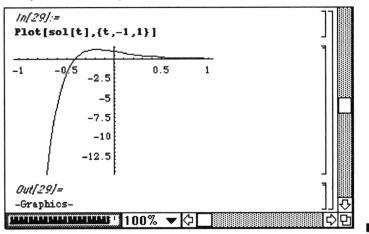

```
In[29]:=
Plot[sol[t],{t,-1,1}]
```

```
Out[29]=
-Graphics-
```

Chapter 9: Applications of the Laplace Transform

Mathematica commands used in **Chapter 9** include:

Clear	If	PlotRange
Collect	Integrate	PlotStyle
Dashing	InverseLaplaceTrans	ReplaceAll
Delta	form	Show
DisplayFunction	LaplaceTransform	Simplify
DSolve	NIntegrate	Solve
Expand	Part	Table
GraphicsArray	Partition	
GrayLevel	Plot	

§9.1 Spring-Mass Systems Revisited

Laplace transforms are useful in solving the spring-mass systems which were discussed in earlier sections. Although the method of Laplace transforms can be used to solve all problems discussed in the section on applications of higher order equations, this method is most useful in alleviating the difficulties associated with problems of this type which involve piecewise-defined forcing functions. Hence, we investigate the use of Laplace transforms to solve the second order initial value problem which models the motion of a mass attached to the end of a spring:

$$m x'' + c x' + k x = f(t)$$

$$x(0) = x_0, \ x'(0) = v_0$$

where m represents the mass, c the damping coefficient, and k the spring constant determined by Hooke's law.

We demonstrate below how the method of Laplace transforms is used to solve this type of initial value problem in which the forcing function is continuous.

■ EXAMPLE 9.1

An object with mass m = 4 is attached to the end of a spring with spring constant k=1. Determine the motion of the mass if c=2, x(0) = −1, x'(0)=1, and the forcing function f(t) is given by:

(a) f(t) = 0

(b) f(t) = 1

(c) f(t) = −1.

Solution:

After loading the **LaplaceTransform** package, we define the left-hand side of the equation in **ode**. As in previous examples, we also define the rules needed to apply the Laplace transform to the differential equation. This is done in **rule**.

```
≣□≣▬▬▬▬▬▬▬▬▬▬ SpringsAgain ▬▬▬▬▬▬▬▬▬▬ ⊡≣
In[1]:=
<<Calculus`LaplaceTransform`
In[2]:=
ode=4 x''[t]+2 x'[t]+x[t];
In[3]:=
rule={x[t]->lx,x'[t]->s lx -x[0],
      x''[t]->s^2 lx -s x[0]-x'[0]};
```

We apply the Laplace transform in **eqx** and determine the Laplace transform of the solution of this homogeneous equation, **lx**, in **soln**.

```
In[4]:=
eqx=ode/.rule
Out[4]=
                         2
lx + 2 (lx s - x[0]) + 4 (lx s  - s x[0] - x'[0])
In[5]:=
soln=Solve[eqx==0,lx]
Out[5]=
             -2 x[0] - 4 s x[0] - 4 x'[0]
{{lx -> -(----------------------------)}}
                              2
                  1 + 2 s + 4 s
```

The initial conditions are applied in **conds**

```
In[6]:=
conds=soln/.{x[0]->-1,x'[0]->1}
Out[6]=
              -2 + 4 s
{{lx -> -(--------------)}}
                        2
            1 + 2 s + 4 s
```

so that the inverse Laplace transform of this result may be found. This is done as follows by completing the square in the denominator.

$$lx = L\{x\} = \frac{-4s+2}{4s^2+2s+1} = -\frac{s-\frac{1}{2}}{s^2+\frac{s}{2}+\frac{1}{4}} = -\frac{s-\frac{1}{2}}{\left(s+\frac{1}{4}\right)^2+\frac{3}{16}} = -\frac{s+\frac{1}{4}}{\left(s+\frac{1}{4}\right)^2+\frac{3}{16}}+\frac{\frac{3}{4}}{\left(s+\frac{1}{4}\right)^2+\frac{3}{16}}.$$

Hence, we apply the inverse Laplace transform to yield

$$x(t) = -e^{-t/4}\left[\cos\left(\frac{\sqrt{3}}{4}t\right)+\sqrt{3}\,\sin\left(\frac{\sqrt{3}}{4}t\right)\right]$$ which is defined below.

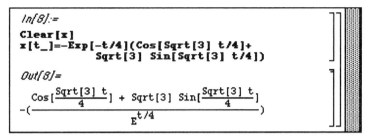

```
In[8]:=
Clear[x]
x[t_]=-Exp[-t/4](Cos[Sqrt[3] t/4]+
            Sqrt[3] Sin[Sqrt[3] t/4])
Out[8]=
```

$$-(\frac{Cos[\frac{Sqrt[3]\ t}{4}]\ +\ Sqrt[3]\ Sin[\frac{Sqrt[3]\ t}{4}]}{E^{t/4}})$$

The solution is plotted below to show that the mass eventually returns to its equilibrium position, x = 0.

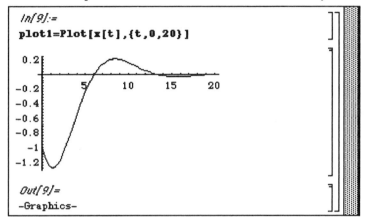

```
In[9]:=
plot1=Plot[x[t],{t,0,20}]
```

```
Out[9]=
-Graphics-
```

We follow the same procedure to solve the problem in part (b) with the exception that the Laplace transform of the forcing function is determined in **1pt** below.

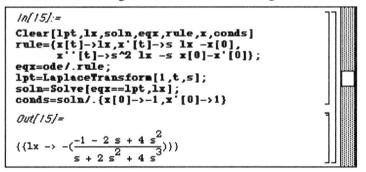

```
In[15]:=
Clear[lpt,lx,soln,eqx,rule,x,conds]
rule={x[t]->lx,x'[t]->s lx -x[0],
        x''[t]->s^2 lx -s x[0]-x'[0]};
eqx=ode/.rule;
lpt=LaplaceTransform[1,t,s];
soln=Solve[eqx==lpt,lx];
conds=soln/.{x[0]->-1,x'[0]->1}
Out[15]=
```

$$\{\{lx\ ->\ -(\frac{-1\ -\ 2\ s\ +\ 4\ s^2}{s\ +\ 2\ s^2\ +\ 4\ s^3})\}\}$$

In this case, we apply **Apart** to the result above to assist in determining the inverse Laplace transform.

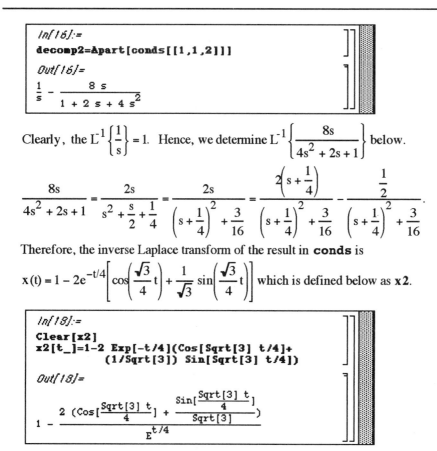

```
In[16]:=
decomp2=Apart[conds[[1,1,2]]]
Out[16]=
1      8 s
─  - ─────────────
s    1 + 2 s + 4 s²
```

Clearly, the $L^{-1}\left\{\dfrac{1}{s}\right\} = 1$. Hence, we determine $L^{-1}\left\{\dfrac{8s}{4s^2 + 2s + 1}\right\}$ below.

$$\frac{8s}{4s^2 + 2s + 1} = \frac{2s}{s^2 + \dfrac{s}{2} + \dfrac{1}{4}} = \frac{2s}{\left(s + \dfrac{1}{4}\right)^2 + \dfrac{3}{16}} = \frac{2\left(s + \dfrac{1}{4}\right)}{\left(s + \dfrac{1}{4}\right)^2 + \dfrac{3}{16}} - \frac{\dfrac{1}{2}}{\left(s + \dfrac{1}{4}\right)^2 + \dfrac{3}{16}}.$$

Therefore, the inverse Laplace transform of the result in **conds** is

$$x(t) = 1 - 2e^{-t/4}\left[\cos\left(\frac{\sqrt{3}}{4}t\right) + \frac{1}{\sqrt{3}}\sin\left(\frac{\sqrt{3}}{4}t\right)\right]$$ which is defined below as **x2**.

```
In[18]:=
Clear[x2]
x2[t_]=1-2 Exp[-t/4](Cos[Sqrt[3] t/4]+
        (1/Sqrt[3]) Sin[Sqrt[3] t/4])
Out[18]=
                          Sqrt[3] t
             Sqrt[3] t  Sin[─────────]
     2 (Cos[─────────] +       4
               4          ─────────────)
1 -                         Sqrt[3]
    ───────────────────────────────────
                  E^{t/4}
```

We plot the solution to the nonhomogeneous equation in **plot2** and then display it with the solution to the corresponding homogeneous equation. Note that the forcing function causes the solution to approach the value of the constant forcing function.

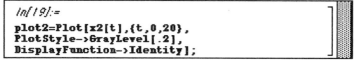

```
In[19]:=
plot2=Plot[x2[t],{t,0,20},
PlotStyle->GrayLevel[.2],
DisplayFunction->Identity];
```

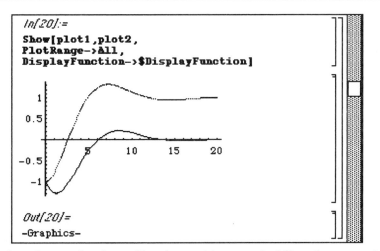

```
In[20]:=
Show[plot1,plot2,
PlotRange->All,
DisplayFunction->$DisplayFunction]
```

```
Out[20]=
-Graphics-
```

In order to solve the problem in part (c), we again follow the same solution procedure. In this case, the only difference is that the forcing function is f(t)=−1. This affects the outcome of the value of **lx** given below.

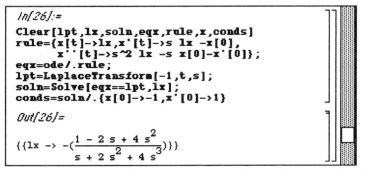

```
In[26]:=
Clear[lpt,lx,soln,eqx,rule,x,conds]
rule={x[t]->lx,x'[t]->s lx -x[0],
      x''[t]->s^2 lx -s x[0]-x'[0]};
eqx=ode/.rule;
lpt=LaplaceTransform[-1,t,s];
soln=Solve[eqx==lpt,lx];
conds=soln/.{x[0]->-1,x'[0]->1}
```

$$Out[26]=$$

$$\{\{lx \;\rightarrow\; -(\frac{1 - 2\,s + 4\,s^2}{s + 2\,s^2 + 4\,s^3})\}\}$$

We again use **Apart** to determine the partial fraction decomposition of **lx**, given below in **decomp3**.

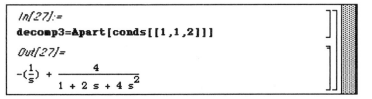

```
In[27]:=
decomp3=Apart[conds[[1,1,2]]]
```

$$Out[27]=$$

$$-(\frac{1}{s}) \;+\; \frac{4}{1 + 2\,s + 4\,s^2}$$

Hence, the inverse Laplace transform is applied in a manner similar to that in parts (a) and (b) to obtain the solution, called **x3**, which is defined below.

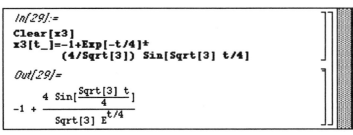

This solution is plotted below in **plot3** and then displayed with the solutions in parts (a) and (b). Notice that the forcing function f(t)=−1 causes the mass to approach the value of x =−1 as t increases.

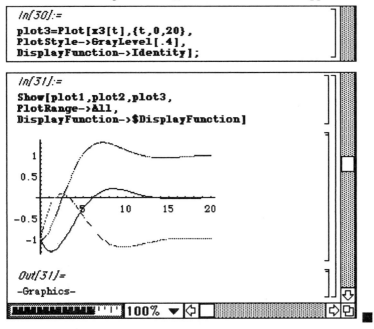

We now consider an initial value problem which involves a discontinuous forcing function.

■ EXAMPLE 9.2

Solve the initial value problem involving no damping

$$x''+x = \begin{cases} \sin t, & 0 \le t < \dfrac{\pi}{2} \\ 0, & t \ge \dfrac{\pi}{2} \end{cases}, \quad x(0) = 0, \ x'(0) = 0.$$

Solution:

We begin by loading the **LaplaceTransform** package and defining the unit step function **u**.

Note that u represents the function $u(t - a) = \begin{cases} 0, t < a \\ 1, t \geq a \end{cases}$ which is useful in defining discontinuous

functions. For example, a function such as $g(t) = \begin{cases} g_1(t), a < t < b \\ 0, \text{otherwise} \end{cases}$ can be rewritten with u as

$g(t) = g_1(t)[u(t - a) - u(t - b)]$.

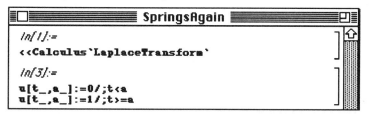

We then define the forcing function using **u** as indicated above and plot this function which is not continuous at t=π/2. Notice that if we only consider positive values of t, u(t−0)=1. Hence, 1 is used in some of the subsequent calculations involving **f** to represent u(t−0).

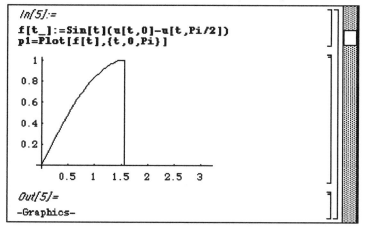

To assist in the determination of the Laplace transform of functions involving the unit step function, we define the function **lpt** with arguments **g** and **a** which computes

$L\{g(t)u(t - a)\} = e^{-as}L\{g(t + a)\}$.

We illustrate its use below with the function sin(t) u(t−π/2).

```
In[6]:=
lpt[g_,a_]:=Exp[-a s]*
              LaplaceTransform[g/.t->t+a,t,s]
In[7]:=
lpt[Sin[t],Pi/2]
Out[7]=
        s
──────────────────
E^(Pi s)/2 (1 + s^2)
```

In order to compute the inverse Laplace transform of functions involving the exponential function, we define **invlpt** below which computes

$$L^{-1}\left\{e^{-as}G(s)\right\} = g(t-a)u(t-a).$$

We illustrate **invlpt** below by using the output of the above example of **lpt**. Note that we simply use the component of *Out[7]* which represents G(s) in the above formula and enter the value of a which in this case is $\pi/2$.

```
In[8]:=
invlpt[lpg_,a_]:=
(InverseLaplaceTransform[lpg,s,t]/.t->t-a)*
               u[t,a]
In[9]:=
invlpt[s/(s^2+1),Pi/2]
Out[9]=
Sin[t] u[t, Pi/2]
```

We are now able to begin solving the problem by, first, entering the left-hand side of the equation.

```
In[10]:=
eq1=x''[t]+x[t];
```

We also define the necessary rules for applying the Laplace transform to the differential equation and apply the rules to the left-hand side of the differential equation in **eqx**.

```
In[11]:=
rule={x[t]->lx,x'[t]->s lx -x[0],
      x''[t]->s^2 lx -s x[0]-x'[0]};
In[12]:=
eqx=eq1/.rule
Out[12]=
lx + lx s^2 - s x[0] - x'[0]
```

The initial conditions are applied below in **ics**.

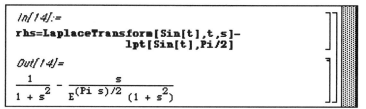

The Laplace transform is applied to the right-hand side function by using **LaplaceTransform** and **lpt** for the component involving the unit step function.

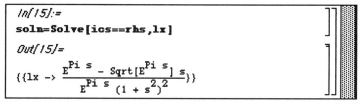

We next equate the Laplace transform of the right- and left-hand sides of the equation and solve for **lx**. Observation of the resulting expression shows that the result can be divided into a component which involves the exponential function and one which does not. We, therefore, use **InverseLaplaceTransform** with the component which does not depend on the exponential function and **invlpt** with the component which depends on the exponential function. The solution is defined as **x** below.

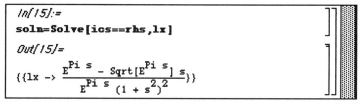

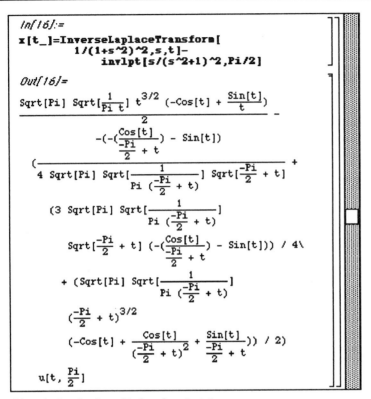

This solution is plotted below in **plot4**.

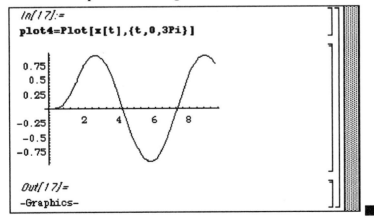

■ EXAMPLE 9.3

Solve the following initial value problem which involves damping in order to note the differences with the previous example:

x"+4x'+13x = f(t), x(0) = 0, x'(0) = 0

where the forcing function is given by: $f(t) = \begin{cases} \sin t, & 0 \le t < \frac{\pi}{2} \\ 0 & , & t \ge \frac{\pi}{2} \end{cases}$.

Solution:

This problem is solved in the same manner as the previous example. In **eq2**, we define the left-hand side of the differential equation along with the transformation rules.

```
SpringsAgain

In[24]:=
Clear[x]

In[25]:=
eq2=x''[t]+4x'[t]+13x[t];

In[26]:=
rule={x[t]->lx,x'[t]->s lx -x[0],
        x''[t]->s^2 lx -s x[0]-x'[0]};
```

We apply the Laplace transform to the left-hand side of the equation and apply the initial conditions to the expression which results.

```
In[27]:=
eqx=eq2/.rule

Out[27]=
               2
13 lx + lx s  + 4 (lx s - x[0]) - s x[0] - x'[0]

In[28]:=
ics=eqx/.{x[0]->0,x'[0]->0}

Out[28]=
                              2
13 lx + 4 lx s + lx s
```

We then use **LaplaceTransform** and **lpt** which was defined in the previous example to compute the Laplace transform of the forcing function. We refer to this result as **rhs** below.

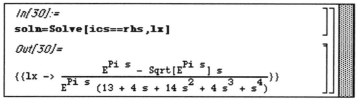

We then equate the Laplace transform of the left- and right-hand sides of the differential equation and solve for the Laplace transform of the solution, **lx**.

In[30]:=
```
soln=Solve[ics==rhs,lx]
```
Out[30]=

$$\{\{lx \rightarrow \frac{E^{Pi\ s} - Sqrt[E^{Pi\ s}]\ s}{E^{Pi\ s}\ (13 + 4\ s + 14\ s^2 + 4\ s^3 + s^4)}\}\}$$

Simplifying this expression, we see that one component involves the exponential function while the other does not. We, therefore, use both InverseLaplaceTransform and invlpt to determine the solution which is called **xd** below.

In[31]:=
```
xd[t_]=InverseLaplaceTransform[
           1/(13+4s+14s^2+4s^3+
           s^4),s,t]-
                 invlpt[s/(13+4s+14s^2+4s^3+
                 s^4),Pi/2]
```
Out[31]=

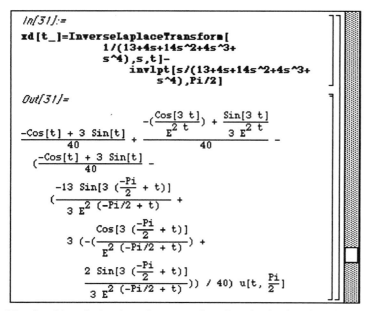

We plot this solution in order to see that after the forcing function becomes zero, the mass returns to its equilibrium position, x=0.

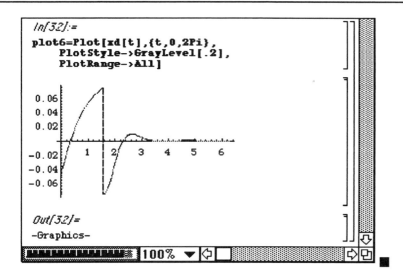

```
In[32]:=
plot6=Plot[xd[t],{t,0,2Pi},
    PlotStyle->GrayLevel[.2],
    PlotRange->All]
```

```
Out[32]=
-Graphics-
```

§9.2 L-R-C Circuits Revisited

Laplace transforms can be used to solve the L-R-C circuits problems which were introduced earlier. Recall that the initial value problem which models this situation is

$$L\frac{d^2q}{dt^2} + R\frac{dq}{dt} + \frac{1}{C}q = v(t)$$

$$q(0) = q_0, \quad q'(0) = i_0,$$

where $\dfrac{dq}{dt} = i$, $q(t) =$ charge, $i(t) =$ current, $L =$ inductance, $R =$ resistance, $C =$ capacitance, and $v(t) =$ voltage supply.

In particular, this method is most useful when the supplied voltage $v(t)$ is piecewise-defined. This is demonstrated through the following example.

■ EXAMPLE 9.4

Suppose that we consider a circuit with a capacitor C, a resistor R, and a voltage supply

$$v(t) = v_0\left[u(t-0) - u(t-1)\right] = v_0\left[1 - u(t-1)\right].$$

Determine the current $i(t)$ in the circuit as a function of q_0, v_0, C, R, and t. Plot solutions for various values of q_0 and v_0. (Note that in this case, L=0.)

Solution:

Using the differential equation which was derived earlier for circuits, we solve the following initial value problem:

$$R\frac{dq}{dt} + \frac{q}{C} = v(t), q(0) = q_0.$$

First, we load the **LaplaceTransform** package which is found in the **Calculus** folder (or directory). Then, the unit step function **u** and the voltage supply function **f** which depends on **u** are defined.

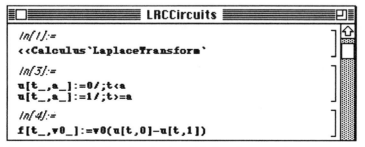

The voltage supply function is plotted below for $v_0=1$ in **plot1**.

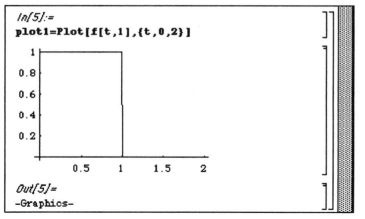

In order to compute the Laplace transform of functions of the form g(t) u(t−a), we define the function **lpt** below. This function properly applies the property

$$L\{g(t)u(t-a)\} = e^{-as}L\{g(t+a)\}$$

when given the function g and the value a. We also need to compute the inverse Laplace transform of functions which involve the exponential function. Hence, we define **invlpt** which computes

$$L^{-1}\{e^{-as}G(s)\} = g(t-a)u(t-a)$$

when G(s)=L{g(t)} and a are given. After these functions are defined, the problem can be solved with Laplace transforms. The differential equation is entered in **eq1** as are the appropriate transformation rules in **rule**.

```
In[6]:=
lpt[g_,a_]:=Exp[-a s]+
            LaplaceTransform[g/.t->t+a,t,s]
In[7]:=
invlpt[lpg_,a_]:=
(InverseLaplaceTransform[
            lpg,s,t]/.t->t-a) u[t,a]
In[8]:=
eq1=r q'[t]+(1/c) q[t];
In[9]:=
rule={q[t]->lq,q'[t]->s lq -q[0]};
```

The transformation rules are applied to the differential equation in **eqq** and the value of **q0** is substituted for q[0] in **ics** for convenience.

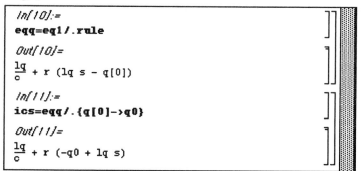

```
In[10]:=
eqq=eq1/.rule
Out[10]=
lq
-- + r (lq s - q[0])
c
In[11]:=
ics=eqq/.{q[0]->q0}
Out[11]=
lq
-- + r (-q0 + lq s)
c
```

The Laplace transform of the right-hand side of the differential equation is computed in **rhs**. Notice that we combine the use of the built-in command **LaplaceTransform** and the user-defined function **lpt** in this computation.

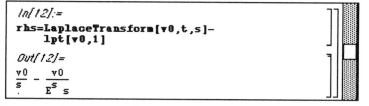

```
In[12]:=
rhs=LaplaceTransform[v0,t,s]-
    lpt[v0,1]
Out[12]=
v0   v0
-- - ----
s     s
     E  s
```

Below, the Laplace transform of the left- and right-hand sides of the differential equation are set equal to one another so that **lq** which represents L{q(t)} can be determined.

In[13]:=
```
solnq=Solve[ics==rhs,lq]
```
Out[13]=
$$\{\{lq \rightarrow -(\frac{-(c\ E^s\ q0\ r\ s)\ +\ c\ v0\ -\ c\ E^s\ v0}{E^s\ s\ +\ c\ E^s\ r\ s^2})\}\}$$

Notice that when simplified, only some components of **lq** involve the exponential function. Because of this, we use the command **InverseLaplaceTransform** to determine the inverse Laplace transform of the portion which does not include the exponential function and **invplt** to determine that of the component which does.

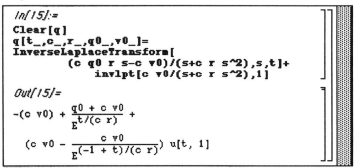

In[15]:=
```
Clear[q]
q[t_,c_,r_,q0_,v0_]=
InverseLaplaceTransform[
        (c q0 r s-c v0)/(s+c r s^2),s,t]+
            invplt[c v0/(s+c r s^2),1]
```
Out[15]=
$$-(c\ v0)\ +\ \frac{q0\ +\ c\ v0}{E^{t/(c\ r)}}\ +$$
$$(c\ v0\ -\ \frac{c\ v0}{E^{(-1\ +\ t)/(c\ r)}})\ u[t,\ 1]$$

The charge q(t) is plotted below for values of c=1, r=1, q_0=1, and v_0=1.

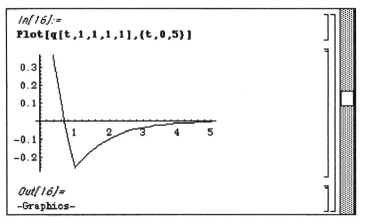

In[16]:=
```
Plot[q[t,1,1,1,1],{t,0,5}]
```

Out[16]=
```
-Graphics-
```

We then plot the solution for values of q_0=0 to q_0=2.5 using increments of 0.5. This is done by defining the function **w** which plots q(t) for the parameters c=1, r=1, q_0=i, and v_0=1. A table of plots is then produced in **wtable** and then displayed as a graphics array after partitioning **wtable** into groups of two in **qlist**.

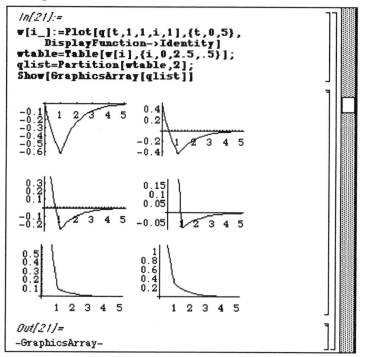

```
In[21]:=
v[i_]:=Plot[q[t,1,1,i,1],{t,0,5},
    DisplayFunction->Identity]
vtable=Table[v[i],{i,0,2.5,.5}];
qlist=Partition[vtable,2];
Show[GraphicsArray[qlist]]
```

```
Out[21]=
-GraphicsArray-
```

Solutions for various values of v_0 are graphed below by defining the function **v** which plots the solution using c=1, r=1, q_0=1, and v_0=i. A table of plots for v_0=0 to v_0=10 using increments of two is then given in **vtable** and displayed as a graphics array.

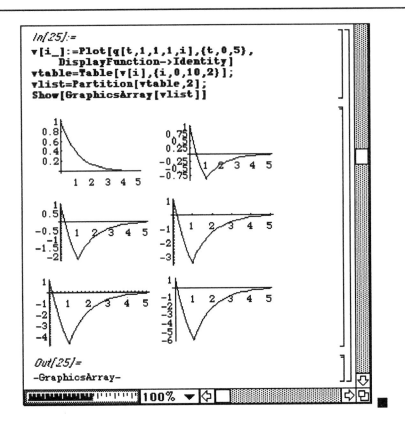

```
In[25]:=
v[i_]:=Plot[q[t,1,1,1,i],{t,0,5},
    DisplayFunction->Identity]
vtable=Table[v[i],{i,0,10,2}];
vlist=Partition[vtable,2];
Show[GraphicsArray[vlist]]
```

```
Out[25]=
-GraphicsArray-
```

100%

§9.3 Population Problems Revisited

Laplace transforms can used to solve the population problems which were discussed as applications of first-order equations. In this case, however, we focus our attention on those problems which include a nonhomogeneous forcing function. Laplace transforms are especially useful when dealing with piecewise-defined forcing functions, but they are useful in many other cases as well. We consider a problem below which involves a continuous forcing function used to describe the presence of immigration or emigration.

■ EXAMPLE 9.5

Let x(t) represent the population of a certain country. The rate at which the population increases depends on the growth rate of the country as well as the rate at which people are being added to or subtracted from the population due to immigration or emigration. Hence, we consider the population problem
$x' - kx = 1000 (1 + a \sin(t))$, $x'(0) = x_0$.

Solve this problem using Laplace transforms with k = 3, x_0=2000, and a = 0.25, 0.75, and 1.25. Plot the solution in each case.

Solution:

We begin by loading the **LaplaceTransform** package, defining the left-hand side of the differential equation in **lhs1**, and entering the transformation rules in **rule1**.

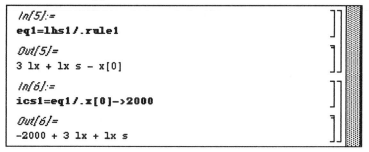

```
In[1]:=
<<Calculus`LaplaceTransform`
In[2]:=
Clear[lhs1,x,lx,eq1,rule1,ics1]
In[3]:=
lhs1=x'[t]+3x[t];
In[4]:=
rule1={x[t]->lx,x'[t]->s lx -x[0]};
```

The Laplace transform of the left-hand side of the equation is computed in **eq1** and the appropriate initial condition applied in **ics1**.

```
In[5]:=
eq1=lhs1/.rule1
Out[5]=
3 lx + lx s - x[0]
In[6]:=
ics1=eq1/.x[0]->2000
Out[6]=
-2000 + 3 lx + lx s
```

The Laplace transform of the right-hand side of the differential equation with a=0.25 is then determined in **rhs**.

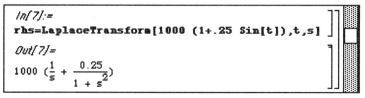

```
In[7]:=
rhs=LaplaceTransform[1000 (1+.25 Sin[t]),t,s]
Out[7]=
          1     0.25
1000 (--- + -------)
          s        2
                1 + s
```

The Laplace transform of the solution is found by equating **ics1** and **rhs1** and then solving for **lx**.

In[8]:=
soln1=Solve[ics1==rhs,lx]

Out[8]=

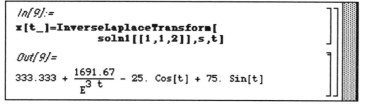

$$\{\{lx \rightarrow -(\frac{-1000 - 2250.\ s - 1000\ s^2 - 2000\ s^3}{3\ s + s^2 + 3\ s^3 + s^4})\}\}$$

Applying the inverse Laplace transform to **lx** (which is extracted from the output list of **soln1** with **soln1[[1,1,2]]**) yields the formula for x(t).

In[9]:=
x[t_]=InverseLaplaceTransform[
                soln1[[1,1,2]],s,t]

Out[9]=

$$333.333 + \frac{1691.67}{E^{3\ t}} - 25.\ Cos[t] + 75.\ Sin[t]$$

The population is then plotted over the interval [0,10]. This graph indicates that the population undergoes periods of increase and decrease on this interval.

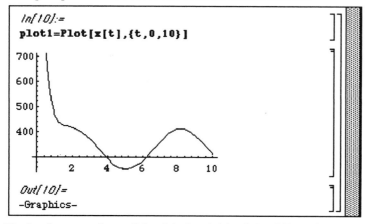

In[10]:=
plot1=Plot[x[t],{t,0,10}]

Out[10]=
-Graphics-

The Laplace transform of the right-hand side of the differential equation with a=0.75 is then computed in **rhs2** and the Laplace transform of the solution determined in **soln2**.

```
In[11]:=
rhs2=LaplaceTransform[
           1000 (1+.75 Sin[t]),t,s]
Out[11]=
            0.75
1000 (1/s + ------)
           1 + s^2

In[12]:=
soln2=Solve[ics1==rhs2,lx]
Out[12]=

            -1000 - 2750. s - 1000 s^2 - 2000 s^3
{{lx -> -(-------------------------------------)}}
            3 s + s^2 + 3 s^3 + s^4
```

The use of **InverseLaplaceTransform** yields the solution for a=0.75 which is defined as **x2**.

```
In[13]:=
x2[t_]=InverseLaplaceTransform[
           soln2[[1,1,2]],s,t]
Out[13]=
            1741.67
333.333 + ------- - 75. Cos[t] + 225. Sin[t]
            E^(3 t)
```

This solution is plotted below in **plot2** with the **GrayLevel** option for later use.

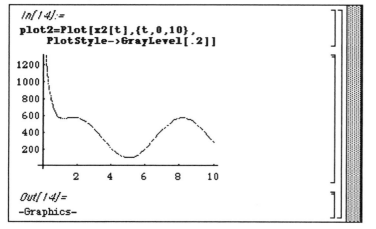

```
In[14]:=
plot2=Plot[x2[t],{t,0,10},
    PlotStyle->GrayLevel[.2]]
```

```
Out[14]=
-Graphics-
```

Finally, we follow similar steps to obtain the solution to the equation with a=1.25.

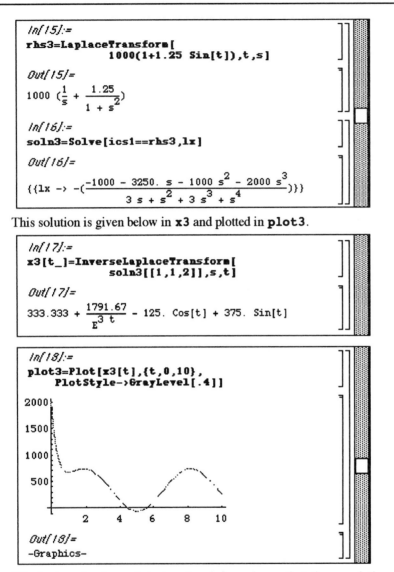

```
In[15]:=
rhs3=LaplaceTransform[
          1000(1+1.25 Sin[t]),t,s]
Out[15]=
             1     1.25
1000 (  --- +  ------- )
             s     1 + s²

In[16]:=
soln3=Solve[ics1==rhs3,lx]
Out[16]=
                    -1000 - 3250. s - 1000 s² - 2000 s³
{{lx -> -( ------------------------------------------ )}}
                      3 s + s² + 3 s³ + s⁴
```

This solution is given below in **x3** and plotted in **plot3**.

```
In[17]:=
x3[t_]=InverseLaplaceTransform[
          soln3[[1,1,2]],s,t]
Out[17]=
                     1791.67
333.333 + --------- - 125. Cos[t] + 375. Sin[t]
                      E³ t
```

```
In[18]:=
plot3=Plot[x3[t],{t,0,10},
    PlotStyle->GrayLevel[.4]]
```

```
Out[18]=
-Graphics-
```

Displaying the three plots together below shows that the population fluctuates more rapidly when a=1.25. This is due to the fact that the forcing function 1000 (1 + a sin(t)) assumes both positive and negative values when a=1.25. In the other cases, a=0.25 and a=0.75, this function is always positive.

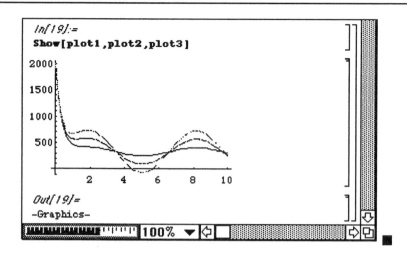

§9.4 The Convolution Theorem

In many cases, we are required to determine the inverse Laplace transform of a product of two functions. Just as in integral calculus when the integral of the product of two functions did not produce the product of the integrals, neither does the inverse Laplace transform of the product yield the product of the inverse Laplace transforms. Therefore, we need the following theorem:

The Convolution Theorem:
Suppose that f(t) and g(t) are piecewise-continuous on $[0,\infty)$ and both of exponential order. Further suppose that the Laplace transform of f is F(s) and that of g is G(s). Then,

$$L^{-1}\{F(s)G(s)\} = L^{-1}\{L\{(f * g)(t)\}\} = (f * g)(t) = \int_0^t f(t-v)g(v)\,dv.$$

Note that $(f * g)(t) = \int_0^t f(t-v)g(v)\,dv$ is called the **convolution integral**.

■ EXAMPLE 9.6

Use the definition of the convolution integral to compute $(f*g)(t)$ if $f(t)=e^{-t}$ and $g(t)=\sin(t)$. Also, verify the Convolution Theorem using these two functions.

Solution:

We make use of the definition and a table of integrals (or Integration By Parts twice) to obtain the following:

$$(f * g)(t) = \int_0^t f(t-v)\, g(v)\, dv = \int_0^t e^{-(t-v)} \sin(v)\, dv = e^{-t} \int_0^t e^v \sin(v)\, dv$$

$$= e^{-t} \left[\frac{e^v}{2} \left(\sin(v) - \cos(v) \right) \right]_0^t = e^{-t} \left[\frac{e^t}{2} \left(\sin(t) - \cos(t) \right) - \frac{1}{2} \left(\sin(0) - \cos(0) \right) \right]$$

$$= \frac{1}{2} \left(\sin(t) - \cos(t) \right) + \frac{1}{2} e^{-t}.$$

Now, according to the Convolution Theorem, $L\{f(t)\}L\{g(t)\}=(f*g)(t)$. In this example, we have

$$F(s)=L\{f(t)\}=L\{e^{-t}\}=\frac{1}{s+1} \quad \text{and} \quad G(s)=L\{g(t)\}=L\{\sin(t)\}=\frac{1}{s^2+1}.$$

Hence, $L^{-1}\left\{ \left(\dfrac{1}{s+1}\right)\left(\dfrac{1}{s^2+1}\right) \right\}$ should be $(f*g)(t)$. We use *Mathematica* to determine this inverse Laplace

transform and, thus, verify the theorem.

The convolution of the two functions is found below with **Integrate**.

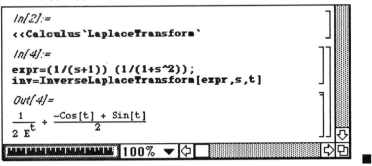

After loading the **LaplaceTransform** package, we compute the inverse Laplace transform of the product F(s) G(s). Note that this result does, in fact, equal the result of (f*g)(t) given above.

```
In[2]:=
<<Calculus`LaplaceTransform`

In[4]:=
expr=(1/(s+1)) (1/(1+s^2));
inv=InverseLaplaceTransform[expr,s,t]

Out[4]=
 1      -Cos[t] + Sin[t]
----  + ----------------
 2 E^t         2
```

100% ▼

■ EXAMPLE 9.7

Use *Mathematica* to define a function which computes the convolution of two functions, and use it to-
determine $(f*g)(t)$ and $(g*f)(t)$ if (a) $f(t)=e^{-t}$ and $g(t)=\sin(t)$; (b) $f(t)=t^3$ and $g(t)=e^{-t}$.

Solution:

We give the definition of the user-defined function **convolution** below. The convolution (f*g)(t) using the functions in part (a), which are the same as those used in the previous example, is then computed to yield the same result as above.

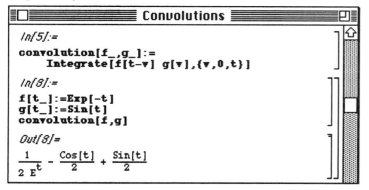

```
                    Convolutions
In[5]:=
convolution[f_,g_]:=
    Integrate[f[t-v] g[v],{v,0,t}]

In[8]:=
f[t_]:=Exp[-t]
g[t_]:=Sin[t]
convolution[f,g]

Out[8]=
 1      Cos[t]    Sin[t]
---- -  ------  + ------
    t      2         2
2 E
```

Next, (g*f)(t) is computed and called **reverse**. Although this result does not appear to equal (f*g)(t), use of **Simplify** with the option setting **Trig->True** shows that *f*g)(t)=(g*f)(t).

```
In[9]:=
reverse=convolution[g,f]

Out[9]=
-(Cos[t] - Sin[t])    Cos[t] (Cos[t] - Sin[t])
------------------  + ------------------------  +
        2                        t
                              2 E
  Sin[t] (Cos[t] + Sin[t])
  ------------------------
               t
            2 E

In[10]:=
Simplify[reverse,Trig->True]

Out[10]=
 1      Cos[t]    Sin[t]
---- -  ------  + ------
    t      2         2
2 E
```

Similar calculations are carried out below with the functions in part(b). First, the functions are defined and (f*g)(t) is computed in **cvl**.

```
In[14]:=
Clear[f,g]
f[t_]:=t^3
g[t_]:=Exp[-t]
cvl=convolution[f,g]

Out[14]=
-6 + 6 t - 3 t² + t³ +
   (6 - 6 t + 3 t² - 3 (-1 + t) t² -
      3 t (-2 + 2 t - t²)) / Eᵗ
```

The result is then expanded and simplified in **one**.

```
In[15]:=
one=Simplify[Expand[cvl]]

Out[15]=
-6 + 6/Eᵗ + 6 t - 3 t² + t³
```

Next, (g*f)(t) is determined in **rcv1**.

```
In[16]:=
rcv1=convolution[g,f]

Out[16]=
6/Eᵗ + Eᵗ (-6/Eᵗ + 6 t/Eᵗ - 3 t²/Eᵗ + t³/Eᵗ)
```

The simplified expression found in **two** is then verified to equal the expression given earlier in **one**.
Hence, (f*g)(t)=(g*f)(t).

```
In[17]:=
two=Simplify[Expand[rcv1]]

Out[17]=
-6 + 6/Eᵗ + 6 t - 3 t² + t³

In[18]:=
one==two

Out[18]=
True
```

□ **EXAMPLE 9.8**

Use the Convolution Theorem to find the Laplace transform of

$$h(t) = \int_0^t \cos(t - v)\sin(v)\,dv.$$

Solution:

Notice that h(t)=(f*g)(t) where f(t)=cos(t) and g(t)=sin(t). Therefore, according to the Convolution Theorem L{(f*g)(t)}=F(s)G(s). In this case, we have

$$L\{h\,(t)\} = L\{f\,(t)\}L\{g\,(t)\} = L\{\cos(t)\}L\{\sin\,(t)\} = \left(\frac{s}{s^2+1}\right)\left(\frac{1}{s^2+1}\right) = \frac{s}{\left(s^2+1\right)^2}.\ \blacksquare$$

The Convolution Theorem is useful in solving numerous problems. In particular, this theorem can be employed to solve **integral equations**, equations which involve an integral of the unknown function. We illustrate this procedure in the following example.

❑ EXAMPLE 9.9

Use the Convolution Theorem to solve the integral equation

$$h(t) = 4t + \int_0^t h(t-v)\sin(v)dv.$$

Solution:

We first note that the integral in the above equation represents (h*g)(t) for g(t)=sin(t). Therefore, if we apply the Laplace transform to both sides of the equation, we obtain

$$L\{h\,(t)\} = L\{4t\} + L\{h\,(t)\}L\{\sin\,(t)\},\ \text{or}\ H(s) = \frac{4}{s^2} + H(s)\frac{1}{s^2+1}.$$

where L{h(t)}=H(s). Solving for H(s), we have

$$H(s)\left(1 - \frac{1}{s^2+1}\right) = \frac{4}{s^2},\ \text{so}\ H(s) = \frac{4\left(s^2+1\right)}{s^4} = \frac{4}{s^2} + \frac{4}{s^4}.$$

Then by computing the inverse Laplace transform, we find that

$$h(t) = 4t + \frac{2}{3}t^3.$$

We verify this result by substituting h(t) into the integral equation. We begin by computing the convolution (h* g)(t), the integral on the right-hand side of the equation, with the user-defined function **convolution** which was introduced in a previous example.

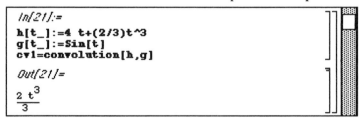

```
In[21]:=

h[t_]:=4 t+(2/3)t^3
g[t_]:=Sin[t]
cv1=convolution[h,g]

Out[21]=

2 t^3
─────
  3
```

The result in **cv1** when added to 4t should equal h(t). Clearly, this is the case. Therefore, we have correctly solved the integral equation through the use of Laplace transforms.

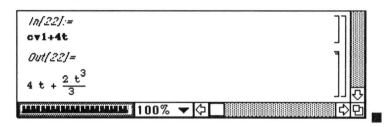

Laplace transforms are helpful in solving problems of other types as well. We illustrate another important problem below.

❑ EXAMPLE 9.10

Recall that the differential equation used to determine the charge q(t) on the capacitor in a L-R-C circuit

is $\bar{L}\dfrac{d^2q}{dt^2} + R\dfrac{dq}{dt} + \dfrac{1}{C}q = v(t)$, $q(0) = 0$, $q'(0) = 0$.

Notice that since $\dfrac{dq}{dt} = i$, this differential equation can be represented as

$\bar{L}\dfrac{di}{dt} + Ri + \dfrac{1}{C}\displaystyle\int_0^t i(u)\,du = v(t)$.   (Note that $\bar{L}$ is being used to denote inductance so that it will

not be confused with the notation used for Laplace transforms.)

Note also that the initial condition $q(0) = 0$ is satisfied since $q(0) = \dfrac{1}{C}\displaystyle\int_0^0 i(u)\,du = 0$.

The condition $q'(0) = 0$ is replaced by $i(0) = 0$.

Solve this **integrodifferential equation**, an equation which involves a derivative as well as an integral of the unknown function, by using the Convolution Theorem.

Solution:

We proceed as in the case of a differential equation by taking the Laplace transform of both sides of the equation. The Convolution Theorem is used in determining the Laplace transform of the integral as follows:

$$L\left\{\int_0^t i(u)\,du\right\} = L\{(1 * i(t))(t)\} = L\{1\}L\{i(t)\} = \dfrac{I(s)}{s}.$$

Therefore, application of the Laplace transform yields

$\bar{L}s\,I(s) - s\,i(0) + R\;I(s) + \dfrac{1}{C}\dfrac{I(s)}{s} = V(t)$.   Since $i(0) = 0$, we have $\bar{L}s\,I(s) + R\;I(s) + \dfrac{1}{C}\dfrac{I(s)}{s} = V(t)$.

(Note that capital letters denote the Laplace transform of the function named with the associated small letter.) Simplifying and solving for I(s), we have

$$I(s) = \frac{V(s)}{\overline{L}Cs^2 + RCs + 1}, \text{ and, hence, } i(t) = L^{-1} \left\{ \frac{V(s)Cs}{\overline{L}Cs^2 + RCs + 1} \right\}. \blacksquare$$

■ EXAMPLE 9.11

Reconsider the previous example with constant values $\overline{L}=C=R=1$, and: (a) $v(t)=1$;

(b) $v(t)=\begin{cases} \sin(t), \ 0 \le t \le \dfrac{\pi}{2} \\ 0, \qquad \text{otherwise} \end{cases}$. Determine $i(t)$ in each case and plot the solution.

Solution:

For (a), we use the formula derived in the previous example.

Since $v(t)=1$, $V(s)=\dfrac{1}{s}$. Therefore, we have $i(t)=L^{-1}\left\{ \dfrac{1}{s^2+s+1} \right\}$. We use *Mathematica* below to

compute this inverse.

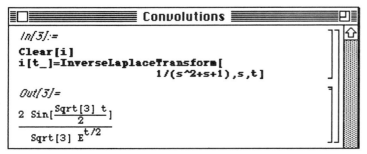

The solution which corresponds to v(t)=1 is then plotted below using a dashed curve.

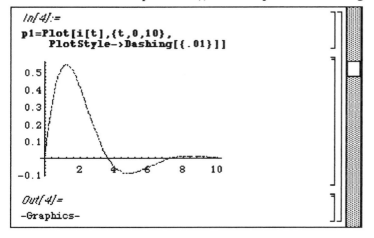

For (b), we make use of the methods indicated in earlier sections involving piecewise-defined forcing
functions. We first define the unit step function **u**.

```
<<Calculus`LaplaceTransform`
In[6]:=
u[t_,a_]:=0/;t<a
u[t_,a_]:=1/;t>=a
```

We then define and plot the forcing function **f** on the interval $[0,\pi]$.

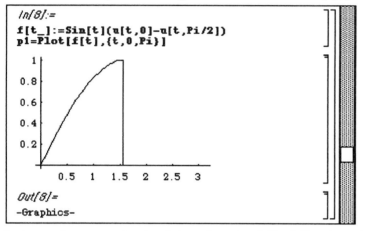

```
In[8]:=
f[t_]:=Sin[t](u[t,0]-u[t,Pi/2])
p1=Plot[f[t],{t,0,Pi}]
```

```
Out[8]=
-Graphics-
```

As in previous examples involving piecewise-defined functions (see **Example 9.2**), we define the
functions **lpt** and **invlpt**.

```
In[9]:=
lpt[g_,a_]:=Exp[-a s]*
            LaplaceTransform[g/.t->t+a,t,s]
In[10]:=
invlpt[lpg_,a_]:=
(InverseLaplaceTransform[lpg,s,t]/.t->t-a)*
                    u[t,a]
```

We, therefore, compute the Laplace transform of v(t) using a combination of **LaplaceTransform**
and **lpt** (since u[1,t]=1). We call this result **capv**.

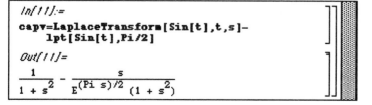

```
In[11]:=
capv=LaplaceTransform[Sin[t],t,s]-
    lpt[Sin[t],Pi/2]
```

$$Out[11]=$$

$$\frac{1}{1+s^2} - \frac{s}{E^{(Pi\ s)/2}\ (1+s^2)}$$

Using the general formula obtained for the Laplace transform of i(t), we note that the denominator in this expression is given by $s^2 + s + 1$ which is entered as **denom** below. Hence, the Laplace transform if i, called **capi**, is given below by the ratio of **capv** and **denom**.

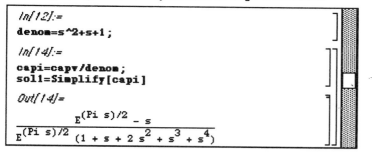

Simplifying the expression given above in **soll**, we notice that one component involves $e^{-\pi s/2}$. Hence, we employ the user-defined function **invlpt** to determine the inverse Laplace transform of this part of **soll** and use **InverseLaplaceTransform** to determine that of the rest of it. The solution which results is defined as **i**.

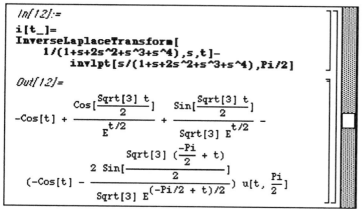

This solution is plotted in **p2** below and displayed with the forcing function in the plot which follows. Notice the effect that the forcing function has on the solution to the differential equation.

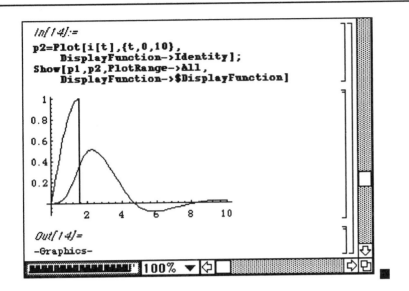

```
In[14]:=
p2=Plot[i[t],{t,0,10},
     DisplayFunction->Identity];
Show[p1,p2,PlotRange->All,
     DisplayFunction->$DisplayFunction]
```

```
Out[14]=
-Graphics-
```

`100%`

§9.5 Differential Equations Involving Impulse Functions

We now consider differential equations of the form $a x'' + b x' + c x = f(t)$ where f is "large" over the short interval $t_0 - \alpha < t < t_0 + \alpha$ and zero otherwise. Hence, we define the impulse delivered by the function $f(t)$ as

the integral $I(t) = \int_{t_0 - a}^{t_0 + a} f(t) \, dt$, or since $f = 0$ on $\left(-\infty, t_0 - \alpha\right) \cup \left(t_0 + \alpha, +\infty\right)$, $I(t) = \int_{-\infty}^{+\infty} f(t) \, dt.$

In order to better understand the impulse function, we let f be defined in the following manner:

$$f(t) = \delta_\alpha\left(t - t_0\right) = \begin{cases} \dfrac{1}{2\alpha} & , t_0 - \alpha < t < t_0 + \alpha \\ 0 & , \text{ otherwise} \end{cases}$$

Therefore, we have

$$I(t) = \int_{t_0 - \alpha}^{t_0 + \alpha} f(t) \, dt = \int_{t_0 - \alpha}^{t_0 + \alpha} \frac{1}{2\alpha} \, dt = \frac{1}{2\alpha}\left(\left(t_0 + \alpha\right) - \left(t_0 - \alpha\right)\right) = \frac{1}{2\alpha}\left(2\alpha\right) = 1.$$

Notice that the value of this integral does not depend on α as long as α is not zero. We now try to create the idealized impulse function by requiring that $\delta_\alpha(t - t_0)$ act on smaller and smaller intervals. From the integral calculation above, we have

$$\lim_{\alpha \to 0} I(t) = 1.$$

We also note that

$$\lim_{\alpha \to 0} \delta_\alpha\left(t - t_0\right) = 0, \, t \neq t_0.$$

We use these properties to now define the idealized unit impulse function as follows:

$$\delta(t - t_0) = 0, t \neq t_0$$

$$\int_{-\infty}^{+\infty} \delta(t - t_0)\ dt = 1.$$

We now state the following useful theorem involving the unit impulse function.

Theorem: Suppose that g is a bounded and continuous function. Then,

$$\int_{-\infty}^{+\infty} \delta(t - t_0)\ g(t)\ dt = g(t_0).$$

The function $\delta(t{-}t_0)$ is known as the **Dirac delta function** and is quite useful in the definition of impulse forcing functions which arise in the differential equations discussed here. The Laplace transform of $\delta(t{-}t_0)$ is found by using the function $\delta_\alpha(t{-}t_0)$ and L'Hopital's rule to be

$$L\{\delta(t - t_0)\} = e^{-st_0}\ .$$

■ EXAMPLE 9.12

Define the function $\delta_\alpha(t{-}t_0)$ with *Mathematica* and plot the function for $t_0 = 0$ values of $\alpha = 0.01$, 0.02, 0.03, 0.04, and 0.05. Determine the area under each curve.

Solution:

We define $\delta_\alpha(t{-}t_0)$ as the function **del** depending on t, t_0, and α below. Notice that we do this on three intervals.

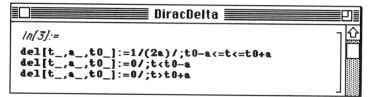

We then define the function **a[i]** which plots **del** on the interval [−2i,2i]. Hence, we generate a table of plots in **deltable** for i = 0.01 to i = 0.05 using increments of 0.01 and increasing levels of gray. These plots are then displayed simultaneously. Notice that smaller and smaller values of α produce a "spike" at the origin.

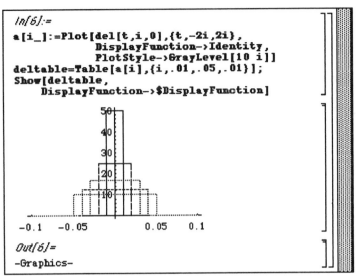

```
In[6]:=
a[i_]:=Plot[del[t,i,0],{t,-2i,2i},
            DisplayFunction->Identity,
            PlotStyle->GrayLevel[10 i]]
deltable=Table[a[i],{i,.01,.05,.01}];
Show[deltable,
     DisplayFunction->$DisplayFunction]
```

```
Out[6]=
-Graphics-
```

Observing the rectangles produced in the plot above, we easily see that each rectangle has area one. We verify this below, however, with **NIntegrate** by producing a table of the area found under each delta function given above.

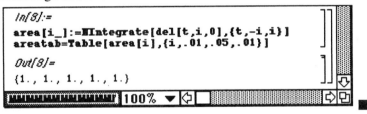

```
In[8]:=
area[i_]:=NIntegrate[del[t,i,0],{t,-i,i}]
areatab=Table[area[i],{i,.01,.05,.01}]
```

```
Out[8]=
{1., 1., 1., 1., 1.}
```

■ **EXAMPLE 9.13**

Verify the formula derived for the Laplace transform of $\delta(t-t_0)$ by using *Mathematica* to compute the following:

(a) $L\{\delta(t-\pi)\}$, (b) $L\{\delta(t-1)\}$, and (c) $L\{\delta(t)\}$.

Solution:

After loading the **LaplaceTransform** package, we compute $L\{\delta(t-\pi)\}=e^{-\pi s}$.

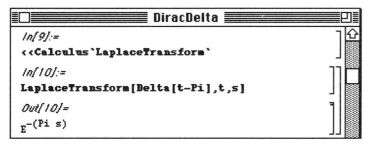

We then determine $L\{\delta(t-1)\}=e^{-s}$.

Finally, we have $L\{\delta(t)\}=e^{0}=1$.

Many applications involve impulse functions. These include L-R-C circuit problems with a voltage surge as well as spring-mass systems with a sudden impact which "excites" the system. We now investigate problems of this type through the use of *Mathematica* and Laplace transforms.

■ EXAMPLE 9.14

Consider a spring-mass system with no damping (c=0), m = 1, and k = 1. Use *Mathematica* and Laplace transforms to solve the initial value problem which models this situation if the forcing function is $\delta(t-\pi)$. That is, a sudden impact occurs at t = π.

Solution:

Using our knowledge of spring-mass systems, we must solve the following initial value problem.

$$x''+x=\delta(t-\pi)$$

$$x(0)=0,\ x'(0)=0.$$

We first compute the Laplace transform of the right-hand side of the equation. This is done with **LaplaceTransform** in **lptdel** below.

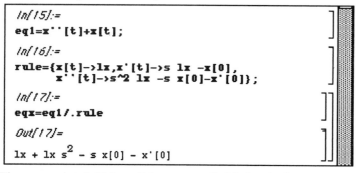

In[14]:=
`lptdel=LaplaceTransform[Delta[t-Pi],t,s]`

Out[14]=
$E^{-(Pi\ s)}$

We then enter the left-hand side of the equation and enter the necessary transformation rules for computing the Laplace transform of the left-hand side with **lx** representing the Laplace transform of x. Hence, **eqx** represents the Laplace transform of the left-hand side of the equation.

In[15]:=
`eq1=x''[t]+x[t];`

In[16]:=
```
rule={x[t]->lx,x'[t]->s lx -x[0],
      x''[t]->s^2 lx -s x[0]-x'[0]};
```

In[17]:=
`eqx=eq1/.rule`

Out[17]=
$lx + lx\ s^2 - s\ x[0] - x'[0]$

The appropriate initial conditions are applied below in **ics**.

In[18]:=
`ics=eqx/.{x[0]->0,x'[0]->0}`

Out[18]=
$lx + lx\ s^2$

The Laplace transform of x is then determined below and extracted from the output list of **soln**.

In[19]:=
`soln=Solve[ics==lptdel,lx]`

Out[19]=
$\{\{lx \rightarrow \frac{1}{E^{Pi\ s} + E^{Pi\ s} s^2}\}\}$

In[20]:=
`soln[[1,1,2]]`

Out[20]=
$\frac{1}{E^{Pi\ s} + E^{Pi\ s} s^2}$

The solution is then calculated with **InverseLaplaceTransform** and called **x**. Notice that this solution is given as

$$x(t) = \begin{cases} -\sin(t), & t \geq \pi \\ 0, & \text{otherwise} \end{cases}$$

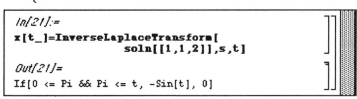

```
In[21]:=
x[t_]=InverseLaplaceTransform[
            soln[[1,1,2]],s,t]

Out[21]=
If[0 <= Pi && Pi <= t, -Sin[t], 0]
```

The solution is shown below in **plot6**.

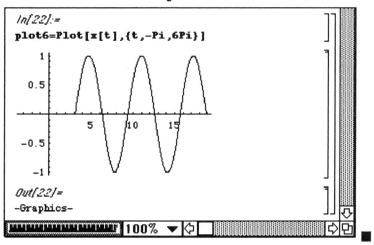

```
In[22]:=
plot6=Plot[x[t],{t,-Pi,6Pi}]
```

```
Out[22]=
-Graphics-
```

■ EXAMPLE 9.15

The forcing function may involve a combination of functions. For example, solve the initial value problem:

$$x''+2x'+x = \delta(t) + h(t-2\pi)$$

$$x(0) = 0, \; x'(0) = 0.$$

Notice that in this case, the system undergoes an initial excitation due to the impulse function. Then, at $t = 2\pi$, a constant force of magnitude one is applied. Notice the effect that this forcing function has on the behavior of the solution by comparing these results with those of the previous example.

Solution:

We begin by defining the functions necessary for working with the piecewis- defined functions. These include the unit step function **u** as well as **lpt**. (See **Example 9.2**.)

```
┌─────────────────────────────────────────────────┐
│ ≡□≡≡≡≡≡≡≡≡≡≡≡ DiracDelta ≡≡≡≡≡≡≡≡≡≡       □│
│ In[41]:=                                        ⇧│
│ u[t_,a_]:=0/;t<a                                 │
│ u[t_,a_]:=1/;t>=a                               ]│
│ In[42]:=                                         │
│ lpt[g_,a_]:=Exp[-a s]*                           │
│        LaplaceTransform[g/.t->t+a,t,s]          ]│
└─────────────────────────────────────────────────┘
```

We then define **invlpt** for computing the inverse Laplace transform of functions involving exponential functions. (See **Example 9.1**.) The left-hand side of the equation and the transformation rules are given in **eq1** and **rule**, respectively.

```
┌─────────────────────────────────────────────────┐
│ In[43]:=                                         │
│ invlpt[lpg_,a_]:=                                │
│ (InverseLaplaceTransform[lpg,s,t]/.t->t-a)*      │
│               u[t,a]                            ]│
│ In[44]:=                                         │
│ eq1=x''[t]+2x'[t]+x[t];                         ]│
│ In[45]:=                                         │
│ rule={x[t]->lx,x'[t]->s lx -x[0],                │
│       x''[t]->s^2 lx -s x[0]-x'[0]};            ]│
└─────────────────────────────────────────────────┘
```

The Laplace transform of the left-hand side is computed below in **eqx**

```
┌─────────────────────────────────────────────────┐
│ In[46]:=                                         │
│ eqx=eq1/.rule                                    │
│ Out[46]=                                         │
│                2                                 │
│ lx + lx s  + 2 (lx s - x[0]) - s x[0] - x'[0]    │
└─────────────────────────────────────────────────┘
```

and the initial conditions applied in **ics**.

```
┌─────────────────────────────────────────────────┐
│ In[47]:=                                         │
│ ics=eqx/.{x[0]->0,x'[0]->0}                      │
│ Out[47]=                                         │
│                    2                             │
│ lx + 2 lx s + lx s                               │
└─────────────────────────────────────────────────┘
```

Next, the Laplace transform of the right-hand side of the equation is computed in **rhs** with **LaplaceTransform** and **lpt**.

```
In[48]:=
rhs=LaplaceTransform[Delta[t],t,s]-
    lpt[1,2Pi]
Out[48]=
        1
1 - ────────
     E^2 Pi s s
```

The Laplace transform of the solution, called **lx**, is then found with **Solve**.

```
In[49]:=
soln=Solve[ics==rhs,lx]
Out[49]=
                          1 - E^2 Pi s
{{lx -> -(─────────────────────────────────────────────────)}}
          E^2 Pi s s + 2 E^2 Pi s s^2 + E^2 Pi s s^3
```

Hence, the solution is determined by using **invlpt** with the portion of **lx** involving $e^{-2\pi s}$ and **InverseLaplaceTransform** with the rest of **lx**.

```
In[34]:=
x[t_]=InverseLaplaceTransform[
         s/(s+2 s^2+s^3),s,t]+
         invlpt[1/(s+2 s^2+s^3),2Pi]
Out[34]=
 t
─── + (1 - E^2 Pi - t - E^2 Pi - t (-2 Pi + t))
E^t

    u[t, 2 Pi]
```

The solution is plotted below.

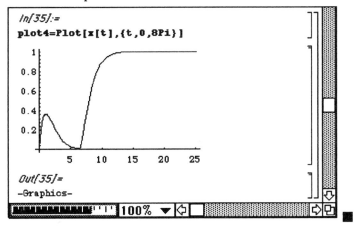

```
In[35]:=
plot4=Plot[x[t],{t,0,8Pi}]
```

```
Out[35]=
-Graphics-
```

100% ▼

Chapter 10: Systems of Ordinary Differential Equations

Mathematica commands used in **Chapter 10** include:

Apart
Cancel
CharacteristicPoly
    nomial
Compiled
ComplexExpand
D
Det
DisplayFunction
DSolve
Eigensystem
Eigenvalues
Eigenvectors
Evaluate
Expand
ExpandNumerator
Factor
Flatten
GraphicsArray
IdentityMatrix
Integrate

Inverse
InverseLaplaceTrans
    form
LaplaceTransform
ListPlot
MatrixForm
MatrixPower
Method
Module
N
NDSolve
NIntegrate
Normal
Numerator
ParametricPlot
ParametricPlot3D
Part
Partition
Plot
PlotPoints
PlotRange

PlotVectorField
PlotVectorField3D
Prolog
ReplaceAll
Series
Short
Show
Simplify
Solve
SolveAlways
Table
TableForm
Ticks
Together
Transpose
VectorHeads

§10.1 Review of Matrix Algebra and Calculus

An **n × m matrix** is an array of the form

$$\begin{pmatrix} a_{11} & a_{12} & \cdots & a_{1m} \\ a_{21} & a_{22} & \cdots & a_{2m} \\ \vdots & \vdots & \ddots & \vdots \\ a_{n1} & a_{n2} & \cdots & a_{nm} \end{pmatrix}$$ with n rows and m columns.

The **transpose** of the $n \times m$ matrix $A = \begin{pmatrix} a_{11} & a_{12} & \cdots & a_{1m} \\ a_{21} & a_{22} & \cdots & a_{2m} \\ \vdots & \vdots & \ddots & \vdots \\ a_{n1} & a_{n2} & \cdots & a_{nm} \end{pmatrix}$ is the $m \times n$ matrix

$$A^T = \begin{pmatrix} a_{11} & a_{21} & \cdots & a_{n1} \\ a_{12} & a_{22} & \cdots & a_{n2} \\ \vdots & \vdots & \ddots & \vdots \\ a_{1m} & a_{2m} & \cdots & a_{nm} \end{pmatrix}.$$

If A and B are $n \times m$ matrices and c is a number, then cA is the matrix obtained by multiplying each element of A by c; A+B is obtained by adding corresponding elements of A and B.

❏ EXAMPLE 10.1

If $A = \begin{pmatrix} -1 & 4 & -2 \\ 6 & 2 & -10 \end{pmatrix}$ and $B = \begin{pmatrix} 2 & -4 & 8 \\ 7 & 4 & 2 \end{pmatrix}$ compute $3A - 9B$.

Solution:

Since $3A = \begin{pmatrix} -3 & 12 & -6 \\ 18 & 6 & -30 \end{pmatrix}$ and $-9B = \begin{pmatrix} -18 & 36 & -72 \\ -63 & -36 & -18 \end{pmatrix}$,

$$3A - 9B = 3A + (-9B) = \begin{pmatrix} -21 & 48 & -78 \\ -45 & -30 & -48 \end{pmatrix}. \quad \blacksquare$$

If $A = \begin{pmatrix} a_{11} & a_{12} & \cdots & a_{1j} \\ a_{21} & a_{22} & \cdots & a_{2j} \\ \vdots & \vdots & \ddots & \vdots \\ a_{n1} & a_{n2} & \cdots & a_{nj} \end{pmatrix}$ is an $n \times j$ matrix and $B = \begin{pmatrix} b_{11} & b_{12} & \cdots & b_{1m} \\ b_{21} & b_{22} & \cdots & b_{2m} \\ \vdots & \vdots & \ddots & \vdots \\ b_{j1} & b_{j2} & \cdots & b_{jm} \end{pmatrix}$ is a $j \times m$ matrix,

$$AB = \begin{pmatrix} a_{11} & a_{12} & \cdots & a_{1j} \\ a_{21} & a_{22} & \cdots & a_{2j} \\ \vdots & \vdots & \ddots & \vdots \\ a_{n1} & a_{n2} & \cdots & a_{nj} \end{pmatrix}\begin{pmatrix} b_{11} & b_{12} & \cdots & b_{1m} \\ b_{21} & b_{22} & \cdots & b_{2m} \\ \vdots & \vdots & \ddots & \vdots \\ b_{j1} & b_{j2} & \cdots & b_{jm} \end{pmatrix}$$ is the unique matrix $C = \begin{pmatrix} c_{11} & c_{12} & \cdots & c_{1m} \\ c_{21} & c_{22} & \cdots & c_{2m} \\ \vdots & \vdots & \ddots & \vdots \\ c_{n1} & c_{n2} & \cdots & c_{nm} \end{pmatrix}$,

where $c_{11} = a_{11}b_{11} + a_{12}b_{21} + \ldots + a_{1j}b_{j1} = \sum_{k=1}^{j} a_{1k}b_{k1}$,

$c_{12} = a_{11}b_{12} + a_{12}b_{22} + \ldots + a_{1j}b_{j2} = \sum_{k=1}^{j} a_{1k}b_{k2}$, and

$c_{uv} = a_{u1}b_{1v} + a_{u2}b_{2v} + ... + a_{uj}b_{jv} = \sum_{k=1}^{j} a_{uk}b_{kv}$. Equivalently, the element of C in the uth row and

vth column, c_{uv}, is obtained by multiplying each member of the uth row of A by the corresponding entry in the vth column of B and adding the result.

❏ EXAMPLE 10.2

Compute AB if $A = \begin{pmatrix} 0 & 4 & 5 \\ -5 & -1 & 5 \end{pmatrix}$ and $B = \begin{pmatrix} -3 & 4 \\ -5 & -4 \\ 1 & -4 \end{pmatrix}$.

Solution:

Since A is a 2 × 3 matrix and B is a 3 × 2 matrix, AB is the 2 × 2 matrix:

$$AB = \begin{pmatrix} 0 & 4 & 5 \\ -5 & -1 & 5 \end{pmatrix}\begin{pmatrix} -3 & 4 \\ -5 & -4 \\ 1 & -4 \end{pmatrix} = \begin{pmatrix} 0\cdot-3+4\cdot-5+5\cdot1 & 0\cdot4+4\cdot-4+5\cdot-4 \\ -5\cdot-3+-1\cdot-5+5\cdot1 & -5\cdot4+-1\cdot-4+5\cdot-4 \end{pmatrix}$$

$$= \begin{pmatrix} -15 & -36 \\ 25 & -36 \end{pmatrix}. \blacksquare$$

The n × n matrix $\begin{pmatrix} 1 & 0 & 0 & 0 \\ 0 & 1 & 0 & 0 \\ \vdots & \vdots & \ddots & \vdots \\ 0 & 0 & 0 & 1 \end{pmatrix}$ is called the **n × n identity matrix**, denoted by I or I_n.

If A is an n × n matrix, then AI=IA=A.

More generally, an n × n matrix is called a **square matrix of order n**.

If $A = (a_{11})$, the **determinant** of A is $\det(A) = a_{11}$; if $A = \begin{pmatrix} a_{11} & a_{12} \\ a_{21} & a_{22} \end{pmatrix}$, the **determinant** of A is

$\det(A) = \begin{vmatrix} a_{11} & a_{12} \\ a_{21} & a_{22} \end{vmatrix} = a_{11}a_{22} - a_{12}a_{21}$. More generally, if $A = \begin{pmatrix} a_{11} & a_{12} & \cdots & a_{1n} \\ a_{21} & a_{22} & \cdots & a_{2n} \\ \vdots & \vdots & \ddots & \vdots \\ a_{n1} & a_{n2} & \cdots & a_{nn} \end{pmatrix}$ is an n × n matrix and A_{ij} denotes the matrix obtained from A by deleting the ith row and jth column from A,

then the **determinant** of A is

$$\det(A) = \begin{vmatrix} a_{11} & a_{12} & \cdots & a_{1n} \\ a_{21} & a_{22} & \cdots & a_{2n} \\ \vdots & \vdots & \ddots & \vdots \\ a_{n1} & a_{n2} & \cdots & a_{nn} \end{vmatrix} = \sum_{j=1}^{n} (-1)^{i+j} a_{ij} \det(A_{ij}) = \sum_{j=1}^{n} (-1)^{i+j} a_{ij} |A_{ij}|.$$

The number $(-1)^{i+j} \det(A_{ij}) = (-1)^{i+j} |A_{ij}|$ is called the **cofactor** of a_{ij}.

The **cofactor matrix**, A^c, of A is the matrix obtained by replacing each element of A by its cofactor:

$$\text{if } A = \begin{pmatrix} a_{11} & a_{12} & \cdots & a_{1n} \\ a_{21} & a_{22} & \cdots & a_{2n} \\ \vdots & \vdots & \ddots & \vdots \\ a_{n1} & a_{n2} & \cdots & a_{nn} \end{pmatrix}, A^c = \begin{pmatrix} |A_{11}| & -|A_{12}| & \cdots & (-1)^{n+1}|A_{1n}| \\ -|A_{21}| & |A_{22}| & \cdots & (-1)^{n}|A_{2n}| \\ \vdots & \vdots & \ddots & \vdots \\ (-1)^{n+1}|A_{n1}| & (-1)^{n}|A_{n2}| & \cdots & |A_{nn}| \end{pmatrix}.$$

❑ EXAMPLE 10.3

Calculate $|A|$ if $A = \begin{pmatrix} -4 & -2 & -1 \\ 5 & -4 & -3 \\ 5 & 1 & -2 \end{pmatrix}$.

Solution:

$$|A| = \begin{vmatrix} -4 & -2 & -1 \\ 5 & -4 & -3 \\ 5 & 1 & -2 \end{vmatrix} = (-4)\begin{vmatrix} -4 & -3 \\ 1 & -2 \end{vmatrix} + (-1)^3 (-2)\begin{vmatrix} 5 & -3 \\ 5 & -2 \end{vmatrix} + (-1)\begin{vmatrix} 5 & -4 \\ 5 & 1 \end{vmatrix}$$

$$= -4\big((-4)(-2) - (-3)(1)\big) + 2\big((5)(-2) - (-3)(5)\big) - \big((5)(1) - (-4)(5)\big) = -59. \quad ■$$

B is an **inverse** of the n × n matrix A means that AB=BA=I.

The **adjoint**, A^a, of an n × n matrix A is the transpose of the cofactor matrix:

$A^a = (A^c)^T$. If $|A| \neq 0$, and $B = \dfrac{1}{|A|}A^a$, then $AB = BA = I$. Therefore, if $|A| \neq 0$, the inverse

of A is given by $A^{-1} = \dfrac{1}{|A|}A^a.$

❑ EXAMPLE 10.4

Find A^{-1} if $A = \begin{pmatrix} 2 & -1 \\ -3 & 1 \end{pmatrix}$.

Solution:

In this case, $\det(A) = \begin{vmatrix} 2 & -1 \\ -3 & 1 \end{vmatrix} = 2 - 3 = -1 \neq 0$ so A^{-1} exists. Moreoever, $A^c = \begin{pmatrix} 1 & 3 \\ 1 & 2 \end{pmatrix}$ so

$A^a = \begin{pmatrix} 1 & 1 \\ 3 & 2 \end{pmatrix}$ and $A^{-1} = \dfrac{1}{|A|} A^a = \begin{pmatrix} -1 & -1 \\ -3 & -2 \end{pmatrix}$. ∎

■ EXAMPLE 10.5

If $A = \begin{pmatrix} 0 & 0 & -2 & 1 \\ 3 & -1 & 7 & 2 \\ -6 & 0 & 5 & -1 \\ -6 & 0 & 1 & -2 \end{pmatrix}$ and $B = \begin{pmatrix} -7 & -6 & -3 & -7 \\ 2 & -3 & 0 & 4 \\ 3 & 4 & 1 & 2 \\ 5 & 6 & 3 & 6 \end{pmatrix}$, compute (i) $3A - 2B$; (ii) B^T; (iii) AB;

(iv) $|A|$, $|B|$, and $|AB|$; and (v) A^{-1}.

Solution:

We begin by defining **matrix** and **matrixb** to correspond to the matrices A and B, respectively.

```
══════════ MatrixOperations ══════════
In[34]:=
matrixa={{0,0,-2,1},{3,-1,7,2},
    {-6,0,5,-1},{-6,0,1,-2}};
matrixb={{-7,-6,-3,-7},{2,-3,0,4},
    {3,4,1,2},{5,6,3,6}};
```

We then compute 3A−2B by computing **3matrixa-2matrixb**, naming the resulting output **parta**, and display **parta** in traditional row-and-column form with the command **MatrixForm**.

```
In[35]:=
parta=3matrixa-2matrixb

Out[35]=
{{14, 12, 0, 17}, {5, 3, 21, -2},
    {-24, -8, 13, -7}, {-28, -12, -3, -18}}

In[36]:=
MatrixForm[parta]

Out[36]//MatrixForm=
14    12    0     17
5     3     21    -2
-24   -8    13    -7
-28   -12   -3    -18
```

Similarly, we use the commands **Transpose** and **MatrixForm** to compute B^T and display the resulting output in traditional row-and-column form.

```
In[37]:=
Transpose[matrixb]//MatrixForm

Out[37]//MatrixForm=
-7   2   3   5
-6  -3   4   6
-3   0   1   3
-7   4   2   6
```

Matrix products are computed using **.**:

```
In[39]:=
product=matrixa.matrixb;
MatrixForm[product]

Out[39]//MatrixForm=
-1  -2   1   2
 8  25   4   1
52  50  20  46
35  28  13  32
```

The determinants are found in **da**, **db**, and **dp** below with **Det**.

```
In[42]:=
da=Det[matrixa]
db=Det[matrixb]
dp=Det[product]

Out[40]=
-36

Out[41]=
-24

Out[42]=
864
```

Finally, the inverse of A is computed with **Inverse** and placed in **MatrixForm**.

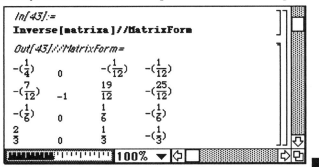

```
In[43]:=
Inverse[matrixa]//MatrixForm

Out[43]//MatrixForm=
```

$$
\begin{array}{cccc}
-(\frac{1}{4}) & 0 & -(\frac{1}{12}) & -(\frac{1}{12}) \\
-(\frac{7}{12}) & -1 & \frac{19}{12} & -(\frac{25}{12}) \\
-(\frac{1}{6}) & 0 & \frac{1}{6} & -(\frac{1}{6}) \\
\frac{2}{3} & 0 & \frac{1}{3} & -(\frac{1}{3})
\end{array}
$$

100%

A nonzero vector $\mathbf{x}$ is an **eigenvector** of the square matrix A means there is a number λ, called an **eigenvalue** of A, so that $A\mathbf{x}=\lambda\mathbf{x}$.

❏ **EXAMPLE 10.6**

Show that $\begin{pmatrix} -1 \\ 2 \end{pmatrix}$ and $\begin{pmatrix} 1 \\ 1 \end{pmatrix}$ are eigenvectors of $\begin{pmatrix} -1 & 2 \\ 4 & -3 \end{pmatrix}$ with corresponding eigenvalues -5 and 1, respectively.

Solution:

Since $\begin{pmatrix} -1 & 2 \\ 4 & -3 \end{pmatrix}\begin{pmatrix} -1 \\ 2 \end{pmatrix} = \begin{pmatrix} 5 \\ -10 \end{pmatrix} = -5\begin{pmatrix} -1 \\ 2 \end{pmatrix}$ and $\begin{pmatrix} -1 & 2 \\ 4 & -3 \end{pmatrix}\begin{pmatrix} 1 \\ 1 \end{pmatrix} = \begin{pmatrix} 1 \\ 1 \end{pmatrix} = 1\begin{pmatrix} 1 \\ 1 \end{pmatrix}$, $\begin{pmatrix} -1 \\ 2 \end{pmatrix}$ and $\begin{pmatrix} 1 \\ 1 \end{pmatrix}$ are eigenvectors

of $\begin{pmatrix} -1 & 2 \\ 4 & -3 \end{pmatrix}$ with corresponding eigenvalues -5 and 1. ∎

If $\mathbf{x}$ is an eigenvector of A with corresponding eigenvalue λ, then $A\mathbf{x}=\lambda\mathbf{x}$ which is equivalent to the equation $(A-\lambda I)\mathbf{x}=0$. Note that $\mathbf{x}$ is an eigenvector if and only if $\det(A-\lambda I)=0$. The equation $\det(A-\lambda I)=0$ is called the **characteristic equation** of A; $\det(A-\lambda I)$ is called the **characteristic polynomial** of A; the roots of the characteristic polynomial of A are the eigenvalues of A. Generally, to find the eigenvectors and corresponding eigenvalues of a square matrix A, we will begin by computing the eigenvalues.

❏ **EXAMPLE 10.7**

Calculate the characteristic polynomial and eigenvalues of the matrices $A = \begin{pmatrix} -4 & 4 & -4 \\ 2 & 3 & -4 \\ 5 & 0 & -1 \end{pmatrix}$ and

$B = \begin{pmatrix} 1 & -3 & 5 \\ 5 & 5 & 0 \\ -5 & -2 & 3 \end{pmatrix}$.

Solution:

The characteristic polynomial of A is

$$\det(A - \lambda I) = \begin{vmatrix} -4 & 4 & -4 \\ 2 & 3 & -4 \\ 5 & 0 & -1 \end{vmatrix} - \begin{pmatrix} \lambda & 0 & 0 \\ 0 & \lambda & 0 \\ 0 & 0 & \lambda \end{pmatrix} = \begin{vmatrix} -4-\lambda & 4 & -4 \\ 2 & 3-\lambda & -4 \\ 5 & 0 & -1-\lambda \end{vmatrix} = -\lambda^3 - 2\lambda^2 - \lambda = -\lambda(\lambda+1)^2$$

and the solutions of $-\lambda(\lambda+1)^2 = 0$ are $\lambda = 0$ and $\lambda = -1$ (with multiplicity two) so the eigenvalues

of $\begin{pmatrix} -4 & 4 & -4 \\ 2 & 3 & -4 \\ 5 & 0 & -1 \end{pmatrix}$ are 0 and -1.

To calculate the characteristic polynomial of B, we use the **Det** command to calculate the determinant of B–λI, name the resulting polynomial, **charpoly**, and then compute the eigenvalues of B by finding the solutions of the characteristic equation.

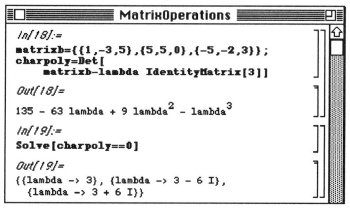

```
In[18]:=
matrixb={{1,-3,5},{5,5,0},{-5,-2,3}};
charpoly=Det[
    matrixb-lambda IdentityMatrix[3]]

Out[18]=
                        2          3
135 - 63 lambda + 9 lambda  - lambda

In[19]:=
Solve[charpoly==0]

Out[19]=
{{lambda -> 3}, {lambda -> 3 - 6 I},
   {lambda -> 3 + 6 I}}
```

Alternatively, *Mathematica* can compute both the characteristic polynomial and eigenvalues of B directly with the commands **CharacteristicPolynomial** and **Eigenvalues**:

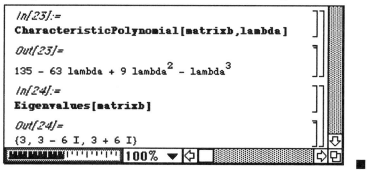

```
In[23]:=
CharacteristicPolynomial[matrixb,lambda]

Out[23]=
                        2          3
135 - 63 lambda + 9 lambda  - lambda

In[24]:=
Eigenvalues[matrixb]

Out[24]=
{3, 3 - 6 I, 3 + 6 I}
```

Once the eigenvalues of a matrix A have been found, the corresponding eigenvectors can be computed.

☐ **EXAMPLE 10.8**

Calculate the eigenvalues and corresponding eigenvectors of $\begin{pmatrix} 4 & -6 \\ 3 & -7 \end{pmatrix}$.

Solution:

The characteristic polynomial of $\begin{pmatrix} 4 & -6 \\ 3 & -7 \end{pmatrix}$ is $\lambda^2 + 3\lambda - 10 = (\lambda + 5)(\lambda - 2)$ so the eigenvalues of

$\begin{pmatrix} 4 & -6 \\ 3 & -7 \end{pmatrix}$ are -5 and 2. Let $\begin{pmatrix} x_1 \\ y_1 \end{pmatrix}$ denote the eigenvector corresponding to the eigenvalue -5. Then,

$\left\{ \begin{pmatrix} 4 & -6 \\ 3 & -7 \end{pmatrix} - (-5)\begin{pmatrix} 1 & 0 \\ 0 & 1 \end{pmatrix} \right\} \begin{pmatrix} x_1 \\ y_1 \end{pmatrix} = 0$. Simplifying yields the system of equations $\begin{cases} 9x_1 - 6y_1 = 0 \\ 3x_1 - 2y_1 = 0 \end{cases}$ so

$y_1 = \frac{3}{2}x_1$. Therefore, if x_1 is any real number $\begin{pmatrix} x_1 \\ \frac{3}{2}x_1 \end{pmatrix}$ is an eigenvector. In particular, if $x_1 = 2$, then

$\begin{pmatrix} 2 \\ 3 \end{pmatrix}$ is an eigenvector of $\begin{pmatrix} 4 & -6 \\ 3 & -7 \end{pmatrix}$ with corresponding eigenvalue -5. Similarly, if we let $\begin{pmatrix} x_2 \\ y_2 \end{pmatrix}$ denote

the eigenvector corresponding to 2, then $\left\{ \begin{pmatrix} 4 & -6 \\ 3 & -7 \end{pmatrix} - 2\begin{pmatrix} 1 & 0 \\ 0 & 1 \end{pmatrix} \right\} \begin{pmatrix} x_2 \\ y_2 \end{pmatrix} = 0$ which yields the system of

equations $\begin{cases} 2x_2 - 6y_2 = 0 \\ 3x_2 - 9y_2 = 0 \end{cases}$ so $y_2 = \frac{1}{3}x_2$. If $x_2 = 3$, then $\begin{pmatrix} 3 \\ 1 \end{pmatrix}$ is an eigenvector of $\begin{pmatrix} 4 & -6 \\ 3 & -7 \end{pmatrix}$ with

corresponding eigenvalue 2. ∎

■ **EXAMPLE 10.9**

Calculate the eigenvalues and corresponding eigenvectors of the matrices (i) $\begin{pmatrix} -2 & 2 & -4 \\ 7 & -7 & -6 \\ -7 & -1 & -2 \end{pmatrix}$; and

(ii) $\begin{pmatrix} 0 & 0 & -1 \\ 0 & 0 & -1 \\ 5 & -3 & -2 \end{pmatrix}$.

Solution:

Proceeding as in the previous examples, we begin by computing the characteristic polynomial of

$$\begin{pmatrix} -2 & 2 & -4 \\ 7 & -7 & -6 \\ -7 & -1 & -2 \end{pmatrix}, \text{ name the result } \mathbf{polya}:$$

```
▤▢▤▤▤▤▤▤▤▤▤ MatrixOperations ▤▤▤▤▤▤▤▤▤▢▤
In[19]:=
matrixa={{-2,2,-4},{7,-7,-6},{-7,-1,-2}};
polya=CharacteristicPolynomial[
                    matrixa,lambda]

Out[19]=
            2          3
320 + 16 lambda - 11 lambda - lambda
```

and then compute the solutions of the characteristic equation. We conclude that the eigenvalues are –8 (with multiplicity two) and 5.

```
In[20]:=
Solve[polya==0]

Out[20]=
{{lambda -> -8}, {lambda -> -8}, {lambda -> 5}}
```

To compute the eigenvector corresponding to –8 we first compute and simplify

$$\left\{ \begin{pmatrix} -2 & 2 & -4 \\ 7 & -7 & -6 \\ -7 & -1 & -2 \end{pmatrix} - (-8) \begin{pmatrix} 1 & 0 & 0 \\ 0 & 1 & 0 \\ 0 & 0 & 1 \end{pmatrix} \right\} \begin{pmatrix} x_1 \\ x_2 \\ x_3 \end{pmatrix}, \text{ name the resulting output } \mathbf{systemone}:$$

```
In[21]:=
systemone=(matrixa-(-8)IdentityMatrix[3]).
                    {x1,x2,x3}

Out[21]=
{6 x1 + 2 x2 - 4 x3, 7 x1 + x2 - 6 x3,
  -7 x1 - x2 + 6 x3}
```

and then solve the equation $\mathbf{systemone}$=0 for x_2 and x_3. The result means that if x_1 is any real

number $\begin{pmatrix} x_1 \\ -x_1 \\ x_1 \end{pmatrix}$ is an eigenvector of $\begin{pmatrix} -2 & 2 & -4 \\ 7 & -7 & -6 \\ -7 & -1 & -2 \end{pmatrix}$ with corresponding eigenvalue – 8. In particular,

if $x_1 = 1$, then $\begin{pmatrix} 1 \\ -1 \\ 1 \end{pmatrix}$ is an eigenvector of $\begin{pmatrix} -2 & 2 & -4 \\ 7 & -7 & -6 \\ -7 & -1 & -2 \end{pmatrix}$ with corresponding eigenvalue – 8.

```
In[22]:=
Solve[systemone==0,{x2,x3}]

Out[22]=
{{x2 -> -x1, x3 -> x1}}
```

Similarly, to find an eigenvector corresponding to the eigenvalue 5 we define **systemtwo** to be the

equation $\left\{ \begin{pmatrix} -2 & 2 & -4 \\ 7 & -7 & -6 \\ -7 & -1 & -2 \end{pmatrix} - 5 \begin{pmatrix} 1 & 0 & 0 \\ 0 & 1 & 0 \\ 0 & 0 & 1 \end{pmatrix} \right\} \begin{pmatrix} x_1 \\ x_2 \\ x_3 \end{pmatrix} = 0$ and then use **Solve** to solve **systemtwo**

for x_2 and x_3 in terms of x_1. The result means that if x_1 is any real number, an eigenvector

corresponding to the eigenvalue 5 is $\begin{pmatrix} x_1 \\ \dfrac{7}{6} x_1 \\ \dfrac{-7}{6} x_1 \end{pmatrix}$. In particular, if $x_1 = 6$, then $\begin{pmatrix} 6 \\ 7 \\ -7 \end{pmatrix}$ is an eigenvector with

corresponding eigenvalue 5.

```
In[23]:=
systemtwo=(matrixa-5IdentityMatrix[3]).
            {x1,x2,x3}==0

Out[23]=
{-7 x1 + 2 x2 - 4 x3, 7 x1 - 12 x2 - 6 x3,
    -7 x1 - x2 - 7 x3} == 0

In[24]:=
Solve[systemtwo,{x2,x3}]

Out[24]=
           7 x1         -7 x1
{{x2 ->  -----, x3 ->  ------}}
            6             6
```

To compute the eigenvalues and corresponding eigenvectors of $\begin{pmatrix} 0 & 0 & -1 \\ 0 & 0 & -1 \\ 5 & -3 & -2 \end{pmatrix}$, we first define

matrixb to be the matrix $\begin{pmatrix} 0 & 0 & -1 \\ 0 & 0 & -1 \\ 5 & -3 & -2 \end{pmatrix}$ and then use **Eigensystem** to compute the

eigenvalues and eigenvectors of **matrixb**. The result means that the eigenvalues of

$\begin{pmatrix} 0 & 0 & -1 \\ 0 & 0 & -1 \\ 5 & -3 & -2 \end{pmatrix}$ are $-1-i$, $-1+i$, and 0 with corresponding eigenvectors $\begin{pmatrix} 1-i \\ 1-i \\ 2 \end{pmatrix}$, $\begin{pmatrix} 1+i \\ 1+i \\ 2 \end{pmatrix}$,

and $\begin{pmatrix} 3 \\ 5 \\ 0 \end{pmatrix}$, respectively.

```
In[26]:=
matrixb={{0,0,-1},{0,0,-1},{5,-3,-2}};
eigs=Eigensystem[matrixb]

Out[26]=
{{-1 - I, -1 + I, 0},
  {{1 - I, 1 - I, 2}, {1 + I, 1 + I, 2},
   {3, 5, 0}}}
```
100% ▼

The **derivative** of an $n \times m$ matrix $A(t) = \begin{pmatrix} a_{11}(t) & a_{12}(t) & \cdots & a_{1m}(t) \\ a_{21}(t) & a_{22}(t) & \cdots & a_{2m}(t) \\ \vdots & \vdots & \ddots & \vdots \\ a_{n1}(t) & a_{n2}(t) & \cdots & a_{nm}(t) \end{pmatrix}$, where $a_{ij}(t)$ is differentiable

for all values of i and j, is $\dfrac{d}{dt} A(t) = \begin{pmatrix} \dfrac{d}{dt} a_{11}(t) & \dfrac{d}{dt} a_{12}(t) & \cdots & \dfrac{d}{dt} a_{1m}(t) \\ \dfrac{d}{dt} a_{21}(t) & \dfrac{d}{dt} a_{22}(t) & \cdots & \dfrac{d}{dt} a_{2m}(t) \\ \vdots & \vdots & \ddots & \vdots \\ \dfrac{d}{dt} a_{n1}(t) & \dfrac{d}{dt} a_{n2}(t) & \cdots & \dfrac{d}{dt} a_{nm}(t) \end{pmatrix}$; the **integral** of $A(t)$, where

$a_{ij}(t)$ is integrable for all values of i and j, is $\int A(t)\, dt = \begin{pmatrix} \int a_{11}(t)\, dt & \int a_{12}(t)\, dt & \cdots & \int a_{1m}(t)\, dt \\ \int a_{21}(t)\, dt & \int a_{22}(t)\, dt & \cdots & \int a_{2m}(t)\, dt \\ \vdots & \vdots & \ddots & \vdots \\ \int a_{n1}(t)\, dt & \int a_{n2}(t)\, dt & \cdots & \int a_{nm}(t)\, dt \end{pmatrix}$.

■ **EXAMPLE 10.10**

Find $\dfrac{d}{dt} A(t)$ and $\int A(t)\, dt$ if $A(t) = \begin{pmatrix} \cos(3t) & \sin(3t) & e^{-t} \\ t & t\sin\left(t^2\right) & e^t \end{pmatrix}$.

Solution:

We begin by defining **matrixa** to be $\begin{pmatrix} \cos(3t) & \sin(3t) & e^{-t} \\ t & t\sin\left(t^2\right) & e^t \end{pmatrix}$. We compute $\dfrac{d}{dt}A(t)$ by using

D to compute derivative of each term of **matrixa**.

```
≡□≡≡≡≡≡≡ MatrixOperations ≡≡≡≡≡≡≡

In[18]:=
matrixa={{Cos[3t],Sin[3t],Exp[-t]},
    {t,t Sin[t^2],Exp[t]}};
In[19]:=
D[matrixa,t]
Out[19]=
{{-3 Sin[3 t], 3 Cos[3 t], -E^-t},
  {1, 2 t^2 Cos[t^2] + Sin[t^2], E^t}}
```

Similarly, we use **Integrate** to compute an anti-derivative of each term of **matrixa** and conclude

$$\int A(t)\,dt = \begin{pmatrix} \dfrac{1}{3}\sin(3t) + c_{11} & \dfrac{-1}{3}\cos(3t) + c_{12} & -e^{-t} + c_{13} \\ \dfrac{1}{2}t^2 + c_{21} & \dfrac{-1}{2}\cos\left(t^2\right) + c_{22} & e^t + c_{23} \end{pmatrix}$$, where each c_{ij} represents an arbitrary

constant

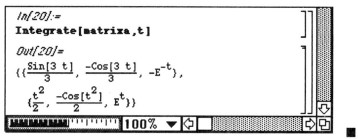

```
In[20]:=
Integrate[matrixa,t]
Out[20]=
{{Sin[3 t]/3, -Cos[3 t]/3, -E^-t},
  {t^2/2, -Cos[t^2]/2, E^t}}
                                100%  ▼
```

§10.2 Preliminary Definitions and Notation

Let $\mathbf{X} = \mathbf{X}(t) = \begin{pmatrix} x_1(t) \\ x_2(t) \\ \vdots \\ x_n(t) \end{pmatrix}$, $\mathbf{A}(t) = \begin{pmatrix} a_{11}(t) & a_{12}(t) & \cdots & a_{1n}(t) \\ a_{21}(t) & a_{22}(t) & \cdots & a_{2n}(t) \\ \vdots & \vdots & \ddots & \vdots \\ a_{n1}(t) & a_{n2}(t) & \cdots & a_{nn}(t) \end{pmatrix}$, and $\mathbf{F}(t) = \begin{pmatrix} f_1(t) \\ f_2(t) \\ \vdots \\ f_n(t) \end{pmatrix}$. Then,

the homogeneous system of first-order linear differential equations

$$\begin{cases} x_1'(t) = a_{11}(t)x_1(t) + a_{12}(t)x_2(t) + \ldots + a_{1n}(t)x_n(t) \\ x_2'(t) = a_{21}(t)x_1(t) + a_{22}(t)x_2(t) + \ldots + a_{2n}(t)x_n(t) \\ \quad\vdots \\ x_n'(t) = a_{n1}(t)x_1(t) + a_{n2}(t)x_2(t) + \ldots + a_{nn}(t)x_n(t) \end{cases}$$ is equivalent to $\mathbf{X}'(t) = \mathbf{A}(t)\mathbf{X}(t)$ and the

nonhomogeneous system $\begin{cases} x_1'(t) = a_{11}(t)x_1(t) + a_{12}(t)x_2(t) + \ldots + a_{1n}(t)x_n(t) + f_1(t) \\ x_2'(t) = a_{21}(t)x_1(t) + a_{22}(t)x_2(t) + \ldots + a_{2n}(t)x_n(t) + f_2(t) \\ \quad\vdots \\ x_n'(t) = a_{n1}(t)x_1(t) + a_{n2}(t)x_2(t) + \ldots + a_{nn}(t)x_n(t) + f_n(t) \end{cases}$ is

equivalent to $\mathbf{X}'(t) = \mathbf{A}(t)\mathbf{X}(t) + \mathbf{F}(t)$.

❑ **EXAMPLE 10.11**

Write the homogeneous system $\begin{cases} x' = -5x + 5y \\ y' = -5x + y \end{cases}$ in matrix form.

Solution:

The homogeneous system $\begin{cases} x' = -5x + 5y \\ y' = -5x + y \end{cases}$ is equivalent to the system $\begin{pmatrix} x' \\ y' \end{pmatrix} = \begin{pmatrix} -5 & 5 \\ -5 & 1 \end{pmatrix}\begin{pmatrix} x \\ y \end{pmatrix}$. ∎

❑ **EXAMPLE 10.12**

Display the nth order differential equation with constant coefficients

$$y^{(n)}(t) + a_{n-1}y^{(n-1)}(t) + \ldots + a_2y^{(2)}(t) + a_1y^{(1)}(t) + a_0y(t) = f(t)$$ as a system.

Solution:

Let $y^{(n)}(t) + a_{n-1}y^{(n-1)}(t) + \ldots + a_2y^{(2)}(t) + a_1y^{(1)}(t) + a_0y(t) = f(t)$ be an nth order differential

equation with constant coefficients. Let $x_1(t) = y(t)$, $x_2 = \dfrac{d}{dt}\left(x_1(t)\right) = y^{(1)}(t)$,

$x_3 = \dfrac{d}{dt}\left(x_2(t)\right) = y^{(2)}(t), \ldots, x_{n-1}(t) = \dfrac{d}{dt}\left(x_{n-2}(t)\right) = y^{(n-2)}$, and $x_n(t) = \dfrac{d}{dt}\left(x_{n-1}(t)\right) = y^{(n-1)}(t)$.

Then the equation $y^{(n)}(t) + a_{n-1}y^{(n-1)}(t) + \ldots + a_2 y^{(2)}(t) + a_1 y^{(1)}(t) + a_0 y(t) = f(t)$ is equivalent to

the system
$$\begin{cases} x_1' = x_2 \\ x_2' = x_3 \\ \vdots \\ x_{n-1}' = x_n \\ x_n' = -a_{n-1}y^{(n-1)}(t) - \ldots - a_2 y^{(2)}(t) - a_1 y^{(1)}(t) - a_0 y(t) + f(t) \end{cases}$$
which can be written in

the matrix form
$$\begin{pmatrix} x_1' \\ x_2' \\ \vdots \\ x_{n-1}' \\ x_n' \end{pmatrix} = \begin{pmatrix} 0 & 1 & 0 & 0 & 0 \\ 0 & 0 & 1 & 0 & 0 \\ \vdots & \vdots & \vdots & \ddots & \vdots \\ 0 & 0 & 0 & 0 & 1 \\ -a_0 & -a_1 & -a_2 & \cdots & -a_n \end{pmatrix} \begin{pmatrix} x_1 \\ x_2 \\ \vdots \\ x_{n-1} \\ x_n \end{pmatrix} + \begin{pmatrix} 0 \\ 0 \\ 0 \\ 0 \\ f(t) \end{pmatrix}. \blacksquare$$

❏ EXAMPLE 10.13

The **Van der Pol equation** is the nonlinear ordinary differential equation
$w'' - \mu\left(1 - w^2\right)w' + w = 0$. Write the Van der Pol equation as a system.

Solution:

Let $x(t) = w(t)$ and $y(t) = x'(t)$.

Then, $y'(t) = x''(t) = w''(t) = \mu\left(1 - w^2\right)w' - w = \mu\left(1 - x^2\right)y - x$

so the Van der Pol equation $w'' - \mu\left(1 - w^2\right)w' + w = 0$ is equivalent to the nonlinear

system $\begin{cases} x' = y \\ y' = \mu\left(1 - x^2\right)y - x \end{cases}$. $\blacksquare$

We now state the following theorems and terminology which are used in establishing the fundamentals of solving systems of differential equations. All proofs are omitted.

Let $\mathbf{X}'(t) = \mathbf{A}(t)\mathbf{X}(t)$ where $\mathbf{X}(t) = \begin{pmatrix} x_1(t) \\ x_2(t) \\ \vdots \\ x_n(t) \end{pmatrix}$ and $\mathbf{A}(t) = \begin{pmatrix} a_{11}(t) & a_{12}(t) & \cdots & a_{1n}(t) \\ a_{21}(t) & a_{22}(t) & \cdots & a_{2n}(t) \\ \vdots & \vdots & \ddots & \vdots \\ a_{n1}(t) & a_{n2}(t) & \cdots & a_{nn}(t) \end{pmatrix}$ where $a_{ij}(t)$ is continuous

for all $1 \le j \le n$ and $1 \le i \le n$.

Let $\left\{\Phi_i\right\}_{i=1}^m = \left\{\begin{pmatrix}\Phi_{1i}\\\Phi_{2i}\\\vdots\\\Phi_{ni}\end{pmatrix}\right\}_{i=1}^m$ be a set of m solutions of $\mathbf{X}'(t) = \mathbf{A}(t)\mathbf{X}(t)$.

The set $\left\{\Phi_i\right\}_{i=1}^m = \left\{\begin{pmatrix}\Phi_{1i}\\\Phi_{2i}\\\vdots\\\Phi_{ni}\end{pmatrix}\right\}_{i=1}^m$ is **linearly dependent** on an interval I means there is a set of

constants $\left\{c_i\right\}_{i=1}^m$ not all zero such that $\sum_{i=1}^m c_i \Phi_i = 0$; otherwise the set is **linearly independent.**

Theorem:

The set $\left\{\Phi_i\right\}_{i=1}^n = \left\{\begin{pmatrix}\Phi_{1i}\\\Phi_{2i}\\\vdots\\\Phi_{ni}\end{pmatrix}\right\}_{i=1}^n$ is linearly independent if and only if the **Wronskian**

$$W\left(\Phi_1, \Phi_2, \ldots \Phi_n\right) = \det\begin{pmatrix}\Phi_{11} & \Phi_{12} & \cdots & \Phi_{1n}\\\Phi_{21} & \Phi_{22} & \cdots & \Phi_{2n}\\\vdots & \vdots & \ddots & \vdots\\\Phi_{n1} & \Phi_{n2} & \cdots & \Phi_{nn}\end{pmatrix} \neq 0.$$

Theorem:
Any n+1 non-trivial solutions of $\mathbf{X}'(t)= \mathbf{A}(t)\mathbf{X}(t)$ are linearly dependent.

Theorem:
There is a set of n non-trivial linearly independent solutions of $\mathbf{X}'(t)=\mathbf{A}(t)\mathbf{X}(t)$.

Theorem:

Let $\left\{\Phi_i\right\}_{i=1}^m = \left\{\begin{pmatrix}\Phi_{1i}\\\Phi_{2i}\\\vdots\\\Phi_{ni}\end{pmatrix}\right\}_{i=1}^n$ be a set of n linearly independent solutions of $\mathbf{X}'(t) = \mathbf{A}(t)\mathbf{X}(t)$.

Then every solution of $\mathbf{X}'(t)=\mathbf{A}(t)\mathbf{X}(t)$ is a linear combination of these solutions.

Definition:

Let $\{\Phi_i\}_{i=1}^m = \left\{\begin{pmatrix} \Phi_{1i} \\ \Phi_{2i} \\ \vdots \\ \Phi_{ni} \end{pmatrix}\right\}_{i=1}^n$ be a set of n linearly independent solutions of $\mathbf{X}'(t) = \mathbf{A}(t)\mathbf{X}(t)$.

Then the matrix $\Phi(t) = \begin{pmatrix} \Phi_1 & \Phi_2 & \cdots & \Phi_n \end{pmatrix} = \begin{pmatrix} \Phi_{11} & \Phi_{12} & \cdots & \Phi_{1n} \\ \Phi_{21} & \Phi_{22} & \cdots & \Phi_{2n} \\ \vdots & \vdots & \ddots & \vdots \\ \Phi_{n1} & \Phi_{n2} & \cdots & \Phi_{nn} \end{pmatrix}$ is called a

fundamental matrix of the system $\mathbf{X}'(t)=\mathbf{A}(t)\mathbf{X}(t)$.

❑ **EXAMPLE 10.14**

Show that $\Phi(t) = \begin{pmatrix} e^{-2t} & -3e^{5t} \\ 2e^{-2t} & e^{5t} \end{pmatrix}$ is a fundamental matrix for the system $\mathbf{X}'(t) = \begin{pmatrix} 4 & -3 \\ -2 & -1 \end{pmatrix}\mathbf{X}(t)$.

Solution:

Since $\begin{pmatrix} e^{-2t} \\ 2e^{-2t} \end{pmatrix}' = \begin{pmatrix} -2e^{-2t} \\ -4e^{-2t} \end{pmatrix} = \begin{pmatrix} 4 & -3 \\ -2 & -1 \end{pmatrix}\begin{pmatrix} e^{-2t} \\ 2e^{-2t} \end{pmatrix}$ and $\begin{pmatrix} -3e^{5t} \\ e^{5t} \end{pmatrix}' = \begin{pmatrix} -15e^{5t} \\ 5e^{5t} \end{pmatrix} = \begin{pmatrix} 4 & -3 \\ -2 & -1 \end{pmatrix}\begin{pmatrix} -3e^{5t} \\ e^{5t} \end{pmatrix}$, both

$\begin{pmatrix} e^{-2t} \\ 2e^{-2t} \end{pmatrix}$ and $\begin{pmatrix} -3e^{5t} \\ e^{5t} \end{pmatrix}$ are solutions of the system $\mathbf{X}'(t) = \begin{pmatrix} 4 & -3 \\ -2 & -1 \end{pmatrix}\mathbf{X}(t)$. The solutions are linearly

independent since $W\left(\begin{pmatrix} e^{-2t} \\ 2e^{-2t} \end{pmatrix}, \begin{pmatrix} -3e^{5t} \\ e^{5t} \end{pmatrix}\right) = \det\begin{pmatrix} e^{-2t} & -3e^{5t} \\ 2e^{-2t} & e^{5t} \end{pmatrix} = 7e^{3t} \neq 0.$ ∎

§10.3 Homogeneous Linear Systems with Constant Coefficients

Let $\mathbf{A} = \mathbf{A}(t) = \begin{pmatrix} a_{11} & a_{12} & \cdots & a_{1n} \\ a_{21} & a_{22} & \cdots & a_{2n} \\ \vdots & \vdots & \ddots & \vdots \\ a_{n1} & a_{n2} & \cdots & a_{nn} \end{pmatrix}$ be an $n \times n$ real matrix and let $\{\lambda_k\}_{k=1}^n$ be the eigenvalues and

$\{\mathbf{v}_k\}_{k=1}^n$ the corresponding eigenvectors of $\mathbf{A}$.

Generally, *Mathematica*, or any mathematical software package, can perform nearly all of the routine linear algebra calculations necessary in this section. For a review of the prerequisite linear algebra, see **Section 10.1**.

Then the general solution of the system $X'(t)=A(t)X(t)$ is determined by the eigenvalues of A. For the moment, we consider the cases when the eigenvalues of **A** are distinct and real or the eigenvalues of **A** are distinct and complex. We will consider the case when **A** has repeated eigenvalues (eigenvalues of multiplicity greater than one) separately.

If the eigenvalues $\{\lambda_k\}_{k=1}^n$ are distinct and real then the general solution of $X'(t) = A(t)X(t)$ is

$$X(t) = c_1 v_1 e^{\lambda_1 t} + c_2 v_2 e^{\lambda_2 t} + ... + c_n v_n e^{\lambda_n t} = \sum_{i=1}^n c_i v_i e^{\lambda_i t}.$$

If the eigenvalues $\{\lambda_k = \alpha_k + \beta_k i\}_{k=1}^m$ where $\beta_k \neq 0$ are complex and the eigenvalues $\{\lambda_k\}_{k=2m+1}^n$ are distinct and real then the general solution of $X'(t) = A(t)X(t)$ is

$$X(t) = c_1 w_{11} e^{\lambda_1 t} + c_2 w_{12} e^{\lambda_1 t} + c_3 w_{21} e^{\lambda_2 t} + c_4 w_{22} e^{\lambda_2 t} + c_{2m-1} w_{m1} e^{\lambda_m t} +$$
$$c_{2m} w_{m2} e^{\lambda_m t} + c_{2m+1} v_{2m+1} e^{\lambda_2 t} ... + c_n v_n e^{\lambda_n t} \text{, where}$$

$$w_{i1} = \frac{1}{2}\left[v_i + \overline{v_i}\right]\cos(\beta_i t) + \frac{i}{2}\left[v_i - \overline{v_i}\right]\sin(\beta_i t) \text{ and } w_{i2} = \frac{i}{2}\left[v_i - \overline{v_i}\right]\cos(\beta_i t) - \frac{1}{2}\left[v_i - \overline{v_i}\right]\sin(\beta_i t).$$

❏ **EXAMPLE 10.15**

Find the general solution of the system $X'(t) = \begin{pmatrix} 5 & -1 \\ 0 & 3 \end{pmatrix} X(t)$.

Solution:

Let $X(t) = \begin{pmatrix} x(t) \\ y(t) \end{pmatrix}$. Then the system $X'(t) = \begin{pmatrix} 5 & -1 \\ 0 & 3 \end{pmatrix} X(t)$ can be rewritten in the form

$\begin{pmatrix} x' \\ y' \end{pmatrix} = \begin{pmatrix} 5 & -1 \\ 0 & 3 \end{pmatrix}\begin{pmatrix} x \\ y \end{pmatrix}$ which is equivalent to the system $\begin{cases} x' = 5x - y \\ y' = 3y \end{cases}$.

The eigenvalues and corresponding eigenvectors of the matrix $\begin{pmatrix} 5 & -1 \\ 0 & 3 \end{pmatrix}$ are $\lambda_1 = 3, v_1 = \begin{pmatrix} 1 \\ 2 \end{pmatrix}$,

$\lambda_2 = 5$, and $v_2 = \begin{pmatrix} 1 \\ 0 \end{pmatrix}$. Therefore, a general solution of the system $X'(t) = \begin{pmatrix} 5 & -1 \\ 0 & 3 \end{pmatrix} X(t)$ is

$$\mathbf{X} = c_1 \begin{pmatrix} 1 \\ 2 \end{pmatrix} e^{3t} + c_2 \begin{pmatrix} 1 \\ 0 \end{pmatrix} e^{5t}. \quad \blacksquare$$

■ EXAMPLE 10.16

Solve $\mathbf{X}'(t) = \begin{pmatrix} 5 & 2 \\ -2 & 2 \end{pmatrix} \mathbf{X}(t)$.

Solution:

We begin by defining **matrix** to be the matrix $\begin{pmatrix} 5 & 2 \\ -2 & 2 \end{pmatrix}$, then computing the eigenvalues

and eigenvectors of $\begin{pmatrix} 5 & 2 \\ -2 & 2 \end{pmatrix}$ with the command **Eigensystem**, and naming the resulting

nested list **eigs**.

```
Systems(ComplexEigenvalues)

In[3]:=
Clear[matrix]
matrix={{5,2},{-2,2}};
MatrixForm[matrix]

Out[3]//MatrixForm=
5    2
-2    2

In[4]:=
eigs=Eigensystem[matrix]

Out[4]=
{{7 - I Sqrt[7]    7 + I Sqrt[7]
  -------------  , -------------},
       2                2
  {{-3 + I Sqrt[7]      -3 - I Sqrt[7]
    -------------, 1}, {-------------, 1}}}
         4                    4
```

The result means that $(7 - i\sqrt{7})/2$, extracted from **eigs** with **eigs[[1, 1]]**, is an eigenvalue of

$\begin{pmatrix} 5 & 2 \\ -2 & 2 \end{pmatrix}$ with corresponding eigenvector $\begin{pmatrix} (-3 + i\sqrt{7})/4 \\ 1 \end{pmatrix}$, extracted from **eigs** with **eigs[[2, 1]]**,

and $(7 + i\sqrt{7})/2$, extracted from **eigs** with **eigs[[1, 2]]**, is an eigenvalue of $\begin{pmatrix} 5 & 2 \\ -2 & 2 \end{pmatrix}$ with

corresponding eigenvector $\begin{pmatrix} (-3 - i\sqrt{7})/4 \\ 1 \end{pmatrix}$, extracted from **eigs** with **eigs[[2, 2]]**.

With the following commands we first extract $\left(7 - i\sqrt{7}\right)/2$ and $\begin{pmatrix} \left(-3 + i\sqrt{7}\right)/4 \\ 1 \end{pmatrix}$ from **eigs**:

```
In[6]:=
eigs[[1,1]]
eigs[[2,1]]

Out[5]=
7 - I Sqrt[7]
―――――――――
    2

Out[6]=
{-3 + I Sqrt[7]
 ―――――――――, 1}
      4
```

and then verify that $\left(7 - i\sqrt{7}\right)/2$ is an eigenvalue of $\begin{pmatrix} 5 & 2 \\ -2 & 2 \end{pmatrix}$ with corresponding eigenvector

$\begin{pmatrix} \left(-3 + i\sqrt{7}\right)/4 \\ 1 \end{pmatrix}$ by computing and simplifying $\left[\begin{pmatrix} 5 & 2 \\ -2 & 2 \end{pmatrix} - \left(7 - i\sqrt{7}\right)/2 \begin{pmatrix} 1 & 0 \\ 0 & 1 \end{pmatrix}\right]\begin{pmatrix} \left(-3 + i\sqrt{7}\right)/4 \\ 1 \end{pmatrix}$.

```
In[7]:=
(matrix-eigs[[1,1]] IdentityMatrix[2]).
    eigs[[2,1]]//Simplify

Out[7]=
{0, 0}
```

In this case the eigenvalues are complex conjugates so we let

$$b_1 = \frac{1}{2}\left[\begin{pmatrix} \left(-3 + i\sqrt{7}\right)/4 \\ 1 \end{pmatrix} + \begin{pmatrix} \left(-3 - i\sqrt{7}\right)/4 \\ 1 \end{pmatrix}\right] = \begin{pmatrix} -3/4 \\ 1 \end{pmatrix} \text{ and}$$

$$b_2 = \frac{i}{2}\left[\begin{pmatrix} \left(-3 + i\sqrt{7}\right)/4 \\ 1 \end{pmatrix} - \begin{pmatrix} \left(-3 - i\sqrt{7}\right)/4 \\ 1 \end{pmatrix}\right] = \begin{pmatrix} -\sqrt{7}/4 \\ 0 \end{pmatrix}.$$

```
In[8]:=
b1=1/2 (eigs[[2,1]]+eigs[[2,2]])//Simplify

Out[8]=
      3
{-(―), 1}
      4

In[9]:=
b2=I/2(eigs[[2,1]]-eigs[[2,2]])//Simplify

Out[9]=
{-Sqrt[7]
 ―――――――, 0}
     4
```

Therefore the general solution is

$$\mathbf{X}(t) = c_1\left[\binom{-3/4}{1}\cos\left(\frac{\sqrt{7}\,t}{2}\right)+\binom{-\sqrt{7}/4}{0}\sin\left(\frac{\sqrt{7}\,t}{2}\right)\right]e^{7t/2} +$$

$$c_2\left[\binom{-3/4}{1}\cos\left(\frac{\sqrt{7}\,t}{2}\right)+\binom{-\sqrt{7}/4}{0}\sin\left(\frac{\sqrt{7}\,t}{2}\right)\right]e^{7t/2}.$$

In this case we can use **DSolve** to obtain the same results by noticing that

the matrix equation $\mathbf{X}'(t) = \begin{pmatrix} 5 & 2 \\ -2 & 2 \end{pmatrix}\mathbf{X}(t)$ is equivalent to the system $\begin{cases} x' = 5x + 2y \\ y' = -2x + 2y \end{cases}$.

In the following command, we use **DSolve** to solve the system and name the resulting output **sol**.

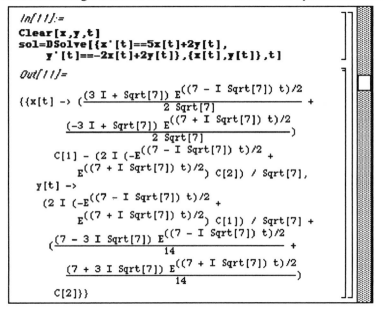

The general solution for x(t) is obtained with **sol[[1,1,2]]**. **ComplexExpand** is used to simplify the result assuming that all the variables are real.

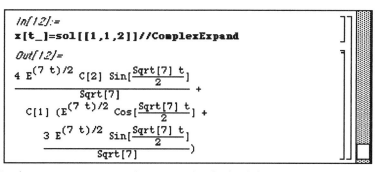

In[12]:=
x[t_]=sol[[1,1,2]]//ComplexExpand

Out[12]=
$$\frac{4\ E^{(7\ t)/2}\ C[2]\ Sin[\frac{Sqrt[7]\ t}{2}]}{Sqrt[7]} +$$
$$C[1]\ (E^{(7\ t)/2}\ Cos[\frac{Sqrt[7]\ t}{2}] +$$
$$\frac{3\ E^{(7\ t)/2}\ Sin[\frac{Sqrt[7]\ t}{2}]}{Sqrt[7]})$$

In the same manner as above, y(t) is obtained from **sol** with **sol[[1,2,2]]** and simplified with **ComplexExpand**.

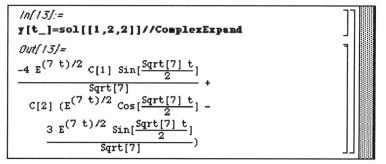

In[13]:=
y[t_]=sol[[1,2,2]]//ComplexExpand

Out[13]=
$$\frac{-4\ E^{(7\ t)/2}\ C[1]\ Sin[\frac{Sqrt[7]\ t}{2}]}{Sqrt[7]} +$$
$$C[2]\ (E^{(7\ t)/2}\ Cos[\frac{Sqrt[7]\ t}{2}] -$$
$$\frac{3\ E^{(7\ t)/2}\ Sin[\frac{Sqrt[7]\ t}{2}]}{Sqrt[7]})$$

The solution is graphed for various values of C[1] and C[2] by first creating a table of {x[t],y[t]} with C[1] replaced by i and C[2] replaced by j for i=−6,−4,−2,0,2,4,6 and j=−6,−4,−2,0,2,4,6 and naming the resulting table **funarray**. In order to convert **funarray** to a list of ordered-functions to be graphed with **ParametricPlot**, **Flatten** is used to remove parentheses from **funarray** and the resulting list of ordered-functions is named **tograph**. **tograph** is displayed in an abbreviated two-line form with **Short**.

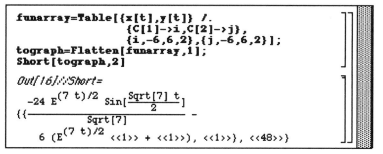

```
funarray=Table[{x[t],y[t]} /.
              {C[1]->i,C[2]->j},
              {i,-6,6,2},{j,-6,6,2}];
tograph=Flatten[funarray,1];
Short[tograph,2]
```

Out[16]//Short=
$$\{\{\frac{-24\ E^{(7\ t)/2}\ Sin[\frac{Sqrt[7]\ t}{2}]}{Sqrt[7]} -$$
$$6\ (E^{(7\ t)/2}\ \ll1\gg + \ll1\gg), \ll1\gg\}, \ll48\gg\}$$

ParametricPlot is then used to graph **tograph** on [−1,1] and the resulting graph is named **graphone** for later use.

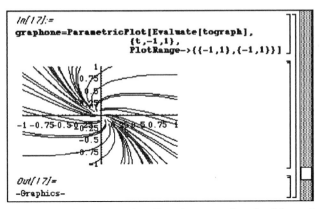

```
In[17]:=
graphone=ParametricPlot[Evaluate[tograph],
                        {t,-1,1},
                        PlotRange->{{-1,1},{-1,1}}]
```

```
Out[17]=
-Graphics-
```

In order to determine the direction associated with each solution, we make use of *Mathematica*'s
PlotField package which is loaded below. Note that in a first-order system in two variables:
X'=X(x,y), Y'=Y(x,y),
each solution must satisfy the relationship

$$\frac{dy}{dx}=\frac{dy/dt}{dx/dt}=\frac{Y(x,y)}{X(x,y)}.$$

Therefore, the direction field associated with the system of differential equations represents a collection
of vectors which are tangent to the family of solutions to the system. The direction field for this example
is plotted with **PlotVectorField** below to illustrate numerous tangent lines to each solution.

```
In[18]:=
<<Graphics`PlotField`
```

```
In[19]:=
graphtwo=PlotVectorField[{5x+2y,-2x+2y},
                         {x,-1,1},{y,-1,1}]
```

```
Out[19]=
-Graphics-
```

By displaying the family of solutions with the direction field below, we see that each solution is directed away from the origin as t increases.

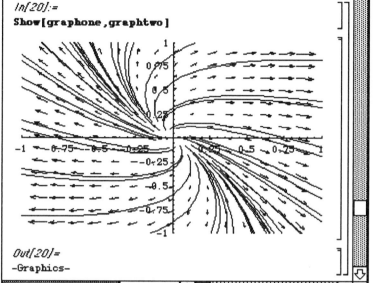

We now consider the case of repeated eigenvalues. Let us first restrict our attention to a system with the repeated eigenvalue $\lambda_1 = \lambda_2$. Assume that all other eigenvalues are distinct. Hence, we obtain one solution to the system $\mathbf{X}_1 = \mathbf{v}_1 e^{\lambda_1 t}$ which corresponds to λ_1. We now seek a second solution corresponding λ_1 in a manner similar to that considered in the case of repeated characteristic values of higher order equations. In this case, however, suppose that the second linearly independent solution corresponding λ_1 is of the form

$$\mathbf{X}_2 = (\mathbf{v}_2 t + \mathbf{w}_2) e^{\lambda_1 t}.$$

Substitution of this solution into the system of equations yields the relationships $\mathbf{v}_2 = \mathbf{v}_1$ and $(A - \lambda_1 I)\mathbf{w}_2 = \mathbf{v}_1$.

Hence, the second solution is $\mathbf{X}_2 = (\mathbf{v}_1 t + \mathbf{w}_2) e^{\lambda_1 t}$ which is found by solving the system $(A - \lambda_1 I)\mathbf{w}_2 = \mathbf{v}_1$.

In the case of three equal eigenvalues $\lambda_1 = \lambda_2 = \lambda_3$, we assume that

$$\mathbf{X}_1 = \mathbf{v}_1 e^{\lambda_1 t}, \ \mathbf{X}_2 = (\mathbf{v}_2 t + \mathbf{w}_2) e^{\lambda_1 t}, \text{ and } \mathbf{X}_3 = (\mathbf{v}_3 t^2 + \mathbf{w}_3 t + \mathbf{u}_3) e^{\lambda_1 t}.$$

Substitution of these solutions into the system of differential equations yields the system of equations which are solved for the unknown vectors $\mathbf{v}_2$, $\mathbf{w}_2$, $\mathbf{v}_3$, $\mathbf{w}_3$, and $\mathbf{u}_3$:

$$\lambda_1 \mathbf{v}_2 = A\mathbf{v}_2, (A - \lambda_1 I)\mathbf{w}_2 = \mathbf{v}_2, \lambda_1 \mathbf{v}_3 = A\mathbf{v}_3, (A - \lambda_1 I)\mathbf{w}_3 = \mathbf{v}_3, \text{ and } (A - \lambda_1 I)\mathbf{u}_3 = \mathbf{w}_3.$$

Similar to the previous case, $v_3=v_2=v_1$ and the vectors w_2, w_3, and u_3 are found by solving the appropriate system. Notice that this method is easily generalized for instances when the multiplicity of the repeated eigenvalue is greater than three. The case of two equal eigenvalues is demonstrated in the following example while that of three is investigated in the applications chapter which follows.

■ **EXAMPLE 10.17**

Solve $X'(t) = \begin{pmatrix} 5 & 3 & -3 \\ 2 & 4 & -5 \\ -4 & 2 & -3 \end{pmatrix} X(t)$.

Solution:

As in the previous example, we define **matrix** to be the matrix $\begin{pmatrix} 5 & 3 & -3 \\ 2 & 4 & -5 \\ -4 & 2 & -3 \end{pmatrix}$ and then use

Eigensystem to compute the eigenvalues and corresponding eigenvectors of $\begin{pmatrix} 5 & 3 & -3 \\ 2 & 4 & -5 \\ -4 & 2 & -3 \end{pmatrix}$.

```
╔═════════════ Systems(RepeatedEigenvalues) ═════════════╗
In[40]:=

Clear[matrix]
matrix={{5,3,-3},{2,4,-5},{-4,2,-3}};
MatrixForm[matrix]

Out[40]//MatrixForm=
5    3    -3
2    4    -5
-4   2    -3

In[41]:=
eigs=Eigensystem[matrix]

Out[41]=
{{-1, -1, 8}, {{0, 1, 1}, {0, 0, 0},
   {-(9/2), -(7/2), 1}}}
```

In this case, the eigenvalue -1 has multiplicity two and corresponding eigenvector $\begin{pmatrix} 0 \\ 1 \\ 1 \end{pmatrix}$ and

the eigenvalue 8 has corresponding eigenvector $\begin{pmatrix} -9/2 \\ -7/2 \\ 1 \end{pmatrix}$. In the following, we verify that -1

is an eigenvalue with corresponding eigenvector $\begin{pmatrix} 0 \\ 1 \\ 1 \end{pmatrix}$ by first extracting -1 and $\begin{pmatrix} 0 \\ 1 \\ 1 \end{pmatrix}$ from **eigs**

with the commands **eigs[[1,1]]** and **eigs[[2,1]]**:

```
In[43]:=
eigs[[1,1]]
eigs[[2,1]]
Out[42]=
-1
Out[43]=
{0, 1, 1}
```

and then computing and simplifying $\left[\begin{pmatrix} 5 & 3 & -3 \\ 2 & 4 & -5 \\ -4 & 2 & -3 \end{pmatrix} - \begin{pmatrix} 1 & 0 & 0 \\ 0 & 1 & 0 \\ 0 & 0 & 1 \end{pmatrix} \right] \begin{pmatrix} 0 \\ 1 \\ 1 \end{pmatrix}$.

```
In[44]:=
(matrix-eigs[[1,1]] IdentityMatrix[3]).
    eigs[[2,1]]//Simplify
Out[44]=
{0, 0, 0}
```

Since $\begin{pmatrix} 0 \\ 1 \\ 1 \end{pmatrix}$ is the only eigenvector corresponding to -1, there is a solution of the form

$$sol_1(t) = \begin{pmatrix} 0 \\ 1 \\ 1 \end{pmatrix} e^{-t} = \begin{pmatrix} 0 \\ e^{-t} \\ e^{-t} \end{pmatrix} \text{ defined below:}$$

```
In[45]:=
sol1[t_]=eigs[[2,1]] Exp[eigs[[1,1]] t]
Out[45]=
{0, E^-t, E^-t}
```

and another linearly independent solution of the form $\text{sol}_2(t) = t\,\text{sol}_1(t) + \begin{pmatrix} a \\ b \\ c \end{pmatrix} e^{-t} = \begin{pmatrix} a\,e^{-t} \\ (b+t)\,e^{-t} \\ (c+t)\,e^{-t} \end{pmatrix}$,

where a, b, and c are constants to be determined, also defined below.

```
In[46]:=
sol2[t_]=t sol1[t]+{a,b,c} Exp[eigs[[1,1]] t]
Out[46]=
{a/E^t, b/E^t + t/E^t, c/E^t + t/E^t}
```

Since $\text{sol}_2(t)$ is a solution of $\mathbf{X}'(t) = \begin{pmatrix} 5 & 3 & -3 \\ 2 & 4 & -5 \\ -4 & 2 & -3 \end{pmatrix}\mathbf{X}(t)$, $\text{sol}_2'(t) = \begin{pmatrix} 5 & 3 & -3 \\ 2 & 4 & -5 \\ -4 & 2 & -3 \end{pmatrix}\text{sol}_2(t)$.

We compute this below and name the resulting system of equations **second**:

```
In[47]:=
second=sol2'[t]==matrix.sol2[t]//Cancel
Out[47]=
{-(a/E^t), E^-t - b/E^t - t/E^t, E^-t - c/E^t - t/E^t} ==
 {5 a/E^t + 3 (b/E^t + t/E^t) - 3 (c/E^t + t/E^t),
  2 a/E^t + 4 (b/E^t + t/E^t) - 5 (c/E^t + t/E^t),
  -4 a/E^t + 2 (b/E^t + t/E^t) - 3 (c/E^t + t/E^t)}
```

and then solve the system of equations **second** with **Solve** and name the resulting solution **vals**.

```
In[48]:=
vals=Solve[second]
Out[48]=
{{b -> (1 + 4 c)/4, a -> -(1/8)}}
```

Then the values of a, b, and c that make sol_2 (t) a solution are $a = \dfrac{-1}{8}$, $b = \dfrac{1+4c}{4}$, and c any real

number. In particular if $c = 0$, then $a = \dfrac{-1}{8}$ and $b = \dfrac{1}{4}$.

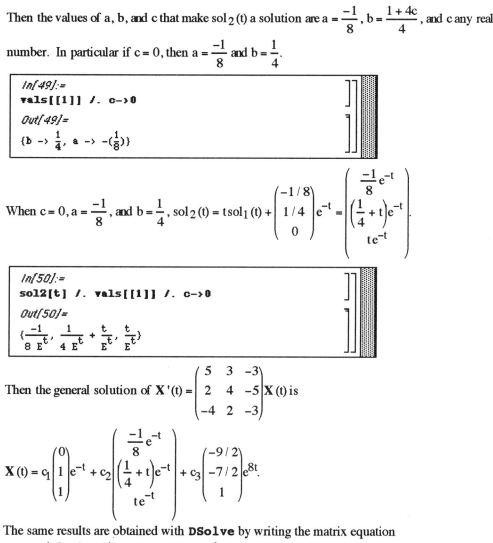

```
In[49]:=
vals[[1]] /. c->0
Out[49]=
{b -> 1/4, a -> -(1/8)}
```

When $c = 0$, $a = \dfrac{-1}{8}$, and $b = \dfrac{1}{4}$, sol_2 (t) $= t\text{sol}_1$ (t) $+ \begin{pmatrix} -1/8 \\ 1/4 \\ 0 \end{pmatrix} e^{-t} = \begin{pmatrix} \dfrac{-1}{8} e^{-t} \\ \left(\dfrac{1}{4}+t\right)e^{-t} \\ te^{-t} \end{pmatrix}$.

```
In[50]:=
sol2[t] /. vals[[1]] /. c->0
Out[50]=
{-1/(8 E^t), 1/(4 E^t) + t/E^t, t/E^t}
```

Then the general solution of $X'(t) = \begin{pmatrix} 5 & 3 & -3 \\ 2 & 4 & -5 \\ -4 & 2 & -3 \end{pmatrix} X(t)$ is

$$X(t) = c_1 \begin{pmatrix} 0 \\ 1 \\ 1 \end{pmatrix} e^{-t} + c_2 \begin{pmatrix} \dfrac{-1}{8} e^{-t} \\ \left(\dfrac{1}{4}+t\right)e^{-t} \\ te^{-t} \end{pmatrix} + c_3 \begin{pmatrix} -9/2 \\ -7/2 \\ 1 \end{pmatrix} e^{8t}.$$

The same results are obtained with **DSolve** by writing the matrix equation

$X'(t) = \begin{pmatrix} 5 & 3 & -3 \\ 2 & 4 & -5 \\ -4 & 2 & -3 \end{pmatrix} X(t)$ as the system $\begin{cases} x' = 5x + 3y - 3z \\ y' = 2x + 4y - 5z. \\ z' = -4x + 2y - 3z \end{cases}$

```
In[2]:=
Clear[x,y,z,t]
sol=DSolve[{x'[t]==5x[t]+3y[t]-3z[t],
    y'[t]==2x[t]+4y[t]-5z[t],
    z'[t]==-4x[t]+2y[t]-3z[t]},
      {x[t],y[t],z[t]},t]
```

$$Out[2]=$$

$$\{\{x[t] \to (C[1] + 2\ E^{9\,t}\ C[1] - C[2] + E^{9\,t}\ C[2] +$$
$$\quad C[3] - E^{9\,t}\ C[3]) \,/\, (3\ E^{t}),$$
$$\ y[t] \to$$
$$\quad (-14\ C[1] + 14\ E^{9\,t}\ C[1] - 72\ t\ C[1] +$$
$$\quad\quad 20\ C[2] + 7\ E^{9\,t}\ C[2] + 72\ t\ C[2] + 7\ C[3] -$$
$$\quad\quad 7\ E^{9\,t}\ C[3] - 72\ t\ C[3]) \,/\, (27\ E^{t}),$$
$$\ z[t] \to$$
$$\quad (4\ C[1] - 4\ E^{9\,t}\ C[1] - 72\ t\ C[1] + 2\ C[2] -$$
$$\quad\quad 2\ E^{9\,t}\ C[2] + 72\ t\ C[2] + 25\ C[3] +$$
$$\quad\quad 2\ E^{9\,t}\ C[3] - 72\ t\ C[3]) \,/\, (27\ E^{t})\}\}$$

x(t) is extracted from **sol** with **sol[[1,1,2]]**:

```
In[3]:=
x[t_]=sol[[1,1,2]]
```

$$Out[3]=$$

$$(C[1] + 2\ E^{9\,t}\ C[1] - C[2] + E^{9\,t}\ C[2] + C[3] -$$
$$\quad E^{9\,t}\ C[3]) \,/\, (3\ E^{t})$$

y(t) extracted with **sol[[1,2,2]]**:

```
In[4]:=
y[t_]=sol[[1,2,2]]
```

$$Out[4]=$$

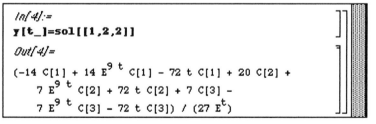

$$(-14\ C[1] + 14\ E^{9\,t}\ C[1] - 72\ t\ C[1] + 20\ C[2] +$$
$$\quad 7\ E^{9\,t}\ C[2] + 72\ t\ C[2] + 7\ C[3] -$$
$$\quad 7\ E^{9\,t}\ C[3] - 72\ t\ C[3]) \,/\, (27\ E^{t})$$

and z(t) is extracted with **sol[[1,3,2]]**.

```
In[5]:=
z[t_]=sol[[1,3,2]]

Out[5]=
(4 C[1] - 4 E^9 t C[1] - 72 t C[1] + 2 C[2] -
   2 E^9 t C[2] + 72 t C[2] + 25 C[3] +
   2 E^9 t C[3] - 72 t C[3]) / (27 E^t)
```

A table is produced for certain constant values in **funarray**. This table is then flattened to level two so that the elements of **tograph** are of the form {x(t), y(t), z(t)}.

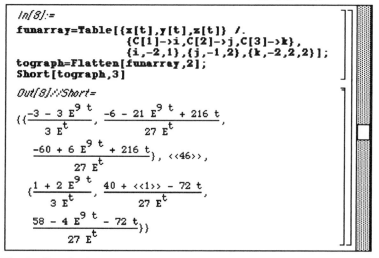

```
In[8]:=
funarray=Table[{x[t],y[t],z[t]} /.
               {C[1]->i,C[2]->j,C[3]->k},
               {i,-2,1},{j,-1,2},{k,-2,2,2}];
tograph=Flatten[funarray,2];
Short[tograph,3]

Out[8]//Short=
{{ (-3 - 3 E^9 t)/(3 E^t) , (-6 - 21 E^9 t + 216 t)/(27 E^t) ,

  (-60 + 6 E^9 t + 216 t)/(27 E^t) }, <<46>>,

 { (1 + 2 E^9 t)/(3 E^t) , (40 + <<1>> - 72 t)/(27 E^t) ,

  (58 - 4 E^9 t - 72 t)/(27 E^t) }}
```

The family of solutions in **tograph** are then plotted with **Plot3D** below.

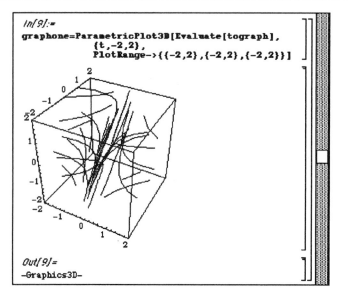

```
In[9]:=
graphone=ParametricPlot3D[Evaluate[tograph],
              {t,-2,2},
              PlotRange->{{-2,2},{-2,2},{-2,2}}]
```

```
Out[9]=
-Graphics3D-
```

The direction associated with these three-dimensional solutions can be determined with **PlotField3D**.

```
In[10]:=
<<Graphics`PlotField3D`
```

```
In[13]:=
graphtwo=PlotVectorField3D[
         {5x+3y-3z,2x+4y-5z,-4x+2y-3z},
         {x,-2,2},{y,-2,2},{z,-2,2},
         VectorHeads->True]
```

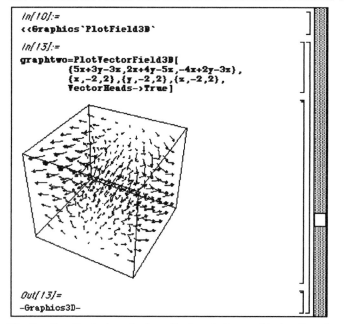

```
Out[13]=
-Graphics3D-
```

The family of solutions is then displayed along with the direction field.

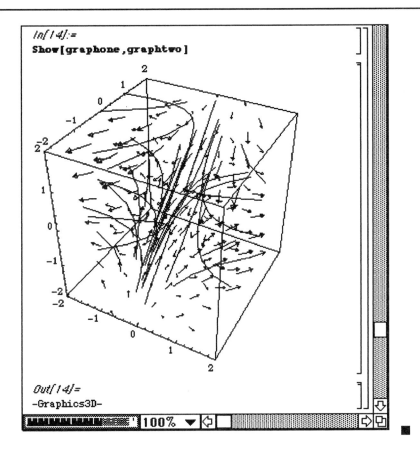

In[14]:=
Show[graphone,graphtwo]

Out[14]=
-Graphics3D-

100%

■ EXAMPLE 10.18

Solve $\mathbf{X}'(t) = \begin{pmatrix} 0 & 0 & 3 \\ 1 & -4 & 2 \\ 0 & -4 & 1 \end{pmatrix} \mathbf{X}(t)$.

Solution:

Proceeding as in the previous examples, we first define **matrix** to be the matrix

$\begin{pmatrix} 0 & 0 & 3 \\ 1 & -4 & 2 \\ 0 & -4 & 1 \end{pmatrix}$ and then compute the eigenvalues and corresponding eigenvectors of $\begin{pmatrix} 0 & 0 & 3 \\ 1 & -4 & 2 \\ 0 & -4 & 1 \end{pmatrix}$.

```
▤□▤▤▤▤  Systems(ReandComEigenvalues) ▤▤▤▤▣▤

In[3]:=

Clear[matrix]
matrix={{0,0,3},{1,-4,2},{0,-4,1}};
MatrixForm[matrix]

Out[3]//MatrixForm=
0     0     3
1    -4     2
0    -4     1

In[4]:=

eigs=Eigensystem[matrix]

Out[4]=

{{-3, -2 I, 2 I}, {{-1, 1, 1}, {3 I/2, 1/4 + I/2, 1},

    {-3 I/2, 1/4 - I/2, 1}}}
```

In this case, one eigenvalue is real and the remaining two are complex conjugates. (Notice that the components of the last two eigenvectors are complex conjugates as well.) Then we define

$$b_1 = \frac{1}{2}\left[\begin{pmatrix} \frac{3i}{2} \\ \frac{1}{4}+\frac{i}{2} \\ 1 \end{pmatrix} + \begin{pmatrix} \frac{-3i}{2} \\ \frac{1}{4}-\frac{i}{2} \\ 1 \end{pmatrix}\right] = \begin{pmatrix} 0 \\ 1/4 \\ 1 \end{pmatrix} \text{ and } b_2 = \frac{i}{2}\left[\begin{pmatrix} \frac{3i}{2} \\ \frac{1}{4}+\frac{i}{2} \\ 1 \end{pmatrix} - \begin{pmatrix} \frac{-3i}{2} \\ \frac{1}{4}-\frac{i}{2} \\ 1 \end{pmatrix}\right] = \begin{pmatrix} -3/2 \\ -1/2 \\ 0 \end{pmatrix}.$$

```
In[5]:=
b1=1/2(eigs[[2,2]]+eigs[[2,3]])

Out[5]=
{0, 1/4, 1}

In[6]:=
b2=I/2(eigs[[2,2]]-eigs[[2,3]])

Out[6]=
{-(3/2), -(1/2), 0}
```

Then the general solution of $X'(t) = \begin{pmatrix} 0 & 0 & 3 \\ 1 & -4 & 2 \\ 0 & -4 & 1 \end{pmatrix} X(t)$ is

$$\mathbf{X}(t) = c_1 \begin{pmatrix} -1 \\ 1 \\ 1 \end{pmatrix} e^{-3t} + c_2 \left[b_1 \cos(2t) + b_2 \sin(2t) \right] + c_3 \left[b_2 \cos(2t) - b_1 \sin(2t) \right]$$

$$= c_1 \begin{pmatrix} -1 \\ 1 \\ 1 \end{pmatrix} e^{-3t} + c_2 \left[\begin{pmatrix} 0 \\ 1/4 \\ 1 \end{pmatrix} \cos(2t) + \begin{pmatrix} -3/2 \\ -1/2 \\ 0 \end{pmatrix} \sin(2t) \right] + c_3 \left[\begin{pmatrix} -3/2 \\ -1/2 \\ 0 \end{pmatrix} \cos(2t) - \begin{pmatrix} 0 \\ 1/4 \\ 1 \end{pmatrix} \sin(2t) \right],$$

defined below as **gensol**.

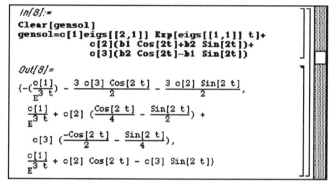

Three solutions are then determined below in **one**, **two**, and **three** by replacing the arbitrary constants in the general solution with those indicated in the commands below. **ParametricPlot3D** is then used to plot these three-dimensional solutions. The option setting **Ticks->False** eliminates all tick marks from the axes. Also, the **PlotRange** option is used to improve the view of the solutions. All three plots are initially suppressed and then shown as a graphics array.

```
In[43]:=
one=gensol /. {c[1]->1,c[2]->-2,c[3]->2};
two=gensol /. {c[1]->2,c[2]->2,c[3]->-2};
three=gensol /. {c[1]->-2,c[2]->2,c[3]->1};

plotone=ParametricPlot3D[one,{t,-1,4},
    PlotRange->{{-6,6},{-6,6},{-5,5}},
    Ticks->None,
    DisplayFunction->Identity];

plottwo=ParametricPlot3D[two,{t,-1,4},
    PlotRange->{{-6,6},{-6,6},{-5,5}},
    Ticks->None,
    DisplayFunction->Identity];

plotthree=ParametricPlot3D[three,{t,-1,4},
    PlotRange->{{-6,6},{-6,6},{-5,5}},
    Ticks->None,
    DisplayFunction->Identity];

Show[GraphicsArray[
        {plotone,plottwo,plotthree}]]
```

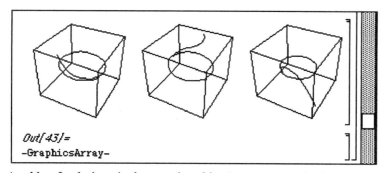

Out[43]=
-GraphicsArray-

A table of solutions is then produced in **funarray**. Again, **Flatten** is used to create a list of the appropriate form so that these solutions may be plotted with **ParametricPlot3D**.

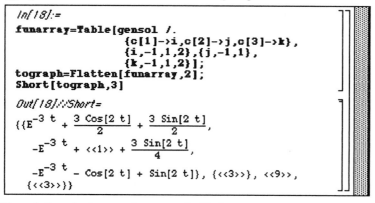

```
In[18]:=
funarray=Table[gensol /.
               {c[1]->i,c[2]->j,c[3]->k},
               {i,-1,1,2},{j,-1,1},
               {k,-1,1,2}];
tograph=Flatten[funarray,2];
Short[tograph,3]

Out[18]//Short=
{{E^-3 t + 3 Cos[2 t]/2 + 3 Sin[2 t]/2 ,
   -E^-3 t + <<1>> + 3 Sin[2 t]/4 ,
   -E^-3 t - Cos[2 t] + Sin[2 t]}, {<<3>>}, <<9>>,
 {<<3>>}}
```

The solutions in **tograph** are graphed below.

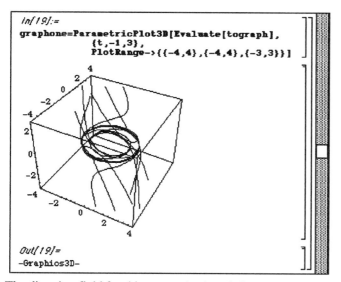

```
In[19]:=
graphone=ParametricPlot3D[Evaluate[tograph],
        {t,-1,3},
        PlotRange->{{-4,4},{-4,4},{-3,3}}]
```

```
Out[19]=
-Graphics3D-
```

The direction field for this system is given below.

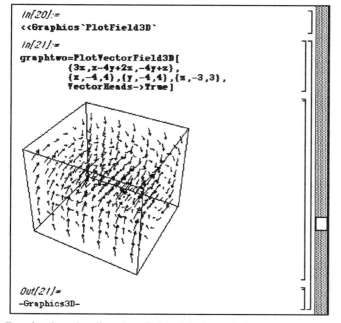

```
In[20]:=
<<Graphics`PlotField3D`
```

```
In[21]:=
graphtwo=PlotVectorField3D[
        {3z,x-4y+2z,-4y+z},
        {x,-4,4},{y,-4,4},{z,-3,3},
        VectorHeads->True]
```

```
Out[21]=
-Graphics3D-
```

By viewing the direction field with the solutions, the "spiraling" motion which accompanies solutions involving sines and cosines is revealed.

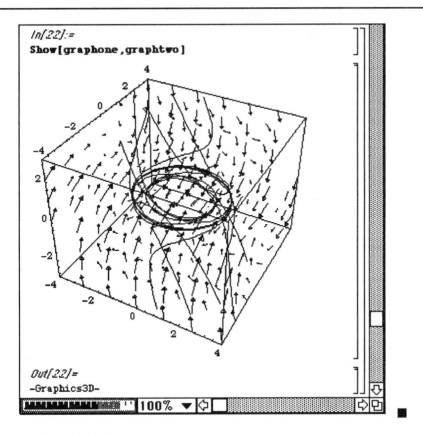

```
In[22]:=
Show[graphone,graphtwo]
```

```
Out[22]=
-Graphics3D-
```

`100%`

■ EXAMPLE 10.19

Solve the systems (a) $\begin{cases} x' = 2x - y \\ y' = -x + 3y \end{cases}$; (b) $\begin{cases} x' = 2x \\ y' = 3x + 2y \end{cases}$ subject to $x(0) = 1$ and $y(0) = 1$; and

(c) $\begin{cases} x' = x + 4y \\ y' = -2x - y \end{cases}$.

Solution:

We use these three examples to illustrate that *Mathematica* can solve every system of the form

$\begin{cases} x' = ax + by \\ y' = cx + dy \end{cases}$, where a, b, c, and d are numbers.

The first system is entered below as **eqone** and **eqtwo** and then solved with **DSolve**. The output is called **sols** for convenience.

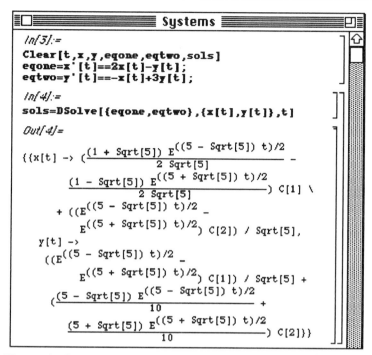

Hence, the first component of the solution is extracted with **sols[[1,1,2]]**

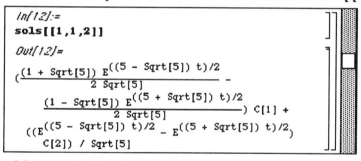

and the second with **sols[[1,2,2]]**.

```
In[13]:=
sols[[1,2,2]]

Out[13]=
((E^((5 - Sqrt[5]) t)/2 - E^((5 + Sqrt[5]) t)/2)
    C[1]) / Sqrt[5] +
  ((5 - Sqrt[5]) E^((5 - Sqrt[5]) t)/2      +
  -----------------------------------------
                    10
    (5 + Sqrt[5]) E^((5 + Sqrt[5]) t)/2
  -------------------------------------) C[2]
                    10
```

A set of ordered pairs for use with the general solution is entered as **pairs** below.

```
In[30]:=
pairs={{0,0},{1,1},{-1,1},{-1,-1},{1,-1},
       {0,1},{1,0},{0,2},{2,0},
       {2,1},{1,2},{-1,2},{1,-2},
       {2,-1},{-2,1}};
```

These numbers are then used for {C[1],C[2]} in the general solution to produce a list of solutions of the system called **funs** below. These solutions are then plotted with **ParametricPlot**.

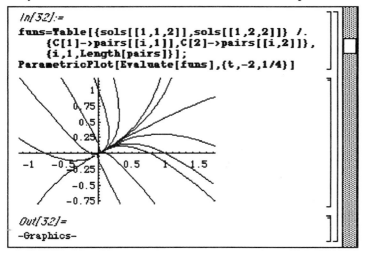

```
In[32]:=
funs=Table[{sols[[1,1,2]],sols[[1,2,2]]} /.
    {C[1]->pairs[[i,1]],C[2]->pairs[[i,2]]},
    {i,1,Length[pairs]}];
ParametricPlot[Evaluate[funs],{t,-2,1/4}]
```

```
Out[32]=
-Graphics-
```

In part (b), we are asked to find the solution which satisfies the system of equations as well as the initial conditions x(0)=1 and y(0)=1. This is accomplished below by including these conditions in the **DSolve** command.

```
In[35]:=
Clear[t,x,y,eqone,eqtwo,sols]
eqone=x'[t]==2x[t];
eqtwo=y'[t]==3x[t]+2y[t];

In[37]:=
sols=DSolve[{eqone,eqtwo,x[0]==1,y[0]==1},
    {x[t],y[t]},t]

Out[37]=
{{x[t] -> E^(2 t), y[t] -> E^(2 t) (1 + 3 t)}}
```

Notice that the result is not a general solution but rather a unique solution (the solution that satisfies the system of equations and passes through the point (1,1) in the x,y-plane). This solution is plotted parametrically below.

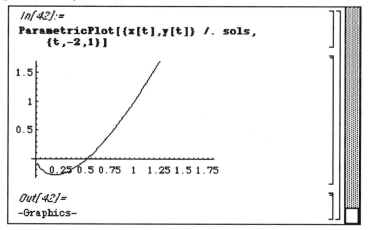

```
In[42]:=
ParametricPlot[{x[t],y[t]} /. sols,
    {t,-2,1}]
```

```
Out[42]=
-Graphics-
```

We solve (c) in the same manner as (a). Note, however, that the resulting solution, **sols**, is expressed in terms of complex exponential functions. To see that the solution is actually a real-valued function we use **ComplexExpand** to expand each part of **sols** assuming that all variables are real.

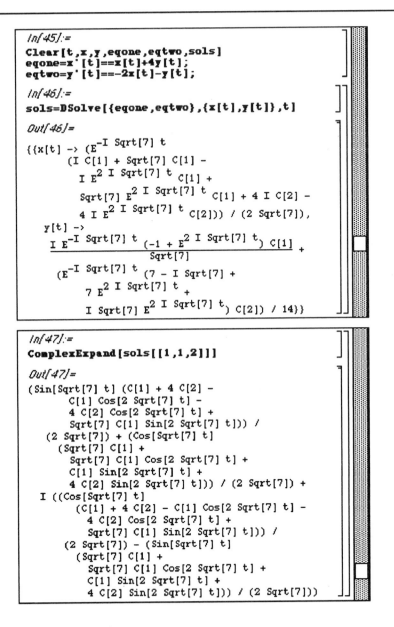

```
In[45]:=
Clear[t,x,y,eqone,eqtwo,sols]
eqone=x'[t]==x[t]+4y[t];
eqtwo=y'[t]==-2x[t]-y[t];

In[46]:=
sols=DSolve[{eqone,eqtwo},{x[t],y[t]},t]

Out[46]=
{{x[t] -> (E^(-I Sqrt[7] t)
     (I C[1] + Sqrt[7] C[1] -
        I E^(2 I Sqrt[7] t) C[1] +
      Sqrt[7] E^(2 I Sqrt[7] t) C[1] + 4 I C[2] -
        4 I E^(2 I Sqrt[7] t) C[2])) / (2 Sqrt[7]),
  y[t] ->
    I E^(-I Sqrt[7] t) (-1 + E^(2 I Sqrt[7] t)) C[1]
    --------------------------------------------------  +
                     Sqrt[7]
    (E^(-I Sqrt[7] t) (7 - I Sqrt[7] +
       7 E^(2 I Sqrt[7] t) +
      I Sqrt[7] E^(2 I Sqrt[7] t)) C[2]) / 14}}
```

```
In[47]:=
ComplexExpand[sols[[1,1,2]]]

Out[47]=
(Sin[Sqrt[7] t] (C[1] + 4 C[2] -
      C[1] Cos[2 Sqrt[7] t] -
      4 C[2] Cos[2 Sqrt[7] t] +
      Sqrt[7] C[1] Sin[2 Sqrt[7] t])) /
  (2 Sqrt[7]) + (Cos[Sqrt[7] t]
    (Sqrt[7] C[1] +
      Sqrt[7] C[1] Cos[2 Sqrt[7] t] +
      C[1] Sin[2 Sqrt[7] t] +
      4 C[2] Sin[2 Sqrt[7] t])) / (2 Sqrt[7]) +
  I ((Cos[Sqrt[7] t]
      (C[1] + 4 C[2] - C[1] Cos[2 Sqrt[7] t] -
        4 C[2] Cos[2 Sqrt[7] t] +
        Sqrt[7] C[1] Sin[2 Sqrt[7] t])) /
    (2 Sqrt[7]) - (Sin[Sqrt[7] t]
      (Sqrt[7] C[1] +
        Sqrt[7] C[1] Cos[2 Sqrt[7] t] +
        C[1] Sin[2 Sqrt[7] t] +
        4 C[2] Sin[2 Sqrt[7] t])) / (2 Sqrt[7]))
```

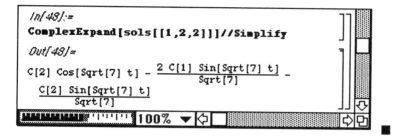

```
In[48]:=
ComplexExpand[sols[[1,2,2]]]//Simplify
Out[48]=
```
$$C[2] \; Cos[Sqrt[7] \; t] - \frac{2 \; C[1] \; Sin[Sqrt[7] \; t]}{Sqrt[7]} -$$
$$\frac{C[2] \; Sin[Sqrt[7] \; t]}{Sqrt[7]}$$

```
100%
```

§10.4 Variation of Parameters

Let $X = X(t) = \begin{pmatrix} x_1(t) \\ x_2(t) \\ \vdots \\ x_n(t) \end{pmatrix}$, $A = A(t) = \begin{pmatrix} a_{11}(t) & a_{12}(t) & \cdots & a_{1n}(t) \\ a_{21}(t) & a_{22}(t) & \cdots & a_{2n}(t) \\ \vdots & \vdots & \ddots & \vdots \\ a_{n1}(t) & a_{n2}(t) & \cdots & a_{nn}(t) \end{pmatrix}$, $F(t) = \begin{pmatrix} f_1(t) \\ f_2(t) \\ \vdots \\ f_n(t) \end{pmatrix}$, and $\Phi(t)$

be a fundamental matrix of the system $X'(t)=A(t)X(t)$. Then the general solution of the homogeneous

system $X'(t) = A(t)X(t)$ is $X = \Phi(t)C$ where $C = \begin{pmatrix} c_1 \\ c_2 \\ \vdots \\ c_n \end{pmatrix}$ is an $n \times 1$ constant vector.

A particular solution of the nonhomogeneous equation $X'(t)=A(t)X(t)+F(t)$ is given by
$X_p = \Phi(t) \int \Phi^{-1}(t) F(t) \, dt$ and the general solution of the system is

$$X = \Phi(t)C + X_p = \Phi(t)C + \Phi(t) \int \Phi^{-1}(t)F(t) \, dt.$$

❑ EXAMPLE 10.20

Solve $X'(t) = \begin{pmatrix} 2 & 5 \\ -4 & -2 \end{pmatrix} X(t) + \begin{pmatrix} \cos(4t) \\ \sin(4t) \end{pmatrix}$.

Solution:

The eigenvalues and corresponding eigenvectors of $\begin{pmatrix} 2 & 5 \\ -4 & -2 \end{pmatrix}$ are $-4i$ and $\begin{pmatrix} \frac{-1}{2} + i \\ 1 \end{pmatrix}$ and $4i$ and

$\begin{pmatrix} \frac{-1}{2} - i \\ 1 \end{pmatrix}$. Let $b_1 = \frac{1}{2}\left[\begin{pmatrix} \frac{-1}{2} + i \\ 1 \end{pmatrix} + \begin{pmatrix} \frac{-1}{2} - i \\ 1 \end{pmatrix} \right] = \begin{pmatrix} \frac{-1}{2} \\ 1 \end{pmatrix}$ and $b_2 = \frac{i}{2}\left[\begin{pmatrix} \frac{-1}{2} + i \\ 1 \end{pmatrix} - \begin{pmatrix} \frac{-1}{2} - i \\ 1 \end{pmatrix} \right] = \begin{pmatrix} -1 \\ 0 \end{pmatrix}$.

Then the general solution of $\mathbf{X}'(t) = \begin{pmatrix} 2 & 5 \\ -4 & -2 \end{pmatrix} \mathbf{X}(t)$ is

$c_1 \left[b_1 \cos(4t) + b_2 \sin(4t) \right] + c_2 \left[b_2 \cos(4t) - b_1 \sin(4t) \right] =$

$$c_1 \begin{pmatrix} \dfrac{-\cos(4t)}{2} - \sin(4t) \\ \cos(4t) \end{pmatrix} + c_2 \begin{pmatrix} -\cos(4t) + \dfrac{\sin(4t)}{2} \\ -\sin(4t) \end{pmatrix} ; \text{ a fundamental matrix is}$$

$$\Phi(t) = \begin{pmatrix} -\cos(4t) - 2\sin(4t) & -2\cos(4t) + \sin(4t) \\ 2\cos(4t) & -2\sin(4t) \end{pmatrix}.$$

By variation of parameters, there is a particular solution of $\mathbf{X}'(t) = \begin{pmatrix} 2 & 5 \\ -4 & -2 \end{pmatrix} \mathbf{X}(t) + \begin{pmatrix} \cos(4t) \\ \sin(4t) \end{pmatrix}$

of the form $\mathbf{X}_p(t) = \Phi(t) \int \Phi^{-1}(t) \begin{pmatrix} \cos(4t) \\ \sin(4t) \end{pmatrix} dt.$ Calculating,

$$\Phi^{-1}(t) = \begin{pmatrix} \dfrac{-\sin(4t)}{2\left(\sin^2(4t) - \cos^2(4t)\right)} & \dfrac{-2\cos(4t) - \sin(4t)}{4\left(\sin^2(4t) - \cos^2(4t)\right)} \\ \dfrac{-\cos(4t)}{2\left(\sin^2(4t) - \cos^2(4t)\right)} & \dfrac{-\cos(4t) - 2\sin(4t)}{4\left(\sin^2(4t) - \cos^2(4t)\right)} \end{pmatrix}$$

$$= \begin{pmatrix} \dfrac{\sec(8t)\sin(4t)}{2} & \dfrac{\sec(8t)\left(2\cos(4t) + \sin(4t)\right)}{4} \\ \dfrac{\cos(4t)\sec(8t)}{2} & \dfrac{\sec(8t)\left(\cos(4t) + 2\sin(4t)\right)}{4} \end{pmatrix};$$

$$\Phi^{-1}(t)\begin{pmatrix} \cos(4t) \\ \sin(4t) \end{pmatrix} = \begin{pmatrix} \dfrac{\sec(8t)\sin(4t)}{2} & \dfrac{\sec(8t)\left(2\cos(4t) + \sin(4t)\right)}{4} \\ \dfrac{\cos(4t)\sec(8t)}{2} & \dfrac{\sec(8t)\left(\cos(4t) + 2\sin(4t)\right)}{4} \end{pmatrix} \begin{pmatrix} \cos(4t) \\ \sin(4t) \end{pmatrix}$$

$$= \begin{pmatrix} \dfrac{\sec(8t)\sin(4t)\left(4\cos(4t) + \sin(4t)\right)}{4} \\ \dfrac{\sec(8t)\left(2\cos^2(4t) + \cos(4t)\sin(4t) + 2\sin^2(4t)\right)}{4} \end{pmatrix} = \begin{pmatrix} \dfrac{-1}{8} + \dfrac{\sec(8t)}{8} + \dfrac{\tan(8t)}{8} \\ \dfrac{\sec(8t)}{2} + \dfrac{\tan(8t)}{8} \end{pmatrix};$$

$$\int \Phi^{-1}(t)\begin{pmatrix} \cos(4t) \\ \sin(4t) \end{pmatrix} dt = \int \begin{pmatrix} \dfrac{-1}{8} + \dfrac{\sec(8t)}{8} + \dfrac{\tan(8t)}{8} \\[2mm] \dfrac{\sec(8t)}{2} + \dfrac{\tan(8t)}{8} \end{pmatrix} dt$$

$$= \begin{pmatrix} \dfrac{-t}{8} - \dfrac{5\mathrm{Ln}\,|\cos(4t) - \sin(4t)|}{64} - \dfrac{3\mathrm{Ln}\,|\cos(4t) + \sin(4t)|}{64} \\[3mm] \dfrac{-5\mathrm{Ln}\,|\cos(4t) - \sin(4t)|}{64} + \dfrac{3\mathrm{Ln}\,|\cos(4t) + \sin(4t)|}{64} \end{pmatrix}; \text{ and}$$

$$\mathbf{X}_p(t) = \Phi(t)\int \Phi^{-1}(t)\begin{pmatrix} \cos(4t) \\ \sin(4t) \end{pmatrix} dt$$

$$= \begin{pmatrix} -\cos(4t) - 2\sin(4t) & -2\cos(4t) + \sin(4t) \\ 2\cos(4t) & -2\sin(4t) \end{pmatrix}\begin{pmatrix} \dfrac{-t}{8} - \dfrac{5\mathrm{Ln}\,|\cos(4t) - \sin(4t)|}{64} - \dfrac{3\mathrm{Ln}\,|\cos(4t) + \sin(4t)|}{64} \\[3mm] \dfrac{-5\mathrm{Ln}\,|\cos(4t) - \sin(4t)|}{64} + \dfrac{3\mathrm{Ln}\,|\cos(4t) + \sin(4t)|}{64} \end{pmatrix}.$$

Consequently, the general solution of $\mathbf{X}'(t) = \begin{pmatrix} 2 & 5 \\ -4 & -2 \end{pmatrix}\mathbf{X}(t) + \begin{pmatrix} \cos(4t) \\ \sin(4t) \end{pmatrix}$ is

$$\mathbf{X}(t) = \begin{pmatrix} -\cos(4t) - 2\sin(4t) & -2\cos(4t) + \sin(4t) \\ 2\cos(4t) & -2\sin(4t) \end{pmatrix}\begin{pmatrix} c_1 \\ c_2 \end{pmatrix} +$$

$$\begin{pmatrix} -\cos(4t) - 2\sin(4t) & -2\cos(4t) + \sin(4t) \\ 2\cos(4t) & -2\sin(4t) \end{pmatrix}\begin{pmatrix} \dfrac{-t}{8} - \dfrac{5\mathrm{Ln}\,|\cos(4t) - \sin(4t)|}{64} - \dfrac{3\mathrm{Ln}\,|\cos(4t) + \sin(4t)|}{64} \\[3mm] \dfrac{-5\mathrm{Ln}\,|\cos(4t) - \sin(4t)|}{64} + \dfrac{3\mathrm{Ln}\,|\cos(4t) + \sin(4t)|}{64} \end{pmatrix}. \quad\blacksquare$$

■ EXAMPLE 10.21

Solve $\mathbf{X}'(t) = \begin{pmatrix} -5 & 3 \\ 2 & -10 \end{pmatrix}\mathbf{X}(t) + \begin{pmatrix} e^{-2t} \\ 1 \end{pmatrix}$.

Solution:

In order to apply variation of parameters, we first calculate a fundamental matrix for the associated homogeneous equation $\mathbf{X}'(t) = \begin{pmatrix} -5 & 3 \\ 2 & -10 \end{pmatrix}\mathbf{X}(t)$. We begin by defining **matrix** to be the matrix

$\begin{pmatrix} -5 & 3 \\ 2 & -10 \end{pmatrix}$ and computing the eigenvalues and corresponding eigenvectors of $\begin{pmatrix} -5 & 3 \\ 2 & -10 \end{pmatrix}$.

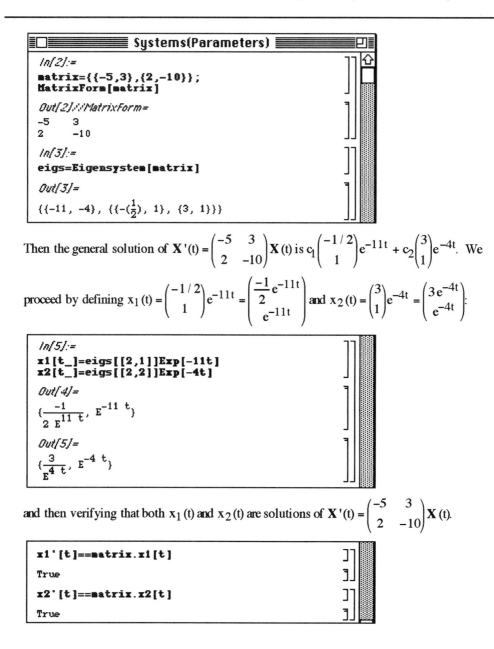

$$In[2]:=$$

$$\mathbf{matrix}=\{\{-5,3\},\{2,-10\}\};$$
$$\mathbf{MatrixForm[matrix]}$$

$$Out[2]//MatrixForm=$$

$$\begin{array}{cc} -5 & 3 \\ 2 & -10 \end{array}$$

$$In[3]:=$$

$$\mathbf{eigs=Eigensystem[matrix]}$$

$$Out[3]=$$

$$\{\{-11,\ -4\},\ \{\{-(\tfrac{1}{2}),\ 1\},\ \{3,\ 1\}\}\}$$

Then the general solution of $\mathbf{X}'(t) = \begin{pmatrix} -5 & 3 \\ 2 & -10 \end{pmatrix} \mathbf{X}(t)$ is $c_1 \begin{pmatrix} -1/2 \\ 1 \end{pmatrix} e^{-11t} + c_2 \begin{pmatrix} 3 \\ 1 \end{pmatrix} e^{-4t}$. We

proceed by defining $x_1(t) = \begin{pmatrix} -1/2 \\ 1 \end{pmatrix} e^{-11t} = \begin{pmatrix} \dfrac{-1}{2} e^{-11t} \\ e^{-11t} \end{pmatrix}$ and $x_2(t) = \begin{pmatrix} 3 \\ 1 \end{pmatrix} e^{-4t} = \begin{pmatrix} 3e^{-4t} \\ e^{-4t} \end{pmatrix}$:

$$In[5]:=$$

$$\mathbf{x1[t\_]=eigs[[2,1]]Exp[-11t]}$$
$$\mathbf{x2[t\_]=eigs[[2,2]]Exp[-4t]}$$

$$Out[4]=$$

$$\{\frac{-1}{2\ E^{11\ t}},\ E^{-11\ t}\}$$

$$Out[5]=$$

$$\{\frac{3}{E^{4\ t}},\ E^{-4\ t}\}$$

and then verifying that both $x_1(t)$ and $x_2(t)$ are solutions of $\mathbf{X}'(t) = \begin{pmatrix} -5 & 3 \\ 2 & -10 \end{pmatrix} \mathbf{X}(t)$.

$$\mathbf{x1'[t]==matrix.x1[t]}$$

True

$$\mathbf{x2'[t]==matrix.x2[t]}$$

True

A fundamental matrix of $\mathbf{X}'(t) = \begin{pmatrix} -5 & 3 \\ 2 & -10 \end{pmatrix} \mathbf{X}(t)$ is $\Phi(t) = \begin{pmatrix} \dfrac{-1}{2}e^{-11t} & 3e^{-4t} \\ e^{-11t} & e^{-4t} \end{pmatrix}$ which we

compute below and name **capphi**. Note that the matrix $\{\mathbf{x1[t]},\mathbf{x2[t]}\}$ corresponds to the matrix

$\begin{pmatrix} \dfrac{-1}{2}e^{-11t} & e^{-11t} \\ 3e^{-4t} & e^{-4t} \end{pmatrix}$; the transpose of $\begin{pmatrix} \dfrac{-1}{2}e^{-11t} & e^{-11t} \\ 3e^{-4t} & e^{-4t} \end{pmatrix}$ is $\begin{pmatrix} \dfrac{-1}{2}e^{-11t} & 3e^{-4t} \\ e^{-11t} & e^{-4t} \end{pmatrix}$.

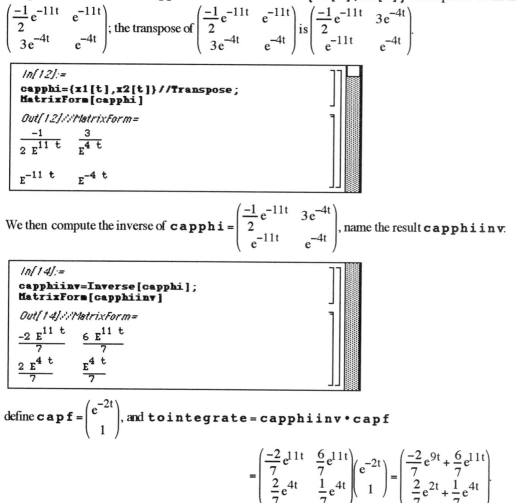

```
In[12]:=
capphi={x1[t],x2[t]}//Transpose;
MatrixForm[capphi]

Out[12]//MatrixForm=

 -1         3
----      ---- t
2 E^11 t   E^4 t

E^-11 t    E^-4 t
```

We then compute the inverse of $\mathbf{capphi} = \begin{pmatrix} \dfrac{-1}{2}e^{-11t} & 3e^{-4t} \\ e^{-11t} & e^{-4t} \end{pmatrix}$, name the result **capphiinv**.

```
In[14]:=
capphiinv=Inverse[capphi];
MatrixForm[capphiinv]

Out[14]//MatrixForm=

-2 E^11 t    6 E^11 t
---------    ---------
   7            7

2 E^4 t      E^4 t
--------     -----
   7           7
```

define $\mathbf{capf} = \begin{pmatrix} e^{-2t} \\ 1 \end{pmatrix}$, and **tointegrate = capphiinv • capf**

$$= \begin{pmatrix} \dfrac{-2}{7}e^{11t} & \dfrac{6}{7}e^{11t} \\ \dfrac{2}{7}e^{4t} & \dfrac{1}{7}e^{4t} \end{pmatrix} \begin{pmatrix} e^{-2t} \\ 1 \end{pmatrix} = \begin{pmatrix} \dfrac{-2}{7}e^{9t} + \dfrac{6}{7}e^{11t} \\ \dfrac{2}{7}e^{2t} + \dfrac{1}{7}e^{4t} \end{pmatrix}.$$

In[16]:=

```
capf={Exp[-2t],1};
tointegrate=capphiinv.capf
```

Out[16]=

$$\{\frac{-2\ E^{9\ t}}{7}+\frac{6\ E^{11\ t}}{7},\ \frac{2\ E^{2\ t}}{7}+\frac{E^{4\ t}}{7}\}$$

We then compute $\int \mathbf{tointegrate}\, d t = \int \begin{pmatrix} \dfrac{-2}{7}e^{9t}+\dfrac{6}{7}e^{11t} \\ \dfrac{2}{7}e^{2t}+\dfrac{1}{7}e^{4t} \end{pmatrix} dt = \begin{pmatrix} \dfrac{-6}{23}e^{9t}+\dfrac{6}{77}e^{11t} \\ \dfrac{1}{7}e^{2t}+\dfrac{1}{28}e^{4t} \end{pmatrix}.$

In[17]:=

```
capu=Integrate[tointegrate,t]
```

Out[17]=

$$\{\frac{-2\ E^{9\ t}}{63}+\frac{6\ E^{11\ t}}{77},\ \frac{E^{2\ t}}{7}+\frac{E^{4\ t}}{28}\}$$

Then, by variation of parameters, $\mathbf{X}_p(t) = \Phi(t)\int \Phi^{-1}(t)\begin{pmatrix} e^{-2t} \\ 1 \end{pmatrix} dt = \mathbf{capphi \cdot capu}$

$$= \begin{pmatrix} \dfrac{-1}{2}e^{-11t} & 3e^{-4t} \\ e^{-11t} & e^{-4t} \end{pmatrix} \begin{pmatrix} \dfrac{-6}{23}e^{9t}+\dfrac{6}{77}e^{11t} \\ \dfrac{1}{7}e^{2t}+\dfrac{1}{28}e^{4t} \end{pmatrix} = \begin{pmatrix} \dfrac{3}{44}+\dfrac{4}{9}e^{-2t} \\ \dfrac{5}{44}+\dfrac{1}{9}e^{-2t} \end{pmatrix}$$

In[19]:=

```
particular=capphi.capu//Simplify
```

Out[19]=

$$\{\frac{3}{44}+\frac{4}{9\ E^{2\ t}},\ \frac{5}{44}+\frac{1}{9\ E^{2\ t}}\}$$

is a particular solution of $\mathbf{X}'(t) = \begin{pmatrix} -5 & 3 \\ 2 & -10 \end{pmatrix}\mathbf{X}(t) + \begin{pmatrix} e^{-2t} \\ 1 \end{pmatrix}$ and the general solution is

$$\mathbf{X}(t) = \begin{pmatrix} \dfrac{-1}{2}e^{-11t} & 3e^{-4t} \\ e^{-11t} & e^{-4t} \end{pmatrix}\begin{pmatrix} c_1 \\ c_2 \end{pmatrix} + \begin{pmatrix} \dfrac{3}{44}+\dfrac{4}{9}e^{-2t} \\ \dfrac{5}{44}+\dfrac{1}{9}e^{-2t} \end{pmatrix}.$$

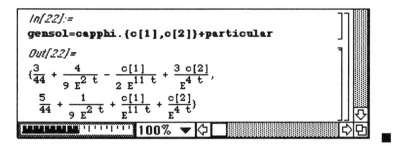

```
In[22]:=
gensol=capphi.{c[1],c[2]}+particular

Out[22]=
```

$$\{\frac{3}{44} + \frac{4}{9\ E^{2\ t}} - \frac{c[1]}{2\ E^{11\ t}} + \frac{3\ c[2]}{E^{4\ t}},$$
$$\frac{5}{44} + \frac{1}{9\ E^{2\ t}} + \frac{c[1]}{E^{11\ t}} + \frac{c[2]}{E^{4\ t}}\}$$

```
100% ▼
```

■

■ EXAMPLE 10.22

Solve $X'(t) = \begin{pmatrix} -4 & 1 & 5 \\ -2 & 0 & 2 \\ 0 & 2 & -4 \end{pmatrix} X(t) + \begin{pmatrix} e^{-t} \\ \cos(t) \\ 1 \end{pmatrix}$.

Solution:

Proceeding as in the previous example, we first solve the homogeneous system

$X'(t) = \begin{pmatrix} -4 & 1 & 5 \\ -2 & 0 & 2 \\ 0 & 2 & -4 \end{pmatrix} X(t)$. We begin by defining **matrix** to be the matrix $\begin{pmatrix} -4 & 1 & 5 \\ -2 & 0 & 2 \\ 0 & 2 & -4 \end{pmatrix}$ and then

using **Eigensystem** to compute the eigenvalues and corresponding eigenvectors of **matrix**.

```
≣□≣═══════════ Systems(Parameters) ═══════════⊞≣
In[11]:=
matrix={{-4,1,5},{-2,0,2},{0,2,-4}};
MatrixForm[matrix]

Out[11]//MatrixForm=
-4   1    5
-2   0    2
0    2   -4

In[12]:=
eigs=Eigensystem[matrix]

Out[12]=
{{-6, -1 - I, -1 + I},
   {{-2, -1, 1}, {4 + I, 3 - I, 2},
    {4 - I, 3 + I, 2}}}
```

Since the eigenvector corresponding to the real eigenvalue -6 is $\begin{pmatrix} -2 \\ -1 \\ 1 \end{pmatrix}$, one solution vector is

$x_1(t) = \begin{pmatrix} -2 \\ -1 \\ 1 \end{pmatrix} e^{-6t}$, defined below.

```
In/13/:=
x1[t_]=eigs[[2,1]]Exp[-6t]
Out/13/=
{ -2/E^6 t , -E^-6 t , E^-6 t }
```

Since the remaining two eigenvalues are complex conjugates, we define $b_1 = \dfrac{1}{2}\left(\begin{pmatrix} 4+i \\ 3-i \\ 2 \end{pmatrix} + \begin{pmatrix} 4-i \\ 3+i \\ 2 \end{pmatrix} \right) = \begin{pmatrix} 4 \\ 3 \\ 2 \end{pmatrix}$

and $b_2 = \dfrac{i}{2}\left(\begin{pmatrix} 4+i \\ 3-i \\ 2 \end{pmatrix} - \begin{pmatrix} 4-i \\ 3+i \\ 2 \end{pmatrix} \right) = \begin{pmatrix} -1 \\ 1 \\ 0 \end{pmatrix}$.

```
In/15/:=
b1=1/2(eigs[[2,2]]+eigs[[2,3]])
b2=I/2(eigs[[2,2]]-eigs[[2,3]])
Out/14/=
{4, 3, 2}
Out/15/=
{-1, 1, 0}
```

Then two other linearly independent solution vectors are

$x_2(t) = \left(b_1 \cos(t) + b_2 \sin(t) \right) e^{-t} = \left(\begin{pmatrix} 4 \\ 3 \\ 2 \end{pmatrix} \cos(t) + \begin{pmatrix} -1 \\ 1 \\ 0 \end{pmatrix} \sin(t) \right) e^{-t} = \begin{pmatrix} 4\cos(t) - \sin(t) \\ 3\cos(t) + \sin(t) \\ 2\cos(t) \end{pmatrix} e^{-t}$ and

$x_3(t) = \left(b_2 \cos(t) - b_1 \sin(t) \right) e^{-t} = \left(\begin{pmatrix} -1 \\ 1 \\ 0 \end{pmatrix} \cos(t) + \begin{pmatrix} 4 \\ 3 \\ 2 \end{pmatrix} \sin(t) \right) e^{-t} = \begin{pmatrix} -\cos(t) - 4\sin(t) \\ \cos(t) - 3\sin(t) \\ -2\sin(t) \end{pmatrix} e^{-t}$, also defined

below.

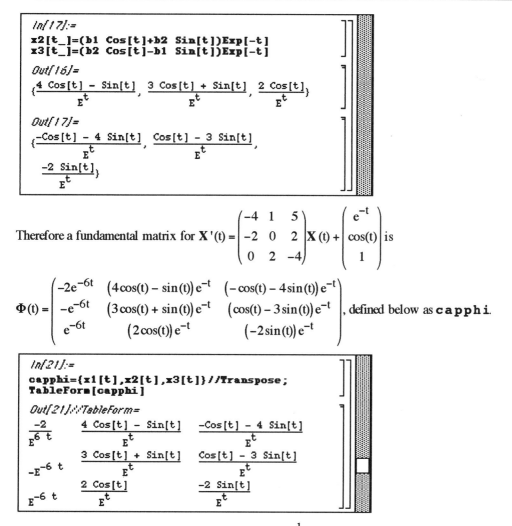

```
In[17]:=
x2[t_]=(b1 Cos[t]+b2 Sin[t])Exp[-t]
x3[t_]=(b2 Cos[t]-b1 Sin[t])Exp[-t]

Out[16]=
 {4 Cos[t] - Sin[t]   3 Cos[t] + Sin[t]   2 Cos[t]
  ─────────────────,  ─────────────────,  ────────}
        E^t                  E^t             E^t

Out[17]=
 {-Cos[t] - 4 Sin[t]   Cos[t] - 3 Sin[t]
  ──────────────────,  ─────────────────,
        E^t                   E^t

   -2 Sin[t]
   ─────────}
      E^t
```

Therefore a fundamental matrix for $\mathbf{X}'(t) = \begin{pmatrix} -4 & 1 & 5 \\ -2 & 0 & 2 \\ 0 & 2 & -4 \end{pmatrix} \mathbf{X}(t) + \begin{pmatrix} e^{-t} \\ \cos(t) \\ 1 \end{pmatrix}$ is

$$\Phi(t) = \begin{pmatrix} -2e^{-6t} & \left(4\cos(t) - \sin(t)\right)e^{-t} & \left(-\cos(t) - 4\sin(t)\right)e^{-t} \\ -e^{-6t} & \left(3\cos(t) + \sin(t)\right)e^{-t} & \left(\cos(t) - 3\sin(t)\right)e^{-t} \\ e^{-6t} & \left(2\cos(t)\right)e^{-t} & \left(-2\sin(t)\right)e^{-t} \end{pmatrix}, \text{ defined below as } \mathbf{capphi}.$$

```
In[21]:=
capphi={x1[t],x2[t],x3[t]}//Transpose;
TableForm[capphi]

Out[21]//TableForm=
   -2         4 Cos[t] - Sin[t]    -Cos[t] - 4 Sin[t]
  ─────       ─────────────────    ──────────────────
  E^6 t             E^t                   E^t

                3 Cos[t] + Sin[t]    Cos[t] - 3 Sin[t]
  -E^-6 t       ─────────────────    ─────────────────
                      E^t                  E^t

                   2 Cos[t]            -2 Sin[t]
  E^-6 t           ────────            ─────────
                     E^t                  E^t
```

We then use *Mathematica* to compute and simplify $\Phi^{-1}(t)$ and name the resulting output
capphiinv.

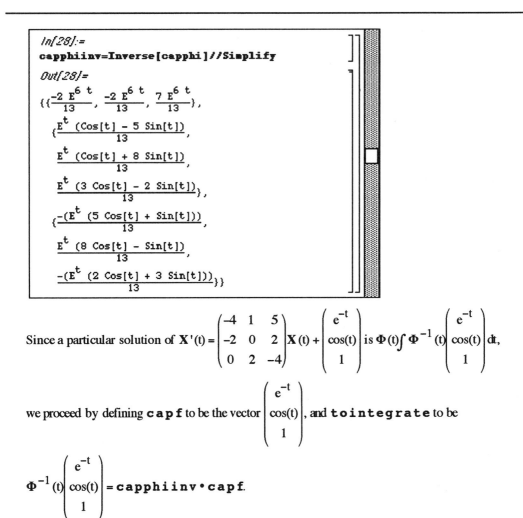

Since a particular solution of $\mathbf{X}'(t) = \begin{pmatrix} -4 & 1 & 5 \\ -2 & 0 & 2 \\ 0 & 2 & -4 \end{pmatrix} \mathbf{X}(t) + \begin{pmatrix} e^{-t} \\ \cos(t) \\ 1 \end{pmatrix}$ is $\Phi(t) \int \Phi^{-1}(t) \begin{pmatrix} e^{-t} \\ \cos(t) \\ 1 \end{pmatrix} dt$,

we proceed by defining **c a p f** to be the vector $\begin{pmatrix} e^{-t} \\ \cos(t) \\ 1 \end{pmatrix}$, and **tointegrate** to be

$$\Phi^{-1}(t) \begin{pmatrix} e^{-t} \\ \cos(t) \\ 1 \end{pmatrix} = \mathbf{capphiinv \cdot capf}.$$

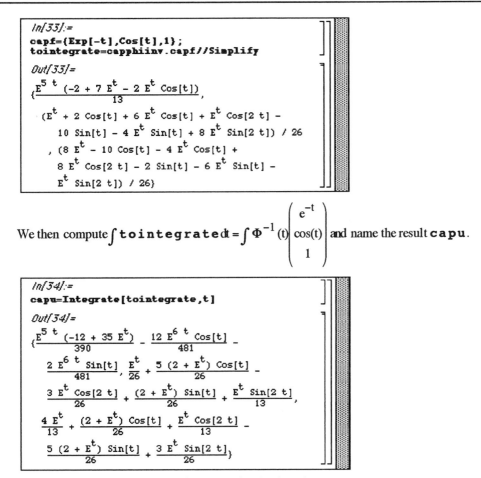

We then compute $\int \mathbf{tointegrated} t = \int \Phi^{-1}(t) \begin{pmatrix} e^{-t} \\ \cos(t) \\ 1 \end{pmatrix}$ and name the result $\mathbf{capu}$.

Therefore a particular solution of the equation is given by

$$\Phi(t) \int \Phi^{-1}(t) \begin{pmatrix} e^{-t} \\ \cos(t) \\ 1 \end{pmatrix} dt = \mathbf{capphi \cdot capu},\ \text{defined as}\ \mathbf{particular}\ \text{below:}$$

```
In[15]:=
particular=capphi.capu//Simplify
```

$$Out[15]=$$

$$\{\frac{43}{78} + \frac{99}{65\,E^t} - \frac{309\,Cos[t]}{481} - \frac{514\,Sin[t]}{481},$$

$$\frac{41}{78} + \frac{82}{65\,E^t} + \frac{86\,Cos[t]}{481} - \frac{257\,Sin[t]}{481},$$

$$\frac{37}{78} + \frac{48}{65\,E^t} - \frac{86\,Cos[t]}{481} - \frac{224\,Sin[t]}{481}\}$$

and the general solution of $\mathbf{X}'(t) = \begin{pmatrix} -4 & 1 & 5 \\ -2 & 0 & 2 \\ 0 & 2 & -4 \end{pmatrix} \mathbf{X}(t) + \begin{pmatrix} e^{-t} \\ \cos(t) \\ 1 \end{pmatrix}$ is $\Phi(t)\begin{pmatrix} c_1 \\ c_2 \\ c_3 \end{pmatrix} + \Phi(t)\int \Phi^{-1}(t)\begin{pmatrix} e^{-t} \\ \cos(t) \\ 1 \end{pmatrix} dt$,

defined below as **gensol**.

```
In[36]:=
gensol=capphi.{c[1],c[2],c[3]}+particular
```

$$Out[36]=$$

$$\{\frac{43}{78} + \frac{99}{65\,E^t} - \frac{2\,c[1]}{E^{6\,t}} - \frac{309\,Cos[t]}{481} +$$

$$\frac{c[3]\,(-Cos[t] - 4\,Sin[t])}{E^t} +$$

$$\frac{c[2]\,(4\,Cos[t] - Sin[t])}{E^t} - \frac{514\,Sin[t]}{481},$$

$$\frac{41}{78} + \frac{82}{65\,E^t} - \frac{c[1]}{E^{6\,t}} + \frac{86\,Cos[t]}{481} +$$

$$\frac{c[3]\,(Cos[t] - 3\,Sin[t])}{E^t} - \frac{257\,Sin[t]}{481} +$$

$$\frac{c[2]\,(3\,Cos[t] + Sin[t])}{E^t},$$

$$\frac{37}{78} + \frac{48}{65\,E^t} + \frac{c[1]}{E^{6\,t}} - \frac{86\,Cos[t]}{481} + \frac{2\,c[2]\,Cos[t]}{E^t} -$$

$$\frac{224\,Sin[t]}{481} - \frac{2\,c[3]\,Sin[t]}{E^t}\}$$

```
|'''''''''''''''''|'''''|'''''|  100%  ▼ ◁ □                 ▷ ◱
```

§10.5 Laplace Transforms

In many cases, Laplace transforms can be used to solve a system of linear differential equations. We illustrate the techniques needed to accomplish this with *Mathematica* in the following examples.

■ EXAMPLE 10.23

Use Laplace transforms to solve the system $\begin{cases} x'(t) - y(t) = e^{-t} \\ y'(t) + 5x(t) + 2y(t) = \sin(3t) \end{cases}$.

Solution:

We begin by loading the package **LaplaceTransform** contained in the **Calculus** folder (or directory) for later use, defining $f(t) = e^{-t}$, $g(t) = \sin(3t)$, computing $L\{f\}(s) = \dfrac{1}{1+s}$

and $L\{g\}(s) = \dfrac{3}{9+s^2}$, and naming the results **lf** and **lg**, respectively.

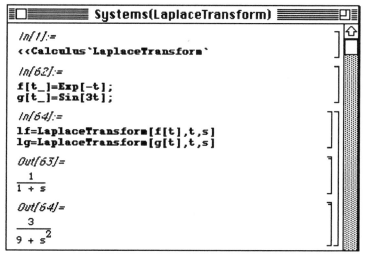

```
═□═══════ Systems(LaplaceTransform) ═════════□═

In[1]:=
<<Calculus`LaplaceTransform`

In[62]:=
f[t_]=Exp[-t];
g[t_]=Sin[3t];

In[64]:=
lf=LaplaceTransform[f[t],t,s]
lg=LaplaceTransform[g[t],t,s]

Out[63]=
1
───
1 + s

Out[64]=
3
─────
9 + s²
```

We then define **lhseqs** to be a list corresponding to the left-hand side of each equation in the system

$\begin{cases} x'(t) - y(t) = e^{-t} \\ y'(t) + 5x(t) + 2y(t) = \sin(3t) \end{cases}$ and **ruleone** to be a list of rules which when applied to a given object

replace each element of the form **x'[t]** by **s lx-x[0]**, **x[t]** by **lx**, **y'[t]** by **s ly-y[0]**, and **y[t]** by **ly**.

```
In[65]:=
lhseqs={x'[t]-y[t],
y'[t]+5x[t]+2y[t]};

In[66]:=
ruleone={x'[t]->s lx-x[0], x[t]->lx,
         y'[t]->s ly-y[0],y[t]->ly};
```

Let **1x** and **1y** denote the Laplace transform of x(t) and y(t), respectively. Then, to compute the Laplace transform of the left-hand side of each equation in the system

$$\begin{cases} x'(t) - y(t) = e^{-t} \\ y'(t) + 5x(t) + 2y(t) = \sin(3t) \end{cases}$$

we apply **ruleone** to **lhseqs** and name the resulting output **laplhs**.

```
In[67]:=
laplhs=lhseqs /. ruleone
Out[67]=
{-ly + lx s - x[0], 5 lx + 2 ly + ly s - y[0]}
```

Thus computing the Laplace transform of each equation in the system $\begin{cases} x'(t) - y(t) = e^{-t} \\ y'(t) + 5x(t) + 2y(t) = \sin(3t) \end{cases}$

yields the system $\begin{cases} -1\,y + 1\,x\,s - x(0) = \dfrac{1}{1+s} \\ 51\,x + 21\,y + 1\,y\,s - y(0) = \dfrac{3}{9+s^2} \end{cases}$. Consequently, to compute $x(t)$ and $y(t)$

we must compute the inverse Laplace transform of **1x** and **1y**. We begin by solving the system

$$\begin{cases} -1\,y + 1\,x\,s - x(0) = \dfrac{1}{1+s} \\ 51\,x + 21\,y + 1\,y\,s - y(0) = \dfrac{3}{9+s^2} \end{cases}$$ for **1 x** and **1 y** and naming the resulting output **solution**.

The formulas for **1x** and **1y** specified in **solution** are extracted from **solution** with the commands **solution[[1,1,2]]** and **solution[[1,2,2]]**, respectively.

```
In[68]:=
solution=Solve[laplhs=={lf,lg},{lx,ly}]
Out[68]=
                      2    3
           (18 + 9 s + 2 s + s )
          (-1 - x[0] - s x[0])
{{lx -> -(─────────────────────────────────
           45 + 63 s + 32 s + 16 s + 3 s + s
                          2      3     4    5
                                2
          (1 + s) (-3 - 9 y[0] - s  y[0])
      ) - ──────────────────────────────── ,
           45 + 63 s + 32 s + 16 s + 3 s + s
                          2      3     4    5
   ly -> -(
                 2
         (-45 - 5 s ) (-1 - x[0] - s x[0])
         ───────────────────────────────── ) -
          45 + 63 s + 32 s + 16 s + 3 s + s
                         2      3     4    5
                2              2
        (s + s ) (-3 - 9 y[0] - s  y[0])
        ────────────────────────────────}}
         45 + 63 s + 32 s + 16 s + 3 s + s
                        2      3     4    5
```

In order to compute the inverse Laplace transform of **1x**, we first extract **solution[[1,1,2]]** from **solution**, display the result as a single fraction, expand the numerator, and name the result **stepone**.

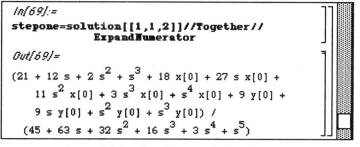

```
In[69]:=
stepone=solution[[1,1,2]]//Together//
              ExpandNumerator

Out[69]=

(21 + 12 s + 2 s^2 + s^3 + 18 x[0] + 27 s x[0] +
    11 s^2 x[0] + 3 s^3 x[0] + s^4 x[0] + 9 y[0] +
    9 s y[0] + s^2 y[0] + s^3 y[0]) /
  (45 + 63 s + 32 s^2 + 16 s^3 + 3 s^4 + s^5)
```

To compute the partial fraction decomposition of **stepone**, we first factor

$$45 + 63s + 32s^2 + 16s^3 + 3s^4 + s^5 = (1+s)\left(9+s^2\right)\left(5+2s+s^2\right)$$ and then define

$$\mathbf{steptwo} = \frac{a}{1+s} + \frac{bs+c}{9+s^2} + \frac{ds+e}{5+2s+s^2}.$$

```
In[70]:=
Factor[45+63s+32s^2+16s^3+3s^4+s^5]

Out[70]=

(1 + s) (9 + s^2) (5 + 2 s + s^2)

In[71]:=
steptwo=a/(1+s)+(b s+c)/(9+s^2)+
          (d s+e)/(5+2s+s^2)

Out[71]=

  a       c + b s       e + d s
----- + --------- + -------------
1 + s     9 + s^2    5 + 2 s + s^2
```

We then write **steptwo** as a single fraction and name the result **stepthree**.

```
In[72]:=
stepthree=Together[steptwo]

Out[72]=

(45 a + 5 c + 9 e + 18 a s + 5 b s + 7 c s +
    9 d s + 9 e s + 14 a s^2 + 7 b s^2 + 3 c s^2 +
    9 d s^2 + e s^2 + 2 a s^3 + 3 b s^3 + c s^3 +
    d s^3 + e s^3 + a s^4 + b s^4 + d s^4) /
  ((1 + s) (9 + s^2) (5 + 2 s + s^2))
```

Finally, we use **SolveAlways** to find the values of a, b, c, d, and e so that the numerator of

stepthree is equal to the numerator of **stepone** for all values of s. The option **Method->3** is used to guarantee that the results are solved explicitly for a, b, c, d, and e.

```
In[73]:=
valsx=SolveAlways[Numerator[stepthree]==
                  Numerator[stepone],s,
                  Method->3]

Out[73]=
{{c -> -(3/13), e -> (63 + 104 x[0] + 52 y[0])/52 ,

  d -> (-7 + 52 x[0])/52 , b -> -(3/26), a -> 1/4}}
```

The partial fraction decomposition of **stepone** is given by replacing a, b, c, d, and e in **stepthree** by the values obtained in **valsx**. We name the resulting output **stepfour**.

```
In[74]:=
stepfour=steptwo /. valsx

Out[74]=
```
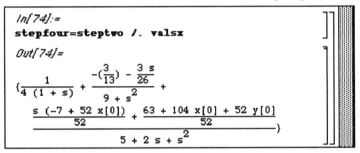

In the same manner as above, to compute the partial fraction decomposition of **ly**, we extract **solution[[1,2,2]]** from **solution**, write the result as a single fraction, expand the numerator, and name the result **stepfive**.

```
In[75]:=
stepfive=solution[[1,2,2]]//Together//
                 ExpandNumerator

Out[75]=
                      2
(-45 + 3 s - 2 s  - 45 x[0] - 45 s x[0] -

     2             3
  5 s  x[0] - 5 s  x[0] + 9 s y[0] +

     2        3         4
  9 s  y[0] + s  y[0] + s  y[0]) /

                  2         3        4     5
 (45 + 63 s + 32 s  + 16 s  + 3 s  + s )
```

To compute the partial fraction decomposition of **stepfive**, **SolveAlways** is used to find the values of a, b, c, d, and e so that the numerator of **stepfive** is equal to the numerator of **steptwo** for all values of a, b, c, d, and e.

```
In[76]:=
valsy=SolveAlways[Numerator[stepthree]==
                  Numerator[stepfive],s,
                  Method->3]
Out[76]=
```

$$\{\{b \to -(\frac{3}{13}),\ d \to \frac{77 + 52\ y[0]}{52},\ c \to \frac{27}{26},$$

$$e \to \frac{35 - 260\ x[0]}{52},\ a \to -(\frac{5}{4})\}\}$$

The partial fraction decomposition of **stepfive** is given by replacing a, b, c, d, and e in **steptwo** by the values obtained in **valsy**. The result is named **stepsix**.

```
In[77]:=
stepsix=steptwo /. valsy
Out[77]=
```

$$\{\frac{-5}{4\ (1 + s)} + \frac{\frac{27}{26} - \frac{3\ s}{13}}{9 + s^2} +$$

$$\frac{\frac{35 - 260\ x[0]}{52} + \frac{s\ (77 + 52\ y[0])}{52}}{5 + 2\ s + s^2}\}$$

To compute x(t) and y(t) we must compute the inverse Laplace transform of **stepfour** and **stepsix**, respectively. We begin by rewriting **stepfour** in the form

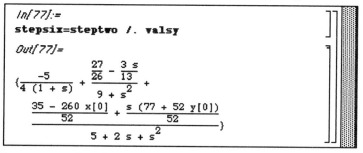

$$\frac{1}{4}\frac{1}{1+s} - \frac{3}{26}\left[\frac{s}{s^2+9} + \frac{2}{s^2+9}\right] - \frac{7}{52}\left[\frac{s+1}{5+2s+s^2} - \frac{10}{5+2s+s^2}\right] +$$

$$x(0)\left[\frac{s+1}{5+2s+s^2} + \frac{1}{5+2s+s^2}\right] + \frac{y(0)}{5+2s+s^2}.$$

Since the inverse Laplace transform of $\frac{1}{1+s}$ is e^{-t}, $\frac{s}{s^2+9}$ is $\cos(3t)$, $\frac{1}{s^2+9}$ is $\frac{2}{3}\sin(3t)$,

$\frac{s+1}{5+2s+s^2}$ is $\cos(2t)\,e^{-t}$, and $\frac{1}{5+2s+s^2}$ is $\frac{1}{2}\sin(2t)\,e^{-t}$, the inverse Laplace transform

of **stepfour** is

$$x(t) = e^{-t}\left[\frac{1}{4} - \frac{7}{52}\left(\cos(2t) - 5\sin(t)\right) + x(0)\left(\cos(2t) + \frac{1}{2}\sin(2t)\right) + \frac{y(0)}{2}\sin(2t)\right] -$$

$$\frac{3}{26}\left(\cos(3t) + \frac{4}{3}\sin(3t)\right)$$

$$= e^{-t}\left[\frac{1}{4} + \left(x(0) - \frac{7}{52}\right)\cos(2t) + \left(\frac{x(0)}{2} + \frac{y(0)}{2} + \frac{35}{52}\right)\sin(2t)\right] - \frac{3}{26}\left(\cos(3t) + \frac{4}{3}\sin(3t)\right),$$

defined below as x(t).

```
In[19]:=
solx[t_]=Exp[-t](1/4+(x[0]-7/52)Cos[2t]+
    (x[0]/2+y[0]/2+35/52)Sin[2t])-
    3/26(Cos[3t]+4/3 Sin[3t])

Out[19]=
```

$$-3\ (\text{Cos}[3\ t]\ +\ \frac{4\ \text{Sin}[3\ t]}{3}) \over 26 \quad +$$

$$(\tfrac{1}{4}\ +\ \text{Cos}[2\ t]\ (-(\tfrac{7}{52})\ +\ x[0])\ +$$

$$\text{Sin}[2\ t]\ (\tfrac{35}{52}\ +\ \tfrac{x[0]}{2}\ +\ \tfrac{y[0]}{2}))\ /\ E^t$$

Similarly, to compute the inverse Laplace transform of **stepsix** we rewrite **stepsix** as

$$\frac{-5}{4}\frac{1}{1+s}-\frac{3}{13}\frac{s}{9+s^2}+\frac{27}{26}\frac{1}{9+s^2}+\frac{35-260x(0)}{52}\frac{1}{5+2s+s^2}+$$

$$\frac{77+52y(0)}{52}\left[\frac{s+1}{5+2s+s^2}-\frac{1}{5+2s+s^2}\right].$$

Then the inverse Laplace transform of **stepsix** is

$$y(t)=\frac{-5}{4}e^{-t}-\frac{3}{13}\cos(3t)+\frac{27}{26}\frac{2}{3}\sin(3t)+\frac{35-260x(0)}{52}\frac{1}{2}\sin(2t)e^{-t}+$$

$$\frac{77+52y(0)}{52}\left[\cos(2t)e^{-t}-\frac{1}{2}\sin(t)e^{-t}\right]$$

$$=\frac{3}{13}\left[3\sin(3t)-\cos(3t)\right]+e^{-t}\left[\frac{-5}{4}+\frac{77+52y(0)}{52}\cos(2t)-\left(\frac{21}{52}+\frac{5}{2}x(0)+\frac{1}{2}y(0)\right)\sin(2t)\right],$$

defined below.

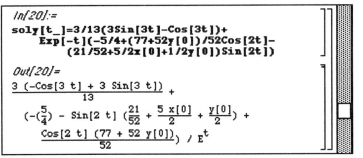

```
In[20]:=
soly[t_]=3/13(3Sin[3t]-Cos[3t])+
    Exp[-t](-5/4+(77+52y[0])/52Cos[2t]-
        (21/52+5/2x[0]+1/2y[0])Sin[2t])

Out[20]=
```

$$\frac{3\ (-\text{Cos}[3\ t]\ +\ 3\ \text{Sin}[3\ t])}{13}\ +$$

$$(-(\tfrac{5}{4})\ -\ \text{Sin}[2\ t]\ (\tfrac{21}{52}\ +\ \tfrac{5\ x[0]}{2}\ +\ \tfrac{y[0]}{2})\ +$$

$$\frac{\text{Cos}[2\ t]\ (77\ +\ 52\ y[0])}{52})\ /\ E^t$$

To graph x(t) and y(t) for various values of x(0) and y(0), we first define a list of ordered pairs **inits** and then define a function **pp[j]** which graphs (x(t),y(t)) when x(0) is replaced by the first element of the jth element of **inits** and y(0) is replaced by the second element of the jth element of **inits**. Then a table, **graphs**, is created of **pp[j]** for j=1, 2, ... , 12 which is then partitioned into a four-by-three array of graphics cells, **array**, and finally displayed as a graphics array.

```
In[21]:=
inits={{0,0},{0,.25},{0,-.25},
       {.15,0},{0,1},{1,1},
       {1,0},{1,-1},{0,-1},
       {-1,-1},{-1,0},{-1,1}};
In[30]:=
pp[j_]:=ParametricPlot[
       {solx[t],soly[t]}/.
           {x[0]->inits[[j,1]],
                y[0]->inits[[j,2]]},
       {t,0,7},
       Ticks->{None,Automatic},
       Compiled->False,
       DisplayFunction->Identity];
graphs=Table[pp[j],{j,1,12}];
array=Partition[graphs,3];
Show[GraphicsArray[array]]
```

The results are displayed below.

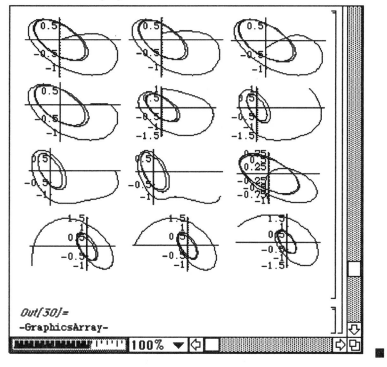

```
Out[30]=
-GraphicsArray-
```

100%

■ **EXAMPLE 10.24**

Use Laplace transforms to solve $\begin{cases} 2x''-5y'=0 \\ -y''-3y-x'=\sin(t) \end{cases}$ subject to the conditions $x(0)=1$, $x'(0)=0$,

$y(0)=1$, and $y'(0)=1$.

Solution:

Proceeding as in the previous examples, we first load the package **LaplaceTransform** and define
lhs to be $2x''-5y'$ and **rhs** to be $-y''-3y-x'$.

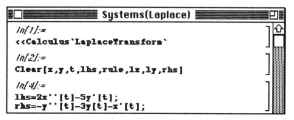

Let **1x** and **1y** denote the Laplace transform of $x(t)$ and $y(t)$, respectively. Then, the Laplace transform
of $x(t)$ is **1x**, $x'(t)$ is $s\,\mathbf{1x}-x(0)$, $x''(t)$ is $s^2\,\mathbf{1x}-sx(0)-x'(0)$, $y(t)$ is **1y**, $y'(t)$ is $s\,\mathbf{1y}-y(0)$,

and $y''(t)$ is $s^2\,\mathbf{1y}-sy(0)-y'(0)$.

In the following commands we first define **rule** so that when **rule** is applied to an object, each
occurrence of **x[t]** is replaced by **1x**, **x'[t]** is replaced by **s 1x-x[0]**, **x''[t]** is replaced by
s^2 1x-s x[0]-x'[0], **y[t]** is replaced by **1y**, **y'[t]** is replaced by **s 1y-y[0]**, and **y''[t]** is
replaced by **s^2 1y-s y[0]-y'[0]**, and then **rule** is applied to **{lhs,rhs}**, naming the resulting
output **eqs**.

The resulting output corresponds to computing the Laplace transform of each member of the left-hand

side of the system $\begin{cases} 2x''-5y'=0 \\ -y''-3y-x'=\sin(t) \end{cases}$.

```
In[5]:=
rule={x[t]->lx,x'[t]->s lx-x[0],
      x''[t]->s^2 lx-s x[0]-x'[0],
      y[t]->ly,y'[t]->s ly-y[0],
      y''[t]->s^2 ly-s y[0]-y'[0]}

Out[5]=
{x[t] -> lx, x'[t] -> lx s - x[0],
              2
   x''[t] -> lx s  - s x[0] - x'[0], y[t] -> ly,
   y'[t] -> ly s - y[0],
              2
   y''[t] -> ly s  - s y[0] - y'[0]}

In[6]:=
eqs={lhs,rhs} /. rule

Out[6]=
                               2
{-5 (ly s - y[0]) + 2 (lx s  - s x[0] - x'[0]),
                             2
   -3 ly - lx s - ly s  + x[0] + s y[0] + y'[0]}
```

Since the Laplace transform of $\sin(t)$ is $\dfrac{1}{s^2+1}$ and the Laplace transform of 0 is 0, we must solve the

system of equations $\begin{cases} -5\big(s\,\mathbf{1}\,\mathbf{y} - y(0)\big) + 2\big(s^2\,\mathbf{1}\,\mathbf{x} - s\,x(0) - x'(0)\big) = 0 \\[2mm] -3\,\mathbf{1}\,\mathbf{y} - s\,\mathbf{1}\,\mathbf{x} - s^2\mathbf{1}\,\mathbf{y} + x(0) + s\,y(0) + y'(0) = \dfrac{1}{s^2+1} \end{cases}$ for $\mathbf{1}\,\mathbf{x}$ and $\mathbf{1}\,\mathbf{y}$ and then

compute the inverse Laplace transform of $\mathbf{1x}$ and $\mathbf{1y}$ to solve the problem. With the following
command, we solve the system of equations, above, and name the resulting output $\mathbf{sols}$.

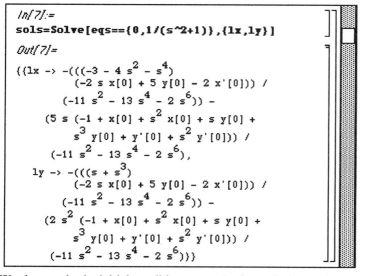

```
In[7]:=
sols=Solve[eqs=={0,1/(s^2+1)},{lx,ly}]

Out[7]=
{{lx -> -(((-3 - 4 s^2 - s^4)
            (-2 s x[0] + 5 y[0] - 2 x'[0])) /
          (-11 s^2 - 13 s^4 - 2 s^6)) -
     (5 s (-1 + x[0] + s^2 x[0] + s y[0] +
            s^3 y[0] + y'[0] + s^2 y'[0])) /
     (-11 s^2 - 13 s^4 - 2 s^6),
   ly -> -(((s + s^3)
            (-2 s x[0] + 5 y[0] - 2 x'[0])) /
          (-11 s^2 - 13 s^4 - 2 s^6)) -
     (2 s^2 (-1 + x[0] + s^2 x[0] + s y[0] +
            s^3 y[0] + y'[0] + s^2 y'[0])) /
     (-11 s^2 - 13 s^4 - 2 s^6)}}
```

We then apply the initial conditions to $\mathbf{sols}$ by replacing each occurrence of $\mathbf{x[0]}$, $\mathbf{y[0]}$, $\mathbf{x'[0]}$ and $\mathbf{y'[0]}$ in $\mathbf{sols}$ by 1, 1, 0, and 1, respectively. $\mathbf{1x}$ is extracted from $\mathbf{conds}$ with $\mathbf{conds[[1,1,2]]}$; $\mathbf{1y}$ with $\mathbf{conds[[1,2,2]]}$.

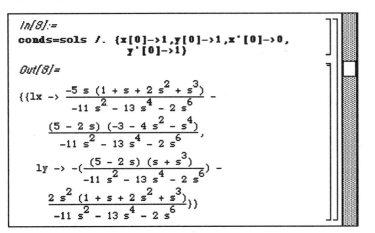

Then x(t) is the inverse Laplace transform of **1x**, computed below:

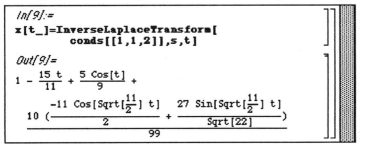

and y(t) is the inverse Laplace transform of **1y**, also computed below:

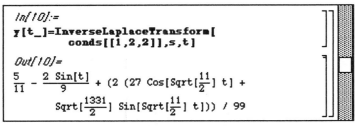

Finally, we graph (x(t),y(t)) on the interval [0,4π].

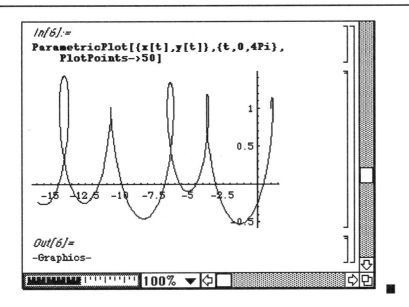

```
In[6]:=
ParametricPlot[{x[t],y[t]},{t,0,4Pi},
    PlotPoints->50]
```

```
Out[6]=
-Graphics-
```

§10.6 Nonlinear Systems, Linearization, and Classification of Equilibrium Points

We now turn our attention to systems of nonlinear equations. The general form of an autonomous system (in which there is no dependence on the independent variable t) is

$$\begin{cases} x'=f(x,y) \\ y'=g(x,y) \end{cases}$$

In order to understand problems of this type, however, we must first introduce some new terminology:

A point (x_0,y_0) is an **equilibrium point** of the system $\begin{cases} x'=f(x,y) \\ y'=g(x,y) \end{cases}$ means that both

$f(x_0,y_0) = 0$ and $g(x_0,y_0) = 0.$

❑ EXAMPLE 10.25

Find the equilibrium points of the system $\begin{cases} x'= 2x - y \\ y'= -x + 3y \end{cases}.$

Solution:

The equilibrium points of the system $\begin{cases} x' = 2x - y \\ y' = -x + 3y \end{cases}$ are the solutions of the system of

equations $\begin{cases} 2x - y = 0 \\ -x + 3y = 0 \end{cases}$. The only solution of the system $\begin{cases} 2x - y = 0 \\ -x + 3y = 0 \end{cases}$ is $(0,0)$ so

the only equilibrium point of the system $\begin{cases} x' = 2x - y \\ y' = -x + 3y \end{cases}$ is $(0,0)$. ∎

When working with nonlinear systems, we can often gain a great deal of information concerning the system by making a linear approximation near each equilibrium point of the nonlinear system and solving the linear system. Although the solution to the linearized system only approximates the actual solution to the nonlinear system, the general behavior of solutions to the nonlinear system near each equilibrium is the same as that of the corresponding linear system in most cases. The first step towards approximating a nonlinear system near each equilibrium point is to find the equilibrium points of the system and the matrix for linearization near each point as defined below:

An equilibrium point (x_0, y_0) of the system $\begin{cases} x' = f(x, y) \\ y' = g(x, y) \end{cases}$ is classified by the eigenvalues

of the matrix $\begin{pmatrix} f_x(x_0, y_0) & f_y(x_0, y_0) \\ g_x(x_0, y_0) & g_y(x_0, y_0) \end{pmatrix}$.

After determining the matrix for linearization for each equilibrium point, the eigenvalues for the matrix must be found. Then, we classify each equilibrium point according to the following criteria:

Let (x_0, y_0) be an equilibrium point of the system $\begin{cases} x' = f(x, y) \\ y' = g(x, y) \end{cases}$ and let λ_1 and λ_2 be the eigenvalues

of the matrix $\begin{pmatrix} f_x(x_0, y_0) & f_y(x_0, y_0) \\ g_x(x_0, y_0) & g_y(x_0, y_0) \end{pmatrix}$.

If $\lambda_2 > \lambda_1$ are distinct, non - zero and positive, (x_0, y_0) is an **unstable node**;

if $\lambda_2 > \lambda_1$ are distinct, non - zero and negative, (x_0, y_0) is a **stable node**; and if

$\lambda_2 > \lambda_1$ are distinct, non - zero and have opposite signs, (x_0, y_0) is a **saddle**.

If $\lambda_1 = \overline{\lambda_2} = \alpha + \beta i$, $\beta \neq 0$, are complex, then (x_0, y_0) is a **stable spiral** if $\alpha < 0$; an

unstable spiral if $\alpha > 0$; and a **center** if $\alpha = 0$.

We will not discuss the case when the eigenvalues are the same or one eigenvalue is zero.

■ **EXAMPLE 10.26**

Find and classify the equilibrium points of the system $\begin{cases} \dfrac{dx}{dt} = 5x + 3y - 4 \\ \dfrac{dy}{dt} = -4x - 3y + 2 \end{cases}$.

Solution:

We locate the critical points by defining f(x,y)=5x+3y−4, g(x,y)=−4x−3y+2 and then solving the system $\begin{cases} f(x,y) = 0 \\ g(x,y) = 0 \end{cases}$ for x and y. The resulting output means that the only equilibrium point of the system is (2,−2).

```
≡▢≡══════════════ EquilibriumPoints ═══════════════▣≡
In[57]:=
Clear[f,g,matrix]
f[x_,y_]=5x+3y-4
g[x_,y_]=-4x-3y+2

Out[56]=
-4 + 5 x + 3 y

Out[57]=
2 - 4 x - 3 y

In[58]:=
Solve[{f[x,y]==0,g[x,y]==0}]

Out[58]=
{{x -> 2, y -> -2}}
```

We then define **matrix** to be the matrix $\begin{pmatrix} f_x(x,y) & f_y(x,y) \\ g_x(x,y) & g_y(x,y) \end{pmatrix}$ and compute the eigenvalues of **matrix**.

```
In[60]:=
matrix={{D[f[x,y],x],D[f[x,y],y]},
    {D[g[x,y],x],D[g[x,y],y]}};
MatrixForm[matrix]

Out[60]//MatrixForm=
 5    3
-4   -3

In[61]:=
Eigenvalues[matrix]

Out[61]=
{-1, 3}
```

Since the eigenvalues have opposite signs, we conclude that (2,–2) is a saddle. In order to graph several solution curves and the direction fields of the system near the point (2,–2), we first use **DSolve** to find

the general solution of the system $\begin{cases} \dfrac{dx}{dt} = 5x + 3y - 4 \\ \dfrac{dy}{dt} = -4x - 3y + 2 \end{cases}$ and then name the resulting output **sols**.

```
In[66]:=
eqone=x'[t]==5x[t]+3y[t]-4;
eqtwo=y'[t]==-4x[t]-3y[t]+2;

In[67]:=
sols=DSolve[{eqone,eqtwo},{x[t],y[t]},t]

Out[67]=
```

$$\left\{\left\{x[t] \to 2 - \frac{C[1]}{2 \, E^t} - \frac{3 \, E^{3\,t} \, C[2]}{2},\right.\right.$$

$$\left.\left. y[t] \to -2 + \frac{C[1]}{E^t} + E^{3\,t} \, C[2]\right\}\right\}$$

We then create a table of ordered functions, **funs**, by replacing each occurrence of **C[1]** and **C[2]** in **{x[t],y[t]}/.sols[[1]]** by **i** and **j**, respectively, for **i**=–4, –3, ... , 3, 4 and **j**=–3, –2, ... , 2, 3. Since **funs** is actually an array of ordered functions, we use **Flatten** to remove parentheses from **funs** and then name the resulting list of ordered functions **tograph**. **Short** is used to display an abbreviated two-line form of **tograph**.

Note that entering the command **{sols[[1,1,2]],sols[[1,2,2]]}** produces the same result as entering the command **{x[t],y[t]}/.sols[[1]]**.

```
In[20]:=
funs=Table[{x[t],y[t]} /. sols[[1]] /.
            {C[1]->i,C[2]->j},
            {i,-4,4},{j,-3,3}];
tograph=Flatten[funs,1];
Short[tograph,2]

Out[20]//Short=
```

$$\left\{\left\{2 + \frac{2}{E^t} + \frac{9 \, E^{3\,t}}{2}, \; -2 - \frac{4}{E^t} - 3 \, E^{3\,t}\right\}, \; \langle\langle 61 \rangle\rangle,\right.$$

$$\left. \left\{2 + \langle\langle 1 \rangle\rangle - \frac{9 \, E^{3\,t}}{2}, \; \langle\langle 1 \rangle\rangle\right\}\right\}$$

We then use **ParametricPlot** to graph the list of ordered functions **tograph** for $-2 \le t \le 2$ and name the resulting graphics object **graphs**. The option **PlotRange->{{-2,4},{-4,2}}** is used to guarantee that the displayed region consist of the rectangle [–2,4] × [–4,2].

In[21]:=
```
graphs=ParametricPlot[Evaluate[tograph],
    {t,-2,2},
    PlotRange->{{-2,4},{-4,2}}]
```

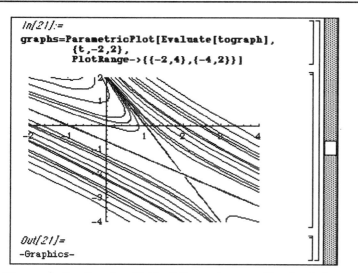

Out[21]=
-Graphics-

To graph the direction field of the system, we first load the package **PlotField** contained in the **Graphics** folder (or directory) and then use the command **PlotVectorField** to graph the direction field of the system on the rectangle $[-2,4] \times [-4,2]$ and name the resulting graphics object **vectors**.

In[22]:=
```
<<Graphics`PlotField`
```

In[23]:=
```
vectors=PlotVectorField[{f[x,y],g[x,y]},
    {x,-2,4},{y,-4,2}]
```

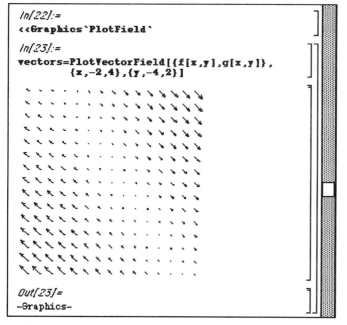

Out[23]=
-Graphics-

Finally, we display both **graphs** and **vectors** simultaneously.

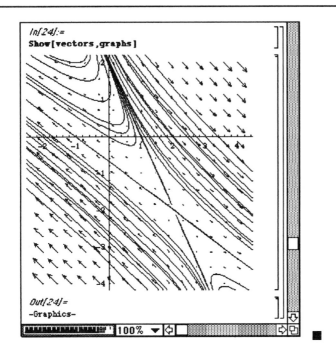

```
In[24]:=
Show[vectors,graphs]
```

```
Out[24]=
-Graphics-
```

We state the following useful theorem:

Theorem:

Suppose that (x_0, y_0) is an equilibrium point of the nonlinear system $\begin{cases} x'=f(x,y) \\ y'=g(x,y) \end{cases}$. Then, the

following relationships hold for the classification of (x_0, y_0) in the nonlinear system and that

in the associated linearized system.

Associated Linearized System	**Nonlinear System**
Stable Node	Stable Node
Unstable Node	Unstable Node
Stable Spiral	Stable Spiral
Unstable Spiral	Unstable Spiral
Saddle	Saddle
Center	No Conclusion

■ **EXAMPLE 10.27**

    The **Bonhoeffer-Van der Pol (BVP) oscillator** is the system of ordinary differential equations

$$\begin{cases} \dfrac{dx}{dt} = x - \dfrac{x^3}{3} - y + I(t) \\ \dfrac{dy}{dt} = c\left(x + a - by\right) \end{cases}$$. Find and calculate the equilibrium point of the BVP oscillator when $I(t) = 0$,

a=1, b=3, and c=1.

Solution:

In the same manner as in the preceding example, we begin by defining a=1, b=3, c=1,

$f(x,y) = x - \dfrac{x^3}{3} - y, g(x,y) = x\left(x + a - by\right)$:

```
≡□≡═══════════════ BVPOscillator ═══════════════□≡

In[11]:=
Clear[a,b,c,x,y,f,g,matrix]
a=1;b=3;c=1;
f[x_,y_]=x-x^3 /3-y
g[x_,y_]=c (x+a-b y)

Out[10]=
     x³
x - ── - y
     3

Out[11]=
1 + x - 3 y
```

and then solving the system $\begin{cases} f(x,y) = 0 \\ g(x,y) = 0 \end{cases}$ for x and y. We name the resulting list of solutions

roots. To obtain numerical approximations of the equilibrium points, we use **N** and then express the approximations in **TableForm**.

```
In[12]:=
roots=Solve[{f[x,y]==0,g[x,y]==0}]

Out[12]=
                2         -1 + Sqrt[5]
{{x -> 1, y -> ─}, {x -> ────────────,
                3              2

      3 + 3 Sqrt[5]
 y -> ────────────},
          18

      -1 - Sqrt[5]         3 - 3 Sqrt[5]
{x -> ────────────, y -> ────────────}}
           2                  18

In[13]:=
N[roots]//TableForm

Out[13]//TableForm=
x -> 1.           y -> 0.666667
x -> 0.618034     y -> 0.539345
x -> -1.61803     y -> -0.206011
```

To classify each equilibrium point, we begin by defining **matrix** to be the matrix
$\begin{pmatrix} f_x(x,y) & f_y(x,y) \\ g_x(x,y) & g_y(x,y) \end{pmatrix}$ and displaying **matrix** in **MatrixForm**

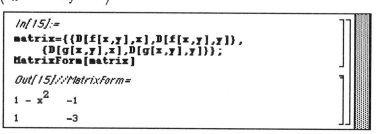

We then compute the eigenvalues of the matrix obtained by replacing each occurrence of x and y in **matrix** by each set of ordered pairs in **pairs**. We begin by computing the eigenvalues of the matrix obtained by replacing each occurrence of x and y in **matrix** by 1 and 2/3, respectively. Since the resulting eigenvalues are negative, (1,2/3) is a stable node.

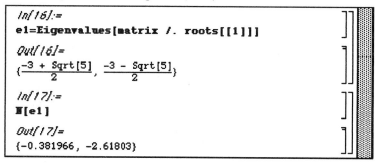

Similarly, since the eigenvalues of the matrix obtained by replacing each occurrence of x and y in the **matrix** by $\dfrac{-1+\sqrt{5}}{2}$ and $\dfrac{3+3\sqrt{5}}{18}$, respectively, have opposite signs, we conclude that $\left(\dfrac{-1+\sqrt{5}}{2}, \dfrac{3+3\sqrt{5}}{18}\right)$ is a saddle.

```
In[18]:=
e2=Eigenvalues[matrix /. roots[[2]]]

Out[18]=
{-7 + Sqrt[5] + Sqrt[1 + Sqrt[5]] Sqrt[9 + Sqrt[5]]
 ----------------------------------------------------- ,
                        4
 -7 + Sqrt[5] - Sqrt[1 + Sqrt[5]] Sqrt[9 + Sqrt[5]]
 ----------------------------------------------------- }
                        4

In[19]:=
N[e2]

Out[19]=
{0.316512, -2.69848}
```

Finally, since the eigenvalues of the matrix obtained by replacing each occurrence of x and y in the

matrix by $\dfrac{-1-\sqrt{5}}{2}$ and $\dfrac{3-3\sqrt{5}}{18}$, respectively, are complex conjugates with the real part negative,

we conclude that $\left(\dfrac{-1-\sqrt{5}}{2}, \dfrac{3-3\sqrt{5}}{18}\right)$ is a stable spiral.

```
In[20]:=
e3=Eigenvalues[matrix /. roots[[3]]]

Out[20]=
{(-7 - Sqrt[5] + Sqrt[-9 + Sqrt[5]]
      Sqrt[-1 + Sqrt[5]]) / 4,
   (-7 - Sqrt[5] - Sqrt[-9 + Sqrt[5]]
      Sqrt[-1 + Sqrt[5]]) / 4}

In[21]:=
N[e3]

Out[21]=
{-2.30902 + 0.722871 I, -2.30902 - 0.722871 I}
```

In order to graph several solution curves of the equation and the direction fields near each equilibrium point, we begin by defining **eqpts** to be a table of ordered pairs corresponding to the three equilibrium points.

```
In[22]:=
eqpts=Table[{x,y} /. roots[[i]],{i,1,3}]
Out[22]=
         2     -1 + Sqrt[5]  3 + 3 Sqrt[5]
{{1,  ---}, {------------, -------------},
         3          2             18
   -1 - Sqrt[5]  3 - 3 Sqrt[5]
  {------------, -------------}}
        2             18
```

We then define **sols** to use **NDSolve** to compute and then graph a numerical solution to the system. **sols[s]** works by

1. Defining the variables **solt**, **x**, **y**, and **t** local to the functions **sols**;

2. Defining **eqone** and **eqtwo** to be the equations $x' = x - \dfrac{x^3}{3} - y$ and $y' = 1 + x - 3y$,

respectively;

3. Defining **solt** to be a numerical solution to the system

$$\left\{ \begin{array}{l} \textbf{eqone} \\ \textbf{eqtwo} \\ x(0) = s\cos(11\pi s) + .8,\, y(0) = s\cos(11\pi s) + .8 \end{array} \right. \quad \text{valid for } 0 \le t \le 5;\text{ and}$$

4. Graphing **solt** for $0 \le t \le 5$. The final graph is contained in the rectangle $|-3,3| \times |-3,3|$.

Note that the graphics object created as a result of executing the command **sols[s]** is not displayed since the option **DisplayFunction->Identity** is included.

```
In[11]:=
Clear[sols]
sols[s_]:=Module[{solt,x,y,t},

    eqone=x'[t]==x[t]-x[t]^3/3-y[t];

    eqtwo=y'[t]==1+x[t]-3y[t];

    solt=NDSolve[{eqone,eqtwo,
        x[0]==s Cos[11 Pi s]+.8,
        y[0]==s Sin[11 Pi s]+.6},
        {x[t],y[t]},{t,0,5}];

    ParametricPlot[{x[t],y[t]} /. solt,
    {t,0,5},Compiled->False,
    PlotRange->{{-3,3},{-3,3}},
    DisplayFunction->Identity]
        ]
```

We then create a table of seventy graphs called **graphs** by computing **sols[s]** for s=.1, .1+2.9/70, ... , 3:

```
In[12]:=
graphs=Table[sols[s],{s,.1,3,2.9/70}];
```

graph the points in the list **eqpts**:

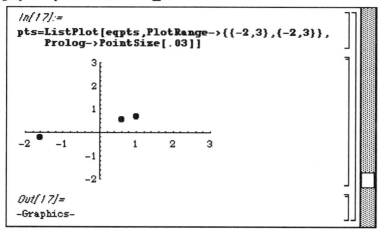

```
In[17]:=
pts=ListPlot[eqpts,PlotRange->{{-2,3},{-2,3}},
    Prolog->PointSize[.03]]
```

```
Out[17]=
-Graphics-
```

and then display both **pnts** and **graphs** simultaneously.

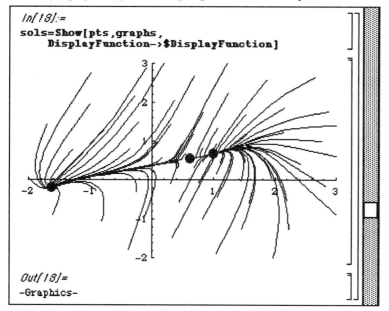

```
In[18]:=
sols=Show[pts,graphs,
    DisplayFunction->$DisplayFunction]
```

```
Out[18]=
-Graphics-
```

To graph the direction fields, we first load the package **PlotField** and then graph the direction field on the rectangle $[-2,3] \times [-2,3]$.

```
In[19]:=
<<Graphics`PlotField`
In[20]:=
vecs=PlotVectorField[{x-x^3/3-y,1+x-3y},
    {x,-2,3},{y,-2,3}]
```

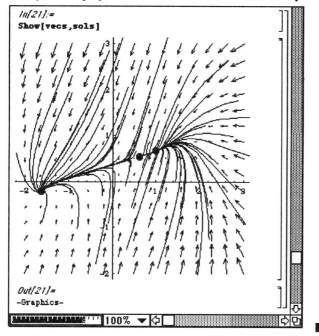

```
Out[20]=
-Graphics-
```

Finally, we display both **vecs** and **sols** simultaneously.

```
In[21]:=
Show[vecs,sols]
```

```
Out[21]=
-Graphics-
```

■ EXAMPLE 10.28

The **Lotka - Volterra system** (Predator - Prey model) is the system $\begin{cases} x' = a_1 x - a_2 x y \\ y' = -b_1 y + b_2 x y \end{cases}$, where

a_1, a_2, b_1, and b_2 are positive constants.

Find and classify the equilibrium points of the Lotka-Volterra equations.

Solution:

In the same manner as in the preceding examples, we begin by defining **eqonerhs** and **eqtworhs** to be $a_1 x - a_2 x y$ and $-b_1 y + b_2 x y$, respectively, and the solving the system of equations

$\begin{cases} a_1 x - a_2 x y = 0 \\ -b_1 y + b_2 x y = 0 \end{cases}$ for x and y to locate the equilibrium points.

```
≡□≡══════════ PredatorPrey ══════════□≡
In[2]:=
eqonerhs=a[1] x - a[2] x y;
eqtworhs=-b[1] y + b[2] x y;

In[3]:=
cps=Solve[{eqonerhs==0,eqtworhs==0},{x,y}]

Out[3]=
{{x -> b[1]/b[2], y -> a[1]/a[2]},
   {x -> 0, y -> 0}}
```

To classify the equilibrium points, we first define **linmatrix** to be the matrix

$$\begin{pmatrix} \dfrac{d}{dx}\left(a_1 x - a_2 x y\right) & \dfrac{d}{dy}\left(a_1 x - a_2 x y\right) \\ \dfrac{d}{dx}\left(-b_1 y + b_2 x y\right) & \dfrac{d}{dy}\left(-b_1 y + b_2 x y\right) \end{pmatrix}$$ and display **linmatrix** in **MatrixForm**

```
In[5]:=
linmatrix={{D[eqonerhs,x],D[eqonerhs,y]},
   {D[eqtworhs,x],D[eqtworhs,y]}};
MatrixForm[linmatrix]

Out[5]//MatrixForm=
a[1] - y a[2]      -(x a[2])
y b[2]             -b[1] + x b[2]
```

We then compute the value of **linmatrix** when $x = \dfrac{b_1}{b_2}$ and $y = \dfrac{a_1}{a_2}$:

```
In[6]:=
linmatrix /. cps[[1]] // MatrixForm

Out[6]//MatrixForm=

0                    -(a[2] b[1])
                        b[2]
a[1] b[2]
  a[2]               0
```

and then compute the eigenvalues. Since the eigenvalues are complex conjugates with the real part equal

to 0, we conclude that the equilibrium point $\left(\dfrac{b_1}{b_2}, \dfrac{a_1}{a_2} \right)$ is a center.

```
In[7]:=
linmatrix /. cps[[1]] // Eigenvalues

Out[7]=
{I Sqrt[a[1]] Sqrt[b[1]],
  -I Sqrt[a[1]] Sqrt[b[1]]}
```

Similarly we compute the value of **linmatrix** when x = 0 and y = 0:

```
In[8]:=
linmatrix /. cps[[2]] // MatrixForm

Out[8]//MatrixForm=
a[1]     0
0       -b[1]
```

and then compute the eigenvalues. Since the eigenvalues are real and have opposite signs, we conclude that the equilibrium point (0,0) is a saddle.

```
In[9]:=
linmatrix /. cps[[2]] // Eigenvalues

Out[9]=
{a[1], -b[1]}
```

Unsuccessfully, we attempt to use **DSolve** to solve the system in the special case when $a_1=2$, $a_2=1$, $b_1=3$, and $b_2=1$:

```
In[11]:=
eqone=x'[t]==2 x[t]-x[t] y[t];
eqtwo=y'[t]==-3 y[t]+x[t] y[t];
```

```
In[12]:=
DSolve[{eqone,eqtwo},{x[t],y[t]},t]

Out[12]=
DSolve[{x'[t] == 2 x[t] - x[t] y[t],
    y'[t] == -3 y[t] + x[t] y[t]},
  {x[t], y[t]}, t]
```

However, we are able to use **NDSolve** to solve the system when x(0)=1 and y(0)=1 for $0 \leq t \leq 3$:

```
In[13]:=
solone=NDSolve[{eqone,eqtwo,x[0]==1,y[0]==1},
      {x[t],y[t]},{t,0,3}]

Out[13]=
{{x[t] ->
    InterpolatingFunction[{0., 3.},
      <>][t], y[t] ->
    InterpolatingFunction[{0., 3.},
      <>][t]}}
```

We then use **ParametricPlot** to graph the solution, **solone**, obtained above.

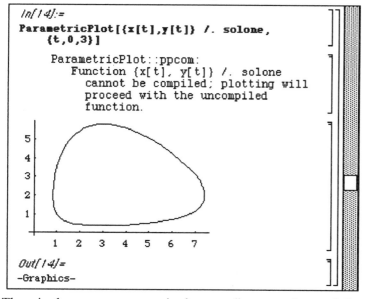

```
In[14]:=
ParametricPlot[{x[t],y[t]} /. solone,
    {t,0,3}]

      ParametricPlot::ppcom:
        Function {x[t], y[t]} /. solone
          cannot be compiled; plotting will
          proceed with the uncompiled
          function.
```

```
Out[14]=
-Graphics-
```

Then, in the same manner as in the preceding example, we define a function **sol** to graph various numerical solutions of the system above.

```
In[15]:=
Clear[sol]

In[17]:=
sol[s_]:=Module[{solt,x,y,t,eqone,eqtwo},

    eqone=x'[t]==2 x[t]-x[t] y[t];

    eqtwo=y'[t]==-3 y[t]+x[t] y[t];

    solt=NDSolve[
        {eqone,eqtwo,x[0]==3s,y[0]==2s},
        {x[t],y[t]},{t,0,4}];

    ParametricPlot[{x[t],y[t]} /. solt,
        {t,0,4},Compiled->False,
        DisplayFunction->Identity]
            ]
```

We plot the solution with **sol** for values of t from t = 1/8 to t = 7/8 using increments of 3/40

```
In[18]:=
graphs=Table[sol[t],{t,1/8,7/8,3/40}]

Out[18]=
{-Graphics-, -Graphics-, -Graphics-,
  -Graphics-, -Graphics-, -Graphics-,
  -Graphics-, -Graphics-, -Graphics-,
  -Graphics-, -Graphics-}
```

and display these graphs below. Notice that all of the solutions oscillate about the center. These solutions reveal the relationship between the two populations: prey, x(t), and predator, y(t). As we follow one cycle counterclockwise beginning, for example, near the point (2,0), we notice that as x(t) increases, then y(t) increases until y(t) becomes overpopulated. Then, since the prey population is too small to supply the predator population, y(t) decreases which leads to an increase in the population of x(t). Then, since the number of predators becomes too small to control the number in the prey population, x(t) becomes overpopulated and the cycle repeats itself.

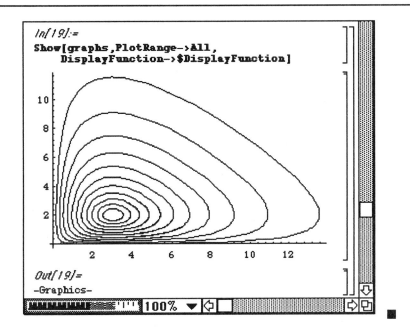

```
In[19]:=
Show[graphs,PlotRange->All,
    DisplayFunction->$DisplayFunction]
```

```
Out[19]=
-Graphics-
```

EXAMPLE 10.29

The Van - der - Pol equation $x'' + \mu\left(x^2 - 1\right)x' + x = 0$ is equivalent to the system

$$\begin{cases} x' = y \\ y' = \mu\left(1 - x^2\right)y - x \end{cases}$$

We find the equilibrium point of this well-known system below in **roots**. The associated linear system is then entered as **linmatrix**.

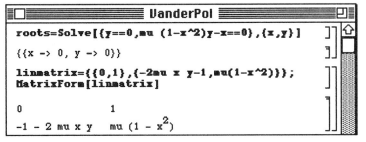

```
roots=Solve[{y==0,mu (1-x^2)y-x==0},{x,y}]

{{x -> 0, y -> 0}}

linmatrix={{0,1},{-2mu x y-1,mu(1-x^2)}};
MatrixForm[linmatrix]

0               1
-1 - 2 mu x y    mu (1 - x )
```

The eigenvalues of the linearized system at the equilibrium point (0,0) are given below.

```
linmatrix /. roots[[1]] // Eigenvalues

{mu + Sqrt[-2 + mu] Sqrt[2 + mu]
 ——————————————————————————————— ,
              2

 mu - Sqrt[-2 + mu] Sqrt[2 + mu]
 ——————————————————————————————— }
              2
```

Notice that these eigenvalues simplify to $\lambda = \dfrac{\mu \pm \sqrt{\mu^2 - 4}}{2}$ which are:

(a) both positive, real if $\mu > 2$ since $\sqrt{\mu^2 - 4} < \mu$. Hence, (0,0) is an unstable node.

(b) complex conjugates with positive real part if $0 < \mu < 2$. Hence, (0,0) is an unstable spiral.

(Note that μ is assumed positive since Van der Pol's equation came about from the study of nonlinear damping. Therefore, μ represents the damping coefficient discussed in previous sections on spring mass systems. We disregard the case with $\mu = 2$ since it results in a repeated eigenvalue.)

We now employ **NDSolve** to determine and plot solutions which correspond to Van der Pol's equation with $\mu=1$. Thus means that the equilibrium point (0,0) is an unstable focus according to our classification above. The two equations are defined in **eqone** and **eqtwo** and then used with **NDSolve** to determine the solution to the initial value problem x[0]=0 on the interval [0,10].

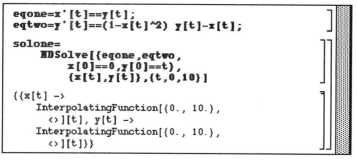

```
eqone=x'[t]==y[t];
eqtwo=y'[t]==(1-x[t]^2) y[t]-x[t];

solone=
    NDSolve[{eqone,eqtwo,
        x[0]==0,y[0]==t},
        {x[t],y[t]},{t,0,10}]
{{x[t] ->
    InterpolatingFunction[{0., 10.},
        <>][t], y[t] ->
    InterpolatingFunction[{0., 10.},
        <>][t]}}
```

The approximate solution is then plotted with **ParametricPlot** below.

```
ParametricPlot[{x[t],y[t]} /. solone,
   {t,0,10}]

   ParametricPlot::ppcom:
      Function {x[t], y[t]} /. solone
         cannot be compiled; plotting will
         proceed with the uncompiled
         function.
```

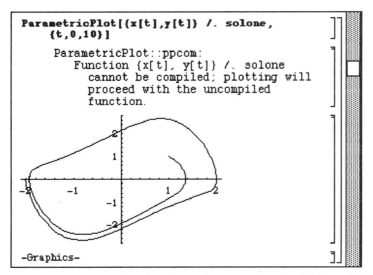

```
-Graphics-
```

We now attempt to plot the phase plane associated with Van der Pol's equation. This is done in a manner similar to the previous example by defining a function **sol** which numerically approximates the solution using the initial conditions x(0)=−s and y(0)=0 and plots the solution parametrically over the interval [0,10].

```
In[1]:=
Clear[sol,eqone,eqtwo,x,y,t]
In[8]:=
sol[s_]:=Module[{solt,x,y,t,eqone,eqtwo},

   eqone=x'[t]==y[t];

   eqtwo=y'[t]==(1-x[t]^2) y[t]-x[t];

   solt=NDSolve[
      {eqone,eqtwo,x[0]==-s,y[0]==0},
      {x[t],y[t]},{t,0,10}];

   ParametricPlot[{x[t],y[t]} /. solt,
      {t,0,10},Compiled->False,
      DisplayFunction->Identity]
         ]
```

A table of solutions is produced in **graphs** below by using x(0)=−s for s = 1/4 to s = 8π using increments of (8π-1/4)/15.

```
In[5]:=
graphs=Table[sol[t],{t,1/4,8Pi,(8Pi-1/4)/15}]

Out[5]=
{-Graphics-, -Graphics-, -Graphics-,
  -Graphics-, -Graphics-, -Graphics-,
  -Graphics-, -Graphics-, -Graphics-,
  -Graphics-, -Graphics-, -Graphics-,
  -Graphics-, -Graphics-, -Graphics-,
  -Graphics-}
```

The solutions which were found are then shown simultaneously to reveal to phase plane. Notice that the solutions seem to approach a closed path as the variable t increases. (Recall that (0,0) is an unstable spiral, so the solutions are directed away from the origin.) This closed path is called a **limit cycle** because all solutions approach it as t increases.

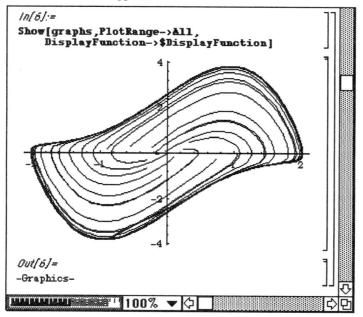

```
In[6]:=
Show[graphs,PlotRange->All,
     DisplayFunction->$DisplayFunction]
```

```
Out[6]=
-Graphics-
```

100% ▼

Several other solutions with initial point outside of the limit cycle are generated below.

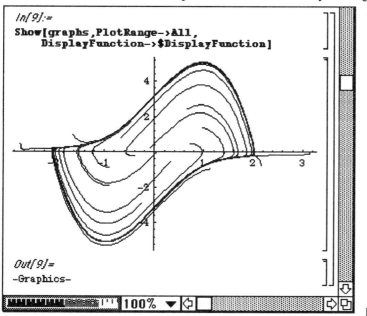

Chapter 11: Applications of Systems of Ordinary Differential Equations

Mathematica commands used in **Chapter 11** include:

AspectRatio	GrayLevel	PlotRange
ComplexExpand	IdentityMatrix	PlotStyle
ComplexToTrig	Integrate	ReplaceAll
DisplayFunction	Inverse	Short
DSolve	Length	Show
Eigensystem	ParametricPlot	Simplify
Eigenvalues	ParametricPlot3D	Solve
Eigenvectors	Part	Table
Evaluate	Partition	
GraphicsArray	Plot	

§11.1 L-R-C Circuits with Loops

As indicated in **Chapter 5**, electrical circuits can be modeled with ordinary differential equations. In this section, we illustrate how a circuit involving loops can be described as a system of linear ordinary differential equations with constant coefficients. This derivation is based on the following principles:

> **Kirchhoff's Current Law:** The current entering a point of the circuit equals the current leaving the point.

> **Kirchhoff's Voltage Law:** The sum of the changes in voltage around each loop in the circuit is zero.

Again, we use the following standard symbols for the components of the circuit:
i(t) = current where i(t) = q'(t), q(t) = charge, R = resistance, C = capacitance, V = voltage, and L = inductance.
The following relationships corresponding to the drops in voltage in the various components of the circuit which were stated in **Chapter 5** are also used:
In a resistor: $V = iR$,

in an inductor: $V = L\dfrac{di}{dt}$,

in a capacitor: $V = \dfrac{1}{C}q$,

and a voltage source offers a voltage drop: $V = -V(t)$.
In determining the drops in voltage around the circuit, we consistently add the voltages in the clockwise

direction. The positive direction is directed from the negative symbol towards the positive symbol associated with the voltage source. In summing the voltage drops encountered in the circuit, a drop across a component is added to the sum if the positive direction through the component agrees with the clockwise direction. Otherwise, this drop is subtracted. In the case of the following L-R-C circuit with one loop involving each type of component, the current is equal around the circuit by Kirchhoff's Current Law as illustrated in the following diagram.

Figure 11.1
A Simple L-R-C Circuit

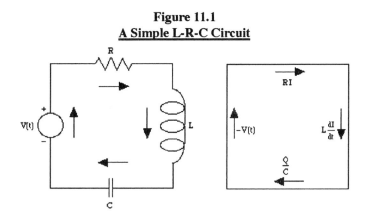

Also, by Kirchoff's Voltage Law, we have the following sum:

$$Ri + L\frac{di}{dt} + \frac{1}{C}q - V(t) = 0.$$

Solving this equation for $\frac{di}{dt}$ and using the relationship between i and q, $\frac{dq}{dt} = i$, we have the following

system of differential equations with indicated initial conditions:

$$\frac{dq}{dt} = i$$

$$\frac{di}{dt} = -\frac{1}{LC}q - \frac{R}{L}i + \frac{V(t)}{L}$$

$$q(0) = q_0, i(0) = i_0.$$

DSolve, the Method of Undetermined Coefficients, or Variation of Parameters can be used to solve problems of this type.

■ **EXAMPLE 11.1**

Consider the L-R-C circuit with L = 1, R = 2, C = 4 and V(t)= e^{-t}. Determine the charge and current using the values given in the list: (0,0), (1,1), (−1,1), (1,−1), (1,2), (1,−2), (−1,2), (−1,−2), (2,1), (2,−1), (3,1), (3,−1), (4,1), (4,−1), (5,1), and (5,−1) where each ordered pair represents (C[1],C[2]) in the general solution. Plot the phase plane for this system using the solutions obtained with this set of constants.

Solution:

First, the general solution of the system, $\begin{cases} q'(t) = i(t) \\ i'(t) = \dfrac{-1}{4}\, q(t) - 2i(t) + e^{-t} \end{cases}$,

is determined in **dsys** below with **DSolve**.

```
═══════════════════════ LRCSystems ═══════════════════════

In[2]:=
Clear[dsys]
dsys=DSolve[{q'[t]==i[t],
        i'[t]==(-1/4)q[t]-2i[t]+Exp[-t]},
        {q[t],i[t]},t]

Out[2]=

{{q[t] -> (-2 (2 - Sqrt[3] +
              ---------
                3 E^t

       E^((-2 + Sqrt[3]) t)/2 C[1])) /
    (2 - Sqrt[3]) -
      (2 (---------------- +      C[2]
            (6 - 3^3/2) E^t    E^((2 + Sqrt[3]) t)/2 )) /
      (2 + Sqrt[3]),
    i[t] ->
     2 - Sqrt[3]            1
     ----------- + ------------------ +
        3 E^t        (6 - 3^3/2) E^t

       E^((-2 + Sqrt[3]) t)/2 C[1] +

            C[2]
       ------------------- }}
       E^((2 + Sqrt[3]) t)/2
```

The solutions q(t) and i(t) are explicitly extracted from dsys with **dsys[[1,1,2]]** and **dsys[[1,2,2]]**, respectively. The constants are then entered in the list named **pairs**.

```
In[4]:=
pairs={{0,0},{1,1},{-1,1},{1,-1},
       {1,2},{1,-2},{-1,2},{-1,-2},
       {2,1},{2,-1},{3,1},{3,-1},
       {4,1},{4,-1},{5,1},{5,-1}};
```

The constants C[1] and C[2] are substituted into the general solution using the elements of **pairs**. This results in the creation of a list of functions called **funcs**. The ith function in the list of functions **funcs** is obtained by replacing each occurrence of C[1] and C[2] in **dsys[[1,1,2]]**, corresponding to the general solution of q(t), and **dsys[[1,2,2]]**, corresponding to the general solution of i(t) by the first member and second member, respectively, of the ith element of **pairs**.

Finally, the phase plane is obtained with **ParametricPlot** by graphing the list of functions **funcs** for $-2 \leq t \leq 2$.

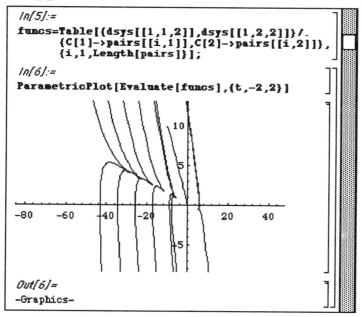

```
In[5]:=
funcs=Table[{dsys[[1,1,2]],dsys[[1,2,2]]}/.
      {C[1]->pairs[[i,1]],C[2]->pairs[[i,2]]},
      {i,1,Length[pairs]}];
In[6]:=
ParametricPlot[Evaluate[funcs],{t,-2,2}]
```

```
Out[6]=
-Graphics-
```

The first element of each component of **funcs** represents $q(t)$. Hence, a function **a** is defined below which plots the charge solutions obtained using some of the constants used earlier. **qtable** evaluates **a** for values of i from i = 1 to i = 8. Hence, a table of the plots of the first nine solutions for charge is generated in **qtable** below.

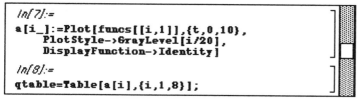

```
In[7]:=
a[i_]:=Plot[funcs[[i,1]],{t,0,10},
      PlotStyle->GrayLevel[i/20],
      DisplayFunction->Identity]
In[8]:=
qtable=Table[a[i],{i,1,8}];
```

These solutions are then displayed with **Show**.

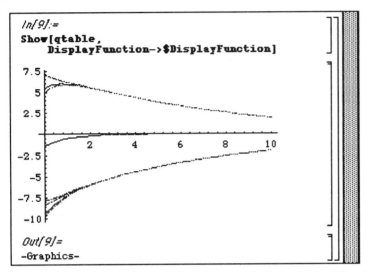

```
In[9]:=
Show[qtable,
    DisplayFunction->$DisplayFunction]
```

```
Out[9]=
-Graphics-
```

A similar procedure is then used to plot the solutions obtained for the current. First, the function **b** is defined in order to plot the solutions. Then, a table of plots is produced with **itable**.

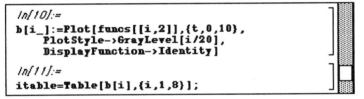

```
In[10]:=
b[i_]:=Plot[funcs[[i,2]],{t,0,10},
    PlotStyle->GrayLevel[i/20],
    DisplayFunction->Identity]
In[11]:=
itable=Table[b[i],{i,1,8}];
```

The solutions which represent the current are shown below.

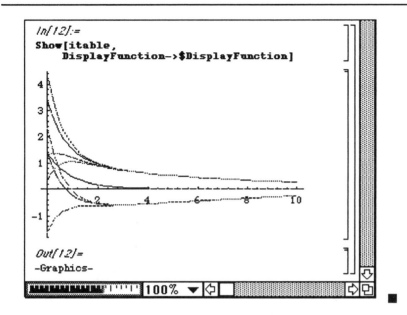

Derivation of the modeling differential equation becomes more complicated as the number of loops in the circuit is increased. For example, consider the circuit below which contains two loops.

Figure 11.2
A Two-Loop Circuit

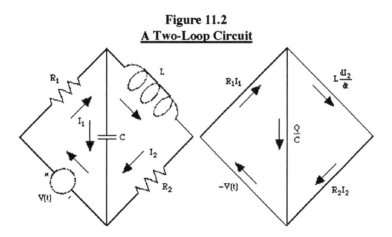

In this case, the current through the capacitor is equivalent to $i_1 - i_2$. Summing the voltage drops around each loop, we have:

$$R_1 i_1 + \frac{1}{C} q - V(t) = 0$$

$$L \frac{d i_2}{dt} + R_2 i_2 - \frac{1}{C} q = 0.$$

Solving the first equation for i_1 and using the relationship

$$\frac{dq}{dt} = i = i_1 - i_2,$$

we have the following system:

$$\frac{dq}{dt} = -\frac{1}{R_1 C} q - i_2 + \frac{V(t)}{R_1}$$

$$\frac{d i_2}{dt} = \frac{1}{LC} q - \frac{R_2}{L} i_2.$$

■ EXAMPLE 11.2

Consider the L-R-C circuit with two loops given that $R_1 = 16$, $L = R_2 = 4$, $C = 1$ and $V(t) = e^{-t}$. Plot the phase plane using the solutions obtained with the following pairs of constants indicating $(C[1], C[2])$: $(0,0)$, $(1,1)$, $(-1,-1)$, $(2,2)$, $(-2,-2)$, $(3,3)$, $(-3,-3)$, $(4,4)$, $(-4,-4)$, $(5,5)$, $(-5,-5)$, $(6,6)$, and $(-6,-6)$. Also, plot $q(t)$ and $i_2(t)$ separately for some of these constants in order to compare the behavior of the solutions. Finally, determine the current $i_1(t)$ and plot the corresponding solutions using the constants above.

Solution:

In a manner similar to that used in the previous example, the general solution is found with **DSolve** in **dsys** below.

```
┌─────────────────────────────────────────────────┐
│ ▤□▦▦▦▦▦▦▦ LRCSystems ▦▦▦▦▦▦▦ ▣▤ │
├─────────────────────────────────────────────────┤
│ In[14]:=                                      ⇧ │
│ Clear[dsys]                                      │
│ dsys=DSolve[{                                    │
│     q'[t]==(-1/16)q[t]-itwo[t]+(1/16)Exp[-t],    │
│     itwo'[t]==(1/4)q[t]-itwo[t]},                │
│     {q[t],itwo[t]},t]                            │
│                                                  │
│ Out[14]=                                         │
```

$$\{\{q[t] \to (15\ E^{((-17\ -\ I\ Sqrt[31])\ t)/32}\ C[1]\ -$$
$$I\ Sqrt[31]\ E^{((-17\ -\ I\ Sqrt[31])\ t)/32}$$
$$C[1]\ +\ 15\ E^{((-17\ +\ I\ Sqrt[31])\ t)/32}$$
$$C[2]\ +\ I\ Sqrt[31]$$
$$E^{((-17\ +\ I\ Sqrt[31])\ t)/32}\ C[2])\ /\ 8,$$
$$itwo[t] \to$$
$$\frac{8}{Sqrt[31]\ (-15\ I\ +\ Sqrt[31])\ E^t}\ +$$
$$\frac{8}{Sqrt[31]\ (15\ I\ +\ Sqrt[31])\ E^t}\ +$$
$$E^{((-17\ -\ I\ Sqrt[31])\ t)/32}\ C[1]\ +$$
$$E^{((-17\ +\ I\ Sqrt[31])\ t)/32}\ C[2]\}\}$$

The constants are entered in the list **pairs** and substituted into the general solution to yield the list of functions, **funcs**. Since the solutions involve complex exponentials, **ComplexExpand** is used to obtain real-valued solutions in **simple**.

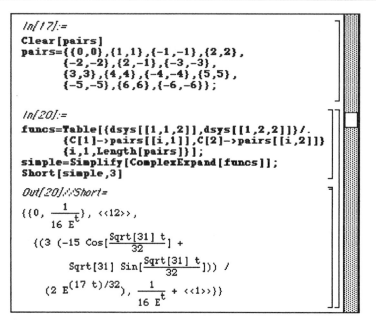

```
In[17]:=
Clear[pairs]
pairs={{0,0},{1,1},{-1,-1},{2,2},
    {-2,-2},{2,-1},{-3,-3},
    {3,3},{4,4},{-4,-4},{5,5},
    {-5,-5},{6,6},{-6,-6}};

In[20]:=
funcs=Table[{dsys[[1,1,2]],dsys[[1,2,2]]}/.
    {C[1]->pairs[[i,1]],C[2]->pairs[[i,2]]}
    {i,1,Length[pairs]}];
simple=Simplify[ComplexExpand[funcs]];
Short[simple,3]

Out[20]//Short=
```

$$\{\{0, \frac{1}{16\ E^t}\},\ \langle\langle12\rangle\rangle,$$

$$\{(3\ (-15\ Cos[\frac{Sqrt[31]\ t}{32}]\ +$$

$$Sqrt[31]\ Sin[\frac{Sqrt[31]\ t}{32}]))\ /$$

$$(2\ E^{(17\ t)/32}),\ \frac{1}{16\ E^t}\ +\ \langle\langle1\rangle\rangle\}\}$$

These solutions are then used with **ParametricPlot** to obtain the phase portrait for the system. Note that even though several error messages are generated, which are not completely displayed, the resulting graphs are correct.

```
In[21]:=
ParametricPlot[Evaluate[simple],{t,-1,20}]

    ParametricPlot::pptr:
        {Comp<<10>>on[{t}, <<2>>][t], <<1>>}
            does not evaluate to a pair of real
            at t = -1..

    ParametricPlot::pptr:
        {Comp<<10>>on[{t}, <<2>>][t], <<1>>}
            does not evaluate to a pair of real
            at t = -0.125.
```

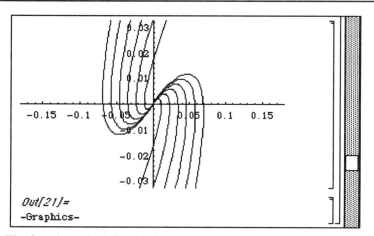

Out[21]=
-Graphics-

The function **b** is defined below which plots the solutions for charge. Several plots are then generated in **qtable1** and **qtable2**. The displays of these plots are given in **q1** and **q2**.

```
In[22]:=
b[i_]:=b[i]=Plot[simple[[i,1]],
{t,0,5},PlotStyle->GrayLevel[i/30],
DisplayFunction->Identity]

In[30]:=
qtable1=Table[b[i],{i,1,5}];
qtable2=Table[b[i],{i,6,10}];
q1=Show[qtable1];
q2=Show[qtable2];
```

Next, a similar function **c** for plotting solutions for the charge, $i_2(t)$, is defined. Tables of plots are created in **ctable1** and **ctable2** with the respective displays given in **c1** and **c2**.

```
In[31]:=
c[i_]:=Plot[simple[[i,2]],
    {t,0,5},PlotStyle->GrayLevel[i/30],
    DisplayFunction->Identity]

In[35]:=
ctable1=Table[c[i],{i,1,5}];
c1=Show[ctable1,PlotRange->All];
ctable2=Table[c[i],{i,6,10}];
c2=Show[ctable2,PlotRange->All];
```

The solution plots for charge and current are shown as a **GraphicsArray** below.

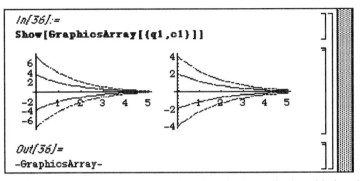

```
In[36]:=
Show[GraphicsArray[{q1,c1}]]
```

```
Out[36]=
-GraphicsArray-
```

Similarly, the respective solutions in **q2** and **c2** are displayed below in a **GraphicsArray**.

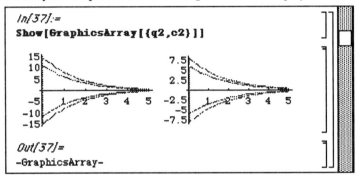

```
In[37]:=
Show[GraphicsArray[{q2,c2}]]
```

```
Out[37]=
-GraphicsArray-
```

The overall current, i(t), is defined as the derivative of the charge, q(t). Hence, the function **curr** below represents this overall current. The list **currtable** gives the list of currents using the constants in **pairs**. Since, $i = i_1 - i_2$, we have that $i_1 = i + i_2$. Therefore, the function **current** defined below represents i_1.

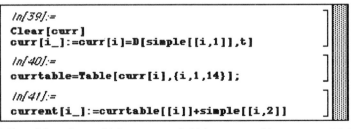

```
In[39]:=
Clear[curr]
curr[i_]:=curr[i]=D[simple[[i,1]],t]
In[40]:=
currtable=Table[curr[i],{i,1,14}];
In[41]:=
current[i_]:=currtable[[i]]+simple[[i,2]]
```

A list of functions which represent $i_1(t)$ is generated in **currentlist** and then plotted simultaneously in **p1**. Note that when the **Plot** command is entered, *Mathematica* produces several error messages, which are not completely displayed here, but the graphs displayed are correct.

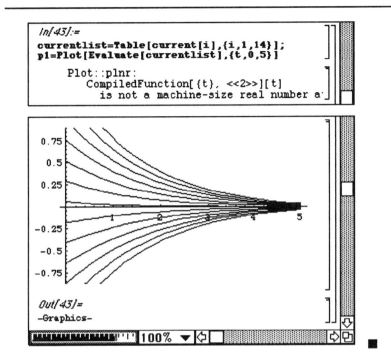

```
In[43]:=
currentlist=Table[current[i],{i,1,14}];
p1=Plot[Evaluate[currentlist],{t,0,5}]

        Plot::plnr:
            CompiledFunction[{t}, <<2>>][t]
                is not a machine-size real number a
```

Out[43]=
-Graphics-

Next, we consider the following circuit which is made up of three loops illustrated in the following diagram.

Figure 11.3
A Three-Loop Circuit

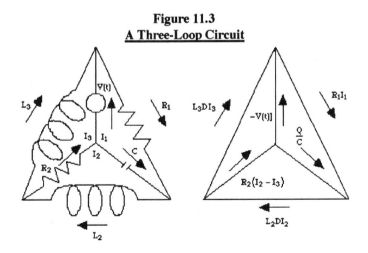

In this circuit, the current through the resistor R_2 is i_2-i_3, and the current through the capacitor is i_1-i_2. Using these quantities in the voltage drop sum equations yields the following three-dimensional system:

$$-V(t)+R_1 i_1 +\frac{1}{C}q=0$$

$$-\frac{1}{C}q+L_2\frac{di_2}{dt}+R_2(i_2-i_3)=0$$

$$V(t)-R_2(i_2-i_1)+L_3\frac{di_3}{dt}=0.$$

Using the relationship

$$\frac{dq}{dt}=i_1-i_2$$

and solving the first equation for i_1, we have the system

$$\frac{dq}{dt}=-\frac{1}{R_1 C}q-i_2+\frac{V(t)}{R_1}$$

$$\frac{di_2}{dt}=-\frac{1}{L_2 C}q-\frac{R_2}{L_2}i_2+\frac{R_2}{L_2}i_3$$

$$\frac{di_3}{dt}=\qquad\frac{R_2}{L_3}i_2-\frac{R_2}{L_3}i_3-\frac{V(t)}{L_3}.$$

■ EXAMPLE 11.3

Consider the circuit with three loops which lead to the previous system of ordinary differential equations. In this case, let $R_1 = R_2 = 1$, $L_1 = L_2 = L_3 = 2$, and $C = 1$. If $V(t) = e^{-t}$, then determine the solution of this nonhomogeneous system. Plot the solution for various initial conditions.

Solution:

The appropriate parameters are defined below where **matrix** represents the coefficient matrix of the corresponding homogeneous system of equations. These parameter values are substituted into the general form of this matrix so that the eigensystem (eigenvalues and eigenvectors) can be determined in **eigens**.

```
≡□≡≡≡≡≡≡≡≡≡≡≡≡≡≡≡ LRCSystems ≡≡≡≡≡≡≡≡≡ 回≡
In[29]:=
r1=1;
r2=1;
l1=2;
l2=2;
l3=2;
c=2;
matrix={{-1/(r1 c),-1,0},
        {-1/(l2 c),-r2/l2,r2/l2},
        {0,r2/l3,-r2/l3}};
eigens=Eigensystem[matrix]

Out[29]=
{{-(1/2), (-1 - Sqrt[2])/2, (-1 + Sqrt[2])/2},
   {{2, 0, 1}, {-2, -Sqrt[2], 1},
    {-2, Sqrt[2], 1}}}
```

The fundamental matrix is defined below as **phi** by extracting the eigenvalues and eigenvectors from the output list of **eigens**.

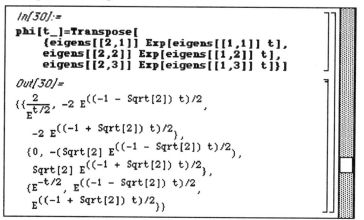

```
In[30]:=
phi[t_]=Transpose[
    {eigens[[2,1]] Exp[eigens[[1,1]] t],
     eigens[[2,2]] Exp[eigens[[1,2]] t],
     eigens[[2,3]] Exp[eigens[[1,3]] t]}]

Out[30]=
{{2/E^(t/2), -2 E^((-1 - Sqrt[2]) t)/2,
    -2 E^((-1 + Sqrt[2]) t)/2},
  {0, -(Sqrt[2] E^((-1 - Sqrt[2]) t)/2),
    Sqrt[2] E^((-1 + Sqrt[2]) t)/2},
  {E^(-t/2), E^((-1 - Sqrt[2]) t)/2,
    E^((-1 + Sqrt[2]) t)/2}}
```

The inverse of the fundamental matrix is then computed and called **invphi**.

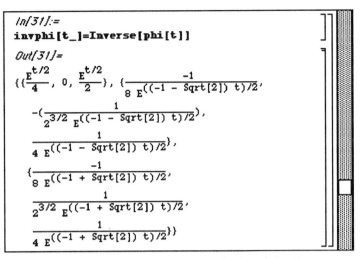

The nonhomogeneous vector of the system is then defined as **f**.

```
In[32]:=
f[t_]={Exp[-t],0,(-1/2) Exp[-t]}

Out[32]=
{E^-t, 0, -1/(2 E^t)}
```

Using the formulas indicated in the previous section on nonhomogeneous systems, the particular solution **solp** and the complementary solution **solh** are determined. Once obtained, these solutions are used to construct the general solution to the nonhomogeneous system. This general solution is given in **solution** below.

```
In[33]:=
solp[t_]=phi[t].Integrate[invphi[u].f[u],
                          {u,0,t}];

In[34]:=
solh[t_]=phi[t].invphi[0].{x0,y0,z0};

In[35]:=
solution=solh[t]+solp[t];
```

The components of the solution vector are extracted for later use and called **x**, **y**, and **z**. Of course, in this case, the three components represent $q(t)$, $i_2(t)$, and $i_3(t)$, respectively. The function **a** is then defined in order to produce a simultaneous plot of the three components using initial values of **x0** = 10, **z0** = 3, and varying the value of **y0**. In **graphs**, an array of plots using **y0** = 1, 2, 3, and 4 is produced and partitioned into groups of two.

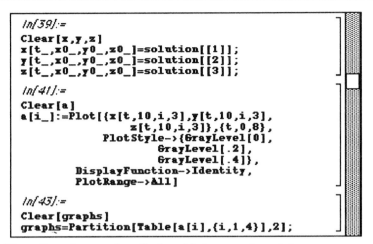

These four plots are then displayed below as a **GraphicsArray**.

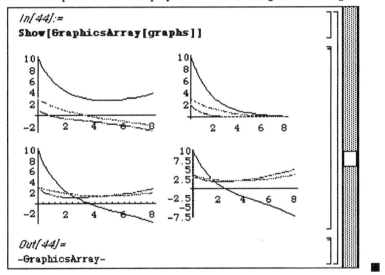

§11.2 Diffusion Problems

Problems involving the determination of the diffusion of a material in a medium also lead to first-order systems of linear ordinary differential equations. For example, consider the situation in which two solutions of a substance are separated by a membrane of permeability P. Assume that the amount of substance which passes through the membrane at any particular time is proportional to the difference between the concentrations of the two solutions. Therefore, if we let x_1 and x_2 represent the two concentrations, and V_1 and V_2 represent the corresponding volume of each solution, the system of differential equations is given by

$$\frac{dx_1}{dt} = P\left(\frac{x_2}{V_2} - \frac{x_1}{V_1}\right)$$

$$\frac{dx_2}{dt} = P\left(\frac{x_1}{V_1} - \frac{x_2}{V_2}\right).$$

■ EXAMPLE 11.4

Consider the system given above with $V_1 = V_2 = V$.

Determine and plot the solution of this system for various initial conditions.

Solution:

With this substitution, the system becomes

$$\frac{dx_1}{dt} = k(x_2 - x_1)$$

$$\frac{dx_2}{dt} = k(x_1 - x_2)$$

where $k = \dfrac{P}{V}$.

DSolve is used to solve this system in **sol** below. Notice that the initial values of x_1 and x_2 are entered as **init1** and **init2**, respectively.

```
≡□□ ════════════════ Diffusion ══════════════ □□≡
 In[69]:=
 sol=DSolve[{x1'[t]==k (x1[t]-x2[t]),
     x2'[t]==k (x2[t]-x1[t]),
     x1[0]==init1,x2[0]==init2},
     {x1[t],x2[t]},t]

 Out[69]=
 {{x1[t] ->
         init1 + E^2 k t init1 + init2 - E^2 k t init2
         ─────────────────────────────────────────────── ,
                              2
     x2[t] ->
         init1 - E^2 k t init1 + init2 + E^2 k t init2
         ───────────────────────────────────────────────}}
                              2
```

The corresponding solutions for x_1 and x_2 which are functions of t, k, **init1**, and **init2** are extracted from the output list of **sol**, with the commands **sol[[1,1,2]]** and **sol[[1,2,2]]**, and called **one** and **two**, respectively. The function **a** which parametrically plots the solution using **init1** = 10, k = 1, and **init2** = i. Hence, a list of plots is generated in **solns** for **i** = 9.75 to **i** = 10.25 using increments of 0.1. Note that this list of plots is partitioned into groups of two.

```
In[72]:=
Clear[one,two]
one[t_,init1_,init2_,k_]=sol[[1,1,2]];
two[t_,init1_,init2_,k_]=sol[[1,2,2]];

In[74]:=
Clear[a]
a[i_]:=ParametricPlot[
    {one[t,10,i,1],two[t,10,i,1]},{t,0,.5},
    DisplayFunction->Identity]

In[75]:=
solns=Partition[Table[a[i],
        {i,9.75,10.25,.1}],2];
```

This collection of plots are displayed below as a **GraphicsArray**. These plots indicate that as x_1 increases, x_2 decreases.

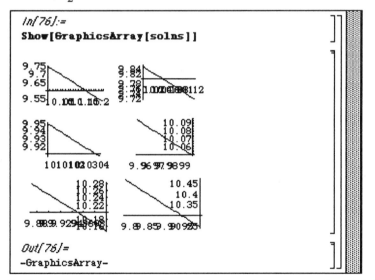

```
In[76]:=
Show[GraphicsArray[solns]]
```

```
Out[76]=
-GraphicsArray-
```

Similarly, the graphs of x_1 and x_2 can be plotted simultaneously. This is done below by first defining the function b which plots these two functions again using **init1** $= 10, k = 1$, and **init2** $=$ **i**.

```
In[78]:=
Clear[b]
b[i_]:=Plot[
    {one[t,10,i,1],two[t,10,i,1]},{t,0,1},
    PlotStyle->{GrayLevel[0],GrayLevel[.4]},
    DisplayFunction->Identity]
```

Below, an array of plots is produced for values of $i = 8$ to $i = 13$ using increments of 1. This list is also partitioned in groups of two and called **simplot**.

```
In[79]:=
simplot=Partition[Table[b[i],{i,8,13}],2];
```

The graphs which are shown below show that when x_1 is initially greater than x_2, then as x_1 increases, x_2 decreases. When the two are initially the same, then they remain unchanged. Finally, when x_1 is initially less than x_2, then as x_1 decreases, x_2 increases. Note that the graph of x_1 is the darker of the two curves.

```
In[80]:=
Show[GraphicsArray[simplot]]
```

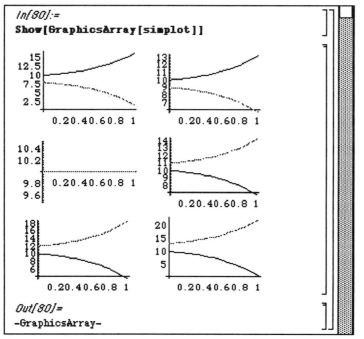

```
Out[80]=
-GraphicsArray-
```

Below, similar calculations are carried out using $k = 0.5$. The function **a** is defined and used to produce an array of plots in **solns2**.

```
In[82]:=
Clear[a]
a[i_]:=ParametricPlot[
    {one[t,10,i,.5],two[t,10,i,.5]},{t,0,1},
    DisplayFunction->Identity]

In[83]:=
solns2=Partition[Table[a[i],
        {i,9.75,10.25,.1}],2]

Out[83]=
{{-Graphics-, -Graphics-},
  {-Graphics-, -Graphics-},
  {-Graphics-, -Graphics-}}
```

These plots are shown below as a **GraphicsArray**.

```
In[84]:=
Show[GraphicsArray[solns2]]
```

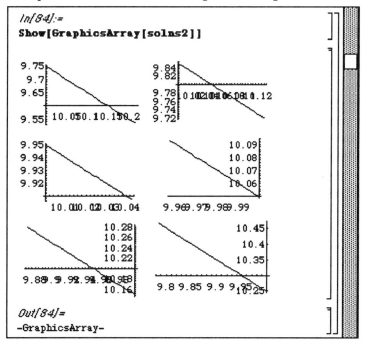

```
Out[84]=
-GraphicsArray-
```

The individual graphs of x_1 and x_2 are then plotted simultaneously below in **simplot2**.

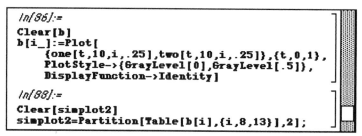

These graphs are then displayed. Note that since k = 0.5 (as opposed to k = 1 in the previous case), then the rate at which the amounts change is not as great as in the previous case.

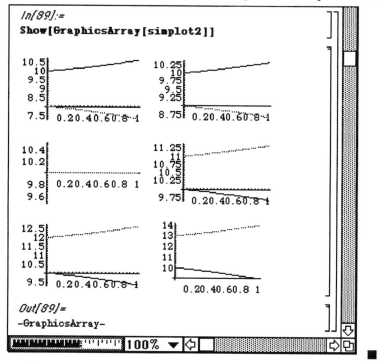

Next, consider the situation in which two solutions are separated by a double-walled membrane, where the inner wall has permeability P_1 and the outer wall has permeability P_2 with $0 < P_1 < P_2$. Suppose that the volume of solution within the inner wall is V_1 and that between the two walls, V_2. Let x represent the concentration of the solution within the inner wall and y, the concentration between the two walls. Assuming that the concentration of the solution outside the outer wall is constantly C, we have the following system of first-order ordinary differential equations.

$$\frac{dx}{dt} = \frac{P_1}{V_1}(y-x)$$

$$\frac{dy}{dt} = \frac{1}{V_2}\left(P_2\,(C-y)+P_1\,(x-y)\right) = \frac{-(P_1+P_2)y}{V_2} + \frac{P_1 x}{V_2} + \frac{P_2 C}{V_2},$$

$$x(0)=x_0,\ y(0)=y_0.$$

■ EXAMPLE 11.5

Consider the system given above with $P_1 = 3$, $P_2 = 8$, and unequal volumes, $V_1 = 2$ and $V_2 = 10$. Determine and plot the solution of this system for various initial conditions.

Solution:

Since this is a nonhomogeneous system, we find the solution without the use of DSolve. Note that when working with nonhomogeneous systems, this procedure tends to decrease the computation time when compared to that of **DSolve**. First, the coefficient matrix of the corresponding homogeneous system is defined in **mat** as are the indicated parameter values. Then, the eigensystem is determined for **mat** using the appropriate parameter values.

```
In[110]:=
p1=3;
p2=8;
vol1=2;
vol2=10;
mat={{-p1/vol1,p1/vol1},
       {(p1/vol2), -(p1+p2)/vol2}}
esys=Eigensystem[mat]

Out[109]=
{{-(3/2), 3/2}, {3/10, -(11/10)}}

Out[110]=
{{-2, -(3/5)}, {{-3, 1}, {5/3, 1}}}
```

The components of **esys**, the eigenvalues and the eigenvectors of **mat**, are extracted from the output list and used below to define the fundamental matrix of the corresponding homogeneous system, **phi**.

```
In[111]:=
phi[t_]=Transpose[
        {esys[[2,1]] Exp[esys[[1,1]] t],
         esys[[2,2]] Exp[esys[[1,2]] t]}]
Out[111]=
```
$$\{\{\frac{-3}{E^{2\ t}},\ \frac{5}{3\ E^{(3\ t)/5}}\},\ \{E^{-2\ t},\ E^{(-3\ t)/5}\}\}$$

The inverse of the fundamental matrix is computed below and named **invphi**.

```
In[112]:=
invphi[t_]=Inverse[phi[t]]
Out[112]=
```
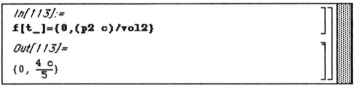

$$\{\{\frac{-3\ E^{2\ t}}{14},\ \frac{5\ E^{2\ t}}{14}\},\ \{\frac{3\ E^{(3\ t)/5}}{14},\ \frac{9\ E^{(3\ t)/5}}{14}\}\}$$

The nonhomogeneous portion of the system is given in the vector-valued function **f** below. Note that this function depends on the concentration **c** . Although this function is constant with respect to t, we define it as a function of t so that this procedure can be applied to other problems in which the nonhomogeneous vector depends on t.

```
In[113]:=
f[t_]={0,(p2 c)/vol2}
Out[113]=
```
$$\{0,\ \frac{4\ c}{5}\}$$

The particular solution is determined below and called **solp**.

```
In[114]:=
solp[t_]=
    phi[t].Integrate[invphi[u].f[u],{u,0,t}]
Out[114]=
```
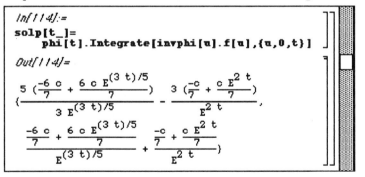

$$\{\frac{5\ (\frac{-6\ c}{7}+\frac{6\ c\ E^{(3\ t)/5}}{7})}{3\ E^{(3\ t)/5}}-\frac{3\ (\frac{-c}{7}+\frac{c\ E^{2\ t}}{7})}{E^{2\ t}},$$

$$\frac{\frac{-6\ c}{7}+\frac{6\ c\ E^{(3\ t)/5}}{7}}{E^{(3\ t)/5}}+\frac{\frac{-c}{7}+\frac{c\ E^{2\ t}}{7}}{E^{2\ t}}\}$$

The complementary solution is then found and called **solh**.

```
In[115]:=
solh[t_]=phi[t].invphi[0].{x0,y0}
```

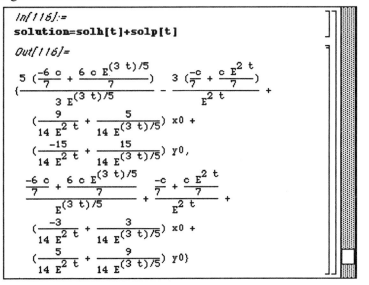

The general solution is, therefore, given as the sum of the complementary and particular solutions. This is given in **solution** below.

```
In[116]:=
solution=solh[t]+solp[t]
```

The solutions are then extracted from the output list in **x** and **y** for later use. Note that these two formulas depend on c, x_0, and y_0, as well as t.

```
In[118]:=
x[t_,x0_,y0_,c_]=solution[[1]];
y[t_,x0_,y0_,c_]=solution[[2]];
```

These solutions can be plotted simultaneously as indicated below in **plot1**.

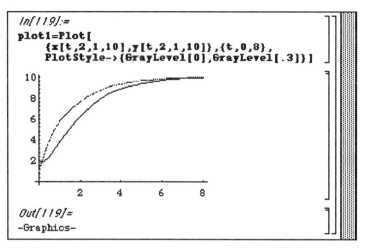

```
In[119]:=
plot1=Plot[
   {x[t,2,1,10],y[t,2,1,10]},{t,0,8},
   PlotStyle->{GrayLevel[0],GrayLevel[.3]}]
```

```
Out[119]=
-Graphics-
```

A similar graph can be produced for various values of the parameter y_0. Hence, the function **a** is defined which plots parametrically using $x_0 = 2$, c = 10, and $y_0 = i$. An array of plots is produced in **graph** and then displayed as a **GraphicsArray**. Notice that all of the graphs approach 10, the value of c.

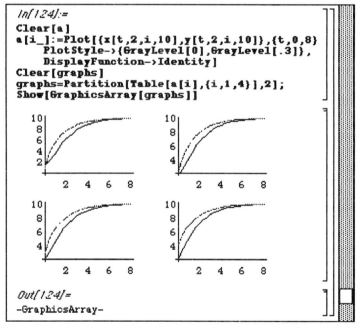

```
In[124]:=
Clear[a]
a[i_]:=Plot[{x[t,2,i,10],y[t,2,i,10]},{t,0,8}
   PlotStyle->{GrayLevel[0],GrayLevel[.3]},
   DisplayFunction->Identity]
Clear[graphs]
graphs=Partition[Table[a[i],{i,1,4}],2];
Show[GraphicsArray[graphs]]
```

```
Out[124]=
-GraphicsArray-
```

Similarly, the graphs are produced with $y_0 = 3$ and $x_0 = 1, 2, 3,$ and 4. The list of plots generated through the use of **a** is then displayed.

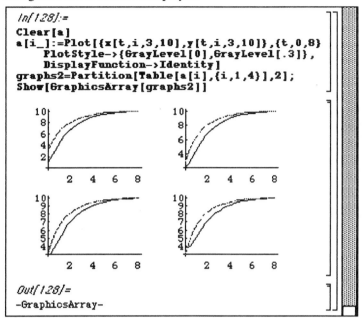

```
In[128]:=
Clear[a]
a[i_]:=Plot[{x[t,i,3,10],y[t,i,3,10]},{t,0,8}
    PlotStyle->{GrayLevel[0],GrayLevel[.3]},
    DisplayFunction->Identity]
graphs2=Partition[Table[a[i],{i,1,4}],2];
Show[GraphicsArray[graphs2]]
```

```
Out[128]=
-GraphicsArray-
```

To investigate the effects of various initial conditions, we determine and plot the solution for other parameter values using $x_0 = 2$, $y_0 = 3$, and $c = i$. An array of plots is generated for $i = 3, 6, 9,$ and 12.

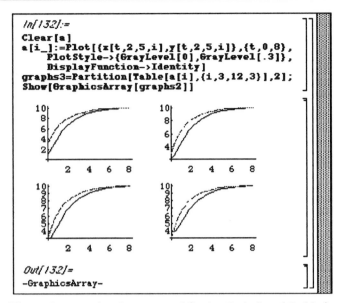

```
In[132]:=
Clear[a]
a[i_]:=Plot[{x[t,2,5,i],y[t,2,5,i]},{t,0,8},
    PlotStyle->{GrayLevel[0],GrayLevel[.3]},
    DisplayFunction->Identity]
graphs3=Partition[Table[a[i],{i,3,12,3}],2];
Show[GraphicsArray[graphs2]]
```

```
Out[132]=
-GraphicsArray-
```

These plots are also demonstrated for $i = 2$, 4, 6, and 8. Notice that in each case, the solutions approach the value of c.

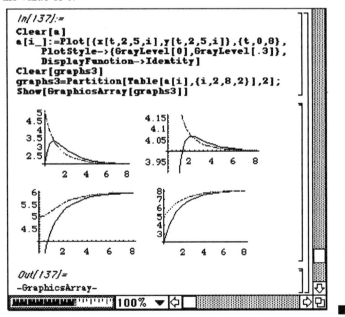

```
In[137]:=
Clear[a]
a[i_]:=Plot[{x[t,2,5,i],y[t,2,5,i]},{t,0,8},
    PlotStyle->{GrayLevel[0],GrayLevel[.3]},
    DisplayFunction->Identity]
Clear[graphs3]
graphs3=Partition[Table[a[i],{i,2,8,2}],2];
Show[GraphicsArray[graphs3]]
```

```
Out[137]=
-GraphicsArray-
```

100%

§11.3 Spring-Mass Systems

The motion of a mass attached to the end of a spring was modeled with a second-order linear differential equation with constant coefficients in **Chapter 5**. This situation can then be expressed as a system of first-order ordinary differential equations as well. Recall the second-order differential equation which models this situation $m\,x'' + c\,x' + k\,x = 0$ where m is the mass attached to the end of the spring, c is the damping coefficient, and k is the spring constant found with Hooke's law. This equation is easily transformed into a system of equations with the following substitution. Let $x' = y$. Hence, $y' = x'' = -k\,x - c\,x'$. Therefore, after substitution, we have the following system:

$x' = y$

$y' = -k\,x - c\,y$

In the previous examples, the motion of each spring was illustrated as a function of time. However, problems of this type may be considered as a phase portrait. In the following example, the phase portraits corresponding to the various situations encountered by spring-mass systems discussed in previous sections (undamped, damped, overdamped, and critically damped) are determined.

■ EXAMPLE 11.6

Consider the spring-mass system modeled by the differential equation $m\,x'' + c\,x' + k\,x = 0$ where m is the mass attached to the end of the spring, c is the damping coefficient, and k is the spring constant found with Hooke's law. Determine the phase portrait for each of the following situations:

(a) $m = 1, c = 0, k = 1$

(b) $m = 1, c = 1, k = 0.5$

(c) $m = 1, c = \sqrt{5}, k = 1$

(d) $m = 1, c = 2, k = 1$

Solution:

This problem is solved in a manner similar to those of previous examples. Since the system is homogeneous, **DSolve** is used below in **spr1** to determine the general solution of the system

$$\begin{cases} x' = y \\ y' = -x \end{cases}.$$

The solutions, which are expressed in terms of complex exponential functions, are extracted from **spr1** with **spr1[[1,1,2]]**, corresponding to x(t), and **spr1[[1,2,2]]**, corresponding to y(t). Note that this system is undamped.

```
========== SpringPortraits ==========

In[5]:=
Clear[spr1]
spr1=DSolve[{x'[t]==y[t],y'[t]==-x[t]},
                      {x[t],y[t]},t]

Out[5]=
{{x[t] -> (E^-I t (C[1] + E^2 I t C[1] + I C[2] -
          I E^2 I t C[2])) / 2,
   y[t] ->
      (E^-I t (-I C[1] + I E^2 I t C[1] + C[2] +
          E^2 I t C[2])) / 2}}
```

Since the solution involves complex exponentials, the **Trigonometry** package which is located in the
Algebra folder (directory) is loaded and contains the command **ComplexToTrig**. Then, the
ComplexToTrig command is used to obtain a real-valued solution by applying Euler's formula. The
result is then simplified and the resulting output is named **sim**. The solution is extracted from **sim** with
the commands **sim[[1,1,2]]**, corresponding to x(t), and **sim[[1,2,2]]**, corresponding to y(t).

```
In[7]:=
<<Algebra`Trigonometry`

In[8]:=
sim=Simplify[ComplexToTrig[spr1]]

Out[8]=
{{x[t] -> C[1] Cos[t] + C[2] Sin[t],
   y[t] -> C[2] Cos[t] - C[1] Sin[t]}}
```

A list of constants is entered below in pairs where each represents (C[1],C[2]). These constants are
substituted into the solution in **funcs** to form a list of solutions. The ith member of functions is
obtained by replacing each occurrence of C[1] and C[2] in **sim[[1,1,2]]** and **sim[[1,2,2]]**
by the first member of the ith element of **pairs** and the second member of the ith element of **pairs**,
respectively.

```
In[10]:=
pairs={{0,0},{1,1},{-1,1},{1,-1},
       {1,2},{1,-2},{-1,2},{-1,-2},
       {2,1},{2,-1},{3,1},{3,-1},
       {4,1},{4,-1},{5,1},{5,-1}};

In[11]:=
funcs=Table[{sim[[1,1,2]],sim[[1,2,2]]}/.
       {C[1]->pairs[[i,1]],C[2]->pairs[[i,2]]},
       {i,1,Length[pairs]}];
```

The elements of **funcs** are then plotted below with **ParametricPlot** to form the phase plane for this system. Note that the direction associated with the trajectories can be determined by using the first differential equation of the system x'=y. Hence, x increases in the first and second quadrants where y> 0 and decreases in the third and fourth quadrants where y< 0. Therefore, the motion is clockwise.

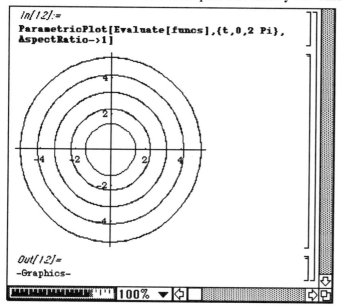

```
In[12]:=
ParametricPlot[Evaluate[funcs],{t,0,2 Pi},
 AspectRatio->1]

Out[12]=
-Graphics-
```

(b) The general solution for this underdamped system is determined with **DSolve** in **spr2** below.

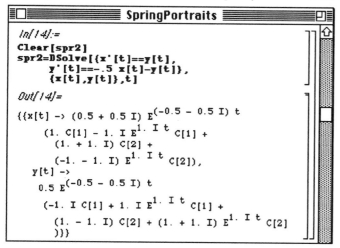

```
SpringPortraits

In[14]:=
Clear[spr2]
spr2=DSolve[{x'[t]==y[t],
     y'[t]==-.5 x[t]-y[t]},
     {x[t],y[t]},t]

Out[14]=
{{x[t] -> (0.5 + 0.5 I) E^(-0.5 - 0.5 I) t
     (1. C[1] - 1. I E^(1. I t) C[1] +
     (1. + 1. I) C[2] +
     (-1. - 1. I) E^(1. I t) C[2]),
  y[t] ->
     0.5 E^(-0.5 - 0.5 I) t
     (-1. I C[1] + 1. I E^(1. I t) C[1] +
     (1. - 1. I) C[2] + (1. + 1. I) E^(1. I t) C[2]
     )}}
```

Again, the **Trigonometry** package is loaded and used to eliminate the complex exponentials in the solution. The simplified solution is called **sim2**.

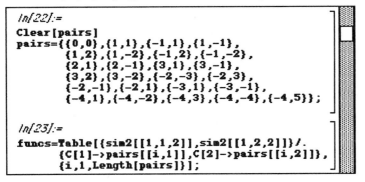

Values for (C[1], C[2]) are taken from the list **pairs** below to form the list of solutions **funcs**.

```
In[22]:=
Clear[pairs]
pairs={{0,0},{1,1},{-1,1},{1,-1},
       {1,2},{1,-2},{-1,2},{-1,-2},
       {2,1},{2,-1},{3,1},{3,-1},
       {3,2},{3,-2},{-2,-3},{-2,3},
       {-2,-1},{-2,1},{-3,1},{-3,-1},
       {-4,1},{-4,-2},{-4,3},{-4,-4},{-4,5}};

In[23]:=
funcs=Table[{sim2[[1,1,2]],sim2[[1,2,2]]}/.
      {C[1]->pairs[[i,1]],C[2]->pairs[[i,2]]},
      {i,1,Length[pairs]}];
```

ParametricPlot is then used to plot the phase plane for this system. Using x'=y, we see that the trajectories are directed inward. This is also determined by observing the exponential terms in the denominators in the solution. Hence, both x and y approach a value of zero as t increases.

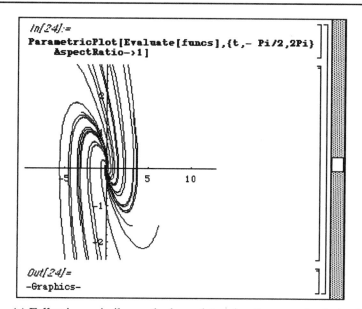

```
In[24]:=
ParametricPlot[Evaluate[funcs],{t,- Pi/2,2Pi}
    AspectRatio->1]
```

```
Out[24]=
-Graphics-
```

(c) Following a similar method, we determine the general solution in **spr3** with **DSolve**. Note that this system is overdamped.

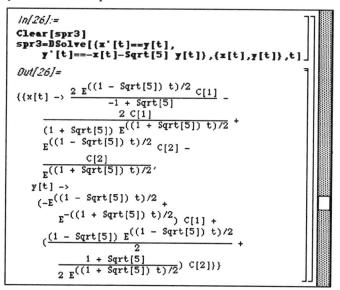

```
In[26]:=
Clear[spr3]
spr3=DSolve[{x'[t]==y[t],
    y'[t]==-x[t]-Sqrt[5] y[t]},{x[t],y[t]},t]
```

$$Out[26]=$$

$$\left\{\left\{x[t] \to \frac{2\, E^{((1\,-\,Sqrt[5])\,t)/2}\, C[1]}{-1\,+\,Sqrt[5]}\,-\right.\right.$$

$$\frac{2\,C[1]}{(1\,+\,Sqrt[5])\,E^{((1\,+\,Sqrt[5])\,t)/2}}\,+$$

$$E^{((1\,-\,Sqrt[5])\,t)/2}\,C[2]\,-$$

$$\frac{C[2]}{E^{((1\,+\,Sqrt[5])\,t)/2}},$$

$$y[t] \to$$

$$\left(-E^{((1\,-\,Sqrt[5])\,t)/2}\,+\right.$$

$$\left.E^{-((1\,+\,Sqrt[5])\,t)/2}\right)\,C[1]\,+$$

$$\left(\frac{(1\,-\,Sqrt[5])\,E^{((1\,-\,Sqrt[5])\,t)/2}}{2}\,+\right.$$

$$\left.\left.\frac{1\,+\,Sqrt[5]}{2\,E^{((1\,+\,Sqrt[5])\,t)/2}}\right)\,C[2]\right\}\right\}$$

The list of ordered pairs in **pairs** is then used to generate a table of solutions through substitution of these values for the arbitrary constants in the general solution.

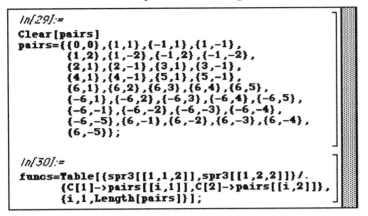

```
In[29]:=
Clear[pairs]
pairs={{0,0},{1,1},{-1,1},{1,-1},
      {1,2},{1,-2},{-1,2},{-1,-2},
      {2,1},{2,-1},{3,1},{3,-1},
      {4,1},{4,-1},{5,1},{5,-1},
      {6,1},{6,2},{6,3},{6,4},{6,5},
      {-6,1},{-6,2},{-6,3},{-6,4},{-6,5},
      {-6,-1},{-6,-2},{-6,-3},{-6,-4},
      {-6,-5},{6,-1},{6,-2},{6,-3},{6,-4},
      {6,-5}};

In[30]:=
funcs=Table[{spr3[[1,1,2]],spr3[[1,2,2]]}/.
      {C[1]->pairs[[i,1]],C[2]->pairs[[i,2]]},
      {i,1,Length[pairs]}];
```

Again, **ParametricPlot** is employed to display the phase plane. The trajectories are directed towards the origin because both x and y approach zero as t increases. The first differential equation of the system, x'=y, also supports this idea.

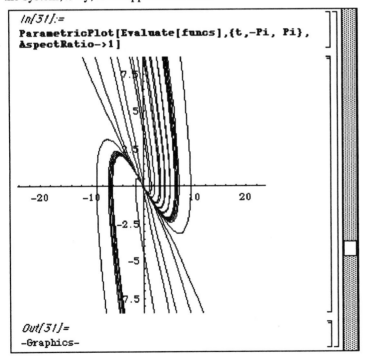

```
In[31]:=
ParametricPlot[Evaluate[funcs],{t,-Pi, Pi},
AspectRatio->1]
```

```
Out[31]=
-Graphics-
```

(d) Once again, **DSolve** is used to find the general solution of this system. Note that this system is critically damped, so there is only one eigenvalue $\lambda = -1$ which is repeated.

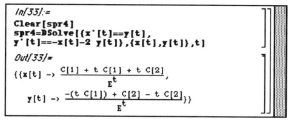

```
In[33]:=
Clear[spr4]
spr4=DSolve[{x'[t]==y[t],
y'[t]==-x[t]-2 y[t]},{x[t],y[t]},t]
```

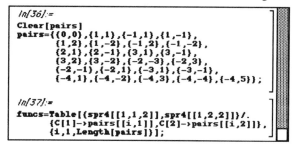

```
Out[33]=
{{x[t] -> (C[1] + t C[1] + t C[2])/E^t,

    y[t] -> (-(t C[1]) + C[2] - t C[2])/E^t}}
```

The list of solutions **funcs** is created below using the constant values given in **pairs**.

```
In[36]:=
Clear[pairs]
pairs={{0,0},{1,1},{-1,1},{1,-1},
       {1,2},{1,-2},{-1,2},{-1,-2},
       {2,1},{2,-1},{3,1},{3,-1},
       {3,2},{3,-2},{-2,-3},{-2,3},
       {-2,-1},{-2,1},{-3,1},{-3,-1},
       {-4,1},{-4,-2},{-4,3},{-4,-4},{-4,5}};

In[37]:=
funcs=Table[{spr4[[1,1,2]],spr4[[1,2,2]]}/.
       {C[1]->pairs[[i,1]],C[2]->pairs[[i,2]]},
       {i,1,Length[pairs]}];
```

The phase plane is then produced with **ParametricPlot**. Again, the trajectories are directed towards the origin.

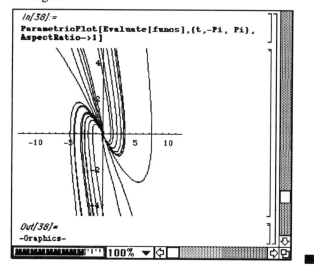

```
In[38]:=
ParametricPlot[Evaluate[funcs],{t,-Pi, Pi},
AspectRatio->1]
```

```
Out[38]=
-Graphics-
```

§11.4 Population Problems

In previous chapters, population problems were discussed which were based on the simple principle that the rate at which a population grows (or decays) is proportional to the number present in the population at any time t. These ideas can be extended to other examples involving more than one population which lead to systems of ordinary differential equations. We illustrate several situations through the following examples.

■ EXAMPLE 11.7

Suppose that the population of two neighboring territories x and y depends on several factors. The birth rate of x is a_1 while that of y is b_1. The rate at which citizens of x move to y is a_2 while that at which citizens move from y to x is b_2. Finally, the mortality rate of each territory is disregarded. Determine the respective populations of these two territories for any time t.

Solution:

Using the simple principles of previous examples, we have that the rate at which population x changes is

$$\frac{dx}{dt} = a_1 x - a_2 x + b_1 y = (a_1 - a_2)x + b_1 y$$

while the rate at which population y changes is

$$\frac{dy}{dt} = b_1 y - b_2 y + a_2 x = (b_1 - b_2)y + a_2 x.$$

Therefore, the system of equations which must be solved is:

$$\frac{dx}{dt} = (a_1 - a_2)x + b_1 y$$

$$\frac{dy}{dt} = a_2 x + (b_1 - b_2)y$$

$x(0) = x_0, y(0) = y_0$ where x_0 and y_0 are the initial populations of territories x and y, respectively.

This homogeneous system of differential equations is solved below with **DSolve**. By defining the function **pop** in terms of the parameters, the system does not have to be reentered to compare the affects that the variation of parameter values has on solutions. We first investigate the solution with $a_1=2$, $a_2=3$, $b_1=2$, and $b_2=1$.

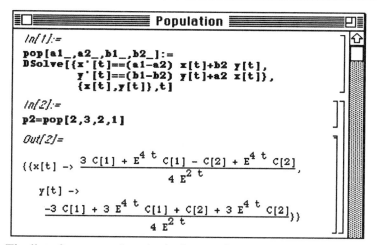

```
                    ══════ Population ═══════
In[1]:=
pop[a1_,a2_,b1_,b2_]:=
DSolve[{x'[t]==(a1-a2) x[t]+b2 y[t],
        y'[t]==(b1-b2) y[t]+a2 x[t]},
        {x[t],y[t]},t]

In[2]:=
p2=pop[2,3,2,1]

Out[2]=
               4 t                    4 t
      3 C[1] + E   C[1] - C[2] + E   C[2]
{{x[t] -> ─────────────────────────────────── ,
                      2 t
                   4 E

  y[t] ->
                4 t                     4 t
    -3 C[1] + 3 E   C[1] + C[2] + 3 E   C[2]
    ───────────────────────────────────────── }}
                      2 t
                   4 E
```

The list of constants in pairs is then used to generate the table of solutions found in **funcs**. The ith member of **funcs** is obtained by replacing each occurrence of C[1] and C[2] in **p2[[1,1,2]]**, corresponding to x(t), and **p2[[1,2,2]]**, corresponding to y(t), by the first member of the ith element of **pairs** and the second member of the ith element of **pairs**, respectively.

```
In[4]:=
pairs={{0,0},{1,1},{1,2},{1,3},{1,4},
       {1,5},{2,1},{2,2},{3,1},{3,2},
       {3,4},{3,5},{4,1},{4,2},{5,1},{5,2}};

In[5]:=
funcs=Table[{p2[[1,1,2]],p2[[1,2,2]]}/.
       {C[1]->pairs[[i,1]],C[2]->pairs[[i,2]]},
       {i,1,Length[pairs]}];
```

The phase plane is then viewed with **ParametricPlot**. Note that the trajectories are directed away from the origin and that only the first quadrant is considered since both populations are nonnegative.

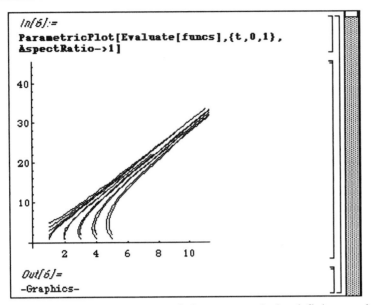

```
In[6]:=
ParametricPlot[Evaluate[funcs],{t,0,1},
AspectRatio->1]
```

```
Out[6]=
-Graphics-
```

We further investigate the populations separately by defining a subset of the list of ordered pairs in **pairssub** below and producing a list of solutions in **funcssub**. Note that these solutions could be extracted from the list **funcs**. The function **a** is then defined so that the two population solutions can be plotted simultaneously.

```
In[8]:=
pairssub={{1,4},{1,5},{4,1},{4,2},{5,1},
                              {5,2}};
```

```
In[9]:=
funcssub=Table[{p2[[1,1,2]],p2[[1,2,2]]}/.
     {C[1]->pairssub[[i,1]],
      C[2]->pairssub[[i,2]]},
      {i,1,Length[pairssub]}];
```

```
In[11]:=
Clear[a]
a[i_]:=
    Plot[{funcssub[[i,1]],funcssub[[i,2]]},
    {t,0,1},
    PlotStyle->{GrayLevel[0],GrayLevel[.2]},
    PlotRange->{0,13},
    DisplayFunction->Identity]
```

After creating the table of plots, they are viewed as a **GraphicsArray**. Notice that under certain initial conditions, both populations increase while under others, a population decreases initially and then increases.

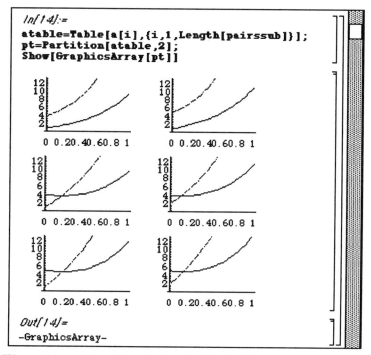

```
In[14]:=
atable=Table[a[i],{i,1,Length[pairssub]}];
pt=Partition[atable,2];
Show[GraphicsArray[pt]]
```

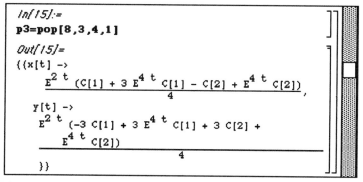

```
Out[14]=
-GraphicsArray-
```

We next investigate the solution obtained with $a_1 = 8$, $a_2 = 3$, $b_1 = 4$, and $b_2 = 1$. The general solution is determined below with **DSolve**.

```
In[15]:=
p3=pop[8,3,4,1]

Out[15]=
{{x[t] ->
      E^(2 t) (C[1] + 3 E^(4 t) C[1] - C[2] + E^(4 t) C[2])
      ─────────────────────────────────────────────────── ,
                              4
  y[t] ->
      E^(2 t) (-3 C[1] + 3 E^(4 t) C[1] + 3 C[2] +
          E^(4 t) C[2])
      ───────────────────────────────────────────
                              4
  }}
```

The constants given in pairs are used to generate the list of solutions in **funcs**.

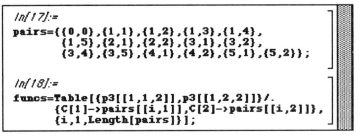

In[17]:=
```
pairs={{0,0},{1,1},{1,2},{1,3},{1,4},
     {1,5},{2,1},{2,2},{3,1},{3,2},
     {3,4},{3,5},{4,1},{4,2},{5,1},{5,2}};
```

In[18]:=
```
funcs=Table[{p3[[1,1,2]],p3[[1,2,2]]}/.
     {C[1]->pairs[[i,1]],C[2]->pairs[[i,2]]},
     {i,1,Length[pairs]}];
```

The phase plane is then produced with **ParametricPlot**. Notice the differences in the shape of the trajectories of this problem and that of the previous case.

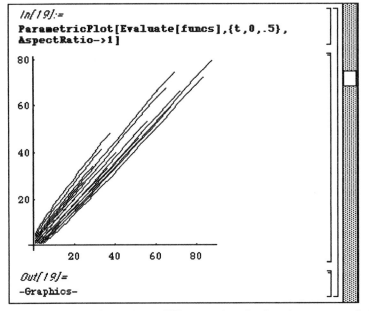

In[19]:=
```
ParametricPlot[Evaluate[funcs],{t,0,.5},
AspectRatio->1]
```

Out[19]=
-Graphics-

We further investigate these differences by plotting the two populations simultaneously by using the function **a** and a subset of the list of solutions given in **funcs**.

```
In[21]:=
pairssub={{1,4},{1,5},{4,1},{4,2},{5,1},
                            {5,2}};
```

```
In[22]:=
funcssub=Table[{p3[[1,1,2]],p3[[1,2,2]]}/.
     {C[1]->pairssub[[i,1]],
      C[2]->pairssub[[i,2]]},
      {i,1,Length[pairssub]}];
```

```
In[24]:=
Clear[a]
a[i_]:=
Plot[{funcssub[[i,1]],funcssub[[i,2]]},
     {t,0,.5},
     PlotStyle->{GrayLevel[0],GrayLevel[.2]},
     DisplayFunction->Identity]
```

The plots are displayed below. Notice that in this case, the population which is initially larger remains larger over the time interval used here. This is in contrast to the previous example.

```
In[27]:=
atable=Table[a[i],{i,1,Length[pairssub]}];
pt=Partition[atable,2];
Show[GraphicsArray[pt]]
```

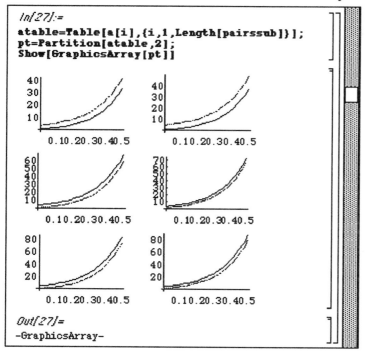

```
Out[27]=
-GraphicsArray-
```

We next investigate the situation with $a_1=1$, $a_2=2$, $b_1=1$, and $b_2=4$. The general solution is determined below with **DSolve**.

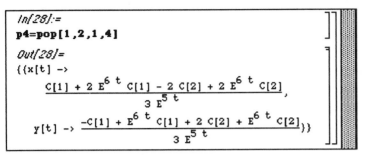

```
In[28]:=
p4=pop[1,2,1,4]
Out[28]=
```
$$\{\{x[t] \to \frac{C[1] + 2\,E^{6\,t}\,C[1] - 2\,C[2] + 2\,E^{6\,t}\,C[2]}{3\,E^{5\,t}},$$

$$y[t] \to \frac{-C[1] + E^{6\,t}\,C[1] + 2\,C[2] + E^{6\,t}\,C[2]}{3\,E^{5\,t}}\}\}$$

Similarly, the list of constants in **pairs** is used to generate the solutions in **funcs**.

```
In[30]:=
pairs={{0,0},{1,1},{1,2},{1,3},{1,4},
       {1,5},{2,1},{2,2},{3,1},{3,2},
       {3,4},{3,5},{4,1},{4,2},{5,1},{5,2}};

In[31]:=
funcs=Table[{p4[[1,1,2]],p4[[1,2,2]]}/.
       {C[1]->pairs[[i,1]],C[2]->pairs[[i,2]]},
       {i,1,Length[pairs]}];
```

The phase plane is then produced with **ParametricPlot**.

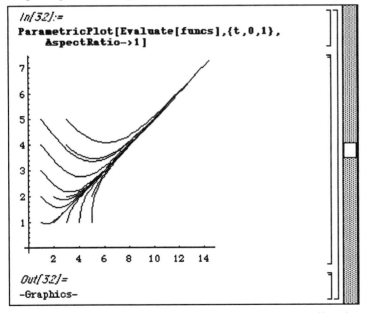

```
In[32]:=
ParametricPlot[Evaluate[funcs],{t,0,1},
    AspectRatio->1]
```

```
Out[32]=
-Graphics-
```

The behavior of the individual populations is more easily viewed by plotting the two solutions

simultaneously. Following the same procedure as was used in the previous two cases, we define a
function **a** which is used to generate these plots for a subset of the set of solutions of this system.

```
In[34]:=
pairssub={{1,4},{1,5},{4,1},{4,2},{5,1},
                              {5,2}};

In[35]:=
funcssub=Table[{p4[[1,1,2]],p4[[1,2,2]]}/.
     {C[1]->pairssub[[i,1]],
      C[2]->pairssub[[i,2]]},
     {i,1,Length[pairssub]}];

In[37]:=
Clear[a]
a[i_]:=Plot[{funcssub[[i,1]],
   funcssub[[i,2]]},{t,0,1},
   PlotStyle->{GrayLevel[0],GrayLevel[.2]},
   DisplayFunction->Identity]
```

An array of plots is created in **atable** and displayed as a **GraphicsArray** below. Notice how the
large value of b_2 affects the population of x (given by the darker of the two curves below). If the value
of x is initially smaller than that of y, then x eventually becomes greater than y. On the other hand, if x
is initially larger than y, then it remains larger than y.

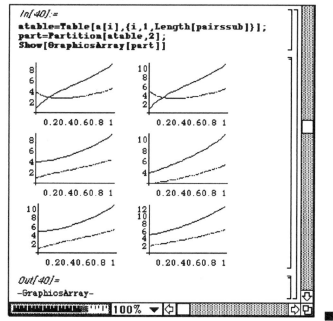

```
In[40]:=
atable=Table[a[i],{i,1,Length[pairssub]}];
part=Partition[atable,2];
Show[GraphicsArray[part]]
```

```
Out[40]=
-GraphicsArray-
```

■ EXAMPLE 11.8

Suppose that the population of three neighboring territories x, y, and z depends on several factors. The birth rates of x, y, and z are a_1, b_1, and c_1, respectively. The rate at which citizens of x move to y is a_2 while that at which citizens move from x to z is a_3. Similarly, the rate at which citizens of y move to x is b_2 while that at which citizens move from y to z is b_3. Also, the rate at which citizens of z move to x is c_2 while that at which citizens move from z to y is c_3. Suppose that the mortality rate of each territory is ignored in the model. Determine the respective populations of the three territories for any time t.

Solution:

The system of equations in this case is similar to that derived in the previous example. The rate at which population x changes is

$$\frac{dx}{dt} = a_1 x - a_2 x - a_3 x + b_1 y + c_1 z = \left(a_1 - a_2 - a_3\right) x + b_1 y + c_1 z$$

while the rate at which population y changes is

$$\frac{dy}{dt} = b_1 y - b_2 y - b_3 y + a_2 x + c_2 z = \left(b_1 - b_2 - b_3\right) y + a_2 x + c_2 z,$$

and that of z is

$$\frac{dz}{dt} = c_1 z - c_2 z - c_3 z + a_3 x + b_3 y = \left(c_1 - c_2 - c_3\right) z + a_3 x + b_3 y.$$

Hence, we must solve the 3×3 system:

$$\frac{dx}{dt} = \left(a_1 - a_2 - a_3\right) x + b_1 y + c_1 z$$

$$\frac{dy}{dt} = a_2 x + \left(b_1 - b_2 - b_3\right) y + c_2 z$$

$$\frac{dz}{dt} = a_3 x + b_3 y + \left(c_1 - c_2 - c_3\right) z$$

$x(0) = x_0$, $y(0) = y_0$, $z(0) = z_0$ where x_0, y_0, and z_0 are the initial populations of territories x, y, and z respectively.

This problem is approached in a manner similar to the previous example except that we now have three dimensions. The function **pop** is defined below in terms of the parameters of the system of equations so that all systems of this form can be solved.

```
╔══════════════════ Population ══════════════════╗
║ In[41]:=                                        ║
║ pop[a1_,a2_,a3_,b1_,b2_,b3_,c1_,c2_,c3_]:=      ║
║ DSolve[{x'[t]==(a1-a2-a3) x[t]+b2 y[t]+         ║
║                                 c2 z[t],        ║
║            y'[t]==(b1-b2-b3) y[t]+a2 x[t]+      ║
║                                 c3 z[t],        ║
║            z'[t]==(c1-c2-c3) z[t]+b3 y[t]+      ║
║                                 a3 x[t]},       ║
║        {x[t],y[t],z[t]},t]                      ║
╚═════════════════════════════════════════════════╝
```

We consider the particular situation with $a_1=4$, $a_2=10$, $a_3=5$, $b_1=8$, $b_2=4$, $b_3=8$, $c_1=5$, $c_2=6$, and $c_3=5$.

The general solution is determined below with **DSolve** and called **p**. This solution involves complex exponentials so **ComplexExpand** is employed to obtain real-valued solutions. The three components of the solution vector are then extracted with the commands **p[[1,1,2]]**, **p[[1,2,2]]**, and **p[[1,3,2]]** and then named **x1**, **x2**, and **x3** for convenience.

```
In[42]:=
p=pop[4,10,5,8,4,8,5,6,5];

In[45]:=
x1=Simplify[ComplexExpand[p[[1,1,2]]]];
x2=Simplify[ComplexExpand[p[[1,2,2]]]];
x3=Simplify[ComplexExpand[p[[1,3,2]]]];
```

The list of ordered triples is then entered below.

```
In[47]:=
Clear[triples]
triples={{1,1,1},{1,1,2},{1,1,3},
         {1,2,1},{1,2,2},{1,2,3},
         {1,2,2},{1,3,2},{1,3,1},
         {3,1,1},{3,2,2},{3,2,3},
         {2,1,1},{2,2,1},{2,3,1},
         {1,3,3},{1,4,1},{1,5,1}};
```

A table of solutions is generated below using the constants in **triples**. The ith member of the list of functions **funcs** is obtained by replacing **C[1]** by the first member of the ith ordered triple of **triples**, **C[2]** by the second member of the ith ordered triple of **triples**, and **C[3]** by the third member of the ith ordered triple of **triples**.

```
In[49]:=
Clear[funcs]
funcs=Table[{x1,x2,x3}/.
       {C[1]->triples[[i,1]],
        C[2]->triples[[i,2]],
        C[3]->triples[[i,3]]},
       {i,1,Length[triples]}];
```

The three-dimensional phase portrait is then produced with **ParametricPlot3D**.

```
In[50]:=
ParametricPlot3D[Evaluate[funcs],{t,0,.1},
    AspectRatio->1]
```

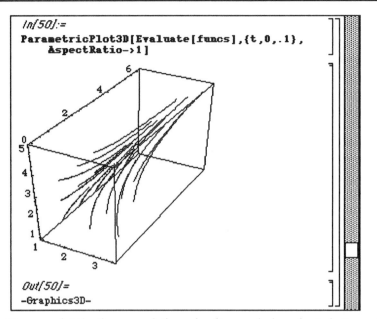

```
Out[50]=
-Graphics3D-
```

We now plot the three populations simultaneously in order to better understand their behavior. In order to generate a subset of the family of solutions produced earlier, a list of six ordered triples called **tripsub** is entered.

```
In[2]:=
tripsub={{1,1,1},{1,1,2},{2,1,1},{2,2,1},
{3,1,1},{3,2,2}};
```

This list is used to produce a table of solutions which are plotted with the function **a** below.

```
In[53]:=
funcssub=Table[{x1,x2,x3}/.
      {C[1]->tripsub[[i,1]],
       C[2]->tripsub[[i,2]],
       C[3]->tripsub[[i,3]]},
      {i,1,Length[tripsub]}];
In[55]:=
Clear[a]
a[i_]:=
Plot[{funcssub[[i,1]],funcssub[[i,2]],
                    funcssub[[i,3]]},
      {t,0,.1},
      PlotStyle->{GrayLevel[0],
           GrayLevel[.2],GrayLevel[.4]},
    DisplayFunction->Identity]
```

After generating a table of plots, the graphs are displayed as a **GraphicsArray**. Note that the darkest of the three curves represents x and the lightest represents z. Notice how the initial populations affect the behavior of the population curves.

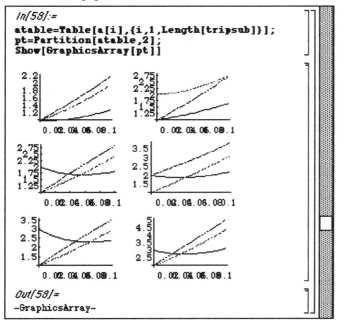

We now consider the system with $a_1=4$, $a_2=6$, $a_3=1$, $b_1=2$, $b_2=6$, $b_3=4$, $c_1=2$, $c_2=6$, and $c_3=3$. This solution is determined below and called **p1**.

```
In[59]:=
p1=pop[4,6,1,2,6,4,2,6,3]

Out[59]=
{{x[t] -> (6 C[1] + 7 E^(13 t) C[1] - 6 C[2] +
        6 E^(13 t) C[2] - 6 C[3] + 6 E^(13 t) C[3]) /
      (13 E^(10 t)), y[t] ->
    (-273 C[1] + 210 E^t C[1] + 63 E^(14 t) C[1] +
        338 C[2] - 210 E^t C[2] + 54 E^(14 t) C[2] +
        156 C[3] - 210 E^t C[3] + 54 E^(14 t) C[3]) /
      (182 E^(11 t)), z[t] ->
    (273 C[1] - 308 E^t C[1] + 35 E^(14 t) C[1] -
        338 C[2] + 308 E^t C[2] + 30 E^(14 t) C[2] -
        156 C[3] + 308 E^t C[3] + 30 E^(14 t) C[3]) /
      (182 E^(11 t))}}
```

As in the previous case, a list of ordered triples is entered and used for the arbitrary constants to yield a table of solutions to the system.

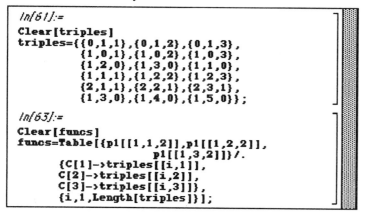

```
In[61]:=
Clear[triples]
triples={{0,1,1},{0,1,2},{0,1,3},
        {1,0,1},{1,0,2},{1,0,3},
        {1,2,0},{1,3,0},{1,1,0},
        {1,1,1},{1,2,2},{1,2,3},
        {2,1,1},{2,2,1},{2,3,1},
        {1,3,0},{1,4,0},{1,5,0}};

In[63]:=
Clear[funcs]
funcs=Table[{p1[[1,1,2]],p1[[1,2,2]],
                    p1[[1,3,2]]}/.
       {C[1]->triples[[i,1]],
        C[2]->triples[[i,2]],
        C[3]->triples[[i,3]]},
        {i,1,Length[triples]}];
```

The phase portrait is then produced via **ParametricPlot3D**.

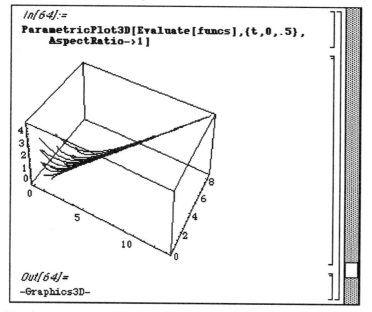

```
In[64]:=
ParametricPlot3D[Evaluate[funcs],{t,0,.5},
    AspectRatio->1]
```

```
Out[64]=
-Graphics3D-
```

We then investigate the three populations separately by using a subset of the constants used earlier and producing a collection of six solutions with the constants in **tripsub**.

```
In[66]:=
tripsub={{0,1,1},{0,1,2},{1,0,2},
        {1,0,3},{1,2,0},{1,3,0}};

In[67]:=
funcssub=Table[{p1[[1,1,2]],p1[[1,2,2]],
                         p1[[1,3,2]]}/.
      {C[1]->tripsub[[i,1]],
       C[2]->tripsub[[i,2]],
       C[3]->tripsub[[i,3]]},
       {i,1,Length[tripsub]}];
```

The function **a** plots the three populations simultaneously.

```
In[69]:=
Clear[a]
a[i_]:=Plot[{funcssub[[i,1]],funcssub[[i,2]],
            funcssub[[i,3]]},
            {t,0,.5},
            PlotStyle->{GrayLevel[0],
               GrayLevel[.2],GrayLevel[.4]},
            DisplayFunction->Identity]
```

A table of plots is produced below in a table and then viewed as a **GraphicsArray**. We notice the effects of the initial conditions on the behavior of the curves.

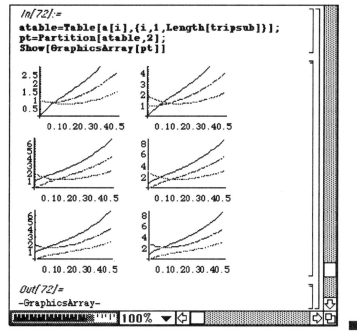

```
In[72]:=
atable=Table[a[i],{i,1,Length[tripsub]}];
pt=Partition[atable,2];
Show[GraphicsArray[pt]]
```

```
Out[72]=
-GraphicsArray-
```

§11.5 Applications Using Laplace Transforms

Coupled Spring-Mass Systems

The motion of a mass attached to the end of a spring was modeled with a second-order linear differential equation with constant coefficients in **Chapter 3**. Similarly, if a second spring and mass are attached to the end of the first mass, then the model becomes that of a system of second-order equations. To more precisely state the problem, let masses m_1 and m_2 be attached to the ends of springs S_1 and S_2 having spring constants k_1 and k_2, respectively. Then, spring S_2 is attached to the base of mass m_1. Suppose that $x(t)$ and $y(t)$ represent the vertical displacement from equilibrium of springs S_1 and S_2, respectively. Since spring S_2 undergoes both elongation and compression when the system is in motion, then according to Hooke's law, S_2 exerts the force $k_2(y-x)$ while S_2 exerts the force $-k_1 x$. Therefore, the force acting on mass m_1 is the sum: $-k_1 x + k_2(y-x)$ and that acting on m_1 is : $-k_2(y-x)$. Hence, by Newton's second law, we have the system

$$m_1 \frac{d^2x}{dt^2} = -k_1 x + k_2 (y - x)$$

$$m_2 \frac{d^2y}{dt^2} = -k_2 (y - x).$$

The initial position and velocity of the two masses m_1 and m_2 are given by $x(0)$, $x'(0)$, $y(0)$, and $y'(0)$, respectively. Hence, the method of Laplace transforms can be used to solve problems of this type. Recall the following property of the Laplace transform: $L\{f(t)\} = s^2 F(s) - s f(0) - f'(0)$ where $F(s)$ is the Laplace transform of $f(t)$. This property is of great use in solving this problem since both equations involve second derivatives. We illustrate the solution of higher order systems of differential equations with Laplace transforms in the following examples.

■ EXAMPLE 11.9

Consider the problem with $m_1 = m_2 = 1$, $k_1 = 3$, and $k_2 = 2$. Suppose that $x(0) = 0$, $x'(0) = 1$, $y(0) = 1$, and $y'(0) = 0$. Determine $x(t)$ and $y(t)$ using Laplace transforms, and plot the solutions if, at equilibrium, both springs have length 1.

Solution:

Using the system derived above, we have the following initial value problem:

$$\frac{d^2x}{dt^2} + x - 2y = 0$$

$$\frac{d^2y}{dt^2} - 2x - 2y = 0$$

$x(0) = 0, x'(0) = 1, \ y(0) = 1, y'(0) = 0.$

After loading the LaplaceTransform package, we define the left-hand side of the differential equations in **eq1** and **eq2**, respectively. We then define the appropriate transformation rules in **rule**.

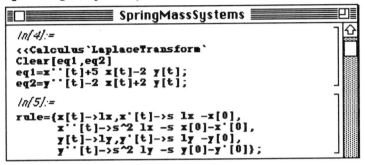

Below, **eqs** represents the left-hand side of the differential equations after application of the Laplace transform.

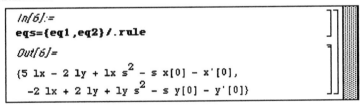

Next, **sols** sets the two equations in **eqs** equal to zero and solves the system for **1x** and **1y**, the Laplace transform of x and y, respectively. In **conds**, the initial conditions are substituted into the results obtained for **1x** and **1y**. Finally, the Laplace transform of x and y are extracted from the output list of conds so that the inverse Laplace transform may be applied to determine x and y. This is accomplished below in the definition of **x** and **y**.

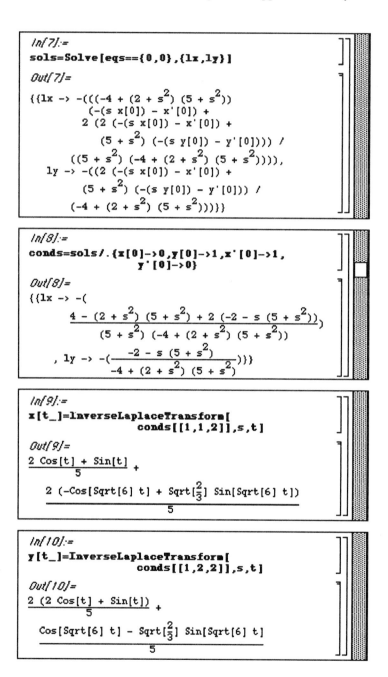

In[7]:=
```
sols=Solve[eqs=={0,0},{1x,1y}]
```

Out[7]=

$$\{\{1x \to -(((-4 + (2 + s^2)(5 + s^2)) \\ (-(s\,x[0]) - x'[0]) + \\ 2\,(2\,(-(s\,x[0]) - x'[0]) + \\ (5 + s^2)(-(s\,y[0]) - y'[0]))) / \\ ((5 + s^2)(-4 + (2 + s^2)(5 + s^2)))), \\ 1y \to -((2\,(-(s\,x[0]) - x'[0]) + \\ (5 + s^2)(-(s\,y[0]) - y'[0])) / \\ (-4 + (2 + s^2)(5 + s^2)))\}\}$$

In[8]:=
```
conds=sols/.{x[0]->0,y[0]->1,x'[0]->1,
             y'[0]->0}
```

Out[8]=

$$\{\{1x \to -(\\ \frac{4 - (2 + s^2)(5 + s^2) + 2\,(-2 - s\,(5 + s^2))}{(5 + s^2)(-4 + (2 + s^2)(5 + s^2))}) \\ , 1y \to -(\frac{-2 - s\,(5 + s^2)}{-4 + (2 + s^2)(5 + s^2)})\}\}$$

In[9]:=
```
x[t_]=InverseLaplaceTransform[
          conds[[1,1,2]],s,t]
```

Out[9]=

$$\frac{2\,\mathrm{Cos}[t] + \mathrm{Sin}[t]}{5} + \\ \frac{2\,(-\mathrm{Cos}[\mathrm{Sqrt}[6]\,t] + \mathrm{Sqrt}[\frac{2}{3}]\,\mathrm{Sin}[\mathrm{Sqrt}[6]\,t])}{5}$$

In[10]:=
```
y[t_]=InverseLaplaceTransform[
          conds[[1,2,2]],s,t]
```

Out[10]=

$$\frac{2\,(2\,\mathrm{Cos}[t] + \mathrm{Sin}[t])}{5} + \\ \frac{\mathrm{Cos}[\mathrm{Sqrt}[6]\,t] - \mathrm{Sqrt}[\frac{2}{3}]\,\mathrm{Sin}[\mathrm{Sqrt}[6]\,t]}{5}$$

In the plot below, the motion of the first spring is represented by the darker curve while the motion of the second spring is given by the lighter curve.

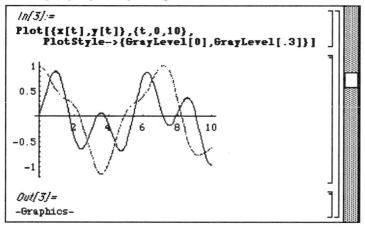

```
In[3]:=
Plot[{x[t],y[t]},{t,0,10},
    PlotStyle->{GrayLevel[0],GrayLevel[.3]}]
```

```
Out[3]=
-Graphics-
```

Since the relative position of the springs is difficult to see from the above plot, we attempt to plot the actual (up-and-down) motion of the two springs simultaneously. We do this by, first, defining that **zigzag** function below. When given two points (a,b) and (c,d), a positive integer **n**, and a small real number **eps**, this function joins the two points with **n** short line segments. The value of **eps** is used to obtain x-values which are slightly larger than and slightly smaller than x =a, the x-coordinate of the first point.

```
In[5]:=
Clear[spring,zigzag,length,points,pairs]
zigzag[{a_,b_},{c_,d_},n_,eps_]:=
    Module[{length,points,pairs,zigzag},
    length=d-b;
    points=Table[b+i length/n,
        {i,1,n-1}];
    pairs=Table[{a+(-1)^i eps,points[[i]]},
        {i,1,n-1}];
    PrependTo[pairs,{a,b}];
    AppendTo[pairs,{c,d}];
    Line[pairs]
    ]
```

We then define the function **spring2** below which displays the graphics of the two springs. This function depends on time **t**, the length of the first spring **len1**, and the length of the second spring **len2**. We use the y-axis to center the graphics. Hence, we take the x-coordinate to be zero in all points. **spring2** first uses **zigzag** to produce the graphics of the first spring from the point {0,x[t]} to the point {0,len1}. It then shows the graphics of a point at the top of the first spring and the subsequent position at the end of the spring. **spring2** then uses **zigzag** to create the graphics of the second spring from this point to a point **len2** units below. Finally, **spring2** shows the graphics of a point to represent the mass at the end of the second spring. Note that **spring2** produces the graphics at a particular value of time. Hence, a list of graphics may be produced with this function and then animated to view the motion of the springs.

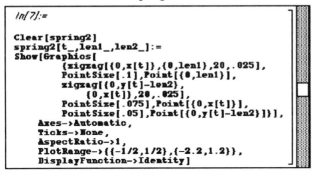

```
In[7]:=

Clear[spring2]
spring2[t_,len1_,len2_]:=
Show[Graphics[
        {zigzag[{0,x[t]},{0,len1},20,.025],
         PointSize[.1],Point[{0,len1}],
         zigzag[{0,y[t]-len2},
             {0,x[t]},20,.025],
         PointSize[.075],Point[{0,x[t]}],
         PointSize[.05],Point[{0,y[t]-len2}]}],
     Axes->Automatic,
     Ticks->None,
     AspectRatio->1,
     PlotRange->{{-1/2,1/2},{-2.2,1.2}},
     DisplayFunction->Identity]
```

Below, we produce an array of graphics cells by partitioning the output of **graphs** into groups of three. We then display the plots as a graphics array. Notice that the large point at the top of each plot represents the point at which the first spring is attached to the rigid support. The smallest point represents the mass at the end of the second spring. At some times below, the two masses cannot be distinguished due to their closeness. An alternative to displaying the resulting graphics cells as a graphics array is to use the command **Do[spring2[t,1,1],{t,0,8}]** and then animate the result.

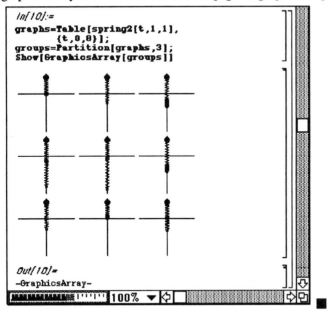

```
In[10]:=
graphs=Table[spring2[t,1,1],
        {t,0,8}];
groups=Partition[graphs,3];
Show[GraphicsArray[groups]]
```

```
Out[10]=
-GraphicsArray-
```

Suppose that external forces $F_1(t)$ and $F_2(t)$ are applied to the masses. This causes the system of equations to become

$$m_1 \frac{d^2x}{dt^2} = -k_1 x + k_2(y-x) + F_1(t)$$

$$m_2 \frac{d^2y}{dt^2} = -k_2(y-x) + F_2(t).$$

We investigate the effects of these external forcing functions in the following example which is again solved through the method of Laplace transforms.

■ EXAMPLE 11.10

Solve the same problem as in the previous example with forcing functions $F_1(t) = 1$ and $F_2(t) = \cos(t)$. Again, determine and plot the solutions.

Solution:

Hence, we have the following initial value problem:

$$\frac{d^2x}{dt^2} + x - 2y = 1$$

$$\frac{d^2y}{dt^2} - 2x - 2y = \cos(t)$$

$$x(0) = 0, x'(0) = 1, \ y(0) = 1, y'(0) = 0.$$

This problem is solved in the same manner as the previous example. The only difference is that the Laplace transform of the forcing functions must be determined. In the previous example, these functions were both zero, so the Laplace transform was also zero for each. Since the Laplace transform of the forcing functions is easily found for each, we simply enter them as **rhs1** and **rhs2** instead of using **LaplaceTransform** to determine them. The output of **sols** represents the Laplace transform of x and y.

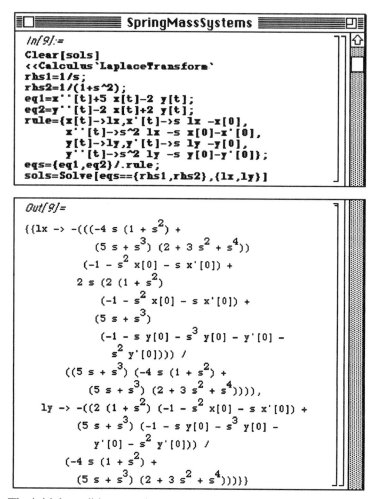

The initial conditions are then applied to these results to obtain two functions of s so that the solution can be determined by applying the inverse Laplace transform.

```
In[10]:=
conds=sols/.{x[0]->0,y[0]->1,x'[0]->1,
              y'[0]->0}

Out[10]=
{{lx -> -((2 s (2 (-1 - s) (1 + s^2) +
              (-1 - s - s^3) (5 s + s^3)) +
          (-1 - s) (-4 s (1 + s^2) +
              (5 s + s^3) (2 + 3 s^2 + s^4))) /
        ((5 s + s^3) (-4 s (1 + s^2) +
              (5 s + s^3) (2 + 3 s^2 + s^4))))),
   ly -> -((2 (-1 - s) (1 + s^2) +
          (-1 - s - s^3) (5 s + s^3)) /
        (-4 s (1 + s^2) +
          (5 s + s^3) (2 + 3 s^2 + s^4)))}}
```

This is accomplished by extracting the appropriate elements of the output list of **conds** and using **InverseLaplaceTransform**. The solution for x, the position of the first spring, is given below in **x**.

```
In[11]:=
x[t_]=InverseLaplaceTransform[
             conds[[1,1,2]],s,t]

Out[11]=
1   t Cos[t]   Sin[t]   5 Cos[t] + 3 Sin[t]
- - ------- + ------ + ------------------- +
3      5         5              25
  (2 (-20 Cos[Sqrt[6] t] +
       11 Sqrt[3/2] Sin[Sqrt[6] t])) / 75
```

and that of y is similarly given in **y**.

```
In[12]:=
y[t_]=InverseLaplaceTransform[
             conds[[1,2,2]],s,t]

Out[12]=
1   2 t Cos[t]   2 Sin[t]
- - --------- + -------- +
3       5           5
  10 Cos[t] + 11 Sin[t]
  --------------------- +
           25
  20 Cos[Sqrt[6] t] - 11 Sqrt[3/2] Sin[Sqrt[6] t]
  -----------------------------------------------
                      75
```

The position functions are plotted below with the darker curve representing the first spring and the lighter curve the second.

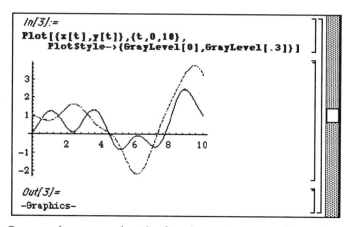

```
In[3]:=
Plot[{x[t],y[t]},{t,0,10},
    PlotStyle->{GrayLevel[0],GrayLevel[.3]}]
```

```
Out[3]=
-Graphics-
```

Once again, we employ the functions **zigzag** and **spring2** which were defined in the previous example to graph to motion of the two springs at various times. Notice that the **PlotRange** setting in **spring2** has been changed in this case.

```
In[5]:=
Clear[spring,zigzag,length,points,pairs]
zigzag[{a_,b_},{c_,d_},n_,eps_]:=
    Module[{length,points,pairs,zigzag},
    length=d-b;
    points=Table[b+i length/n,
        {i,1,n-1}];
    pairs=Table[{a+(-1)^i eps,points[[i]]},
        {i,1,n-1}];
    PrependTo[pairs,{a,b}];
    AppendTo[pairs,{c,d}];
    Line[pairs]
    ]
```

```
In[7]:=

Clear[spring2]
spring2[t_,len1_,len2_]:=
Show[Graphics[
    {zigzag[{0,x[t]},{0,len1},20,.025],
    PointSize[.1],Point[{0,len1}],
    zigzag[{0,y[t]-len2},
        {0,x[t]},20,.025],
    PointSize[.075],Point[{0,x[t]}],
    PointSize[.05],Point[{0,y[t]-len2}]}],
    Axes->Automatic,
    Ticks->None,
    AspectRatio->1,
    PlotRange->{{-1/2,1/2},{-3.2,1.2}},
    DisplayFunction->Identity]
```

The springs are plotted for values of t from t =0 and t =8 using increments of one. The list which results is then partitioned into groups of three and then displayed as a graphics array. As stated in the previous example, an alternative approach is to use the command **Do[spring2[t,1,1],{t,0,8}]** and then animate the resulting graphics cells.

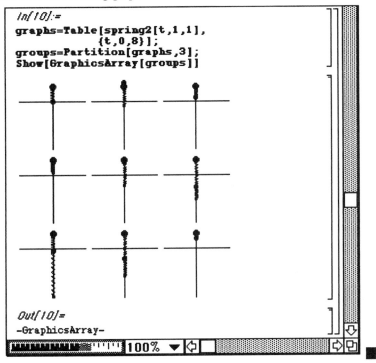

```
In[10]:=
graphs=Table[spring2[t,1,1],
        {t,0,8}];
groups=Partition[graphs,3];
Show[GraphicsArray[groups]]
```

```
Out[10]=
-GraphicsArray-
```

The previous situation can be modified by including a third spring with spring constant k_3 between the base of the mass m_2 and a lower support. Hence, the motion of the spring-mass system is affected by the third spring. Using the techniques of the earlier case, this model becomes:

$$m_1 \frac{d^2 x}{dt^2} = -k_1 x + k_2 (y - x) + F_1(t)$$

$$m_2 \frac{d^2 y}{dt^2} = k_3 y - k_2 (y - x) + F_2(t).$$

■ EXAMPLE 11.11

Consider the problem with $m_1 = m_2 = 1$, $k_1 = k_2 = k_3 = 1$. Suppose that $x(0) = 0$, $x'(0) = -1$, $y(0) = 0$, and $y'(0) = 1$. Determine $x(t)$ and $y(t)$ using Laplace transforms, and plot the solutions if, at equilibrium, all three springs have length 1 and the forcing functions are both zero.

Solution:

Using the above derivation, we have

$$\frac{d^2x}{dt^2} + 2x - y = 0$$

$$\frac{d^2y}{dt^2} - x + 2y = 0$$

$x(0) = 0, x'(0) = -1, \ y(0) = 0, y'(0) = 1.$

We solve this problem in a manner similar to the previous two examples. We begin by defining the left-hand side of the equations in **eq1** and **eq2** as well as the transformation rules needed to apply the Laplace transform method.

```
                   SpringMassSystems
In[4]:=

Clear[x,y,eqs,eq1,eq2]
<<Calculus`LaplaceTransform`
eq1=x''[t]+2 x[t]- y[t];
eq2=y''[t]- x[t]+2 y[t];

In[5]:=

rule={x[t]->lx,x'[t]->s lx -x[0],
      x''[t]->s^2 lx -s x[0]-x'[0],
      y[t]->ly,y'[t]->s ly -y[0],
      y''[t]->s^2 ly -s y[0]-y'[0]};
```

Once again, the Laplace transform is applied in **eqs** by **rule**.

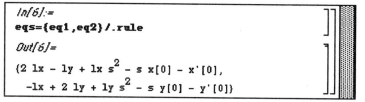

```
In[6]:=

eqs={eq1,eq2}/.rule

Out[6]=
                  2
{2 lx - ly + lx s  - s x[0] - x'[0],
                   2
  -lx + 2 ly + ly s  - s y[0] - y'[0]}
```

Since this system is homogeneous, each of the components of **eqs** are equated to zero and solved for **lx** and **ly**, the Laplace transform of x and y, respectively.

```
In[8]:=
Clear[sols,lx,ly,x,y]
sols=Solve[eqs=={0,0},{lx,ly}]

Out[8]=
{{lx -> -((-(s x[0]) +
           (-1 + (2 + s^2)^2) (-(s x[0]) - x'[0]) -
           x'[0] + (2 + s^2) (-(s y[0]) - y'[0])) /
          ((2 + s^2) (-1 + (2 + s^2)^2))),
  ly -> -((-(s x[0]) - x'[0] +
           (2 + s^2) (-(s y[0]) - y'[0])) /
          (-1 + (2 + s^2)^2))}}
```

Next, the initial conditions are applied in **conds**

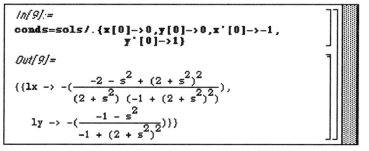

and the solutions are determined with **InverseLaplaceTransform** as before.

```
In[10]:=
x[t_]=InverseLaplaceTransform[
              conds[[1,1,2]],s,t]

Out[10]=
-(Sin[Sqrt[3] t] / Sqrt[3])
```

```
In[11]:=
y[t_]=InverseLaplaceTransform[
              conds[[1,2,2]],s,t]

Out[11]=
-Sin[Sqrt[3] t] / (2 Sqrt[3])  +  Sqrt[3] Sin[Sqrt[3] t] / 2
```

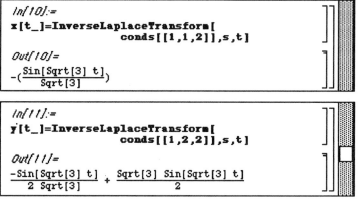

The position functions are plotted simultaneously below.

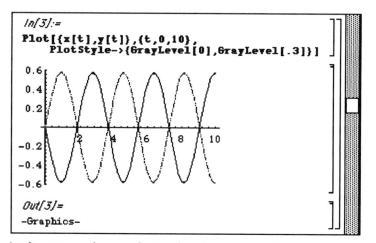

```
In[3]:=
Plot[{x[t],y[t]},{t,0,10},
    PlotStyle->{GrayLevel[0],GrayLevel[.3]}]
```

```
Out[3]=
-Graphics-
```

Again, we use zigzag to better view the motion of these two masses. We also define a function called **spring3**, similar to **spring2**, which incorporates the third spring. Notice that **spring3** also depends on the length of the third spring, **len3**, and uses **zigzag** to draw the third spring.

```
In[5]:=
Clear[spring,zigzag,length,points,pairs]
zigzag[{a_,b_},{c_,d_},n_,eps_]:=
    Module[{length,points,pairs,zigzag},
    length=d-b;
    points=Table[b+i length/n,
            {i,1,n-1}];
    pairs=Table[{a+(-1)^i eps,points[[i]]},
            {i,1,n-1}];
    PrependTo[pairs,{a,b}];
    AppendTo[pairs,{c,d}];
    Line[pairs]
    ]
```

```
In[7]:=

Clear[spring3]
spring3[t_,len1_,len2_,len3_]:=
Show[Graphics[
        {zigzag[{0,x[t]},{0,len1},20,.025],
        PointSize[.025],Point[{0,len1}],
        zigzag[{0,y[t]-len2},
            {0,x[t]},20,.025],
        PointSize[.075],Point[{0,x[t]}],
        PointSize[.075],Point[{0,y[t]-len2}],
        zigzag[{0,-(len2+len3)},
            {0,y[t]-len2},20,.025],
        PointSize[.025],
            Point[{0,-(len2+len3)}]}]
        ],
    Axes->Automatic,
    Ticks->None,
    AspectRatio->1,
    PlotRange->{{-1/2,1/2},
        {-(len2+len3),len1}},
    DisplayFunction->Identity]
```

Again, the springs are plotted for t-values for t =0 to t =8 using increments of one, partitioned into groups of three, and displayed as a graphics array.

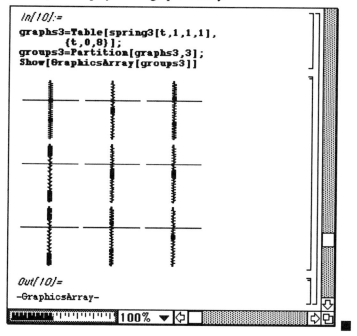

```
In[10]:=
graphs3=Table[spring3[t,1,1,1],
        {t,0,8}];
groups3=Partition[graphs3,3];
Show[GraphicsArray[groups3]]
```

```
Out[10]=
-GraphicsArray-
```

100% ▼

The Double Pendulum

In a method similar to that of the simple pendulum in Chapter 3 and that of the coupled spring system in the previous section, the motion of a double pendulum is modeled by the following system of equations using the approximation $\sin \theta = \theta$ for small displacements:

$$(m_1 + m_2)\, \ell_1^2\, \theta_1'' + m_2\, \ell_1\, \ell_2\, \theta_2'' + (m_1 + m_2)\, \ell_1\, g\, \theta_1 = 0$$

$$m_2\, \ell_2^2\, \theta_2'' + m_2\, \ell_1\, \ell_2\, \theta_1'' + m_2\, \ell_2\, g\, \theta_2 = 0,$$

where θ_1 represents the displacement of the upper pendulum, and θ_2 that of the lower pendulum. Also, m_1 and m_2 represent the mass attached to the upper and lower pendulums, respectively, while the length of each is given by ℓ_1 and ℓ_2.

■ EXAMPLE 11.12

Suppose that $m_1 = 3$, $m_2 = 1$, and each pendulum has length 16. If $\theta_1(0) = 1$, $\theta_1'(0) = 0$, $\theta_2(0) = -1$, and $\theta_2'(0) = 0$, then solve the double pendulum problem using $g = 32$. Also, plot the solution.

Solution:

Application of the system of equations given above yields the following system of second-order equations

$$4(16)^2\, \theta_1'' + (16)^2\, \theta_2'' + 4(16)(32)\, \theta_1 = 0$$

$$(16)^2\, \theta_2'' + (16)^2\, \theta_1'' + (16)(32)\, \theta_2 = 0,$$

$\theta_1(0) = 1$, $\theta_1'(0) = 0$, $\theta_2(0) = 1$, $\theta_2'(0) = 0$.

This system is solved similarly to the prior examples. For convenience, we refer to θ_1 as **x** and θ_2 as **y** in the commands below. We define the left-hand sides of the equations in **eq1** and **eq2** as well as the transformation rules in **rule**.

```
▤▱◼════════ DoublePendulum ══════▱◲▤
In[2]:=
Clear[x,rule,eq1,eq2]
<<Calculus`LaplaceTransform`

In[4]:=
eq1=4 16^2x''[t]+16^2 y''[t]+64 32x[t];
eq2=16^2 y''[t]+16^2x''[t]+16 32y[t];

In[5]:=
rule={x[t]->lx,x'[t]->s lx -x[0],
      x''[t]->s^2 lx -s x[0]-x'[0],
      y[t]->ly,y'[t]->s ly -y[0],
      y''[t]->s^2 ly -s y[0]-y'[0]};
```

The Laplace transform is applied below in **eqs**.

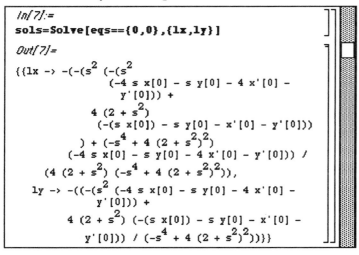

```
In[6]:=
eqs={eq1,eq2}/.rule

Out[6]=
{2048 lx + 1024 (lx s² - s x[0] - x'[0]) +
    256 (ly s² - s y[0] - y'[0]),
  512 ly + 256 (lx s² - s x[0] - x'[0]) +
    256 (ly s² - s y[0] - y'[0])}
```

Since the system is homogeneous, each of the components of **eqs** is equated to zero and the Laplace transform of x and y, **lx** and **ly**, are found.

```
In[7]:=
sols=Solve[eqs=={0,0},{lx,ly}]

Out[7]=
{{lx -> -(-(s² (-(s²
                (-4 s x[0] - s y[0] - 4 x'[0] -
                 y'[0])) +
              4 (2 + s²)
                (-(s x[0]) - s y[0] - x'[0] - y'[0]))
          ) + (-s⁴ + 4 (2 + s²)²)
          (-4 s x[0] - s y[0] - 4 x'[0] - y'[0])) /
      (4 (2 + s²) (-s⁴ + 4 (2 + s²)²)),
  ly -> -((-(s² (-4 s x[0] - s y[0] - 4 x'[0] -
            y'[0])) +
          4 (2 + s²) (-(s x[0]) - s y[0] - x'[0] -
            y'[0])) / (-s⁴ + 4 (2 + s²)²))}}
```

Below, the initial conditions are applied.

```
In[8]:=
conds=sols/.{x[0]->1,y[0]->-1,
             x'[0]->0,y'[0]->0}

Out[8]=
                  -(-3 s⁵ - 3 s (-s⁴ + 4 (2 + s²)²))
{{lx -> ─────────────────────────────────────── ,
             4 (2 + s²) (-s⁴ + 4 (2 + s²)²)

               -3 s³
  ly -> ─────────────── }}
         -s⁴ + 4 (2 + s²)²
```

As in previous examples, **InverseLaplaceTransform** is used to obtain the formulas for x and y. These are given below.

```
In[9]:=
x[t_]=InverseLaplaceTransform[
              conds[[1,1,2]],s,t]
Out[9]=
3 Cos[2 t]     Cos[ 2 t / Sqrt[3] ]
─────────  +  ─────────────────
    4                  4
```

```
In[10]:=
y[t_]=InverseLaplaceTransform[
              conds[[1,2,2]],s,t]
Out[10]=
-3 Cos[2 t]     Cos[ 2 t / Sqrt[3] ]
──────────  +  ─────────────────
     2                   2
```

The position functions are then plotted with the lighter curve representing the second spring and the darker curve the first spring.

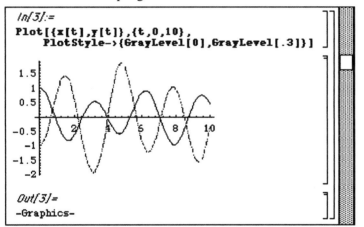

```
In[3]:=
Plot[{x[t],y[t]},{t,0,10},
    PlotStyle->{GrayLevel[0],GrayLevel[.3]}]
```

```
Out[3]=
-Graphics-
```

As with the coupled spring problems above, we can generate the graphics for more easily viewing the motion of the double pendulum. We do this by defining the function **pen2** which depends on the time t as well as the length of the two pendulums, **len1** and **len2**, below. Since the angles are measured from the vertical axis, polar coordinates with the reference angle at $3\pi/2$ are used. In this function, **pt1** represents the position of the mass attached to the end of the first pendulum and **pt2** that of the mass at the end of the second spring. **pen2** uses **Line** to produce the graphics of the lines joining the points representing the masses.

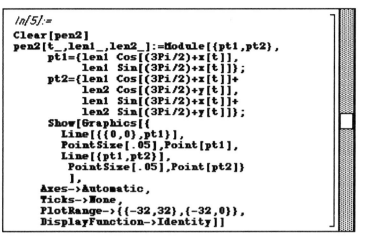

```
In[5]:=
Clear[pen2]
pen2[t_,len1_,len2_]:=Module[{pt1,pt2},
    pt1={len1 Cos[(3Pi/2)+x[t]],
         len1 Sin[(3Pi/2)+x[t]]};
    pt2={len1 Cos[(3Pi/2)+x[t]]+
         len2 Cos[(3Pi/2)+y[t]],
         len1 Sin[(3Pi/2)+x[t]]+
         len2 Sin[(3Pi/2)+y[t]]};
    Show[Graphics[{
      Line[{{0,0},pt1}],
      PointSize[.05],Point[pt1],
      Line[{pt1,pt2}],
       PointSize[.05],Point[pt2]}
       ],
    Axes->Automatic,
    Ticks->None,
    PlotRange->{{-32,32},{-32,0}},
    DisplayFunction->Identity]]
```

Below, we generate the graphics of the pendulum for t =0 to t =8 using increments of one. These graphics are partitioned into groups of three and displayed as a graphics array. In the same manner as in the previous examples, an alternative to producing an array of graphics cells is to use the command **Do[pen2[t,1,6,16],{t,0,8}]** and then animate the resulting graphics cells.

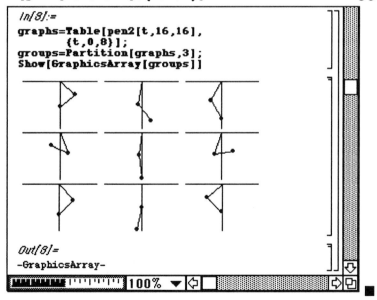

```
In[8]:=
graphs=Table[pen2[t,16,16],
    {t,0,8}];
groups=Partition[graphs,3];
Show[GraphicsArray[groups]]
```

```
Out[8]=
-GraphicsArray-
```

100%

Chapter 12: Fourier Series and Applications to Partial Differential Equations

Mathematica commands used in **Chapter 12** include:

BesselJ	NIntegrate	Simplify
ContourPlot	ParametricPlot3D	Solve
Contours	Part	Table
ContourShading	Partition	TableForm
D	Plot	
DisplayFunction	Plot3D	
DSolve	PlotPoints	
Evaluate	PlotRange	
GraphicsArray	ReplaceAll	
NDSolve	Show	

§12.1 Orthogonal Functions and Sturm-Liouville Problems

Definition:
Consider the second-order homogeneous equation with parameter λ independent of x

$$a_2(x)y''(x) + a_1(x)y'(x) + \left[a_0(x) + \lambda\right]y(x) = 0 \text{ and let } p(x) = e^{\int \frac{a_1(x)}{a_2(x)}}, \quad q(x) = \frac{a_0(x)}{a_2(x)}p(x), \text{ and}$$

$s(x) = \dfrac{p(x)}{a_2(x)}$. Then the equation $a_2(x)y''(x) + a_1(x)y'(x) + \left[a_0(x) + \lambda\right]y(x) = 0$ is the same as

the equation $\dfrac{d}{dx}\left(p\dfrac{dy}{dx}\right) + \left(q + \lambda s\right)y = 0.$ The equation $\dfrac{d}{dx}\left(p\dfrac{dy}{dx}\right) + \left(q + \lambda s\right)y = 0$ is called a

Sturm-Liouville equation.

The Sturm - Liouville equation $\dfrac{d}{dx}\left(p\dfrac{dy}{dx}\right) + \left(q + \lambda s\right)y = 0$ subject to the conditions

$a_1 y(a) + a_2 y'(a) = 0$ and $b_1 y(b) + b_2 y'(b) = 0$, where a_1 and a_2 are constants not both zero

and b_1 and b_2 are constants not both zero, is called a **Sturm-Liouville problem**. A number λ for which the Sturm-Liouville problem has a non-trivial solution is called an **eigenvalue** of the problem; and the non-trivial solution associated with the eigenvalue is called the **eigenfunction** of the problem (corresponding to the eigenvalue λ).

❑ EXAMPLE 12.1

Find the eigenvalues and eigenfunctions of the Sturm-Liouville problem

$y'' + \lambda y = 0$ subject to $y(0) = 0$ and $y\left(\dfrac{\pi}{2}\right) = 0.$

Solution:

If λ is negative, then $\lambda = -\mu^2$ for some $\mu > 0$ and $y'' - \mu^2 y = 0$ has general solution

$y = c_1 e^{\mu x} + c_2 e^{-\mu x}$. Since $y(0) = 0$, $c_1 + c_2 = 0$ and since $y\left(\dfrac{\pi}{2}\right) = 0$, $c_1 e^{\mu \pi / 2} + c_2 e^{-\mu \pi / 2} = 0$

so we must have that $c_1 = c_2 = 0$. Since $y = 0$ is not an eigenfuction, λ is not an eigenvalue. In a similar manner, $\lambda = 0$ is not an eigenvalue.

If λ is positive, then $\lambda = \mu^2$ for some $\mu > 0$ and $y'' + \mu^2 y = 0$ has general solution

$y = c_1 \cos(\mu x) + c_2 \sin(\mu x)$. Since $y(0) = 0$, $c_1 = 0$ and since $y\left(\dfrac{\pi}{2}\right) = 0$, $c_2 \sin\left(\dfrac{\pi \mu}{2}\right) = 0$ so

$\dfrac{\pi \mu}{2} = 0, \pm \pi, \pm 2\pi, \dots$ and $\mu = 0, \pm 2, \pm 4, \dots$. Since $y = 0$ is not an eigenfunction, $\mu \neq 0$ so

$\mu = \pm 2, \pm 4, \dots$.

Then $\lambda = \mu^2 = 4, 16, 36, \dots$ are the eigenvalues and $\sin(2x)$, $\sin(4x), \dots$ are the corresponding eigenfunctions. ∎

Definition:

Let $f(x)$ and $g(x)$ be integrable on the interval (a,b) (note that a could be $-\infty$ and b could be $+\infty$) and $w(x) > 0$ on (a,b). The **inner product** of $f(x)$ and $g(x)$ with respect to the weight function $w(x)$ on the interval (a,b) is $\int_a^b f(x) g(x) w(x) \, dx$. f and g are **orthogonal** on (a, b) with respect to the weight function $w(x)$ means $\int_a^b f(x) g(x) w(x) \, dx = 0.$

❑ EXAMPLE 12.2

Verify the orthogonality property for the eigenfunctions in the previous example.

Solution:

In the previous example, the family of eigenfunctions was $\left\{\sin(2nx)\right\}_{n=1}^{\infty}$. Let $n \neq m$ be two integers.

Then, $\displaystyle\int_0^{\pi/2} \sin(2nx) \sin(2mx) \, dx = \left(\dfrac{n \cos(2nx) \sin(2mx)}{2(m^2 - n^2)} + \dfrac{m \cos(2mx) \sin(2nx)}{2(n^2 - m^2)} \right)\Bigg|_{x=0}^{x=\pi/2} = 0.$

Note that *Mathematica* can be used to both compute an antiderivative of $\sin(2nx)\sin(2mx)$:

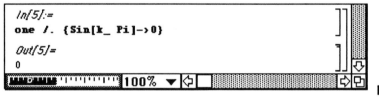

and compute the value of the definite integral $\int_0^{\pi/2}\sin(2nx)\sin(2mx)\,dx$:

```
In[4]:=
one=Integrate[Sin[2 n x] Sin[2 m x],{x,0,Pi/2}]
Out[4]=
n Cos[n Pi] Sin[m Pi]     m Cos[m Pi] Sin[n Pi]
───────────────────── + ─────────────────────
      2    2                   2    2
   2 (m  - n )              2 (-m  + n )
```

Since the sine of integer multiples of π is zero, we enter the following to verify the orthogonality property.

```
In[5]:=
one /. {Sin[k_ Pi]->0}
Out[5]=
0
```

Theorem:
Let p, q, and s be continuous functions on the interval [a,b] in the Sturm-Liouville problem

$$\frac{d}{dx}\left(p\frac{dy}{dx}\right) + \left(q + \lambda s\right)y = 0$$

subject to the conditions $a_1 y(a) + a_2 y'(a) = 0$ and $b_1 y(b) + b_2 y'(b) = 0$

be continuous on the interval [a, b] and let y_j and y_k be the eigenfunctions corresponding to the eigenvalues λ_j and λ_k, respectively. If y_j and y_k are continuously differentiable on [a,b], y_j and y_k are orthogonal with respect to the weight function s on [a, b].

❑ EXAMPLE 12.3
Verify the orthogonality property for the eigenfunctions of Legendre's equation:

$$\left(1-x^2\right)y''-2xy'+n\left(n+1\right)y = 0.$$

Solution:

The eigenfunctions of Legendre's equation are the Legendre polynomials. A table of these polynomials of order 0 to order 9 are given below in **lptable**.

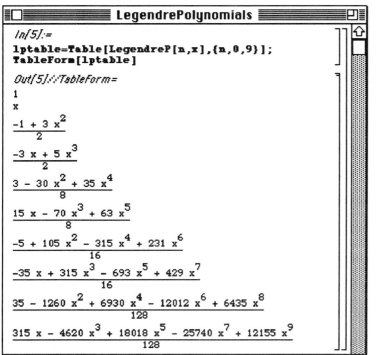

```
                        LegendrePolynomials
In[5]:=
lptable=Table[LegendreP[n,x],{n,0,9}];
TableForm[lptable]

Out[5]//TableForm=
1
x
-1 + 3 x²
─────────
    2
-3 x + 5 x³
──────────
     2
3 - 30 x² + 35 x⁴
─────────────────
        8
15 x - 70 x³ + 63 x⁵
────────────────────
         8
-5 + 105 x² - 315 x⁴ + 231 x⁶
─────────────────────────────
              16
-35 x + 315 x³ - 693 x⁵ + 429 x⁷
────────────────────────────────
               16
35 - 1260 x² + 6930 x⁴ - 12012 x⁶ + 6435 x⁸
────────────────────────────────────────────
                    128
315 x - 4620 x³ + 18018 x⁵ - 25740 x⁷ + 12155 x⁹
────────────────────────────────────────────────
                      128
```

Since Legendre's equation has singularities at $x = 1$ and $x = -1$, convergence of the power series method of solution which was used to obtain these solutions is valid only on the interval $-1 < x < 1$. Notice also that Legendre's equation is placed in the form of the previous theorem (known as **self-adjoint** form or a Sturm-Liouville equation) as follows:

$$\frac{d}{dx}\left[\left(1-x^2\right)y'\right]+n(n+1)y=0.$$

By letting $\lambda = n(n+1)$, then we see that $s = 1$. Hence, the orthogonality of the Legendre polynomials is verified with the following:

$$\int_{-1}^{1} P_n(x)P_m(x)\,dx = 0, m \neq n.$$

This is done in the table below by calculating this integral using the polynomials in **lptable**. Notice that only the elements along the diagonal (when $i = j$) are nonzero. Hence, the orthogonality is verified for these polynomials.

```
In[9]:=
orth=Table[Integrate[lptable[[i]] lptable[[j]],
        {x,-1,1}],
        {i,1,10},{j,1,10}]//MatrixForm

Out[9]//MatrixForm=
```

2	0	0	0	0	0	0	0	0	0
0	$\frac{2}{3}$	0	0	0	0	0	0	0	0
0	0	$\frac{2}{5}$	0	0	0	0	0	0	0
0	0	0	$\frac{2}{7}$	0	0	0	0	0	0
0	0	0	0	$\frac{2}{9}$	0	0	0	0	0
0	0	0	0	0	$\frac{2}{11}$	0	0	0	0
0	0	0	0	0	0	$\frac{2}{13}$	0	0	0
0	0	0	0	0	0	0	$\frac{2}{15}$	0	0
0	0	0	0	0	0	0	0	$\frac{2}{17}$	0
0	0	0	0	0	0	0	0	0	$\frac{2}{19}$

`100%`

§12.2 Introduction to Fourier Series

In calculus, we discussed the power series expansion of functions. We now investigate series of the form,

$$f(x) = \sum_{n=1}^{\infty} c_n \phi_n(x)$$ where $\{\phi_n(x)\}$ represents the set of eigenfunctions of a particular eigenvalue

problem. When the eigenfunctions are sines and cosines, the expansion is known as a **Fourier series**. We begin by discussing series of this form.

Fourier Sine Series

The family of functions $\left\{ \sin\left(\frac{n\pi x}{a}\right) \right\}_{n=1}^{+\infty}$ satisfies the orthogonality relation

$$\int_0^a \sin\left(\frac{n\pi x}{a}\right) \sin\left(\frac{m\pi x}{a}\right) dx = \begin{cases} 0, n \neq m \\ a/2, n = m \end{cases}.$$

If $f(x) = \sum\limits_{n=1}^{\infty} \alpha_n \sin\left(\dfrac{n\pi x}{a}\right)$, then multiplying the equation by $\sin\left(\dfrac{m\pi x}{a}\right)$ results in

$f(x)\sin\left(\dfrac{m\pi x}{a}\right) = \sum\limits_{n=1}^{\infty} \alpha_n \sin\left(\dfrac{n\pi x}{a}\right)\sin\left(\dfrac{m\pi x}{a}\right)$. Assuming this series can be integrated

term - by - term, we obtain $\int_0^a f(x)\sin\left(\dfrac{m\pi x}{a}\right)dx = \sum\limits_{n=1}^{\infty}\int_0^a \alpha_n \sin\left(\dfrac{n\pi x}{a}\right)\sin\left(\dfrac{m\pi x}{a}\right)dx$ so that

$\int_0^a f(x)\sin\left(\dfrac{m\pi x}{a}\right)dx = \alpha_m \dfrac{a}{2}$ and consequently $\alpha_m = \dfrac{2}{a}\int_0^a f(x)\sin\left(\dfrac{m\pi x}{a}\right)dx$.

■ EXAMPLE 12.4

Determine the Fourier Sine series of the function $f(x) = x + x^2$ on the interval $[0,2]$.

Solution:

We begin by defining the coefficient α_n according the integral formula above in **alpha**. Note that we choose to use the index n in these calculations.

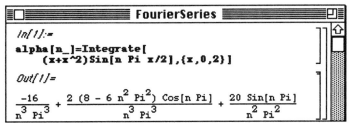

```
In[1]:=
alpha[n_]=Integrate[
     (x+x^2)Sin[n Pi x/2],{x,0,2}]
Out[1]=
 -16      2 (8 - 6 n^2 Pi^2) Cos[n Pi]    20 Sin[n Pi]
------- + ------------------------------ + --------------
 n^3 Pi^3          n^3 Pi^3                   n^2 Pi^2
```

Below, we simplify the expression which results by replacing $\cos(n\pi)$ with $(-1)^n$ and $\sin(n\pi)$ with zero. We call this result **coeff** and use this formula to calculate the necessary coefficients of the Fourier Sine series.

```
In[2]:=
coeff[n_]=alpha[n]/.
          Cos[n Pi]->(-1)^n/.Sin[n Pi]->0
Out[2]=
 -16      2 (-1)^n (8 - 6 n^2 Pi^2)
------- + --------------------------
 n^3 Pi^3        n^3 Pi^3
```

The **nth** term of the series expansion is defined as **fterm**

```
In[3]:=
fterm[x_,n_]:=coeff[n] Sin[n Pi x/2]
```

so that approximations of the function $f(x)=x + x^2$ on the interval [0,2] using the sum of the first n terms for n=2 to n=8 using increments of 2 are determined in **fapproxtable** below. To save space, only an abbreviated output list of this table is given.

```
In[5]:=
fapproxtable=Table[
    Sum[fterm[x,n],{n,1,i}],{i,2,8,2}];
Short[fapproxtable,6]

Out[5]//Short=
{(-16   2 (8 - 6 Pi^2)       Pi x
 (---- - -------------) Sin[----] +
  Pi^3      Pi^3              2

 (-2   8 - 24 Pi^2
 (--- + ----------) Sin[Pi x], <<2>>,
  Pi^3    4 Pi^3

 (-16   2 (8 - 6 Pi^2)       Pi x
 (---- - -------------) Sin[----] +
  Pi^3      Pi^3              2

 (-2   8 - 24 Pi^2
 (--- + ----------) Sin[Pi x] +
  Pi^3    4 Pi^3

  -16    2 (8 + <<1>>)       3 Pi x
 (----- - ------------) Sin[------] + <<4>> +
  27 Pi^3    27 Pi^3            2

   -1     8 - 384 Pi^2
 (------ + ------------) Sin[4 Pi x]}
  32 Pi^3    256 Pi^3
```

The elements of **fapproxtable** are all rather complicated and would be difficult to plot by hand. Therefore, we use *Mathematica* to observe the accuracy of the approximation with a Fourier Sine Series. This is done by defining the function fplot which plots the ith element of **fapproxtable** and function $f(x)=x + x^2$ simultaneously on the interval [0,2]. The plot of f(x) appears as the lighter curve because of the level of gray used. In **gtable**, a list of plots is produced for each of the elements of **fapproxtable**. This list is partitioned into groups of two and displayed as a graphics array. Notice that as more terms of the series are used, the accuracy of the approximation improves.

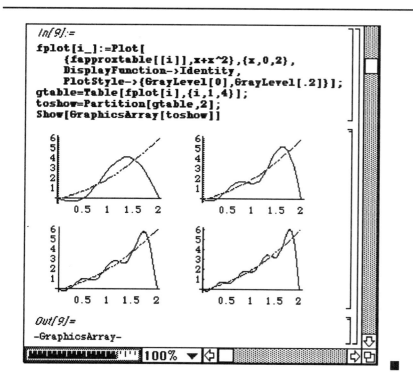

```
In[9]:=
fplot[i_]:=Plot[
    {fapproxtable[[i]],x+x^2},{x,0,2},
    DisplayFunction->Identity,
    PlotStyle->{GrayLevel[0],GrayLevel[.2]}];
gtable=Table[fplot[i],{i,1,4}];
toshow=Partition[gtable,2];
Show[GraphicsArray[toshow]]
```

```
Out[9]=
-GraphicsArray-
```

Fourier Cosine Series

The family of functions $\left\{\cos\left(\dfrac{n\pi x}{a}\right)\right\}_{n=0}^{+\infty}$ satisfies the orthogonality relation

$$\int_0^a \cos\left(\frac{n\pi x}{a}\right)\cos\left(\frac{m\pi x}{a}\right)dx = \begin{cases} 0, n \ne m \\ a/2, n = m \ne 0 \\ a, n = m = 0 \end{cases}$$

If $f(x) = \displaystyle\sum_{n=0}^{\infty} \beta_n \cos\left(\frac{n\pi x}{a}\right)$, then multiplying the equation by $\cos\left(\dfrac{m\pi x}{a}\right)$ results in

$f(x)\cos\left(\dfrac{m\pi x}{a}\right) = \displaystyle\sum_{n=0}^{\infty} \beta_n \cos\left(\frac{n\pi x}{a}\right)\cos\left(\frac{m\pi x}{a}\right)$. Assuming this series can be integrated

term - by - term, we obtain $\int_0^a f(x)\cos\left(\dfrac{m\pi x}{a}\right)dx = \displaystyle\sum_{n=0}^{\infty} \int_0^a \beta_n \cos\left(\frac{n\pi x}{a}\right)\cos\left(\frac{m\pi x}{a}\right)dx$ so that

$$\int_0^a f(x)\,dx = \beta_m a, \text{ if } m = 0, \text{ and } \int_0^a f(x)\cos\left(\frac{m\pi x}{a}\right)dx = \beta_m \frac{a}{2}, \text{ if } m \neq 0, \text{consequently}$$

$$\beta_0 = \frac{1}{a}\int_0^a f(x)\,dx \text{ and } \beta_m = \frac{2}{a}\int_0^a f(x)\cos\left(\frac{m\pi x}{a}\right)dx.$$

■ EXAMPLE 12.5

Determine the Fourier Cosine Series of $f(x)=x + x^2$ on the interval $[0,2]$.

Solution:

We begin by finding β_0 using the integral formula derived above.

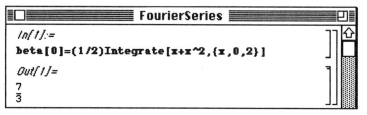

```
━━━━━━━━━━━━━━━━━ FourierSeries ━━━━━━━━━━━━━━━━━
In[1]:=
beta[0]=(1/2)Integrate[x+x^2,{x,0,2}]

Out[1]=
7
─
3
```

We then perform a similar calculation to determine β_n for the series expansion

```
In[2]:=
beta[n_]=Integrate[
          (x+x^2)Cos[n Pi x/2],{x,0,2}]

Out[2]=

  -4       20 Cos[n Pi]     2 (-8 + 6 n^2 Pi^2) Sin[n Pi]
────── + ───────────── + ──────────────────────────────
n^2 Pi^2    n^2 Pi^2                n^3 Pi^3
```

and simplify this expression with the replacements indicated below. Hence, we use this simplified expression, called **coeff2**, to represent the coefficients β_n in subsequent calculations.

```
In[3]:=
coeff2[n_]=beta[n]/.
            Cos[n Pi]->(-1)^n/. Sin[n Pi]->0

Out[3]=

  -4      20 (-1)^n
────── + ──────────
n^2 Pi^2  n^2 Pi^2
```

The **n**th term of the series for n=1, 2, ... is defined in **fterm**

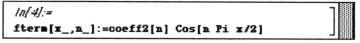

```
In[4]:=
fterm[x_,n_]:=coeff2[n] Cos[n Pi x/2]
```

and the sum of the first n terms is defined as **fsum** below.

```
In[5]:=
fsum[x_,n_]:=beta[0]+Sum[fterm[x,i],{i,1,n}]
```

Therefore, we can determine the approximation of this function on [0,2] with a Fourier Cosine Series using **fsum**. We observe the approximation below through the use of the function **f2plot** which plots the sum of the first **i** terms of the expansion and the function $f(x)=x + x^2$ simultaneously on the interval [0,2]. This is done in **gtable** for i=1 to i=7 using increments of 2. The list of plots which results is partitioned into groups of two and displayed as a graphics array. Notice that this series appears to yield the better approximation since the accuracy is improved using fewer terms than in the case of the Fourier Sine Series which was discussed in the previous example.

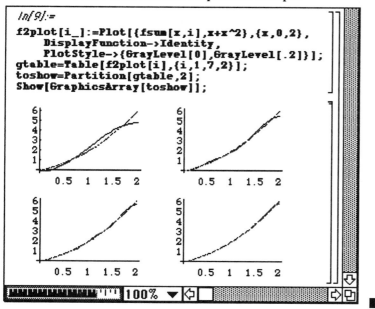

```
In[9]:=
f2plot[i_]:=Plot[{fsum[x,i],x+x^2},{x,0,2},
    DisplayFunction->Identity,
    PlotStyle->{GrayLevel[0],GrayLevel[.2]}];
gtable=Table[f2plot[i],{i,1,7,2}];
toshow=Partition[gtable,2];
Show[GraphicsArray[toshow]];
```

Fourier Trigonometric Series

The family of functions $\left\{1, \sin\left(\dfrac{n\pi x}{a}\right), \cos\left(\dfrac{n\pi x}{a}\right)\right\}_{n=1}^{+\infty}$ satisfies the orthogonality relations

$$\int_{-a}^{a}\cos\left(\frac{n\pi x}{a}\right)\cos\left(\frac{m\pi x}{a}\right)dx = \begin{cases} 0, n \neq m \\ a, n = m \end{cases}, \int_{-a}^{a}\sin\left(\frac{n\pi x}{a}\right)\cos\left(\frac{m\pi x}{a}\right)dx = 0 \text{ for all } n \text{ and } m, \text{ and}$$

$$\int_{-a}^{a}\sin\left(\frac{n\pi x}{a}\right)\sin\left(\frac{m\pi x}{a}\right)dx = \begin{cases} 0, n \neq m \\ a, n = m \end{cases}.$$

If $f(x) = \beta_0 + \sum_{n=1}^{\infty} \left[\alpha_n \sin\left(\frac{n\pi x}{a}\right) + \beta_n \cos\left(\frac{n\pi x}{a}\right) \right]$ and assuming the series can be integrated

term - by - term, yields $\int_{-a}^{a} f(x)dx = \int_{-a}^{a} \beta_0 \, dx + \sum_{n=1}^{\infty} \int_{-a}^{a} \left[\alpha_n \sin\left(\frac{n\pi x}{a}\right) + \beta_n \cos\left(\frac{n\pi x}{a}\right) \right] dx$

so $\beta_0 = \frac{1}{2a} \int_{-a}^{a} f(x)dx$. Multiplying each side of the equation

$f(x) = \beta_0 + \sum_{n=1}^{\infty} \left[\alpha_n \sin\left(\frac{n\pi x}{a}\right) + \beta_n \cos\left(\frac{n\pi x}{a}\right) \right]$ by $\sin\left(\frac{m\pi x}{a}\right)$ and integrating results in

$\int_{-a}^{a} f(x)\sin\left(\frac{m\pi x}{a}\right)dx = \int_{-a}^{a} \beta_0 \sin\left(\frac{m\pi x}{a}\right)dx +$

$$\sum_{n=1}^{\infty} \int_{-a}^{a} \left[\alpha_n \sin\left(\frac{n\pi x}{a}\right)\sin\left(\frac{m\pi x}{a}\right) + \beta_n \cos\left(\frac{n\pi x}{a}\right)\sin\left(\frac{m\pi x}{a}\right) \right] dx$$

so that $\int_{-a}^{a} f(x)\sin\left(\frac{m\pi x}{a}\right)dx = \int_{-a}^{a} \alpha_m \sin\left(\frac{m\pi x}{a}\right)\sin\left(\frac{m\pi x}{a}\right)dx = \alpha_m a$ and thus

$\alpha_m = \frac{1}{a} \int_{-a}^{a} f(x)\sin\left(\frac{m\pi x}{a}\right)dx$. Similarly, multiplying each side of the equation by $\cos\left(\frac{m\pi x}{a}\right)$

and integrating results in

$\int_{-a}^{a} f(x)\cos\left(\frac{m\pi x}{a}\right)dx = \int_{-a}^{a} \beta_0 \cos\left(\frac{m\pi x}{a}\right)dx +$

$$\sum_{n=1}^{\infty} \int_{-a}^{a} \left[\alpha_n \sin\left(\frac{n\pi x}{a}\right)\cos\left(\frac{m\pi x}{a}\right) + \beta_n \cos\left(\frac{n\pi x}{a}\right)\cos\left(\frac{m\pi x}{a}\right) \right] dx$$

so that $\int_{-a}^{a} f(x)\cos\left(\frac{m\pi x}{a}\right)dx = \int_{-a}^{a} \beta_m \cos\left(\frac{m\pi x}{a}\right)\cos\left(\frac{m\pi x}{a}\right)dx = \beta_m a$ and thus

$\beta_m = \frac{1}{a} \int_{-a}^{a} f(x)\cos\left(\frac{m\pi x}{a}\right)dx$.

Convergence of the Fourier Trigonometric Series:
In order to explain the convergence of the Fourier Series, we must give the following definition:
A function is **piecewise continuous** on a finite interval if it is continuous at each point in the interval except
possibly at a finite number of points. Hence, the points of discontinuity are finite.
Hence, we can state the following theorem:

The Convergence Theorem:
If the function f is periodic such that f(x+2a)=f(x) for all x, and both f and f' are piecewise continuous, then

$$f(x) = \beta_0 + \sum_{n=1}^{\infty} \left[\alpha_n \sin\left(\frac{n\pi x}{a}\right) + \beta_n \cos\left(\frac{n\pi x}{a}\right) \right]$$

converges for all values of x. At all points x where the function is continuous, the sum of the series equals the value of f(x). At points of discontinuity, the sum of the series is the average value

$$\frac{f\left(x^+\right) + f\left(x^-\right)}{2}$$ where $f\left(x^+\right)$ represents the value of f to the right of x and $f\left(x^-\right)$ represents the value of f to the left of x.

Therefore, we can expect that the series expansions obtained with the Fourier Series approach the function. Fortunately, with the help of *Mathematica*, we will be able to view this convergence graphically. We illustrate this procedure through the following example by, first, using the **FourierTransform** package, because this package which is found in the Calculus folder (directory)contains commands which compute the Fourier series. In particular, **FourierTrigSeries[f,{x,xmin,xmax},n]** yields the Fourier trigonometric series expansion of order **n** of the function **f** with period of length 2a= **xmax−xmin**.

■ EXAMPLE 12.6

Determine the Fourier Trigonometric Series of the periodic function f(x)=x on the interval (−1,1) such that f(x+2)=f(x) for x>1.

Solution:

We begin by loading the **FourierTransform** package, entering the function f(x)=x, and defining **fstable** to use **FourierTrigSeries** to compute the sum of the first n terms of the Fourier Trigonometric Series (known as the Fourier Trigonometric Series expansion of order n) for n=2 to n=8 using increments of two.

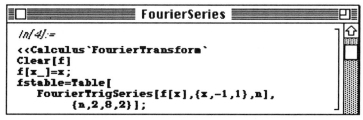

The elements of fstable are printed in **TableForm** below. Note that this list does not contain any cosine terms due to the fact that f(x) is an odd function on (−1,1). Recall that a function is <u>**odd**</u> if f(− x)=−x and <u>**even**</u> if f(−x)=f(x). Hence, the integrals used to calculate β_0 and β_n in the Fourier Trigonometric Series equate to zero since we are multiplying an odd function by an even function (cosine is even) to yield an odd function. The integral over one period of an odd function is zero.

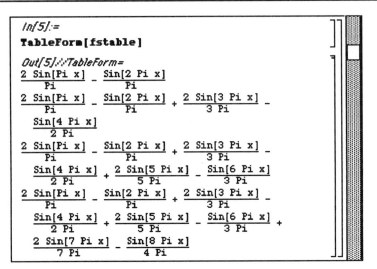

We would like to plot the elements of **fstable** and f(x)=x in order to compare the curves. We do this by defining the periodic extension of f below in **g**. Hence, f merely repeats itself over each subsequent interval which follows (−1,1) of length 2.

```
In[7]:=
g[x_]:=x/;-1<=x<=1
g[x_]:=g[x-2]/;x>1
```

The function **plotfourier** plots the ith element of **fstable** along with **g** on the interval [−1,3], and we perform this function for i=1 to i=4. The resulting list of graphics is partitioned into groups of two and displayed as a graphics array. Notice the improved accuracy of the approximation as more terms are used.   As stated in the convergence theorem above, the Fourier Trigonometric Series converges to f(x) for all x such that the function is continuous. At points such as x=1, the series converges to 0, the average value of $f(x^+)=-1$ and $f(x^-)=1$.

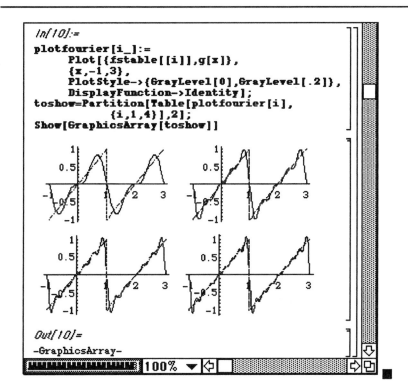

```
In[10]:=
plotfourier[i_]:=
    Plot[{fstable[[i]],g[x]},
    {x,-1,3},
    PlotStyle->{GrayLevel[0],GrayLevel[.2]},
    DisplayFunction->Identity];
toshow=Partition[Table[plotfourier[i],
        {i,1,4}],2];
Show[GraphicsArray[toshow]]
```

```
Out[10]=
-GraphicsArray-
```

■ EXAMPLE 12.7

Determine the Fourier Trigonometric Series of the periodic function $f(x)=x^2$ on the interval $(-1,1)$ such that $f(x+2)=f(x)$.

Solution:

We follow a similar procedure as in the previous example by using the **FourierTransform** command **FourierTrigSeries** to compute the sum of the first n terms of the Fourier Trigonometric Series (known as the Fourier Trigonometric Series expansion of order n) for n=1 to n=4 using increments of one.

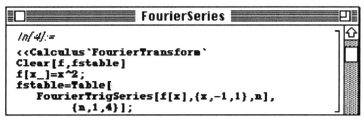

```
FourierSeries
In[4]:=
<<Calculus`FourierTransform`
Clear[f,fstable]
f[x_]=x^2;
fstable=Table[
    FourierTrigSeries[f[x],{x,-1,1},n],
        {n,1,4}];
```

The elements of **fstable** are listed below. Note that no terms involve sine. This is due to the fact that $f(x)=x^2$ is an even function on the interval $(-1,1)$. Hence the integral formula to determine α_n is zero for all n. (The product of an even function and the sine function in the formula is odd. Therefore, the integral over one period is zero.)

```
In[5]:=
TableForm[fstable]

Out[5]//TableForm=
1   4 Cos[Pi x]
─ - ──────────
3       Pi²
1   4 Cos[Pi x]   Cos[2 Pi x]
─ - ────────── + ───────────
3       Pi²          Pi²
1   4 Cos[Pi x]   Cos[2 Pi x]   4 Cos[3 Pi x]
─ - ────────── + ─────────── - ─────────────
3       Pi²          Pi²           9 Pi²
1   4 Cos[Pi x]   Cos[2 Pi x]   4 Cos[3 Pi x]
─ - ────────── + ─────────── - ───────────── +
3       Pi²          Pi²           9 Pi²
   Cos[4 Pi x]
   ──────────
     4 Pi²
```

In order to graph the periodic extension of f, we define **g** below. Note that an **If** statement could be used to define g as well.

```
In[7]:=
g[x_]:=x^2/;-1<=x<=1
g[x_]:=g[x-2]/;x>1
```

Then, **plotfourier** is defined as in the previous example to graph the approximation and g simultaneously on the interval [-1,3]. The results show that the Cosine Series yields a good approximation of this periodic function. We have used as many terms as in the Sine Series in the previous example and achieved greater accuracy. In the last plot, the approximation is difficult to distinguish from the original function.

```
In[10]:=
plotfourier[i_]:=
    Plot[{fstable[[i]],g[x]},
    {x,-1,3},
    PlotStyle->{GrayLevel[0],GrayLevel[.2]},
    DisplayFunction->Identity];
toshow=Partition[Table[plotfourier[i],
    {i,1,4}],2];
Show[GraphicsArray[toshow]]
```

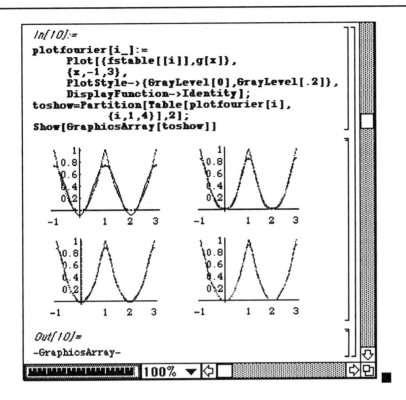

```
Out[10]=
-GraphicsArray-
```

100% ▼

■ EXAMPLE 12.8

Consider the periodic function $f(x)=x$, $f(x+1)=x$. Determine the Fourier Trigonometric Series expansion of this function of order $n = 2, 4, 6,$ and 8. Also, plot the function along with the approximation for each of these values of n.

Solution:

We first define the function and then create a table of Fourier series expansions for the indicated values of n in **fouriertable**. Notice that there are no cosine terms in the series in the **fouriertable** entries. This is due to the fact that the integral used to compute the β_n is zero for all values of n except n=0. Notice that we are only considering the interval [0,1]. Hence, f is neither even nor odd on this interval, so the integral used to compute β_0 does not necessarily have to equal zero.

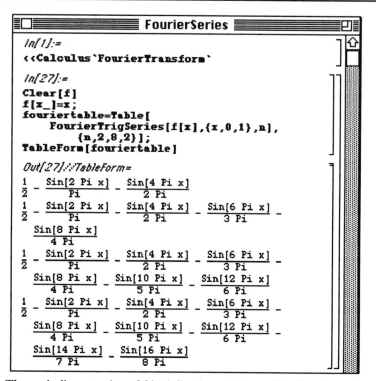

The periodic extension of f is defined as **g** below. The function **plotfourier** is then defined to plot the series expansions given in **fouriertable**.

```
In[29]:=
g[x_]:=x /; x<=1
g[x_]:=g[x-1] /; x>1
In[30]:=
plotfourier[i_]:=
    Plot[{fouriertable[[i]],g[x]},
    {x,0,4},
    PlotStyle->{GrayLevel[0],GrayLevel[.4]},
    DisplayFunction->Identity]
```

Since the plot is initially suppressed in the function **plotfourier**, **Show** is used to display the graphs after they are partitioned into groups of two in **toshow**.

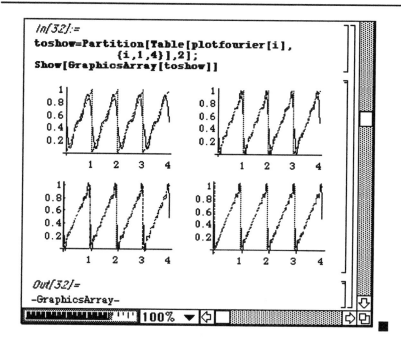

```
In[32]:=
toshow=Partition[Table[plotfourier[i],
          {i,1,4}],2];
Show[GraphicsArray[toshow]]
```

```
Out[32]=
-GraphicsArray-
```

Bessel Functions

We now discuss an eigenfunction expansion of a more general kind. Recall Bessel's equation discussed earlier in **Chapter 6**,

$$x^2 y'' + xy' + \left(x^2 - n^2\right) y = 0.$$

A more generalized form of this equation is given by

$$x^2 y'' + xy' + \left(\lambda^2 x^2 - n^2\right) y = 0.$$

Thus, the change of variable $s = \lambda x$ transforms this equation into the previous form to yield the solutions $y_n(x) = J_n(\lambda x)$. Hence, the set $\left\{J_n(\lambda x)\right\}$ is the set of eigenfunctions of Bessel's equation. Therefore, we can use this set of functions to form the eigenfunction expansion, but we must first discuss some of the properties of Bessel functions.

Let α_{nm} denote the mth zero of the nth Bessel function $J_n(x), n = 0, 1, 2, ...$ and $m = 1, 2, 3, ...$

Then for each n the family of functions $\left\{ J_n\left(\frac{\alpha_{nm}}{R} x\right) \right\}_{m=1}^{\infty}$ satisfies the orthogonality relation

$$\int_0^R x J_n\left(\frac{\alpha_{nm}}{R} x\right) J_n\left(\frac{\alpha_{nk}}{R} x\right) dx = \begin{cases} 0, k \neq m \\ \frac{R^2}{2} \left(J_{n+1}\left(\alpha_{nm}\right)\right)^2, k = m \end{cases}$$

To verify the orthogonality relation, recall that the Bessel function $J_n(x)$ satisfies Bessel's equation

$$t^2 y'' + t y' + \left(t^2 - n^2\right) y = 0 \text{ so that } t^2 \frac{d^2}{dt^2}\left(J_n(t)\right) + t \frac{d}{dt}\left(J_n(t)\right) + \left(t^2 - n^2\right) J_n(t) = 0.$$

If $t = \lambda x$, then $\frac{dx}{dt} = \frac{1}{\lambda}$ and applying the chain rule yields $\frac{d}{dt}\left(J_n(t)\right) = \frac{1}{\lambda} J_n'\left(\lambda x\right)$ and

$\frac{d^2}{dt^2}\left(J_n(t)\right) = \frac{1}{\lambda^2} J_n''\left(\lambda x\right).$ Substituting into the equation

$$t^2 \frac{d^2}{dt^2}\left(J_n(t)\right) + t \frac{d}{dt}\left(J_n(t)\right) + \left(t^2 - n^2\right) J_n(t) = 0 \text{ results in}$$

$x^2 J_n''\left(\lambda x\right) + x J_n'\left(\lambda x\right) + \left(\lambda^2 x^2 - n^2\right) J_n\left(\lambda x\right) = 0.$ Dividing this equation by x and rewriting

yields the equivalent equation $\left(x J_n'\left(\lambda x\right)\right)' + \left(\frac{-n^2}{x} + \lambda^2 x\right) J_n\left(\lambda x\right) = 0.$

Multiplying this equation by $2x J_n'\left(\lambda x\right)$ results in

$$2x J_n'\left(\lambda x\right)\left(x J_n'\left(\lambda x\right)\right)' + \left(\frac{-n^2}{x} + \lambda^2 x\right) 2x J_n'\left(\lambda x\right) J_n\left(\lambda x\right) = 0 \text{ which can be written in the form}$$

$$\left[\left(x J_n'\left(\lambda x\right)\right)^2\right]' + \left(\lambda^2 x^2 - n^2\right)\left[\left(J_n\left(\lambda x\right)\right)^2\right]' = 0 \text{ and integrating yields}$$

$$\left(x J_n'\left(\lambda x\right)\right)^2 \Big|_0^R = -\int_0^R \left(\lambda^2 x^2 - n^2\right)\left[\left(J_n\left(\lambda x\right)\right)^2\right]' dx.$$

The Bessel functions satisfy the recurrence relations

(i) $J_{\mu-1}(x) + J_{\mu+1}(x) = \frac{2\mu}{x} J_\mu(x);$ (ii) $J_{\mu-1}(x) - J_{\mu+1}(x) = 2 J_\mu'(x);$

(iii) $\mu J_\mu(x) + x J_\mu'(x) = x J_{\mu-1}(x);$ and (iv) $\mu J_\mu(x) - x J_\mu'(x) = x J_{\mu+1}(x).$

Replacing μ by n, x by λx, multiplying by (iv) and applying the chain rule to (iv) results in

$x J_n'\left(\lambda x\right) = n J_n\left(\lambda x\right) - \lambda x J_{n+1}\left(\lambda x\right).$

Therefore $\left(x J_n{'}(\lambda x) \right)^2 \Big|_0^R = \left(n J_n(\lambda x) - \lambda x J_{n+1}(\lambda x) \right)^2 \Big|_0^R$. When $\lambda = \dfrac{\alpha_{nm}}{R}$,

$\left(n J_n(\lambda x) - \lambda x J_{n+1}(\lambda x) \right)^2 \Big|_0^R = \lambda^2 R^2 \left(J_{n+1}(\lambda R) \right)^2$.

Using integration by parts we obtain

$-\int_0^R \left(\lambda^2 x^2 - n^2 \right) \left[\left(J_n(\lambda x) \right)^2 \right]' dx = -\left[\left(\lambda^2 x^2 - n^2 \right) \left(J_n(\lambda x) \right)^2 \right] \Big|_0^R + 2\lambda^2 \int_0^R x \left(J_n(\lambda x) \right)^2 dx$. In

the case when $\lambda = \dfrac{\alpha_{nm}}{R}$,

$-\left[\left(\lambda^2 x^2 - n^2 \right) \left(J_n(\lambda x) \right)^2 \right] \Big|_0^R + 2\lambda^2 \int_0^R x \left(J_n(\lambda x) \right)^2 dx = 2\lambda^2 \int_0^R x \left(J_n(\lambda x) \right)^2 dx$. Therefore

when $\lambda = \dfrac{\alpha_{nm}}{R}$, $\lambda^2 R^2 \left(J_{n+1}(\lambda R) \right)^2 = 2\lambda^2 \int_0^R x \left(J_n(\lambda x) \right)^2 dx$ so

$\int_0^R x \left(J_n(\lambda x) \right)^2 dx = \dfrac{R^2}{2} \left(J_{n+1}(\lambda R) \right)^2$.

Having established these properties, we can now investigate the eigenfunction expansion (called a generalized Fourier series) involving Bessel functions. We begin by defining the series expansion and then determining the coefficients a_m.

If $f(x) = \displaystyle\sum_{m=1}^{\infty} a_m J_n\left(\dfrac{\alpha_{nm}}{R} x \right)$ and assuming the series can be integrated term - by - term, multiplying

each term in the equation by $x J_n\left(\dfrac{\alpha_{nk}}{R} x \right)$ and integrating results in

$\int_0^R x f(x) J_n\left(\dfrac{\alpha_{nk}}{R} x \right) dx = \displaystyle\sum_{m=1}^{\infty} \int_0^R a_m x J_n\left(\dfrac{\alpha_{nm}}{R} x \right) J_n\left(\dfrac{\alpha_{nk}}{R} x \right) dx$

$= \int_0^R a_k x \left(J_n\left(\dfrac{\alpha_{nk}}{R} x \right) \right)^2 dx = a_k \dfrac{R^2}{2} \left(J_{n+1}(\alpha_{nk}) \right)^2$ so that

$a_k = \dfrac{2}{R^2 \left(J_{n+1}(\alpha_{nk}) \right)^2} \int_0^R x f(x) J_n\left(\dfrac{\alpha_{nk}}{R} x \right) dx$.

■ EXAMPLE 12.9
Since the zeros of the Bessel functions play an important role in the generalized Fourier series involving

Bessel functions, use *Mathematica* to find the first eight zeros of the Bessel functions of the first kind, $J_n(x)$, of order n=0,1,2,...,6.

Solution:

The Bessel functions of order n = 0, ..., 5 are plotted below as a graphics array.

```
≣□        ════ TheWaveEquation(3D) ════        □≣
In[53]:=
graphs=Table[Plot[BesselJ[n,x],{x,0,33},
    DisplayFunction->Identity],{n,0,5}];
array=Partition[graphs,3];
Show[GraphicsArray[array]]
```

These graphs are used to approximate the values of the zeros for each function. For example, the zeros of the Bessel function of order zero appear to occur at x = 2.5, 5.5, 8.7, 11.8, 15.1, 18.1, 21.2, and 24.4. Hence, these values are entered in the list **az** which is used with **FindRoot** in **a[0]** to supply the initial guess for each of the first eight zeros. A similar list of approximate zeros is given for each function. In general, **a[i]** is a list of the first eight zeros of the Bessel function of order **i**.

```
In[67]:=
az={2.5,5.5,8.7,11.8,15.1,18.1,21.2,24.4};
a[0]=Table[FindRoot[
    BesselJ[0,x]==0,{x,az[[i]]}],{i,1,8}];
at={4,7,10,13.2,16.4,19.6,22.6,26};
a[1]=Table[FindRoot[
    BesselJ[1,x]==0,{x,at[[i]]}],{i,1,8}];
atwo={5.2,8.4,11.8,14.7,18,21.1,24.4,27.4};
a[2]=Table[FindRoot[
    BesselJ[2,x]==0,{x,atwo[[i]]}],{i,1,8}];
athree={6.5,10,13,16.2,19.4,22.5,25.8,29};
a[3]=Table[FindRoot[
    BesselJ[3,x]==0,{x,athree[[i]]}],
                         {i,1,8}];
afour={7.6,11.1,14.5,17.7,20.9,24.1,27.3,
                         30.5};
a[4]=Table[FindRoot[
    BesselJ[4,x]==0,{x,afour[[i]]}],{i,1,8}];
afive={8.9,12.4,15.6,19.1,22.3,25.5,28.7,
                         31.9};
a[5]=Table[FindRoot[
    BesselJ[5,x]==0,{x,afive[[i]]}],{i,1,8}];
asix={10,13.5,17.2,20,23.6,26.9,30,33.3};
a[6]=Table[FindRoot[
    BesselJ[6,x]==0,{x,asix[[i]]}],{i,1,8}];
```

After these lists are obtained, they are combined to form the single list **zeros** which is viewed in **TableForm** below.

```
In[69]:=
zeros=Table[a[i][[j,1,2]],{i,0,6},{j,1,8}];
TableForm[zeros]

Out[69]//TableForm=
2.40483    5.52008    8.65373    11.7915    14.9309
      18.0711   21.2116   24.3525
3.83171    7.01559    10.1735    13.3237    16.4706
      19.6159   22.7601   25.9037
5.13562    8.41724    11.6198    14.796     17.9598
      21.117    24.2701   27.4206
6.38016    9.76102    13.0152    16.2235    19.4094
      22.5827   25.7482   28.9084
7.58834    11.0647    14.3725    17.616     20.8269
      24.019    27.1991   30.371
8.77148    12.3386    15.7002    18.9801    22.2178
      25.4303   28.6266   31.8117
9.93611    13.5893    17.0038    20.3208    23.5861
      26.8202   30.0337   33.233
```

For convenience, this list can be saved as **besseltable**. In doing so, these time-consuming calculations may be avoided each time the list of zeros is needed. Instead, the file **besseltable** may be easily read.

```
In[70]:=
zeros>>besseltable
```

If the calculations have just been done, the α_{mn} which are necessary in the calculation of the series coefficients of the eigenfunction expansion are defined in the following way.

```
In[71]:=
alpha[i_,j_]:=zeros[[i+1,j]]
```

However, if the zeros must first be read in from **besseltable**, the following commands must be performed. These functions will be explained in more detail in future examples.

```
In[72]:=
getzeros=ReadList["besseltable"];
In[73]:=
alpha[i_,j_]:=getzeros[[1]][[i+1,j]]
```

`100%  ▼`

■ EXAMPLE 12.10

Use a Bessel function expansion with $y_n(x) = J_0(\lambda x)$ to approximate the function $f(x) = x$ on $0 < x < 1$ using five, six, seven, and eight terms in the expansion. Plot the approximation as well as $f(x)$ on this interval.

Solution:

First, we compute the first eight zeros of the Bessel function of order zero in **a[0]**. This is done in a manner similar to **besseltable** in the previous example. In fact, the **besseltable** file could be used as opposed to performing the following calculation. However, we are only concerned with the zeros of the Bessel function of order zero in this problem.

```
≡▭▭▭▭▭▭▭▭▭ FourierSeries ≡▭▭▭▭▭▭
In[2]:=
az={2.5,5.5,8.7,11.8,15.1,18.1,21.2,24.4};
a[0]=Table[FindRoot[
    BesselJ[0,x]==0,{x,az[[i]]}],{i,1,8}];
```

Below, we define the integral formula needed to compute a_n for the Bessel function expansion. We choose to use **NIntegrate** to avoid the difficulties associated with integrating the Bessel function.

```
In[5]:=
Clear[alpha,f]
f[x_]:=x
alpha[n_]:=alpha[n]=
    NIntegrate[x f[x]*
        BesselJ[0,a[0][[n,1,2]]x],{x,0,1}]/
    NIntegrate[
        x(BesselJ[0,a[0][[n,1,2]]x])^2,
            {x,0,1}]//N
```

The first eight coefficients are determined below in **atable**.

```
In[6]:=
atable=Table[alpha[n]//Chop,{n,1,8}]

Out[6]=
{0.817455, -1.13349, 0.798285, -0.747008,
 0.631543, -0.597359, 0.535993, -0.512335}
```

These values are used to define the nth term of the series expansion in **fterm** below.

```
In[7]:=
fterm[x_,n_]:=atable[[n]]*
        BesselJ[0,a[0][[n,1,2]]x]
```

Hence, the approximation of the function is given by the sum of the terms defined in **fterm**. Notice that we define **fapprox** recursively so that *Mathematica* remembers all values of **fapprox** as they are computed. We find the Bessel function expansions for n=5 to n=8 in **approxes** below.

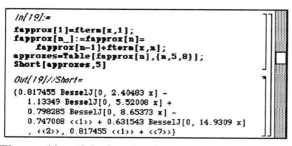

```
In[19]:=
fapprox[1]=fterm[x,1];
fapprox[n_]:=fapprox[n]=
    fapprox[n-1]+fterm[x,n];
approxes=Table[fapprox[n],{n,5,8}];
Short[approxes,5]

Out[19]//Short=
{0.817455 BesselJ[0, 2.40483 x] -
 1.13349 BesselJ[0, 5.52008 x] +
 0.798285 BesselJ[0, 8.65373 x] -
 0.747008 <<1>> + 0.631543 BesselJ[0, 14.9309 x]
 , <<2>>, 0.817455 <<1>> + <<7>>}
```

The graphics of the function f(x)=x is generated in **plotf** below using a shade of gray so that it can be distinguished from the approximation.

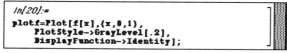

```
In[20]:=
plotf=Plot[f[x],{x,0,1},
    PlotStyle->GrayLevel[.2],
    DisplayFunction->Identity];
```

The function **plotfapprox** produces the graphics of **fapprox[i]** and displays this graph with that of f. This function is carried out for i=5 to i=8 using increments of one, and the list of graphics which results is partitioned into groups of two. This array is then displayed as a graphics array. Notice that in this case as with the Fourier Series, the approximation improves as more terms are used in the expansion. This particular eigenfunction expansion will be of great use in a later section in solving the wave equation on a circular region.

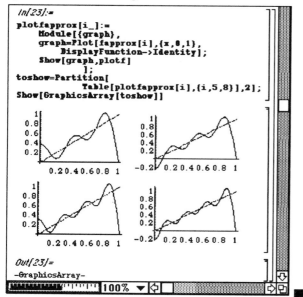

```
In[23]:=
plotfapprox[i_]:=
    Module[{graph},
    graph=Plot[fapprox[i],{x,0,1},
        DisplayFunction->Identity];
    Show[graph,plotf]
        ];
toshow=Partition[
            Table[plotfapprox[i],{i,5,8}],2];
Show[GraphicsArray[toshow]]
```

```
Out[23]=
-GraphicsArray-
```

§12.3 The One-Dimensional Heat Equation

One of the more important differential equations is the heat equation. In one spatial dimension, the solution to the heat equation translates into the temperature(at any position x and any time t) in a thin rod or wire of length a. Since the rate at which heat flows through the rod depends on the material which makes up the rod, the constant k which is related to the thermal diffusivity of the material is included in the heat equation as stated below. The first problem that we will investigate is the situation in which the temperature at the ends of the rod are constantly kept at zero and the initial temperature distribution in the rod is represented as the function f(x). Hence, the fixed end zero temperature is given in the boundary conditions (ii) while the initial heat distribution is given in (iii).

The Heat Equation with Homogeneous Boundary Conditions

(i) $k\dfrac{\partial^2 u}{\partial x^2} = \dfrac{\partial u}{\partial t}, 0 < x < a, t > 0;$

(ii) $u(0,t) = 0, u(a,t) = 0, t > 0;$ and

(iii) $u(x,0) = f(x), 0 < x < a.$

The solution to this problem (and several others that we will discuss later) is determined by a method called **separation of variables**, because we can assume that the solution is the product of two functions, one that depends on x and one that depends on t. In making this assumption, the partial differential equation is transformed into a pair of ordinary differential equations which can be easily solved in many cases. We now illustrate the application of this useful method through the solution of the heat equation.

If $u(x,t) = X(x)T(t)$, then $k\dfrac{\partial^2 u}{\partial x^2} = \dfrac{\partial u}{\partial t}$ is equivalent to $kX''(x)T(t) = X(x)T'(t)$ so $\dfrac{X''(x)}{X(x)} = \dfrac{T'(t)}{kT(t)}$ for

all values of $0 < x < a$ and $t > 0$. Therefore, each of the functions $\dfrac{X''(x)}{X(x)}$ and $\dfrac{T'(t)}{kT(t)}$ is constant.

Then, there is a constant $-\lambda^2$ so that $\dfrac{X''(x)}{X(x)} = -\lambda^2$ and $\dfrac{T'(t)}{kT(t)} = -\lambda^2$ which yields the two ordinary

differential equations $X'' + \lambda^2 X = 0$ and $T' + \lambda^2 kT = 0$ with solutions $X(x) = a_1 \cos(\lambda x) + a_2 \sin(\lambda x)$

and $T(t) = a_3 e^{-\lambda^2 kt}$, respectively. The boundary conditions $u(0,t) = 0$ and $u(a,t) = 0$ require that $X(0) = 0$

and $X(a) = 0$. If $X(0) = 0$, $a_1 = 0$ and if $X(a) = 0$, $a_2 \sin(\lambda a) = 0$ so either $a_2 = 0$ or $\lambda a = n\pi$ for

$n = \pm 1, \pm 2, \ldots$ and thus $\lambda = \dfrac{n\pi}{a}$.

Therefore, $u(x,t) = \sum_{n=1}^{\infty} \alpha_n \sin(\lambda_n x) e^{-\lambda_n^2 k t}$, where $\lambda_n = \dfrac{n\pi}{a}$. Since

$u(x,0) = \sum_{n=1}^{\infty} \alpha_n \sin(\lambda_n x) = f(x)$ is a Fourier series, $\alpha_n = \dfrac{2}{a} \int_0^a f(x) \sin\left(\dfrac{n\pi x}{a}\right) dx.$

■ EXAMPLE 12.11

Solve the following heat equation with homogeneous boundary conditions:

(i) $\dfrac{\partial^2 u}{\partial x^2} = \dfrac{\partial u}{\partial t}, 0 < x < 1, t > 0;$

(ii) $u(0,t) = 0, u(1,t) = 0, t > 0;$ and

(iii) $u(x,0) = \begin{cases} x, & 0 < x < 1/2 \\ 1 - x, & 1/2 < x < 1 \end{cases}$

Solution:

We begin by defining the necessary parameters a and k as well as the piecewise-defined initial temperature distribution function given as **f1** and **f2** below.

```
▊▭▭▭▭▭▭▭▭▭▭  HeatEquation  ▭▭▭▭▭▭▭▭▭▭▭▊
In[5]:=

Clear[a,k,f,alpha,uapprox]
k=1;
a=1;
f1[x_]:=x/;0<=x<=1/2
f2[x_]:=1-x/;1/2<=x<=1
```

Since the integral used to determine α_n differs on the intervals [0,1/2] and [1/2,1], we must enter the two integrals separately. **Chop** is used to round off very small values to zero.

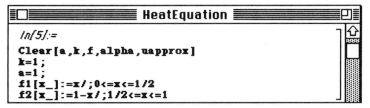

```
In[6]:=

alpha[n_]:=alpha[n]=
    (2/a) (NIntegrate[
    f1[x]* Sin[n Pi x/a],{x,0,1/2}]+
    NIntegrate[
    f2[x]* Sin[n Pi x/a],{x,1/2,1}])//N//Chop;
```

The first five coefficients of the series expansion are determined below. Note that because of the manner in which **alpha** is defined, *Mathematica* remembers these values.

```
In[7]:=
Table[{n,alpha[n]},{n,1,5}]//TableForm

Out[7]//TableForm=
1    0.405285
2    0
3    -0.0450316
4    0
5    0.0162114
```

The **n**th term of the series expansion is defined below in **u**

```
In[8]:=
u[x_,t_,n_]:=alpha[n] Sin[n Pi x/a]*
                Exp[-k t(n Pi/a)^2]
```

so that the approximation of the solution of the heat equation using the first five terms of the expansion is given by **uapprox**.

```
In[9]:=
uapprox[x_,t_]=
    Sum[u[x,t,j],{j,1,5}]

Out[9]=
```

$$\frac{0.405285 \; Sin[Pi \; x]}{E^{Pi^2 \; t}} - \frac{0.0450316 \; Sin[3 \; Pi \; x]}{E^{9 \; Pi^2 \; t}} +$$

$$\frac{0.0162114 \; Sin[5 \; Pi \; x]}{E^{25 \; Pi^2 \; t}}$$

In order to plot this approximate solution to the heat equation, we produce the list of graphics cells below in **graphs**. Several plotting options are used to bring uniformity to the plots. These graphs represent the approximate solution at times t=0 to t=0.35 using increments of .35/8. Since nine graphs are generated, the resulting list is partitioned into groups of three and displayed as a graphics array. Note that the height above the t-axis represents the temperature. Hence, as t increases, the temperature approaches zero. This is to be expected since the ends of the rod are constantly held at zero, so eventually, the entire rod will have this same temperature.

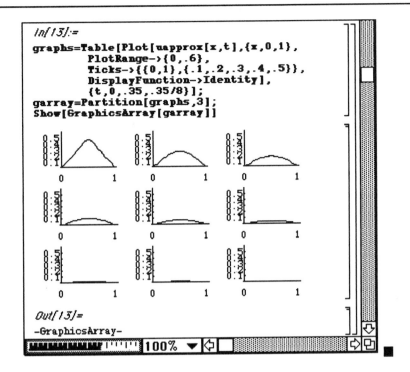

```
In[13]:=
graphs=Table[Plot[uapprox[x,t],{x,0,1},
        PlotRange->{0,.6},
        Ticks->{{0,1},{.1,.2,.3,.4,.5}},
        DisplayFunction->Identity],
        {t,0,.35,.35/8}];
garray=Partition[graphs,3];
Show[GraphicsArray[garray]]
```

```
Out[13]=
-GraphicsArray-
```

The Heat Equation with Nonhomogeneous Boundary Conditions

(i) $k\dfrac{\partial^2 u}{\partial x^2} = \dfrac{\partial u}{\partial t}, 0 < x < a, t > 0;$

(ii) $u(0,t) = T_0, u(a,t) = T_1, t > 0;$ and

(iii) $u(x,0) = f(x), 0 < x < a.$

The method of separation of variables typically fails if one or both of the boundary conditions are nonhomogeneous. Hence, the method must be altered in order to solve problems of this type. In the case of the heat equation, this modification involves the observation that as $t \to \infty$, the temperature $u(x,t)$ approaches a steady - state temperature, $s(x)$, which only depends on x. Therefore, the temperature is made up of the sum of $v(x,t)$, the transient temperature, and $s(x)$. Mathematically, this is represented as $u(x,t) = v(x,t) + s(x)$.

Since $s(x)$ must satisfy the heat equation, we have that $s''(x)=0$ since $\dfrac{\partial s}{\partial t}=0$. We also assume

that the nonhomogeneous boundary conditions are associated with the steady - state solution. We

do this by substituting $x=0$ into the solution:

$u(0,t)=v(0,t)+s(0)=T_0$.

This relationship is satisfied by $v(0,t)=0$ and $s(0)=T_0$. Similarly,

$u(a,t)=v(a,t)+s(a)=T_1$,

which is satisfied by $v(a,t)=0$ and $s(a)=T_1$.

Now, the transient temperature must satisfy the heat equation as well, so

$$k\frac{\partial^2 v}{\partial x^2}=\frac{\partial v}{\partial t}, 0<x<a, t>0,$$

and the initial condition implies that $u(x,0)=v(x,0)+s(x)$. Therefore, we have the initial transient

temperature function $v(x,0)=u(x,0)-s(x)=f(x)-s(x)$.

Therefore, the heat equation with nonhomogeneous boundary conditions is divided into two smaller

problems: (a) the steady-state problem

(i) $s''=0, 0<x<a, t>0$;

(ii) $s(0)=T_0, s(a)=T_1, t>0$

and (b) the transient temperature problem (heat equation with homogeneous boundary conditions)

(i) $k\dfrac{\partial^2 v}{\partial x^2}=\dfrac{\partial v}{\partial t}, 0<x<a, t>0$;

(ii) $v(0,t)=0, v(a,t)=0, t>0$; and

(iii) $v(x,0)=f(x)-s(x), 0<x<a$.

We begin by solving the steady-state problem. Clearly, the solution of the ordinary differential equation
$s''=0$ is $s(x)=Ax+B$, where A and B are arbitrary constants which are found through use of the boundary
conditions:

$s(0)=B=T_0$ and $s(a)=Aa+B=T_1$, so $B=T_0$ and $A=\dfrac{T_1-T_0}{a}$. Therefore, the steady - state

temperature is given by $s(x)=\left(\dfrac{T_1-T_0}{a}\right)x+T_0$.

We now solve the transient temperature problem using the results of the heat equation with homogeneous
boundary conditions. Therefore, the transient temperature is given by

$$v(x,t)=\sum_{n=1}^{\infty}\alpha_n \sin(\lambda_n x)e^{-\lambda_n^2 kt}, \text{ where } \lambda_n=\frac{n\pi}{a}.$$

Applying the initial condition, we have

$$v(x,0) = \sum_{n=1}^{\infty} \alpha_n \sin(\lambda_n x) = f(x) - s(x),$$

the Fourier sine series for the function $f(x) - s(x)$, so $\alpha_n = \dfrac{2}{a}\int_0^a [f(x) - s(x)]\sin\left(\dfrac{n\pi x}{a}\right)dx.$

Therefore, the solution is given by

$$u(x,t) = v(x,t) + s(x) = \left(\frac{T_1 - T_0}{a}\right)x + T_0 + \sum_{n=1}^{\infty} \alpha_n \sin(\lambda_n x)\, e^{-\lambda_n{}^2 kt},$$

where $\lambda_n = \dfrac{n\pi}{a}$ and $\alpha_n = \dfrac{2}{a}\int_0^a [f(x) - s(x)]\sin\left(\dfrac{n\pi x}{a}\right)dx.$

■ EXAMPLE 12.12

Solve the following heat equation with nonhomogeneous boundary conditions:

(i) $\dfrac{\partial^2 u}{\partial x^2} = \dfrac{\partial u}{\partial t}, 0 < x < 1, t > 0;$

(ii) $u(0,t) = 0, u(1,t) = 1/2, t > 0;$ and

(iii) $u(x,0) = x^2,\ 0 < x < 1.$

Solution:

We define the appropriate parameters **a, k, t0,** and **t1** below as well as the steady-state temperature **s** and the initial temperature distribution **f**.

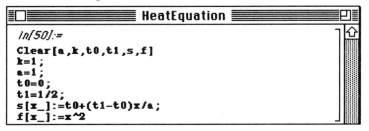

Next, we define the series coefficients **alpha** using **NIntegrate** and **Chop** with the formula derived for the problem with nonhomogeneous boundary conditions above.

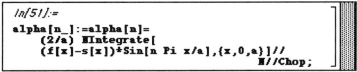

Using **alpha**, we determine the first five coefficients

```
In[52]:=
Table[{n,alpha[n]},{n,1,5}]//TableForm

Out[52]//TableForm=
1    0.405285
2    0
3    -0.0450316
4    0
5    0.0162114
```

and define the **nth** term in transient temperature with these coefficients in **v** below.

```
In[53]:=
v[x_,t_,n_]:=alpha[n] Sin[n Pi x/a]+
             Exp[-k t(n Pi/a)^2]
```

The sum of the first five terms is found in **vapprox** below.

```
In[54]:=
vapprox[x_,t_]=
    Sum[v[x,t,j],{j,1,5}]

Out[54]=
0.405285 Sin[Pi x]   0.0450316 Sin[3 Pi x]
------------------ - ---------------------- +
      Pi^2 t                 9 Pi^2 t
     E                      E
0.0162114 Sin[5 Pi x]
---------------------
     25 Pi^2 t
    E
```

Hence, the approximate solution of the heat equation with nonhomogeneous boundary conditions is the sum of the steady-state solution and the transient solution as defined below in **uapprox**.

```
In[55]:=
uapprox[x_,t_]=s[x]+vapprox[x,t]

Out[55]=
x   0.405285 Sin[Pi x]   0.0450316 Sin[3 Pi x]
- + ------------------ - ---------------------- +
2         Pi^2 t                 9 Pi^2 t
         E                      E
0.0162114 Sin[5 Pi x]
---------------------
     25 Pi^2 t
    E
```

We plot this approximate solution for values of t from t=0 to t=0.35 using increments of 0.35/8 and display these graphs as a graphics array. Notice that as t increases, the solution approaches the steady-state temperature function, the graph of which is a line, as expected. This is because the transient temperature approaches zero as t increases.

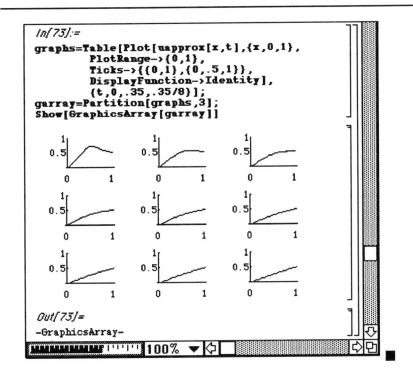

```
In[73]:=
graphs=Table[Plot[uapprox[x,t],{x,0,1},
        PlotRange->{0,1},
        Ticks->{{0,1},{0,.5,1}},
        DisplayFunction->Identity],
        {t,0,.35,.35/8}];
garray=Partition[graphs,3];
Show[GraphicsArray[garray]]
```

```
Out[73]=
-GraphicsArray-
```

§12.4 The One-Dimensional Wave Equation

The one-dimensional wave equation is important in solving a very interesting problem. Suppose that we pluck a string (i.e., a guitar or violin string) of length a and constant mass density that is fixed at each end. The question that we need to answer is: "What is the position of the string at a particular instance of time?". We answer this question by modeling the physical situation with a differential equation, namely the wave equation in one spatial variable. The partial differential equation with necessary initial and boundary conditions are stated below. Notice that the boundary conditions (ii) represent the fixed ends of the string while the initial conditions (iii) represent the initial position and velocity of the string, respectively.

The Wave Equation

(i) $c^2 \dfrac{\partial^2 u}{\partial x^2} = \dfrac{\partial^2 u}{\partial t^2}, 0 < x < a, t > 0;$

(ii) $u(0,t) = 0, u(a,t) = 0, t \geq 0;$ and

(iii) $u(x,0) = f(x), \dfrac{\partial u}{\partial t}\bigg|_{t=0} = g(x), 0 \leq x \leq a.$

We now solve the problem through separation of variables techniques.

If $u(x,t) = X(x) \cdot T(t)$, then (i) becomes $c^2 X''(x)T(t) = X(x)T''(t)$ which results in

$$\frac{X''(x)}{X(x)} = \frac{T''(t)}{c^2 T(t)} = -\lambda^2 \text{ so that } X'' + \lambda^2 X = 0 \text{ and } T'' + \lambda^2 c^2 T = 0. \text{ Thus,}$$

$X(t) = c_1 \cos(\lambda x) + c_2 \sin(\lambda x)$ and $T(t) = c_3 \cos(\lambda ct) + c_4 \sin(\lambda ct)$.

The boundary conditions (ii) yield $X(0) = 0$ and $X(a) = 0$ so that $c_1 = 0$ and $c_2 \sin(\lambda a) = 0$.

Therefore, $\lambda = \dfrac{n\pi}{a}$ for $n = 1, 2, 3, \ldots$ and thus $X(x) = \sin\left(\dfrac{n\pi x}{a}\right)$. Consequently,

$$u_n(x,t) = \left[\alpha_n \sin\left(\frac{n\pi ct}{a}\right) + \beta_n \cos\left(\frac{n\pi ct}{a}\right)\right] \sin\left(\frac{n\pi x}{a}\right) \text{ and}$$

$$u(x,t) = \sum_{n=1}^{\infty} u_n(x,t) = \sum_{n=1}^{\infty}\left[\alpha_n \sin\left(\frac{n\pi ct}{a}\right) + \beta_n \cos\left(\frac{n\pi ct}{a}\right)\right] \sin\left(\frac{n\pi x}{a}\right). \text{ By (iii) } u(x,0) = f(x) \text{ and}$$

$$\text{thus } u(x,0) = \sum_{n=1}^{\infty} \beta_n \sin\left(\frac{n\pi x}{a}\right) = f(x) \text{ so } \beta_n = \frac{2}{a}\int_0^a f(x)\sin\left(\frac{n\pi x}{a}\right)dx.$$

$$\text{Computing } \frac{\partial u}{\partial t} = \sum_{n=1}^{\infty}\left[\alpha_n \frac{n\pi c}{a}\cos\left(\frac{n\pi ct}{a}\right) - \beta_n \frac{n\pi c}{a}\sin\left(\frac{n\pi ct}{a}\right)\right]\sin\left(\frac{n\pi x}{a}\right) \text{ and applying (iii),}$$

$$\left.\frac{\partial u}{\partial t}\right|_{t=0} = \sum_{n=1}^{\infty} \alpha_n \frac{n\pi c}{a}\sin\left(\frac{n\pi x}{a}\right) = g(x) \text{ so } \alpha_n \frac{n\pi c}{a} = \frac{2}{a}\int_0^a g(x)\sin\left(\frac{n\pi x}{a}\right)dx \text{ and thus}$$

$$\alpha_n = \frac{2}{n\pi c}\int_0^a g(x)\sin\left(\frac{n\pi x}{a}\right)dx.$$

■ **EXAMPLE 12.13**

Solve the wave equation with c=1 and a=1 subject to the indicated initial conditions:

(i) $\dfrac{\partial^2 u}{\partial x^2} = \dfrac{\partial^2 u}{\partial t^2}, 0 < x < 1, t > 0;$

(ii) $u(0,t) = 0, u(a,t) = 0, t \geq 0;$ and

(iii) $u(x,0) = \sin(\pi x), \left.\dfrac{\partial u}{\partial t}\right|_{t=0} = 3x + 1, 0 \leq x \leq 1.$

Solution:

The appropriate parameters and initial condition functions are entered below.

```
╔═══════════════════════════════════════════════════╗
║▤□▇▇▇▇▇▇▇▇▇▇▇ TheWaveEquation ▇▇▇▇▇▇▇▇▇▇□▤║
╟───────────────────────────────────────────────────╢
║ Clear[alpha,beta]                              ] ⬆ ║
║                                                    ║
║ In[4]:=                                            ║
║ a=1;                                               ║
║ c=1;                                               ║
║ f[x_]=Sin[Pi x];                                   ║
║ g[x_]=3x+1;                                        ║
╚════════════════════════════════════════════════════╝
```

Next, the functions to determine the coefficients α_n and β_n in the series approximation of the solution $u(x,t)$ are defined in **alpha** and **beta** below. The use of **NIntegrate** in these functions causes the calculations to be performed more quickly in most cases.

```
╔═══════════════════════════════════════════════════╗
║ In[6]:=                                            ║
║                                                    ║
║ alpha[n_]:=alpha[n]=                               ║
║     2/(n Pi c)*NIntegrate[                         ║
║          g[x]Sin[n Pi x/a],{x,0,a}]//N//Chop;      ║
║ beta[n_]:=beta[n]=                                 ║
║     2/a*NIntegrate[f[x]Sin[n Pi x/a],              ║
║          {x,0,a}]//Chop;                           ║
╚═══════════════════════════════════════════════════╝
```

A table of the first five α_n and β_n is found below.

```
╔═══════════════════════════════════════════════════╗
║ Table[{n,alpha[n],beta[n]},                      ] ║
║            {n,1,5}]//TableForm                     ║
║                                                    ║
║ Out[7]//TableForm=                                 ║
║ 1    1.01321       1.0000000000000000001           ║
║ 2   -0.151982      0                                ║
║ 3    0.112579      0                                ║
║ 4   -0.0379954     0                                ║
║ 5    0.0405285     0                                ║
╚═══════════════════════════════════════════════════╝
```

The function **u** defined below represents the nth term in the series expansion.

```
╔═══════════════════════════════════════════════════╗
║ In[9]:=                                            ║
║                                                    ║
║ u[x_,t_,n_]:=                                      ║
║     (alpha[n]Sin[n Pi c t/a]+beta[n]              ║
║     Cos[n Pi c t/a])Sin[n Pi x/a];                 ║
╚═══════════════════════════════════════════════════╝
```

Hence, **uapprox** determines the approximation of order five by summing the first five terms of the expansion.

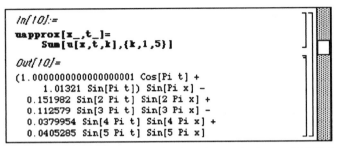

```
In[10]:=
uapprox[x_,t_]=
    Sum[u[x,t,k],{k,1,5}]

Out[10]=
(1.0000000000000000001 Cos[Pi t] +
    1.01321 Sin[Pi t]) Sin[Pi x] -
  0.151982 Sin[2 Pi t] Sin[2 Pi x] +
  0.112579 Sin[3 Pi t] Sin[3 Pi x] -
  0.0379954 Sin[4 Pi t] Sin[4 Pi x] +
  0.0405285 Sin[5 Pi t] Sin[5 Pi x]
```

In **graphs** below, **uapprox** is plotted over the interval [0,2] (in x) for values of t from t=0 to t=2 using increments of 2/19. Hence, this produces a list of 20 plots which is partitioned into groups of four in **garray** and viewed as a graphics array.

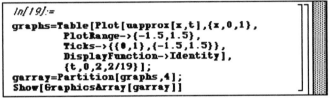

```
In[19]:=
graphs=Table[Plot[uapprox[x,t],{x,0,1},
        PlotRange->{-1.5,1.5},
        Ticks->{{0,1},{-1.5,1.5}},
        DisplayFunction->Identity],
        {t,0,2,2/19}];
garray=Partition[graphs,4];
Show[GraphicsArray[garray]]
```

These plots can be displayed via a **Do** command so that the resulting graphs may be animated to see the motion of the string.

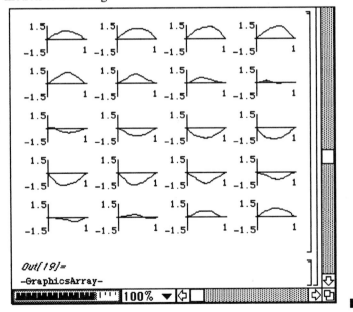

```
Out[19]=
-GraphicsArray-
```

§12.5 Laplace's Equation

Laplace's equation, often called the potential equation, is one of the most useful partial differential equations in that it arises in many fields of study. These include fluid flows as well as electrostatic and gravitational potential. Because the potential does not depend on time, no initial condition is required. Hence, we are faced with solving a pure boundary value problem when working with Laplace's equation. The boundary conditions can be stated in different forms. If the value of the solution is given around the boundary of the region, then the boundary value problem is called a Dirichlet problem whereas if the normal derivative of the solution is given around the boundary, the problem is known as a Neumann problem. We now investigate the solutions to Laplace's equation in a rectangular region by, first, stating and solving the general form of the Dirichlet problem:

Laplace's Equation (The Dirichlet Problem)

(i) $\dfrac{\partial^2 u}{\partial x^2} + \dfrac{\partial^2 u}{\partial y^2} = 0, 0 < x < a, 0 < y < b;$

(ii) $u(x,0) = f_1(x)$ and $u(x,b) = f_2(x), 0 < x < a;$ and

(iii) $u(0,y) = g_1(y)$ and $u(a,y) = g_2(y), 0 < y < b.$

This problem is solved through separation of variables as well:

If $u_1(x,y)$ satisfies the three conditions $\begin{cases} \text{(ia)} \dfrac{\partial^2 u_1}{\partial x^2} + \dfrac{\partial^2 u_1}{\partial y^2} = 0, 0 < x < a, 0 < y < b \\ \text{(iia)}\, u_1(x,0) = f_1(x) \text{ and } u_1(x,b) = f_2(x), 0 < x < a \\ \text{(iiia)}\ u(0,y) = 0 \text{ and } u(a,y) = 0, 0 < y < b \end{cases}$

and $u_2(x,y)$ satisfies the three conditions

$\begin{cases} \text{(ib)} \dfrac{\partial^2 u_2}{\partial x^2} + \dfrac{\partial^2 u_2}{\partial y^2} = 0, 0 < x < a, 0 < y < b \\ \text{(iib)}\ u_2(x,0) = 0 \text{ and } u_2(x,b) = 0, 0 < x < a \qquad \text{then} \\ \text{(iiib)}\ u_2(0,y) = g_1(y) \text{ and } u_2(a,y) = g_2(y), 0 < y < b \end{cases}$

$u(x,y) = u_1(x,y) + u_2(x,y)$ satisfies (i) – (iii).

If $u_1(x,y) = X_1(x)Y_1(y)$, then (ia) and (iiia) become $X_1''(x)Y_1(y) + X_1(x)Y_1''(y) = 0$ and $X_1(0) = 0$ and $X_1(a) = 0$, respectively. Dividing $X_1''(x)Y_1(y) + X_1(x)Y_1''(y) = 0$ by $X_1(x)Y_1(y)$

results in $\dfrac{X_1''(x)}{X_1(x)} = -\dfrac{Y_1''(y)}{Y_1(y)} = -\lambda^2$ which yields the two ordinary differential equations

$X_1'' + \lambda^2 X_1 = 0$ and $Y_1'' - \lambda^2 Y_1 = 0.$ The problem $\begin{cases} X_1'' + \lambda^2 X_1 = 0 \\ X_1(0) = 0 \text{ and } X_1(a) = 0 \end{cases}$ is an eigenvalue problem

with eigenvalues $\lambda_n = \pm \dfrac{n\pi}{a}$ and corresponding eigenfunctions $X_{1n}(x) = \sin(\lambda_n x).$ The corresponding

solutions of $Y_1'' - \lambda^2 Y_1 = 0$ are $Y_{1n} = \alpha_n \cosh(\lambda_n y) + \beta_n \sinh(\lambda_n y).$ Therefore,

$$u_1(x,y) = \sum_{n=1}^{\infty} \left[\alpha_n \cosh(\lambda_n y) + \beta_n \sinh(\lambda_n y) \right] \sin(\lambda_n x). \text{ Since}$$

$$u_1(x,0) = \sum_{n=1}^{\infty} \alpha_n \sin(\lambda_n x) = f_1(x), \alpha_n = \frac{2}{a}\int_0^a f_1(x)\sin(\lambda_n x)\, dx \text{ and since}$$

$$u_1(x,b) = \sum_{n=1}^{\infty} \left[\alpha_n \cosh(\lambda_n b) + \beta_n \sinh(\lambda_n b) \right] \sin(\lambda_n x) = f_2(x),$$

$\alpha_n \cosh(\lambda_n b) + \beta_n \sinh(\lambda_n b) = \dfrac{2}{a}\int_0^a f_2(x)\sin(\lambda_n x)\, dx$ so

$\beta_n = \dfrac{1}{\sinh(\lambda_n b)}\left[\dfrac{2}{a}\int_0^a f_2(x)\sin(\lambda_n x)\, dx - \alpha_n \cosh(\lambda_n b) \right].$

If $u_2(x,y) = X_2(x)Y_2(y)$, then (ib) and (iib) become $X_2''(x)Y_2(y) + X_2(x)Y_2''(y) = 0$ and
$Y_2(0) = 0$ and $Y_2(b) = 0$, respectively. Dividing $X_2''(x)Y_2(y) + X_2(x)Y_2''(y) = 0$ by $X_2(x)Y_2(y)$

results in $-\dfrac{X_2''(x)}{X_2(x)} = \dfrac{Y_2''(y)}{Y_2(y)} = -\mu^2$ which yields the two ordinary differential equations

$X_2'' - \mu^2 X_2 = 0$ and $Y_2'' + \mu^2 Y_2 = 0.$ The problem $\begin{cases} Y_2'' + \mu^2 Y_2 = 0 \\ Y_2(0) = 0 \text{ and } Y_2(b) = 0 \end{cases}$ is an eigenvalue problem

with eigenvalues $\mu_n = \pm \dfrac{n\pi}{b}$ and corresponding eigenfunctions $Y_{2n}(y) = \sin(\mu_n y).$ The corresponding

solutions of $X_2'' - \mu^2 X_2 = 0$ are $X_{2n} = A_n \cosh(\mu_n x) + B_n \sinh(\mu_n x).$ Therefore,

$$u_2(x,y) = \sum_{n=1}^{\infty} \left[A_n \cosh(\mu_n x) + B_n \sinh(\mu_n x) \right] \sin(\mu_n y). \text{ Since}$$

$$u_2(0,y) = \sum_{n=1}^{\infty} A_n \sin(\mu_n y) = g_1(y), A_n = \frac{2}{b}\int_0^b g_1(y)\sin(\mu_n y)\, dy \text{ and since}$$

$$u_2(a,y) = \sum_{n=1}^{\infty}\left[A_n \cosh(\mu_n a) + B_n \sinh(\mu_n a)\right]\sin(\mu_n y) = g_2(y),$$

$$A_n \cosh(\mu_n a) + B_n \sinh(\mu_n a) = \frac{2}{b}\int_0^b g_2(y)\sin(\mu_n y)\, dy \text{ so}$$

$$B_n = \frac{1}{\sinh(\mu_n a)}\left[\frac{2}{b}\int_0^b g_2(y)\sin(\mu_n y)\, dy - A_n \cosh(\mu_n a)\right].$$

■ EXAMPLE 12.14

Solve Laplace's equation over the indicated rectangular region using the indicated boundary conditions.

(i) $\dfrac{\partial^2 u}{\partial x^2} + \dfrac{\partial^2 u}{\partial y^2} = 0, 0 < x < 1, 0 < y < 2;$

(ii) $u(x,0) = x$ and $u(x,2) = 2 + x, 0 < x < 1;$ and

(iii) $u(0,y) = y$ and $u(1,y) = 1 + y, 0 < y < 2.$

Solution:

We begin by defining the parameters a and b as well as the boundary condition functions.

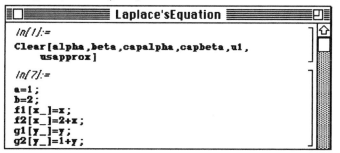

```
Laplace'sEquation
In[1]:=
Clear[alpha,beta,capalpha,capbeta,u1,
    usapprox]
In[7]:=
a=1;
b=2;
f1[x_]=x;
f2[x_]=2+x;
g1[y_]=y;
g2[y_]=1+y;
```

The functions which are needed to determine the series coefficients are defined in **alpha** and **beta**. Notice the manner in which these definitions are made. This causes *Mathematica* to remember the values without having to recalculate them. The function **lambda** is also defined which yields a numerical value for each value of n.

```
In[10]:=
lambda[n_]=n Pi/a//N;
alpha[n_]:=alpha[n]=
    2/a*NIntegrate[f1[x]Sin[lambda[n] x],
                  {x,0,a}]//N//Chop;
beta[n_]:=beta[n]=
    1/Sinh[lambda[n] b]*
    (2/a*NIntegrate[f2[x]Sin[lambda[n] x],
                  {x,0,a}]-
    alpha[n]Cosh[lambda[n] b])//N//Chop;
```

A table of the first 8 values of each coefficient is then determined.

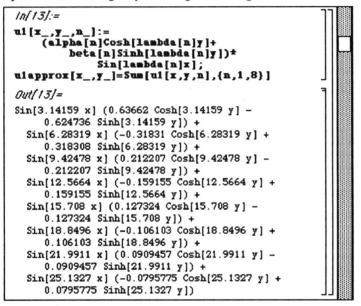

```
In[11]:=
Table[{n,alpha[n],beta[n]},
            {n,1,8}]//TableForm

Out[11]//TableForm=
1    0.63662      -0.624736
2   -0.31831       0.318308
3    0.212207     -0.212207
4   -0.159155      0.159155
5    0.127324     -0.127324
6   -0.106103      0.106103
7    0.0909457    -0.0909457
8   -0.0795775     0.0795775
```

The expression for **u1** represents the nth term of the series expansion, so **u1approx** yields the expansion of order eight by summing the first eight terms.

```
In[13]:=
u1[x_,y_,n_]:=
    (alpha[n]Cosh[lambda[n]y]+
        beta[n]Sinh[lambda[n]y])*
            Sin[lambda[n]x];
u1approx[x_,y_]=Sum[u1[x,y,n],{n,1,8}]

Out[13]=
Sin[3.14159 x] (0.63662 Cosh[3.14159 y] -
    0.624736 Sinh[3.14159 y]) +
  Sin[6.28319 x] (-0.31831 Cosh[6.28319 y] +
    0.318308 Sinh[6.28319 y]) +
  Sin[9.42478 x] (0.212207 Cosh[9.42478 y] -
    0.212207 Sinh[9.42478 y]) +
  Sin[12.5664 x] (-0.159155 Cosh[12.5664 y] +
    0.159155 Sinh[12.5664 y]) +
  Sin[15.708 x] (0.127324 Cosh[15.708 y] -
    0.127324 Sinh[15.708 y]) +
  Sin[18.8496 x] (-0.106103 Cosh[18.8496 y] +
    0.106103 Sinh[18.8496 y]) +
  Sin[21.9911 x] (0.0909457 Cosh[21.9911 y] -
    0.0909457 Sinh[21.9911 y]) +
  Sin[25.1327 x] (-0.0795775 Cosh[25.1327 y] +
    0.0795775 Sinh[25.1327 y])
```

Since this is a function of the two spatial variables, x and y, we use **Plot3D** to plot the solution. We also view the contour plot of this function. The lists of values found in **vals1** and **vals2** are joined to form the list **contourvals**.

```
In[14]:=
threed=Plot3D[ulapprox[x,y],{x,0,1},{y,0,2},
              DisplayFunction->Identity];
In[17]:=
vals1=Table[i,{i,0,.25,.25/11}];
vals2=Table[i,{i,.25,1.5,1.25/10}];
contourvals=Union[vals1,vals2];
```

The list of values in **contourvals** is used to plot the associated contours with the **ContourPlot** option **Contours**. Both plots are then viewed as a graphics array. The contour plot shows how the function increases and decreases over the rectangular region which is sometimes difficult to see on the three-dimensional plot.

```
In[19]:=
contourgraphs=ContourPlot[ulapprox[x,y],
    {x,0,1},{y,0,2},PlotPoints->30,
    Contours->contourvals,
    ContourShading->False,
    DisplayFunction->Identity];
Show[GraphicsArray[{threed,contourgraphs}]]
```

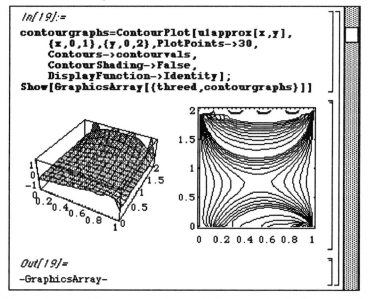

```
Out[19]=
-GraphicsArray-
```

Next, the coefficients A_n and B_n of the second solution $u_2(x,y)$ are found through the use of the functions **capalpha** and **capbeta** below. The function **mu** is also defined which yields a numerical approximation of μ_n for each value of n.

```
In[20]:=
mu[n_]=n Pi/b//N;

In[22]:=
capalpha[n_]:=capalpha[n]=
    2/b*NIntegrate[g1[y]Sin[mu[n] y],
                        {y,0,b}]//N//Chop;
capbeta[n_]:=capbeta[n]=
    1/Sinh[mu[n] a]*
    (2/b*NIntegrate[g2[y]Sin[mu[n] y],
                        {y,0,b}]-
    capalpha[n]Cosh[mu[n] a])//N//Chop;
```

The first twelve values of A_n and B_n are found in the table below.

```
In[23]:=
Table[{n,capalpha[n],capbeta[n]},
            {n,1,12}]//TableForm

Out[23]//TableForm=
1     1.27324      -0.281713
2    -0.63662       0.583877
3     0.424413     -0.40923
4    -0.31831       0.317123
5     0.254648     -0.254253
6    -0.212207      0.212172
7     0.181891     -0.181879
8    -0.159155      0.159154
9     0.141471     -0.141471
10   -0.127324      0.127324
11    0.115749     -0.115749
12   -0.106103      0.106103
```

Therefore, **u2** defined below represents the nth term of the series expansion of $u_2(x,y)$ and **u2approx** gives the sum of the first twelve terms (the expansion of order twelve).

```
In[25]:=
u2[x_,y_,n_]:=
    (capalpha[n]Cosh[mu[n]x]+
        capbeta[n]Sinh[mu[n]x])*
            Sin[mu[n]y];
u2approx[x_,y_]=Sum[u2[x,y,n],{n,1,12}]
```

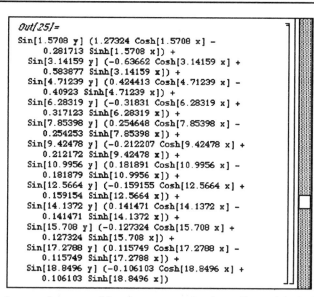

```
Out[25]=
Sin[1.5708 y] (1.27324 Cosh[1.5708 x] -
    0.281713 Sinh[1.5708 x]) +
 Sin[3.14159 y] (-0.63662 Cosh[3.14159 x] +
    0.583877 Sinh[3.14159 x]) +
 Sin[4.71239 y] (0.424413 Cosh[4.71239 x] -
    0.40923 Sinh[4.71239 x]) +
 Sin[6.28319 y] (-0.31831 Cosh[6.28319 x] +
    0.317123 Sinh[6.28319 x]) +
 Sin[7.85398 y] (0.254648 Cosh[7.85398 x] -
    0.254253 Sinh[7.85398 x]) +
 Sin[9.42478 y] (-0.212207 Cosh[9.42478 x] +
    0.212172 Sinh[9.42478 x]) +
 Sin[10.9956 y] (0.181891 Cosh[10.9956 x] -
    0.181879 Sinh[10.9956 x]) +
 Sin[12.5664 y] (-0.159155 Cosh[12.5664 x] +
    0.159154 Sinh[12.5664 x]) +
 Sin[14.1372 y] (0.141471 Cosh[14.1372 x] -
    0.141471 Sinh[14.1372 x]) +
 Sin[15.708 y] (-0.127324 Cosh[15.708 x] +
    0.127324 Sinh[15.708 x]) +
 Sin[17.2788 y] (0.115749 Cosh[17.2788 x] -
    0.115749 Sinh[17.2788 x]) +
 Sin[18.8496 y] (-0.106103 Cosh[18.8496 x] +
    0.106103 Sinh[18.8496 x])
```

As was the case with **u1approx**, the three-dimensional plot of **u2approx** is produced with **Plot3D** as is the contour plot with **ContourPlot**. These two graphs are viewed as a graphics array below. Notice that in this case, the 25 function values used to generate the contour plot are selected automatically by *Mathematica*.

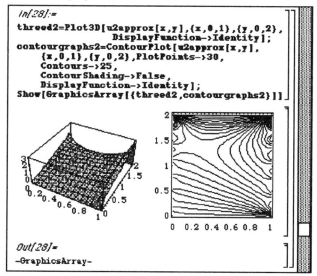

```
In[28]:=
threed2=Plot3D[u2approx[x,y],{x,0,1},{y,0,2},
                DisplayFunction->Identity];
contourgraphs2=ContourPlot[u2approx[x,y],
    {x,0,1},{y,0,2},PlotPoints->30,
    Contours->25,
    ContourShading->False,
    DisplayFunction->Identity];
Show[GraphicsArray[{threed2,contourgraphs2}]]
```

```
Out[28]=
-GraphicsArray-
```

The solution u(x,y) is the sum of $u_1(x,y)$ and $u_2(x,y)$. Thus, **uapprox** represents the sum of these two solutions.

```
In[29]:=
uapprox[x_,y_]=
    u1approx[x,y]+u2approx[x,y];
```

The solution is plotted with **Plot3D** and **ContourPlot**. The contour plot reveals the '"ripples"' which appear to cross the rectangular region in the three-dimensional plot.

```
In[32]:=
threedu=Plot3D[uapprox[x,y],{x,0,1},{y,0,2},
                DisplayFunction->Identity];
contourgraphsu=ContourPlot[uapprox[x,y],
    {x,0,1},{y,0,2},PlotPoints->40,
    Contours->25,
    ContourShading->False,
    DisplayFunction->Identity];
Show[GraphicsArray[{threedu,contourgraphsu}]]
```

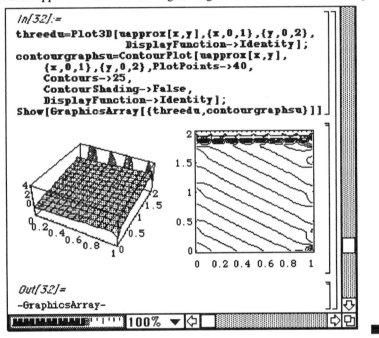

```
Out[32]=
-GraphicsArray-
```

100% ▼

§12.6 The Two-Dimensional Wave Equation in a Circular Region

One of the more interesting problems involving two spatial dimensions (x and y) is the wave equation. This is due to the fact that the solution to this problem represents something that we are all familiar with a drumhead. Since most drumheads are in the shape of a circle, we will investigate the solution of the wave equation in a circular region. In solving this problem as well as numerous problems in applied mathematics, polar or cylindrical coordinate systems are often convenient to use. The two-dimensional wave equation in a circular region which is radially symmetric (no dependence on θ) with boundary and initial conditions is easily expressed in polar coordinates as follows:

(i) $\quad \dfrac{\partial^2 u}{\partial t^2} = c^2 \left(\dfrac{\partial^2 u}{\partial r^2} + \dfrac{1}{r} \dfrac{\partial u}{\partial r} \right) \quad 0 < r < R,\ 0 < t;$

(ii) $\quad u(R,t) = 0 \quad 0 < t;$

(iii) $\quad \big| u(0,t) \big| \quad \text{bounded}, \quad 0 < t;$

(iv) $\quad u(r,0) = f(r) \quad 0 < r < R;$ and

(v) $\quad \dfrac{\partial u}{\partial t}(r,0) = g(r) \quad 0 < r < R.$

If $u(r,t) = G(t)W(r)$, then (i) and (ii) become $G''W = c^2\left(GW'' + \dfrac{1}{r}GW'\right)$ and $W(R) = 0.$ Dividing

by WG results in $\dfrac{G''}{c^2 G} = \dfrac{W''}{W} + \dfrac{W'}{rW} = -k^2$, where $-k^2$ is the constant of separation, which yields

the two ordinary differential equations $\dfrac{d^2 G}{dt^2} + \lambda^2 G = 0$, where $\lambda = ck$; and $\dfrac{d^2 W}{dr^2} + \dfrac{1}{r}\dfrac{dW}{dr} + k^2 W = 0$

$\dfrac{d^2 W}{dr^2} + \dfrac{1}{r}\dfrac{dW}{dr} + k^2 W = 0$ is Bessel's equation of order zero and, thus, has general solution of the form

$W(r) = c_1 J_0(kr) + c_2 Y_0(kr)$ where $J_0(kr)$ and $Y_0(kr)$ are Bessel functions of order zero of the first

and second kind, respectively. By (iii) $W(0)$ is bounded and since $Y_0(kr)$ is unbounded, $c_2 = 0.$

Applying the condition (ii) $W(R) = 0$ leads to the equation $J_0(kr) = 0.$ Hence, $k_m = \dfrac{\alpha_m}{R}$ where

α_m denotes the mth zero of $J_0(x)$ and $m = 1,2,3,\dots$ so that $W_m(r) = J_0(k_m r).$

The corresponding solutions of $\dfrac{d^2 G}{dt^2} + \lambda^2 G = 0$ are $G_m(t) = a_m \cos(\lambda_m t) + b_m \sin(\lambda_m t)$ where

$\lambda_m = ck_m = \dfrac{c}{R}\alpha_m.$ Therefore, the functions

$u_m(r,t) = W_m(t)G_m(t) = \left(a_m \cos(\lambda_m t) + b_m \sin(\lambda_m t)\right)J_0(k_m r)$ are solutions of (i) satisfying

the boundary conditions (ii) and (iii). The general solution of the problem (i) – (v) is of the form

$$u(r,t) = \sum_{m=1}^{\infty} u_m(r,t) = \sum_{m=1}^{\infty}\left(a_m \cos(\lambda_m t) + b_m \sin(\lambda_m t)\right)J_0(k_m r).$$

By (iv), $u(r,0) = \displaystyle\sum_{m=1}^{\infty} a_m J_0(k_m r) = f(r)$ so that $a_m = \dfrac{2}{R^2\left(J_1(\alpha_m)\right)^2}\displaystyle\int_0^R rf(r)J_0(k_m r)\,dr.$

Applying (v) results in $\dfrac{\partial u}{\partial t}(r,0) = \displaystyle\sum_{m=1}^{\infty}\lambda_m b_m J_0(k_m r) = g(r)$ so that

$$\lambda_m \, b_m = \frac{2}{R^2 \left(J_1(\alpha_m) \right)^2} \int_0^R r \, g(r) J_0(k_m \, r) \, dr \text{ and consequently}$$

$$b_m = \frac{2}{c \alpha_m R \left(J_1(\alpha_m) \right)^2} \int_0^R r \, g(r) J_0(k_m \, r) \, dr.$$

As a practical matter, in nearly all cases these formulas are difficult to evaluate. For a limited number of functions, integration by parts using $\dfrac{d}{dr}\left(r^\nu J_\nu(r) \right) = r^{\nu+1} J_\nu(r)$ is possible. However, even when possible, the resulting calculations are lengthy and tedious.

■ EXAMPLE 12.15

Solve the wave equation on the circular region with boundary R=1 and parameter c=1 with initial position f(r)=r(r−1) and initial velocity g(r)=sin(πr).

Solution:

We, first, read in the table of zeros of the Bessel functions which was compiled earlier in **besseltable** and call this table **getzeros**. We then define the function **alpha[i,j]** which represents the **j**th zero of the Bessel function of order **i**.

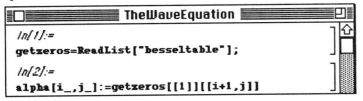

```
In[1]:=
getzeros=ReadList["besseltable"];

In[2]:=
alpha[i_,j_]:=getzeros[[1]][[i+1,j]]
```

Next, we enter the appropriate radius in **capr**, parameter **c**, and initial position and velocity functions, **f** and **g**.

```
In[7]:=
Clear[a,b,k,lambda,um,uapprox,capr,c,f,g]
capr=1;
c=1;
f[r_]=r(r-1);
g[r_]=Sin[Pi r];
```

We define the formula for k_m, in this case, below. This function uses the zeros of the Bessel function of order zero only. The eigenvalue λ_m, which depends on **k**, is then defined in **lambda**. Using the formulas for the coefficients a_m and b_m which were derived above are then defined in **a** and **b** so that an approximate solution may be determined. Note that we use **NIntegrate** in order to avoid the difficulties in integration associated with the presence of the Bessel function of order zero.

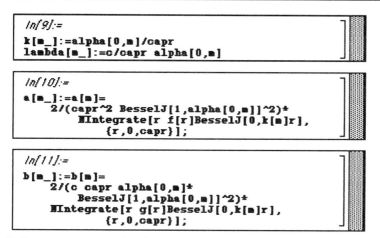

```
In[9]:=
k[m_]:=alpha[0,m]/capr
lambda[m_]:=c/capr alpha[0,m]
```

```
In[10]:=
a[m_]:=a[m]=
    2/(capr^2 BesselJ[1,alpha[0,m]]^2)±
        NIntegrate[r f[r]BesselJ[0,k[m]r],
            {r,0,capr}];
```

```
In[11]:=
b[m_]:=b[m]=
    2/(c capr alpha[0,m]±
        BesselJ[1,alpha[0,m]]^2)±
    NIntegrate[r g[r]BesselJ[0,k[m]r],
            {r,0,capr}];
```

Below, we compute the first six values of a_m and b_m. Because of the manner in which **a** and **b** are defined, *Mathematica* remembers these values.

```
In[12]:=
Table[{m,a[m],b[m]},{m,1,6}]//TableForm

Out[12]//TableForm=
1   -0.323503     0.52118
2    0.208466    -0.145776
3    0.00763767  -0.0134216
4    0.0383536   -0.00832269
5    0.00534454  -0.00250503
6    0.0150378   -0.00208315
```

The **m**th term of the series solution is defined in u below.

```
In[13]:=
u[m_,r_,t_]:=(a[m]Cos[lambda[m] t]+
    b[m]Sin[lambda[m] t])±
        BesselJ[0,k[m] r]
```

Hence, an approximate solution is obtained in **uapprox** by summing the first six terms of **u** given above. The formula of this approximation is printed as output to show that it is of the correct form.

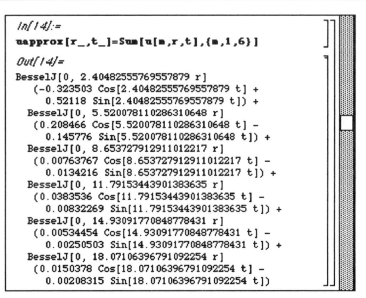

```
In[14]:=
uapprox[r_,t_]=Sum[u[n,r,t],{n,1,6}]

Out[14]=
BesselJ[0, 2.40482555769557879 r]
   (-0.323503 Cos[2.40482555769557879 t] +
    0.52118 Sin[2.40482555769557879 t]) +
BesselJ[0, 5.520078110286310648 r]
   (0.208466 Cos[5.520078110286310648 t] -
    0.145776 Sin[5.520078110286310648 t]) +
BesselJ[0, 8.653727912911012217 r]
   (0.00763767 Cos[8.653727912911012217 t] -
    0.0134216 Sin[8.653727912911012217 t]) +
BesselJ[0, 11.79153443901383635 r]
   (0.0383536 Cos[11.79153443901383635 t] -
    0.00832269 Sin[11.79153443901383635 t]) +
BesselJ[0, 14.93091770848778431 r]
   (0.00534454 Cos[14.93091770848778431 t] -
    0.00250503 Sin[14.93091770848778431 t]) +
BesselJ[0, 18.07106396791092254 r]
   (0.0150378 Cos[18.07106396791092254 t] -
    0.00208315 Sin[18.07106396791092254 t])
```

Without the assistance of the computer, we would be forced to stop at this point. Fortunately, we can use *Mathematica* to produce the graphics associated with this function. Since this function is independent of the angular coordinate θ, we can plot this function over the interval [0,1] to yield a side view of half of the drumhead. This is accomplished in **graphs** below by plotting **uapprox** for values of t from t=0 to t=2 using increments of 2/15. The list of graphics which results is then partitioned into groups of four and displayed as a graphics array. A similar list of graphics can be generated with a **Do** command so that the resulting list may be animated to show the motion of the drumhead. Additionally, **ParametricPlot3D** (located in the **Graphics** directory) may be employed using polar coordinates as the x and y coordinates and **uapprox** as the z coordinate to produce a three-dimensional view of the drumhead. However, the two-dimensional plots below are much easier to produce and since the drum is radially symmetric, the general shape of the drumhead in three dimensions can be obtained from them.

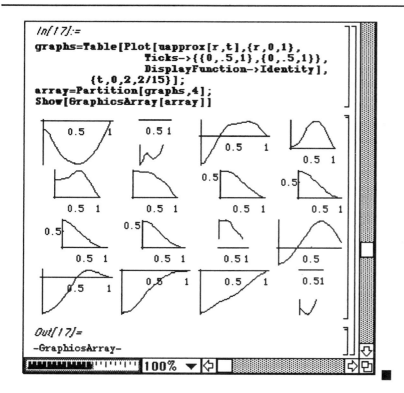

```
In[17]:=
graphs=Table[Plot[uapprox[r,t],{r,0,1},
                 Ticks->{{0,.5,1},{0,.5,1}},
                 DisplayFunction->Identity],
          {t,0,2,2/15}];
array=Partition[graphs,4];
Show[GraphicsArray[array]]
```

```
Out[17]=
-GraphicsArray-
```

■ EXAMPLE 12.16

The problem that describes the displacement of a circular membrane in its general case is:

(i) $\dfrac{1}{r}\dfrac{\partial}{\partial r}\left(r\dfrac{\partial u}{\partial r}\right)+\dfrac{1}{r^2}\dfrac{\partial^2 u}{\partial\theta^2}=\dfrac{1}{c^2}\dfrac{\partial^2 u}{\partial t^2}$, $0<r<a$, $-\pi<\theta\le\pi$;

(ii) $u(a,\theta,t)=0$, $0<t$, $-\pi<\theta\le\pi$;

(iii) $|u(0,\theta,t)|$ bounded, $0<t$, $-\pi<\theta\le\pi$;

(iv) $u(r,-\pi,t)=u(r,\pi,t)$, $0<r<a$, $0<t$;

(v) $\dfrac{\partial u}{\partial\theta}(r,-\pi,t)=\dfrac{\partial u}{\partial\theta}(r,\pi,t)$, $0<r<a$, $0<t$;

(vi) $u(r,\theta,0)=f(r,\theta)$, $0<r<a$, $-\pi<\theta\le\pi$; and

(vii) $\dfrac{\partial u}{\partial t}(r,\theta,0)=g(r,\theta)$, $0<r<a$, $-\pi<\theta\le\pi$.

If $u(r,\theta,t) = R(r)\Theta(\theta)T(t)$ then (i) become $R''\Theta T + \dfrac{1}{r}R'\Theta T + \dfrac{1}{r^2}R\Theta''T = \dfrac{1}{c^2}R\Theta T''$. Dividing by

$R\Theta T$ yields $\dfrac{R''}{R} + \dfrac{1}{r}\dfrac{R'}{R} + \dfrac{1}{r^2}\dfrac{\Theta''}{\Theta} = \dfrac{1}{c^2}\dfrac{T''}{T} = -\lambda^2$, where $-\lambda^2$ is the constant of separation. We then

obtain the two equations $T'' + \lambda^2 c^2 T = 0$ and $\dfrac{R''}{R} + \dfrac{1}{r}\dfrac{R'}{R} + \dfrac{1}{r^2}\dfrac{\Theta''}{\Theta} = -\lambda^2$. Multiplying

$\dfrac{R''}{R} + \dfrac{1}{r}\dfrac{R'}{R} + \dfrac{1}{r^2}\dfrac{\Theta''}{\Theta} = -\lambda^2$ by r^2 and separating results in $r^2\dfrac{R''}{R} + r\dfrac{R'}{R} + r^2\lambda^2 = \dfrac{-\Theta''}{\Theta} = \mu^2$, where

μ^2 is the constant of separation, which yields the two equations $\Theta'' + \mu^2\Theta = 0$ and

$r^2 R'' + rR' + \left(r^2\lambda^2 - \mu^2\right)R = 0.$

The boundary conditions (ii) – (v) become $R(a) = 0, |R(0)|$ bounded, $\Theta(-\pi) = \Theta(\pi)$, and $\Theta'(-\pi) = \Theta'(\pi)$,

respectively. The problem $\begin{cases} \Theta'' + \mu^2\Theta = 0 \\ \Theta(-\pi) = \Theta(\pi),\ \Theta'(-\pi) = \Theta'(\pi) \end{cases}$ is an eigenvalue problem with eigenvalues

$\mu^2 = 0$ and $\mu^2 = m^2$ and corresponding eigenfunctions $\Theta_0 = 1$ and $\Theta_m = \cos(m\theta)$ or $\sin(m\theta)$. The

corresponding solutions of $r^2 R'' + rR' + \left(r^2\lambda^2 - \mu^2\right)R = 0$, which we recognize as Bessel's equation,

since $\mu = m$ are $R_m(r) = c_{1m}J_m(\lambda r) + c_{2m}Y_m(\lambda r)$. Since $|R(0)|$ is bounded, $c_{2m} = 0$ and since

$R(a) = 0, J_m(\lambda a) = 0$ so $\lambda_{mn} = \dfrac{\alpha_{mn}}{a}$, where α_{mn} denotes the nth zero of the Bessel function

$J_m(x)$.

Therefore,

$u(r,\theta,t) = \sum_n a_{0n} J_0(\lambda_{0n} r) \cos(\lambda_{0n} ct) + \sum_{m,n} a_{mn} J_m(\lambda_{mn} r) \cos(m\theta) \cos(\lambda_{mn} ct)$

$\qquad + \sum_{m,n} b_{mn} J_m(\lambda_{mn} r) \sin(m\theta) \cos(\lambda_{mn} ct) + \sum_n A_{0n} J_0(\lambda_{0n} r) \sin(\lambda_{0n} ct)$

$\qquad\qquad + \sum_{m,n} A_{mn} J_m(\lambda_{mn} r) \cos(m\theta) \sin(\lambda_{mn} ct)$

$\qquad\qquad\qquad + \sum_{m,n} B_{mn} J_m(\lambda_{mn} r) \sin(m\theta) \sin(\lambda_{mn} ct).$

$a_{0n} = \dfrac{\int_0^{2\pi}\int_0^a f(r,\theta) J_0(\lambda_{0n} r) r\, dr\, d\theta}{2\pi \int_0^a \left[J_0(\lambda_{0n} r)\right]^2 r\, dr},$

$$a_{mn} = \frac{\int_0^{2\pi} \int_0^a f(r,\theta)\, J_m(\lambda_{mn}\, r)\cos(m\theta)\, r\, dr\, d\theta}{\pi \int_0^a \left[J_m(\lambda_{mn}\, r) \right]^2 r\, dr},$$

$$b_{mn} = \frac{\int_0^{2\pi} \int_0^a f(r,\theta)\, J_m(\lambda_{mn}\, r)\sin(m\theta)\, r\, dr\, d\theta}{\pi \int_0^a \left[J_m(\lambda_{mn}\, r) \right]^2 r\, dr},$$

$$A_{0n} = \frac{\int_0^{2\pi} \int_0^a g(r,\theta)\, J_0(\lambda_{0n}\, r)\, r\, dr\, d\theta}{2\pi \lambda_{0n}\, c \int_0^a \left[J_0(\lambda_{0n}\, r) \right]^2 r\, dr},$$

$$A_{mn} = \frac{\int_0^{2\pi} \int_0^a g(r,\theta)\, J_m(\lambda_{mn}\, r)\cos(m\theta)\, r\, dr\, d\theta}{\pi \lambda_{mn}\, c \int_0^a \left[J_m(\lambda_{mn}\, r) \right]^2 r\, dr}, \quad \text{and}$$

$$B_{mn} = \frac{\int_0^{2\pi} \int_0^a g(r,\theta)\, J_m(\lambda_{mn}\, r)\sin(m\theta)\, r\, dr\, d\theta}{\pi \lambda_{mn}\, c \int_0^a \left[J_m(\lambda_{mn}\, r) \right]^2 r\, dr}.$$

The table of zeros which were found earlier and saved as **besseltable** are read in and called **getzeros**. A function **alpha** is then defined so that these zeros of the Bessel functions can more easily be obtained from the list.

```
≡□≡══════ TheWaveEquation(3D) ══════□≡
In[1]:=
getzeros=ReadList["besseltable"];
In[2]:=
alpha[i_,j_]:=getzeros[[1]][[i+1,j]]
```

The appropriate parameter values as well as the initial condition functions are defined below. Notice that the functions describing the initial position and velocity are defined as the product of functions. This enables the calculations to be carried out in the manner which follows.

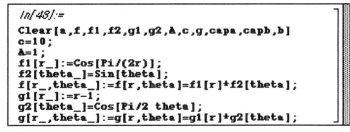

```
In[49]:=
Clear[a,f,f1,f2,g1,g2,A,c,g,capa,capb,b]
c=10;
A=1;
f1[r_]:=Cos[Pi/(2r)];
f2[theta_]=Sin[theta];
f[r_,theta_]:=f[r,theta]=f1[r]*f2[theta];
g1[r_]:=r-1;
g2[theta_]=Cos[Pi/2 theta];
g[r_,theta_]:=g[r,theta]=g1[r]*g2[theta];
```

The coefficients a_{0n} are determined with the function **a** below.

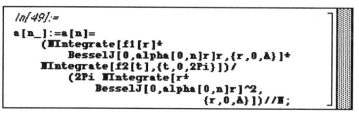

Hence, **as** represents a table of the first five values of a_{0n}.

```
In[50]:=
as=Table[a[n]//Chop,{n,1,5}]

Out[50]=
{0, 0, 0, 0, 0}
```

Since the denominator of each integral formula used to find a_{mn} and b_{mn} are the same, the function **bjmn** which computes this value is defined below. A table of nine values of this coefficient is then determined.

```
In[52]:=
bjmn[m_,n_]:=bjmn[m,n]=
    NIntegrate[r*BesselJ[m,alpha[m,n]r]^2,
                                {r,0,A}]//N
Table[bjmn[m,n]//Chop,{m,1,3},{n,1,3}]

Out[52]=
{{0.0811076, 0.0450347, 0.0311763},
 {0.0576874, 0.0368243, 0.0270149},
 {0.0444835, 0.0311044, 0.0238229}}
```

Since the initial position function f is defined as the product of a function **f1** of r and a function **f2** of θ, we determine the value of the integral of the product of **f1** and the appropriate Bessel function in **fbjmn** and create a table of values. **Chop** is used to round off very small numbers to zero.

```
In[54]:=
fbjmn[m_,n_]:=fbjmn[m,n]=
    NIntegrate[f1[r]*
        BesselJ[m,alpha[m,n]r]r,{r,0,A}]//N
Table[fbjmn[m,n]//Chop,{m,1,3},{n,1,3}]

Out[54]=
{{-0.0462147, 0.100255, -0.00368118},
 {-0.0637856, 0.078574, 0.0145639},
 {-0.070229, 0.0572423, 0.0219089}}
```

Hence, the values of **fbjmn** and **bjmn** which were found with the **Table** commands above are used to determine a_{mn} below

```
In[56]:=
a[m_,n_]:=a[m,n]=
        (fbjmn[m,n]*
        NIntegrate[f2[t]*
                Cos[m t] ,{t,0,2Pi}])/
        (Pi bjmn[m,n])//N;
Table[a[m,n]//Chop,{m,1,3},{n,1,3}]
Out[56]=
{{0, 0, 0}, {0, 0, 0}, {0, 0, 0}}
```

as well as the value of b_{mn}. Note that defining the coefficients in this manner cuts down on unnecessary computation time.

```
In[58]:=
b[m_,n_]:=b[m,n]=
        (fbjmn[m,n]*
        NIntegrate[f2[t]*
                Sin[m t],{t,0,2Pi}])/
        (Pi bjmn[m,n])//N;
Table[b[m,n]//Chop,{m,1,3},{n,1,3}]
Out[58]=
{{-0.569795, 2.22617, -0.118076}, {0, 0, 0},
  {0, 0, 0}}
```

The values of A_{0n} are found similar to those of a_{0n}. After defining the function **capa** to calculate these coefficients, a table of values is then found.

```
In[60]:=
capa[n_]:=capa[n]=
        (NIntegrate[g1[r]*
                BesselJ[0,alpha[0,n]r]r,{r,0,A}]*
        NIntegrate[g2[t],{t,0,2Pi}])/
        (2Pi c alpha [0,n]*
        NIntegrate[r*BesselJ[0,
                        alpha[0,n]r]^2,
                        {r,0,A}])//N;
Table[capa[n]//Chop,{n,1,6}]
Out[60]=
{0.00142231, 0.0000542518, 0.0000267596,
  6.41976 10^-6, 4.95843 10^-6, 1.88585 10^-6}
```

The value of the integral of the component of g, g1, which depends on r and the appropriate Bessel function, is defined as **gbjmn** below.

```
In[62]:=
gbjmn[m_,n_]:=gbjmn[m,n]=
    NIntegrate[g1[r]*
        BesselJ[m,alpha[m,n]r]r,{r,0,A}]//N;
Table[gbjmn[m,n]//Chop,{m,1,3},{n,1,3}]

Out[62]=
{{-0.0743906, -0.019491, -0.00989293},
 {-0.0554379, -0.0227976, -0.013039},
 {-0.0433614, -0.0226777, -0.0141684}}
```

Hence, A_{mn} is found by taking the product of integrals, **gbjmn** depending on r and one depending on θ. A table of coefficient values is generated in this case as well.

```
In[64]:=
capa[m_,n_]:=capa[m,n]=
    (gbjmn[m,n]*
        NIntegrate[g2[t]Cos[m t],{t,0,2Pi}])/
        (Pi alpha[m,n] c bjmn[m,n])//N;
Table[capa[m,n]//Chop,{m,1,3},{n,1,3}]

Out[64]=
{{0.0035096, 0.000904517, 0.000457326},
 {-0.00262692, -0.00103252, -0.000583116},
 {-0.000503187, -0.000246002, -0.000150499}}
```

Similarly, the B_{mn} are determined.

```
In[66]:=
capb[m_,n_]:=capb[m,n]=
    (gbjmn[m,n]*
        NIntegrate[g2[t]Sin[m t],{t,0,2Pi}])/
        (Pi alpha[m,n] c bjmn[m,n])//N;
Table[capb[m,n]//Chop,{m,1,3},{n,1,3}]

Out[66]=
{{0.00987945, 0.00254619, 0.00128736},
 {-0.0147894, -0.00581305, -0.00328291},
 {-0.00424938, -0.00207747, -0.00127095}}
```

Now that the necessary coefficients have been found, we must construct the approximate solution to the wave equation by using our results. Below, **term1** represents those terms of the expansion involving a_{0n}, **term2** those terms involving a_{mn}, **term3** those involving a_{mn}, **term4** those involving A_{0n}, **term5** those involving A_{mn}, and **term6** those involving B_{mn}.

```
In[67]:=
Clear[term1,term2,term3,term4,term5,term6]

In[73]:=
term1[r_,t_,n_]:=a[n]*BesselJ[0,alpha[0,n]r]*
                Cos[alpha[0,n] c t];
term2[r_,t_,th_,m_,n_]:=a[m,n]*
          BesselJ[m,alpha[m,n]r]*Cos[m th]*
                Cos[alpha[m,n] c t];
term3[r_,t_,th_,m_,n_]:=b[m,n]*
                BesselJ[m,alpha[m,n]r]*
                Sin[m th] Cos[alpha[m,n] c t];
term4[r_,t_,n_]:=capa[n]*
                BesselJ[0,alpha[0,n]r]*
                Sin[alpha[0,n] c t];
term5[r_,t_,th_,m_,n_]:=capa[m,n]*
                BesselJ[m,alpha[m,n]r]*
                Cos[m th] Sin[alpha[m,n] c t];
term6[r_,t_,th_,m_,n_]:=capb[m,n]*
                BesselJ[m,alpha[m,n]r]*
                Sin[m th] Sin[alpha[m,n] c t];
```

Therefore, the solution is given as the sum of these terms as indicated in **u** below.

```
In[75]:=
Clear[u]
u[r_,t_,th_]:=
    Sum[term1[r,t,n],{n,1,5}]+
    Sum[term2[r,t,th,m,n],{m,1,3},{n,1,3}]+
    Sum[term3[r,t,th,m,n],{m,1,3},{n,1,3}]+
    Sum[term4[r,t,n],{n,1,5}]+
    Sum[term5[r,t,th,m,n],{m,1,3},{n,1,3}]+
    Sum[term6[r,t,th,m,n],{m,1,3},{n,1,3}];
```

The solution is compiled below in **uc**

```
In[3]:=
uc=Compile[{r,t,th},u[r,t,th]]

Out[3]=
CompiledFunction[{r, t, th}, u[r, t, th],
  -CompiledCode-]
```

and then plotted with the function **tplot** which uses **ParametricPlot3D** to produce the graph of the solution at a particular value of t. Note that **th** represents the angle θ and that the x and y coordinates are given in terms of polar coordinates.

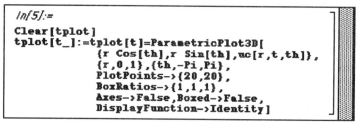

```
In[5]:=
Clear[tplot]
tplot[t_]:=tplot[t]=ParametricPlot3D[
            {r Cos[th],r Sin[th],uc[r,t,th]},
            {r,0,1},{th,-Pi,Pi},
            PlotPoints->{20,20},
            BoxRatios->{1,1,1},
            Axes->False,Boxed->False,
            DisplayFunction->Identity]
```

A table of four plots for t=0 to t=1/2 using increments of 1/6 is produced in **graphs** below. This table is then partitioned into groups of 2 and displayed as a graphics array. These plots can be produced with a **Do** command as well so that the results may be animated.

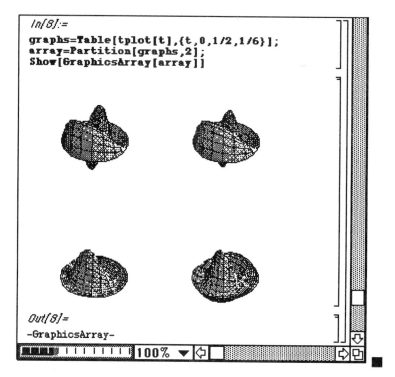

```
In[8]:=
graphs=Table[tplot[t],{t,0,1/2,1/6}];
array=Partition[graphs,2];
Show[GraphicsArray[array]]
```

```
Out[8]=
-GraphicsArray-
```

100%

Appendix: Numerical Methods

Mathematica commands used in the **Appendix** include:

AspectRatio	ListPlot	PlotStyle
Clear	Module	ReplaceAll
Dashing	ParametricPlot	Short
DisplayFunction	Part	Show
DSolve	Plot	Table
GrayLevel	PlotJoined	TableForm
Length	PlotRange	

Euler's Method

Consider the initial value problem

$$\frac{dy}{dx} = f(x,y), \ y(x_0) = y_0.$$

In some cases, an exact solution to this problem cannot be determined. Hence, we must attempt to approximate the solution. We begin by recalling the Taylor series expansion of y about $x = x_0$ given by

$$y(x) = y(x_0) + y'(x_0)(x - x_0) + \frac{y''(x_0)}{2!}(x - x_0)^2 + \cdots \ .$$

Since we know the value of y at the initial value of $x = x_0$, we use this value to approximate y at $x_1 = x_0 + h$ which is near x_0 in the following manner. We first evaluate the Taylor series at $x_1 = x_0 + h$ to yield

$$y(x_0 + h) = y(x_0) + y'(x_0)h + \frac{y''(x_0)}{2!}h^2 + \cdots \ .$$

Substituting $y' = f(x,y)$ into this expansion, using $y(x_0) = y_0$, and calling this new value y_1, we have

$$y_1 = y(x_0 + h) = y(x_0) + f(x_0, y_0)h + \frac{1}{2!}\frac{df}{dx}(x_0, y_0)h^2 + \cdots \ = y_0 + f(x_0, y_0)h + \frac{1}{2!}\frac{df}{dx}(x_0, y_0)h^2 + \cdots \ .$$

Hence, the initial point (x_0, y_0) is used to determine y_1. A first-order approximation is obtained from this series by disregarding the terms of order h^2 and higher. In other words, we determine y_1 from

$$y_1 = y_0 + f(x_0, y_0)h \ .$$

We next use the point (x_1, y_1) to approximate the value of y at $x_2 = x_1 + h$. Calling this value y_2, we have

from $y(x) = y(x_1) + y'(x_1)(x - x_1) + \frac{y''(x_1)}{2!}(x - x_1)^2 + \cdots$ that an approximate value of $y(x_2)$ is

$$y_2 = y(x_1) + f(x_1, y_1)h = y_1 + f(x_1, y_1)h.$$

Hence, we obtain the recursive relationship for the first-order approximate value of y at x_{n+1} given by

$$y_{n+1} = y_n + f(x_n, y_n)h \ , \ n = 0, 1, 2, \dots \ .$$

■ EXAMPLE 1

Use Euler's method to approximate the solution of the following initial value problem

$$\frac{dy}{dx} = xy, \ y(0) = 1$$

for h=0.1, 0.05, and 0.025. Also, determine the exact solution and compare the results.

Solution:

We begin by defining the function f(x,y)=xy as well as the initial condition. The recursive formula for x and y are also given. Notice that due to the manner by which y is defined, all prior values of y are remembered so that future values can be based on them.

```
▤▯▭▭▭▭▭▭▭▭ Euler'sMethod ▭▭▭▭▭▭▭▭▯▤
In[6]:=
Clear[f,x,y,h]
f[x_,y_]=x y;
h=.1;
y[0]=1;
x[n_]=n h;
y[n_]:=y[n]=y[n-1]+h f[x[n-1],y[n-1]]
```

A table of approximate values based on this recursive formula is given in **ytable**.

```
In[7]:=
ytable=Table[y[i],{i,0,10}]

Out[7]=
{1, 1, 1.01, 1.0302, 1.06111, 1.10355, 1.15873,
   1.22825, 1.31423, 1.41937, 1.54711}
```

A list of ordered pairs is formed below in **intpts1**. Notice that the index for x must be one smaller than that of y since **ytable[[i]]** represents the ith element of the list while **x[i]** substitutes **i** into the formula to determine x.

```
In[8]:=
intpts1=Table[{x[i-1],ytable[[i]]},
        {i,1,Length[ytable]}]

Out[8]=
{{0, 1}, {0.1, 1}, {0.2, 1.01}, {0.3, 1.0302},
   {0.4, 1.06111}, {0.5, 1.10355}, {0.6, 1.15873},
   {0.7, 1.22825}, {0.8, 1.31423}, {0.9, 1.41937},
   {1., 1.54711}}
```

These points can be plotted with **ListPlot** so that the accuracy of the approximation can be compared to that of the exact solution. Notice that the problem considered here is separable and, therefore, can be solved. This is done as follows:

$$\frac{dy}{dx} = xy$$

$$\frac{dy}{y} = x \, dx$$

$$\text{Log}(y) = \frac{x^2}{2} + C$$

$$y = ke^{x^2/2}.$$

Applying the initial condition, we have k=1. The approximate solution obtained through Euler's method and the exact solution are plotted below in **plot1** and **exactplot**, respectively.

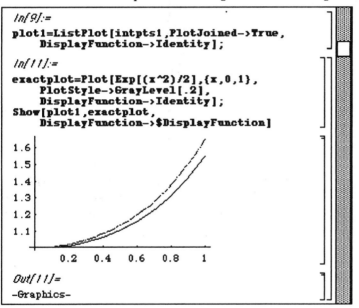

```
In[9]:=
plot1=ListPlot[intpts1,PlotJoined->True,
    DisplayFunction->Identity];

In[11]:=
exactplot=Plot[Exp[(x^2)/2],{x,0,1},
    PlotStyle->GrayLevel[.2],
    DisplayFunction->Identity];
Show[plot1,exactplot,
    DisplayFunction->$DisplayFunction]
```

```
Out[11]=
-Graphics-
```

Since the approximation is not as accurate as x increases, we reduce the increment size, h, and repeat the approximation process.

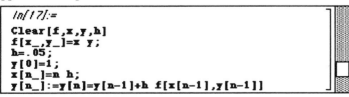

```
In[17]:=
Clear[f,x,y,h]
f[x_,y_]=x y;
h=.05;
y[0]=1;
x[n_]=n h;
y[n_]:=y[n]=y[n-1]+h f[x[n-1],y[n-1]]
```

The approximate values of y using h = 0.05 are determined in **y2table** below. Note that the number of entries in this table is increased so that values are determined from x = 0 to x = 1.

```
In[18]:=
y2table=Table[y[i],{i,0,20}]

Out[18]=
{1, 1, 1.0025, 1.00751, 1.01507, 1.02522,
  1.03803, 1.05361, 1.07204, 1.09348, 1.11809,
  1.14604, 1.17756, 1.21288, 1.2523, 1.29613,
  1.34474, 1.39853, 1.45796, 1.52357, 1.59594}
```

Again, a list of ordered pairs is created in **intpts2**.

```
In[19]:=
intpts2=Table[{x[i-1],y2table[[i]]},
         {i,1,Length[y2table]}]

Out[19]=
{{0, 1}, {0.05, 1}, {0.1, 1.0025},
  {0.15, 1.00751}, {0.2, 1.01507},
  {0.25, 1.02522}, {0.3, 1.03803},
  {0.35, 1.05361}, {0.4, 1.07204},
  {0.45, 1.09348}, {0.5, 1.11809},
  {0.55, 1.14604}, {0.6, 1.17756},
  {0.65, 1.21288}, {0.7, 1.2523},
  {0.75, 1.29613}, {0.8, 1.34474},
  {0.85, 1.39853}, {0.9, 1.45796},
  {0.95, 1.52357}, {1., 1.59594}}
```

These points are plotted below in **plot2** using a dashed line. It is then displayed along with the exact solution in **exactplot** and the results from the first approximation in **plot1**. Notice that the approximation obtained with h = 0.05 is better than the approximation obtained with h = 0.1.

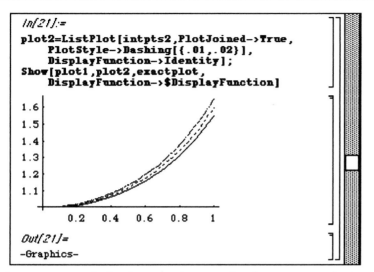

```
In[21]:=
plot2=ListPlot[intpts2,PlotJoined->True,
    PlotStyle->Dashing[{.01,.02}],
    DisplayFunction->Identity];
Show[plot1,plot2,exactplot,
    DisplayFunction->$DisplayFunction]
```

```
Out[21]=
-Graphics-
```

In an attempt to further improve the approximation, we repeat the process with h = 0.025.

```
In[27]:=
Clear[f,x,y,h]
f[x_,y_]=x y;
h=.025;
y[0]=1;
x[n_]=n h;
y[n_]:=y[n]=y[n-1]+h f[x[n-1],y[n-1]]
```

A table of approximate y-values is given in **y3table**. In this case, the index must be increased to 40 to insure that approximate values are determined over the entire interval from x =0 to x =1.

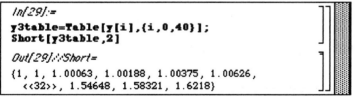

```
In[29]:=
y3table=Table[y[i],{i,0,40}];
Short[y3table,2]
```

```
Out[29]//Short=
{1, 1, 1.00063, 1.00188, 1.00375, 1.00626,
   <<32>>, 1.54648, 1.58321, 1.6218}
```

A list of ordered pairs using the approximate y-values obtained with Euler's formula is created in **intpts3** below.

```
In[31]:=
intpts3=Table[{x[i-1],y3table[[i]]},
        {i,1,Length[y3table]}];
Short[intpts3,2]
```

```
Out[31]//Short=
{{0, 1}, {0.025, 1}, {0.05, 1.00063}, <<36>>,
  {0.975, 1.58321}, {1., 1.6218}}
```

These points are then plotted in **plot3** using a dashed line and displayed with the exact solution (shown in **exactplot**) to reveal that the approximation obtained with h =0.025 is better than the two previous approximations.

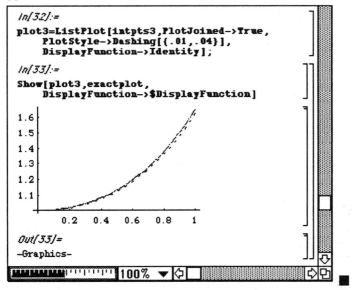

```
In[32]:=
plot3=ListPlot[intpts3,PlotJoined->True,
    PlotStyle->Dashing[{.01,.04}],
    DisplayFunction->Identity];
```

```
In[33]:=
Show[plot3,exactplot,
    DisplayFunction->$DisplayFunction]
```

```
Out[33]=
-Graphics-
```

Euler's method of approximation is improved by making use of the intermediate value of y given by

$$y^*_{n+1} = y_n + f(x_n, y_n)h$$

and then using this value to compute the approximation in the following way:

$$y_{n+1} = y_n + \frac{1}{2}\left[f(x_n, y_n) + f\left(x_{n+1}, y^*_{n+1}\right)\right]h.$$

■ **EXAMPLE 2**

Use the improved Euler's method to approximate the solution of the initial value problem in Example 1:

$$\frac{dy}{dx} = xy, \ y(0) = 1$$

for h=0.1. Compare these results to those from the earlier example as well as the exact solution.

Solution:

First, the recursive formula used with the improved Euler's method is defined. Note that the intermediate value is directly entered this formula below. An alternate formula can be defined with *Mathematica*'s **Module** command as will be seen in subsequent examples.

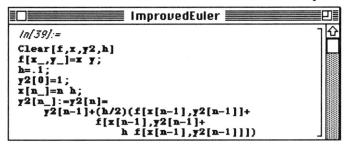

```
                 ImprovedEuler
In[39]:=
Clear[f,x,y2,h]
f[x_,y_]=x y;
h=.1;
y2[0]=1;
x[n_]=n h;
y2[n_]:=y2[n]=
     y2[n-1]+(h/2)(f[x[n-1],y2[n-1]]+
          f[x[n-1],y2[n-1]+
               h f[x[n-1],y2[n-1]]])
```

As in the previous example, a table of approximations is determined using the recursive formula given in **y2**. This table is called **yimptable**, because it is generated using the improved Euler's method.

```
In[41]:=
Clear[yimptable]
yimptable=Table[y2[i],{i,0,10}]

Out[41]=
{1, 1, 1.01005, 1.03045, 1.06183, 1.10515,
   1.16179, 1.23359, 1.32296, 1.43304, 1.56781}
```

In addition to investigating the accuracy graphically, we produce the chart below. The first column represents the x-value, the second the approximation obtained previously with Euler's method (h =0.1), the third that obtained with the improved Euler's method (h =0.1), while the fourth column represents the exact value of y. Notice the slight improvement in the approximation obtained with this method.

```
In[42]:=
Table[{(.1 i),ytable[[i+1]],
     yimptable[[i+1]],Exp[((.1 i)^2)/2]},
          {i,0,10}]//TableForm

Out[42]//TableForm=
0       1          1          1
0.1     1          1          1.00501
0.2     1.01       1.01005    1.0202
0.3     1.0302     1.03045    1.04603
0.4     1.06111    1.06183    1.08329
0.5     1.10355    1.10515    1.13315
0.6     1.15873    1.16179    1.19722
0.7     1.22825    1.23359    1.27762
0.8     1.31423    1.32296    1.37713
0.9     1.41937    1.43304    1.4993
1.      1.54711    1.56781    1.64872
```

We now plot this approximation by, first, generating the list of ordered pairs in **impeulerpts1** below.

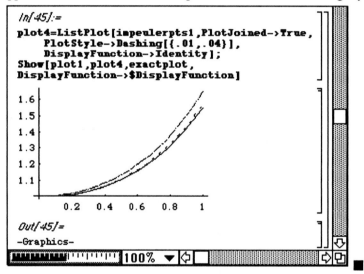

```
In[43]:=
impeulerpts1=Table[{x[i-1],yimptable[[i]]},
        {i,1,Length[yimptable]}]

Out[43]=
{{0, 1}, {0.1, 1}, {0.2, 1.01005},
  {0.3, 1.03045}, {0.4, 1.06183}, {0.5, 1.10515},
  {0.6, 1.16179}, {0.7, 1.23359}, {0.8, 1.32296},
  {0.9, 1.43304}, {1., 1.56781}}
```

These points are plotted with **ListPlot** in **plot4** and joined with a dashed line. When this plot is displayed with the exact solution (which is given in the lighter shade of gray) and the approximation obtained earlier with Euler's method (which is represented as the darker line), we see that the approximation obtained with the improved Euler method does slightly increase the accuracy.

The Runge-Kutta Method

In an attempt to improve on the approximation obtained with Euler's method as well as avoid the analytic differentiation of the function $f(x,y)$ to obtain y'', y''', ..., the Runge-Kutta method which involves many more computations at each step is introduced. Let us begin with the Runge-Kutta method of order two. Suppose that we know the value of y at x_n. We now use the point (x_n, y_n) to approximate the value of y at a nearby value $x = x_n + h$ by assuming that

$y_{n+1} = y_n + A k_1 + B k_2$ where

$k_1 = hf(x_n, y_n)$ and $k_2 = hf(x_n + ah, y_n + bk_1)$.

We can also use the Taylor series expansion of y to obtain another representation of $y_{n+1} = y(x_n + h)$ as follows:

$$y(x_n + h) = y(x_n) + hy'(x_n) + h^2 \frac{y''(x_n)}{2!} + \cdots = y_n + hy'(x_n) + h^2 \frac{y''(x_n)}{2!} + \cdots .$$

Now, since $y_{n+1} = y_n + Ak_1 + Bk_2 = y_n + Ahf(x_n, y_n) + Bhf(x_n + ah, y_n + bhf(x_n, y_n))$, we wish to determine values of A, B, a, and b such that these two representations of y_{n+1} agree. Clearly, if we let A=1 and B=0, then the relationship holds up to order h. However, we can choose these parameters more wisely so that agreement occurs up through terms of order h^2. This is accomplished by considering the Taylor expansion of a function of two variables. In our case, we have

$$f(x_n + ah, y_n + bhf(x_n, y_n)) = f(x_n, y_n) + ah \frac{\partial f}{\partial x}(x_n, y_n) + bhf(x_n, y_n) \frac{\partial f}{\partial x}(x_n, y_n) + O(h^2).$$

The power series is then substituted into the following expression and then simplified to yield:

$$y_{n+1} = y_n + Ahf(x_n, y_n) + bhf(x_n + ah, y_n + bhf(x_n, y_n))$$

$$= y_n + (A + B)hf(x_n, y_n) + aBh^2 \frac{\partial f}{\partial x}(x_n, y_n) + bBh^2 f(x_n, y_n) \frac{\partial f}{\partial x}(x_n, y_n) + O(h^3).$$

Comparing the above expression to the following power series obtained directly from the Taylor series of y

$$y(x_n + h) = y(x_n) + hf(x_n, y_n) + \frac{h^2}{2} \frac{\partial f}{\partial x}(x_n, y_n) + \frac{h^2}{2} f(x_n, y_n) \frac{\partial f}{\partial y}(x_n, y_n) + O(h^3) \quad \text{or}$$

$$y_{n+1} = y_n + hf(x_n, y_n) + \frac{h^2}{2} \frac{\partial f}{\partial x}(x_n, y_n) + \frac{h^2}{2} f(x_n, y_n) \frac{\partial f}{\partial y}(x_n, y_n) + O(h^3),$$

we see that A, B, a, and b must satisfy the following system of nonlinear equations:

$$A + B = 1, \quad aA = \frac{1}{2}, \quad \text{and } bB = \frac{1}{2}.$$

Therefore, choosing a = b =1, the Runge-Kutta method of order two uses the equations :

$$y_{n+1} = y(x_n + h) = y_n + \frac{1}{2} hf(x_n, y_n) + \frac{1}{2} hf(x_n + h, y_n + hf(x_n, y_n))$$

$$= y_n + \frac{1}{2}(k_1 + k_2)$$

where $k_1 = hf(x_n, y_n)$ and $k_2 = hf(x_n + h, y_n + k_1)$.

■ **EXAMPLE 3**

Use the Runge-Kutta method to approximate the solution of the initial value problem from Example 1:

$$\frac{dy}{dx} = xy, \quad y(0) = 1$$

for h=0.1. Compare these results to those from the improved Euler's method from the previous example as well as the exact solution.

Solution:

We begin by defining the function f, the increment size h, and the initial value of y below. Then, the formula to determine x at each step and the function **yrk** which recursively determines the approximate y-values is defined. Notice the manner in which **Module** is used in the definition of yrk. This definition remembers y-values as they are computed.

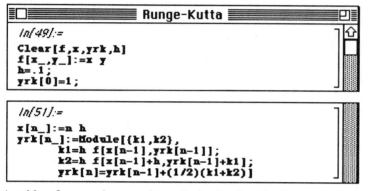

```
≣□□ ≣≣≣≣≣≣≣ Runge-Kutta ≣≣≣≣≣≣≣ □◨≣

In[49]:=
Clear[f,x,yrk,h]
f[x_,y_]:=x y
h=.1;
yrk[0]=1;
```

```
In[51]:=
x[n_]:=n h
yrk[n_]:=Module[{k1,k2},
        k1=h f[x[n-1],yrk[n-1]];
        k2=h f[x[n-1]+h,yrk[n-1]+k1];
        yrk[n]=yrk[n-1]+(1/2)(k1+k2)]
```

A table of approximate values obtained using the Runge-Kutta method in **yrk** is given below in **rktable1**. These values are then used to generate a list of ordered pairs in **rkpts1**.

```
In[52]:=
rktable1=Table[yrk[i],{i,0,10}]

Out[52]=
{1, 1.005, 1.02018, 1.04599, 1.08322, 1.13305,
  1.19707, 1.27739, 1.37677, 1.49876, 1.64788}
```

```
In[53]:=
rkpts1=Table[{x[i-1],rktable1[[i]]},
        {i,1,Length[rktable1]}]

Out[53]=
{{0, 1}, {0.1, 1.005}, {0.2, 1.02018},
  {0.3, 1.04599}, {0.4, 1.08322}, {0.5, 1.13305},
  {0.6, 1.19707}, {0.7, 1.27739}, {0.8, 1.37677},
  {0.9, 1.49876}, {1., 1.64788}}
```

The points are plotted with ListPlot and joined with a dashed line in **plot5**.

```
In[55]:=
Clear[plot5]
    plot5=ListPlot[rkpts1,PlotJoined->True,
    PlotStyle->Dashing[{.01,.04}],
    DisplayFunction->Identity];
```

The graph of the approximation obtained with the Runge-Kutta method, that obtained with Euler's method, and the exact solution are displayed simultaneously below. Recall that the Euler method approximation is represented by the darkest curve. Hence, the Runge-Kutta approximation is very close to the actual solution. Graphically, they appear to be the same curve.

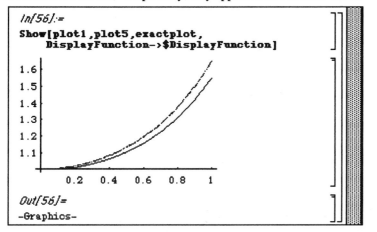

```
In[56]:=
Show[plot1,plot5,exactplot,
      DisplayFunction->$DisplayFunction]
```

```
Out[56]=
-Graphics-
```

To better understand the accuracy of the approximation obtained through the Runge-Kutta method, we produce the table below by adding a third column to the previous table to represent the Runge-Kutta approximation. (Note that the second and third columns represent the approximations obtained by Euler's method and the improved Euler's method, respectively.) Clearly, the Runge-Kutta method yields the best approximation of the three methods. Note, also, that the Runge-Kutta method with h=0.1 gives a better approximation than was achieved with Euler's method and h =0.025.

```
In[57]:=
Table[{.1 i,ytable[[i+1]],
      yimptable[[i+1]],rktable1[[i+1]],
      Exp[((.1 i)^2)/2]},{i,0,10}]//TableForm
```

```
Out[57]//TableForm=
0        1         1         1          1
0.1      1         1         1.005      1.00501
0.2      1.01      1.01005   1.02018    1.0202
0.3      1.0302    1.03045   1.04599    1.04603
0.4      1.06111   1.06183   1.08322    1.08329
0.5      1.10355   1.10515   1.13305    1.13315
0.6      1.15873   1.16179   1.19707    1.19722
0.7      1.22825   1.23359   1.27739    1.27762
0.8      1.31423   1.32296   1.37677    1.37713
0.9      1.41937   1.43304   1.49876    1.4993
1.       1.54711   1.56781   1.64788    1.64872
```

```
100%
```

The terms of the power series expansions used in the derivation of the Runge-Kutta method of order two can be made to match up to order 4. These computations are rather complicated, so they will not be discussed here. However, after much work, the approximation at each step is found to be made with

$$y_{n+1} = y_n + \frac{h}{6}\left[k_1 + 2k_2 + 2k_3 + k_4\right], \ n=0, 1, 2, \ldots \ \text{where}$$

$$k_1 = f(x_n, y_n), \ k_2 = f\left(x_n + \frac{h}{2}, y_n + \frac{hk_1}{2}\right), \ k_3 = f\left(x_n + \frac{h}{2}, y_n + \frac{hk_2}{2}\right), \text{ and } k_4 = f(x_{n+1}, y_n + hk_3).$$

■ EXAMPLE 4

Use the fourth-order Runge-Kutta method to approximate the solution of the problem in the previous example:

$$\frac{dy}{dx} = xy, \ y(0) = 1$$

for h=0.1. Compare these results to those obtained with Euler's method, the Runge-Kutta method of order two, as well as the exact solution.

Solution:

We define the appropriate functions below. Again, **Module** is used in the definition of the recursive formula which is called **yrk4** in this case.

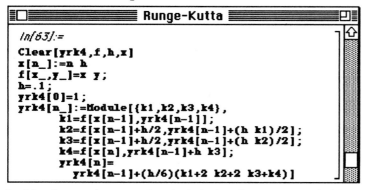

```
In[63]:=
Clear[yrk4,f,h,x]
x[n_]:=n h
f[x_,y_]=x y;
h=.1;
yrk4[0]=1;
yrk4[n_]:=Module[{k1,k2,k3,k4},
    k1=f[x[n-1],yrk4[n-1]];
    k2=f[x[n-1]+h/2,yrk4[n-1]+(h k1)/2];
    k3=f[x[n-1]+h/2,yrk4[n-1]+(h k2)/2];
    k4=f[x[n],yrk4[n-1]+h k3];
    yrk4[n]=
        yrk4[n-1]+(h/6)(k1+2 k2+2 k3+k4)]
```

A table of approximate y-values is then given in **rktable4** below

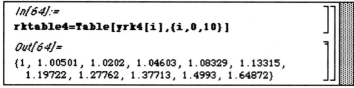

```
In[64]:=
rktable4=Table[yrk4[i],{i,0,10}]

Out[64]=
{1, 1.00501, 1.0202, 1.04603, 1.08329, 1.13315,
  1.19722, 1.27762, 1.37713, 1.4993, 1.64872}
```

and a list of ordered pairs generated in **rkpts4**.

```
In[65]:=
rkpts4=Table[{x[i-1],rktable4[[i]]},
        {i,1,Length[rktable4]}]
Out[65]=
{{0, 1}, {0.1, 1.00501}, {0.2, 1.0202},
  {0.3, 1.04603}, {0.4, 1.08329}, {0.5, 1.13315},
  {0.6, 1.19722}, {0.7, 1.27762}, {0.8, 1.37713},
  {0.9, 1.4993}, {1., 1.64872}}
```

These points are then plotted with **ListPlot** and joined with a dashed line in **plot6**. This graph is displayed with the solution obtained with Euler's method and the exact solution. As with the Runge-Kutta approximation of order two, the method of order four also offers a good approximation. It appears to equal the exact solution graphically.

```
In[66]:=
plot6=ListPlot[rkpts4,PlotJoined->True,
    PlotStyle->Dashing[{.01,.08}],
    DisplayFunction->Identity];
```

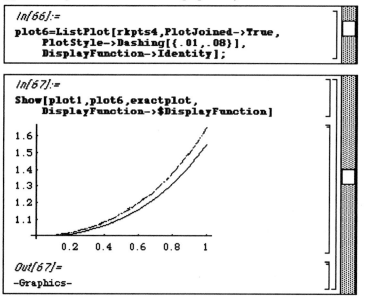

```
In[67]:=
Show[plot1,plot6,exactplot,
    DisplayFunction->$DisplayFunction]
```

```
Out[67]=
-Graphics-
```

We create the following table to better see the accuracy of the approximation obtained with the Runge-Kutta method of order four. These approximate values are given in the third column below while those obtained with the method of order two are given in the second column. The exact values are represented in the last column which match to five decimal places with the fourth-order approximate values.

```
In[68]:=
Table[{.1 i,rktable1[[i+1]],
    rktable4[[i+1]],Exp[((.1 i)^2)/2]},
    {i,0,10}]//TableForm

Out[68]//TableForm=
0       1         1          1
0.1     1.005     1.00501    1.00501
0.2     1.02018   1.0202     1.0202
0.3     1.04599   1.04603    1.04603
0.4     1.08322   1.08329    1.08329
0.5     1.13305   1.13315    1.13315
0.6     1.19707   1.19722    1.19722
0.7     1.27739   1.27762    1.27762
0.8     1.37677   1.37713    1.37713
0.9     1.49876   1.4993     1.4993
1.      1.64788   1.64872    1.64872
```
100%

We now illustrate the Runge-Kutta method of order four on a nonlinear equation.

■ EXAMPLE 5

Use the fourth-order Runge-Kutta method to approximate the solution of the nonlinear initial value problem:

$$\frac{dy}{dx} = x\sqrt{y}, \; y(1)=1$$

for h=0.1. Determine the exact solution to this separable equation and compare the results of the approximate solution to those of the exact solution.

Solution:

We define the necessary functions below. Notice that since the initial x-value is x = 1, the formula to determine x at each step must be modified slightly. Then, the formula for **yrk4** which is the same as that in the previous example is defined.

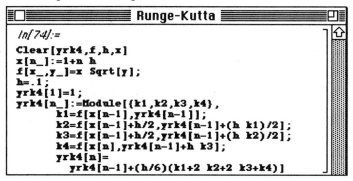

```
Runge-Kutta
In[74]:=
Clear[yrk4,f,h,x]
x[n_]:=1+n h
f[x_,y_]=x Sqrt[y];
h=.1;
yrk4[1]=1;
yrk4[n_]:=Module[{k1,k2,k3,k4},
    k1=f[x[n-1],yrk4[n-1]];
    k2=f[x[n-1]+h/2,yrk4[n-1]+(h k1)/2];
    k3=f[x[n-1]+h/2,yrk4[n-1]+(h k2)/2];
    k4=f[x[n],yrk4[n-1]+h k3];
    yrk4[n]=
        yrk4[n-1]+(h/6)(k1+2 k2+2 k3+k4)]
```

A list of approximate values is given below in **rktable42**

```
In[75]:=
rktable42=Table[yrk4[i],{i,1,10}]

Out[75]=
{1, 1.11831, 1.2544, 1.41016, 1.5876, 1.78891,
   2.0164, 2.27256, 2.56, 2.88151}
```

and a list of ordered pairs in **rkpts42**.

```
In[77]:=
Clear[rkpts42]
rkpts42=Table[{x[i],rktable42[[i+1]]},
       {i,0,9}]

Out[77]=
{{1, 1}, {1.1, 1.11831}, {1.2, 1.2544},
   {1.3, 1.41016}, {1.4, 1.5876}, {1.5, 1.78891},
   {1.6, 2.0164}, {1.7, 2.27256}, {1.8, 2.56},
   {1.9, 2.88151}}
```

These points are plotted in **plot8** with **ListPlot** and joined with the option **PlotJoined->True**.

```
In[78]:=
plot8=ListPlot[rkpts42,PlotJoined->True,
     DisplayFunction->Identity];
```

The exact solution to this separable equation is given by

$$\frac{dy}{dx} = x\sqrt{y}$$

$$\frac{dy}{\sqrt{y}} = x\ dx$$

$$2(y)^{1/2} = \frac{x^2}{2} + C$$

$$y = \left(\frac{x^2}{4} + C\right)^2$$

and applying the initial condition, we have

$$y = \frac{1}{16}\left(\frac{x^2}{4} + 3\right)^2 .$$

This solution, the lighter curve, is displayed below along with the approximate solution.

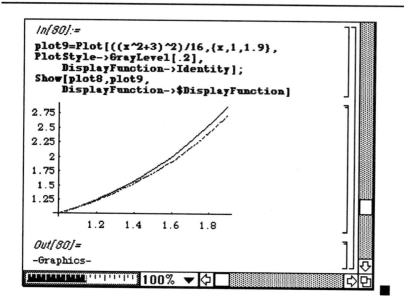

In[80]:=

```
plot9=Plot[((x^2+3)^2)/16,{x,1,1.9},
 PlotStyle->GrayLevel[.2],
     DisplayFunction->Identity];
 Show[plot8,plot9,
     DisplayFunction->$DisplayFunction]
```

Out[80]=

-Graphics-

Systems of Differential Equations

Euler's method for approximation which was discussed for first-order equations may be extended to include systems of first-order equations. Therefore, the initial value problem

$$\begin{cases} \dfrac{dx}{dt} = f(t,x,y) \\[2mm] \dfrac{dy}{dt} = g(t,x,y) \\[2mm] x(t_0) = x_0, \; y(t_0) = y_0 \end{cases}$$

is approximated at each step by the recursive relationship based on the Taylor expansion of x and y up to order h:

$$x_{n+1} = x_n + hf(t_n, x_n, y_n)$$
$$y_{n+1} = y_n + hg(t_n, x_n, y_n)$$

where $t_n = t_0 + nh$, $n = 0,1,2, \dots$.

■ EXAMPLE 6

Use Euler's method to approximate the solution of the initial value problem:

$$\begin{cases} \dfrac{dx}{dt} = x - y + 1 \\[2mm] \dfrac{dy}{dt} = x + 3y + e^{-t} \\[2mm] x(0) = 0,\ y(0) = 1 \end{cases}$$

for h=0.1 and 0.05. Compare these results to those of the exact solution of the system of equations.

Solution:

We begin by defining the functions f and g as well as the increment size, the initial values of x and y, the formula for incrementing t, and the recursive formulas for determining subsequent values of x and y.

```
▤□▤▤▤▤▤▤ Systems ▤▤▤▤▤▤▤▤
In[9]:=

Clear[f,g,t,h,x,y]
f[t_,x_,y_]:=x-y+1;
g[t_,x_,y_]:=x+3y+Exp[-t];
h=.1;
x[0]=0;
y[0]=1;
t[n_]:=h n;
x[n_]:=x[n]=x[n-1]+h(f[t[n-1],x[n-1],y[n-1]])
y[n_]:=y[n]=y[n-1]+h(g[t[n-1],x[n-1],y[n-1]])
```

A table containing the approximate values of x and y at equally spaced values of t using a step-size of h = 0.1 is given below in **txytable**.

```
In[11]:=
Clear[txytable]
txytable=Table[{t[i],x[i],y[i]},{i,0,10}]

Out[11]=
{{0, 0, 1}, {0.1, 0, 1.4}, {0.2, -0.04, 1.91048},
  {0.3, -0.135048, 2.5615},
  {0.4, -0.304703, 3.39053},
  {0.5, -0.574227, 4.44425},
  {0.6, -0.976074, 5.78076},
  {0.7, -1.55176, 7.47226},
  {0.8, -2.35416, 9.60842},
  {0.9, -3.45042, 12.3005}, {1., -4.9255, 15.6862}}
```

We then use **DSolve** to find the exact solution to this nonhomogeneous system of ordinary differential equations. Notice that the unknowns are called **x1** and **y1** in the command below so that there is no confusion with the approximate values found above. These solutions are then extracted from the output list of **de** and called **xexact** and **yexact**, respectively.

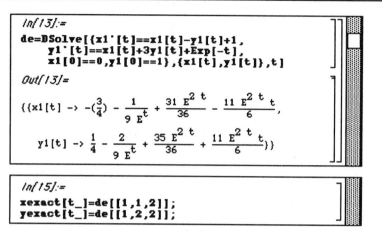

```
In[13]:=
de=DSolve[{x1'[t]==x1[t]-y1[t]+1,
    y1'[t]==x1[t]+3y1[t]+Exp[-t],
    x1[0]==0,y1[0]==1},{x1[t],y1[t]},t]

Out[13]=
```

$$\{\{x1[t] \to -(\frac{3}{4}) - \frac{1}{9 E^t} + \frac{31 E^{2t}}{36} - \frac{11 E^{2t} t}{6},$$

$$y1[t] \to \frac{1}{4} - \frac{2}{9 E^t} + \frac{35 E^{2t}}{36} + \frac{11 E^{2t} t}{6}\}\}$$

```
In[15]:=
xexact[t_]=de[[1,1,2]];
yexact[t_]=de[[1,2,2]];
```

The table called **compare** below can be used to investigate the accuracy of the approximation associated with this method. The first column gives the value of t. The second column represents the approximate value of x and should be compared to the third column which gives the exact value of x. Similarly, the fourth and fifth columns give the approximate and exact values of y, respectively.

```
In[17]:=
compare=Table[{t[i],x[i],xexact[t[i]],y[i],
                yexact[t[i]]},{i,0,10}];
TableForm[compare]

Out[17]//TableForm=
```

0	0	0	1	1
0.1	0	-0.0226978	1.4	1.46032
0.2	-0.04	-0.103346	1.91048	2.06545
0.3	-0.135048	-0.265432	2.5615	2.85904
0.4	-0.304703	-0.540105	3.39053	3.89682
0.5	-0.574227	-0.968408	4.44425	5.24975
0.6	-0.976074	-1.60412	5.78076	7.00806
0.7	-1.55176	-2.51737	7.47226	9.28638
0.8	-2.35416	-3.79926	9.60842	12.23
0.9	-3.45042	-5.56767	12.3005	16.0232
1.	-4.9255	-7.97468	15.6862	20.8987

Clearly, the accuracy of this approximation diminishes as t increases. Hence, we attempt to improve the approximation by decreasing the increment size. We do this below by entering the value h = 0.05 and repeating the procedure which was followed above. Again, to avoid confusion, we call the approximate values obtained in this case, **x2** and **y2**.

```
In[26]:=
Clear[f,g,t,h,x,y]
f[t_,x_,y_]:=x-y+1;
g[t_,x_,y_]:=x+3y+Exp[-t];
h=.05;
x2[0]=0;
y2[0]=1;
t[n_]=h n;
x2[n_]:=x2[n]=x2[n-1]+h(f[t[n-1],
                          x2[n-1],y2[n-1]])
y2[n_]:=y2[n]=y2[n-1]+h(g[t[n-1],
                          x2[n-1],y2[n-1]])
```

After defining the appropriate functions, a table called **txytable2** containing the approximate values of x and y obtained with h=0.05 is generated below.

```
In[28]:=
Clear[txytable2]
txytable2=Table[{t[i],x2[i],y2[i]},{i,0,20}];
```

We then compare the approximate values with the exact values in the table **compare2** below. As with the previous case, the approximate values of x and y are given in the second and fourth columns, respectively, while the exact values are given in the third and fifth. Notice that the accuracy is improved with the decreased step-size h.

```
In[94]:=
compare2=Table[{t[i],x2[i],xexact[t[i]],
               y2[i],yexact[t[i]]},{i,0,20}];
TableForm[compare2]

Out[94]//TableForm=
```

0	0	0	1	1
0.05	0	-0.00532454	1.2	1.21439
0.1	-0.01	-0.0226978	1.42756	1.46032
0.15	-0.0318781	-0.054467	1.68644	1.74231
0.2	-0.0677939	-0.103346	1.98084	2.06545
0.25	-0.120226	-0.172465	2.31552	2.43552
0.3	-0.192013	-0.265432	2.69577	2.85904
0.35	-0.286402	-0.386392	3.12758	3.34338
0.4	-0.407102	-0.540105	3.61763	3.89682
0.45	-0.558338	-0.732029	4.17344	4.52876
0.5	-0.744927	-0.968408	4.80342	5.24975
0.55	-0.972344	-1.25639	5.51701	6.07171
0.6	-1.24681	-1.60412	6.32479	7.00806
0.65	-1.57539	-2.02091	7.23861	8.07394
0.7	-1.96609	-2.51737	8.27174	9.28638
0.75	-2.42798	-3.10558	9.43902	10.6645
0.8	-2.97133	-3.79926	10.7571	12.23
0.85	-3.60776	-4.61405	12.2446	14.0071
0.9	-4.35037	-5.56767	13.9222	16.0232
0.95	-5.214	-6.68027	15.8134	18.3088
1.	-6.21537	-7.97468	17.944	20.8987

100%

Since we would like to be able to improve the approximation without using such a small value for h, we seek to improve the method. As with first-order equations, the Runge-Kutta method can be extended to systems. In this case, the recursive formula at each step is

$$x_{n+1} = x_n + \frac{h}{6}(k_1 + 2k_2 + 2k_3 + k_4)$$

$$y_{n+1} = y_n + \frac{h}{6}(m_1 + 2m_2 + 2m_3 + m_4)$$

where

$$k_1 = f(t_n, x_n, y_n) \qquad m_1 = g(t_n, x_n, y_n)$$

$$k_2 = f\left(t_n + \frac{h}{2}, x_n + \frac{hk_1}{2}, y_n + \frac{hm_1}{2}\right) \qquad m_2 = g\left(t_n + \frac{h}{2}, x_n + \frac{hk_1}{2}, y_n + \frac{hm_1}{2}\right)$$

$$k_3 = f\left(t_n + \frac{h}{2}, x_n + \frac{hk_2}{2}, y_n + \frac{hm_2}{2}\right) \qquad m_3 = g\left(t_n + \frac{h}{2}, x_n + \frac{hk_2}{2}, y_n + \frac{hm_2}{2}\right)$$

$$k_4 = f(t_n + h, x_n + hk_3, y_n + hm_3) \qquad m_4 = g(t_n + h, x_n + hk_3, y_n + hm_3).$$

■ EXAMPLE 7

Use the Runge-Kutta method to approximate the solution of the initial value problem from the previous example:

$$\begin{cases} \dfrac{dx}{dt} = x - y + 1 \\[2mm] \dfrac{dy}{dt} = x + 3y + e^{-t} \\[2mm] x(0) = 0, \ y(0) = 1 \end{cases}$$

for h=0.1. Compare these results to those of the exact solution of the system of equations as well as those obtained with Euler's method.

Solution:

We begin by defining the appropriate functions below. Note that we use **xrk** and **yrk** to represent the approximate values of x and y obtained with this method.

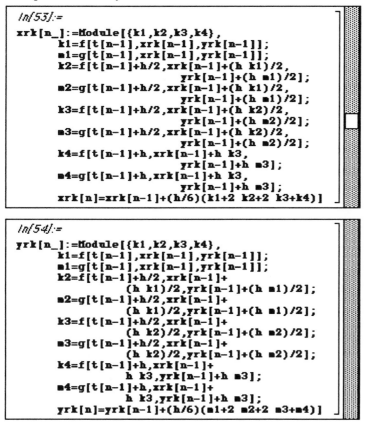

```
▤□▭▭▭▭▭▭ Systems ▭▭▭▭▭▭ 🖳▣
In[45]:=
f[t_,x_,y_]:=x-y+1;
g[t_,x_,y_]:=x+3y+Exp[-t];
In[52]:=
Clear[f,g,t,h,xrk,yrk]
f[t_,x_,y_]:=x-y+1;
g[t_,x_,y_]:=x+3y+Exp[-t];
h=.1;
xrk[0]=0;
yrk[0]=1;
t[n_]=h n;
```

The recursive formulas for **xrk** and **yrk** are defined below using the **Module** command. Notice that
the manner in which these two functions are defined causes previous values to be retained so that
subsequent values may be based on them.

```
In[53]:=
xrk[n_]:=Module[{k1,k2,k3,k4},
     k1=f[t[n-1],xrk[n-1],yrk[n-1]];
     m1=g[t[n-1],xrk[n-1],yrk[n-1]];
     k2=f[t[n-1]+h/2,xrk[n-1]+(h k1)/2,
                     yrk[n-1]+(h m1)/2];
     m2=g[t[n-1]+h/2,xrk[n-1]+(h k1)/2,
                     yrk[n-1]+(h m1)/2];
     k3=f[t[n-1]+h/2,xrk[n-1]+(h k2)/2,
                     yrk[n-1]+(h m2)/2];
     m3=g[t[n-1]+h/2,xrk[n-1]+(h k2)/2,
                     yrk[n-1]+(h m2)/2];
     k4=f[t[n-1]+h,xrk[n-1]+h k3,
                     yrk[n-1]+h m3];
     m4=g[t[n-1]+h,xrk[n-1]+h k3,
                     yrk[n-1]+h m3];
     xrk[n]=xrk[n-1]+(h/6)(k1+2 k2+2 k3+k4)]
```

```
In[54]:=
yrk[n_]:=Module[{k1,k2,k3,k4},
     k1=f[t[n-1],xrk[n-1],yrk[n-1]];
     m1=g[t[n-1],xrk[n-1],yrk[n-1]];
     k2=f[t[n-1]+h/2,xrk[n-1]+
             (h k1)/2,yrk[n-1]+(h m1)/2];
     m2=g[t[n-1]+h/2,xrk[n-1]+
             (h k1)/2,yrk[n-1]+(h m1)/2];
     k3=f[t[n-1]+h/2,xrk[n-1]+
             (h k2)/2,yrk[n-1]+(h m2)/2];
     m3=g[t[n-1]+h/2,xrk[n-1]+
             (h k2)/2,yrk[n-1]+(h m2)/2];
     k4=f[t[n-1]+h,xrk[n-1]+
             h k3,yrk[n-1]+h m3];
     m4=g[t[n-1]+h,xrk[n-1]+
             h k3,yrk[n-1]+h m3];
     yrk[n]=yrk[n-1]+(h/6)(m1+2 m2+2 m3+m4)]
```

A table of approximate values for x and y is created below in **sols**.

```
In[55]:=
sols=Table[{t[i],xrk[i],yrk[i]},{i,0,10}]

Out[55]=
{{0, 0, 1}, {0.1, -0.0226878, 1.46031},
  {0.2, -0.10332, 2.06541},
  {0.3, -0.265382, 2.85897},
  {0.4, -0.540021, 3.8967},
  {0.5, -0.968273, 5.24956},
  {0.6, -1.60391, 7.00778},
  {0.7, -2.51707, 9.28596},
  {0.8, -3.79882, 12.2294},
  {0.9, -5.56704, 16.0223},
  {1., -7.97379, 20.8975}}
```

The approximate and the exact values can then be compared in the table below. This table has the same format as the two previous tables in the example using Euler's method. Notice the vast improvement in the accuracy associated with the Runge-Kutta method. Although this method requires more calculations at each step, the improvement in the approximation favors its use.

```
In[57]:=
comparerk=Table[{t[i],xrk[i],xexact[t[i]],
          yrk[i],yexact[t[i]]},{i,0,10}];
TableForm[comparerk]

Out[57]//TableForm=
0      0              0              1         1
0.1   -0.0226878     -0.0226978     1.46031   1.46032
0.2   -0.10332       -0.103346      2.06541   2.06545
0.3   -0.265382      -0.265432      2.85897   2.85904
0.4   -0.540021      -0.540105      3.8967    3.89682
0.5   -0.968273      -0.968408      5.24956   5.24975
0.6   -1.60391       -1.60412       7.00778   7.00806
0.7   -2.51707       -2.51737       9.28596   9.28638
0.8   -3.79882       -3.79926       12.2294   12.23
0.9   -5.56704       -5.56767       16.0223   16.0232
1.    -7.97379       -7.97468       20.8975   20.8987
```

Since the Runge-Kutta method can be extended to systems of first-order equations, this method can be used to solve higher order differential equations. This is accomplished by transforming the higher order equation into a system of first-order equations. We illustrate this below with an equation, the pendulum equation, that we have solved in many situations.

■ **EXAMPLE 8**

Use the Runge-Kutta method to approximate the solution of the initial value problem

$x'' + \sin(x) = 0$, $x(0) = 0$, $x'(0) = 1$

for h=0.1. Compare these results to those of the exact solution.

Solution:

We begin by transforming the second-order equation into a system of first-order equations. We do this by letting x'=y, so y'=x''=−sin(x). Hence, f(t,x,y)=y and g(t,x,y)=−sin(x). Again, we use **xrk** and **yrk** to represent the approximate values.

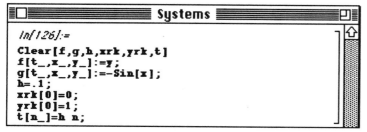

```
In[126]:=

Clear[f,g,h,xrk,yrk,t]
f[t_,x_,y_]:=y;
g[t_,x_,y_]:=-Sin[x];
h=.1;
xrk[0]=0;
yrk[0]=1;
t[n_]=h n;
```

A table of approximate values is given in sols below. Notice that we use enough values of t so that the t-values eventually become larger than 2π.

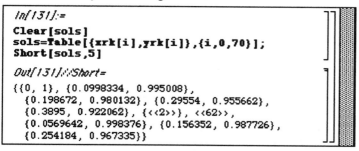

```
In[131]:=

Clear[sols]
sols=Table[{xrk[i],yrk[i]},{i,0,70}];
Short[sols,5]

Out[131]//Short=

{{0, 1}, {0.0998334, 0.995008},
  {0.198672, 0.980132}, {0.29554, 0.955662},
  {0.3895, 0.922062}, {<<2>>}, <<62>>,
  {0.0569642, 0.998376}, {0.156352, 0.987726},
  {0.254184, 0.967335}}
```

These points are plotted with ListPlot below in plot10. In earlier sections, we have found approximate solutions to this system by replacing sin(x) with x in the system. This leads to the solution graphed in **plot11** below.

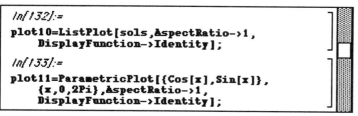

```
In[132]:=

plot10=ListPlot[sols,AspectRatio->1,
    DisplayFunction->Identity];

In[133]:=

plot11=ParametricPlot[{Cos[x],Sin[x]},
    {x,0,2Pi},AspectRatio->1,
    DisplayFunction->Identity];
```

These two approximate solutions are then displayed simultaneously below to reveal that they yield similar approximations.

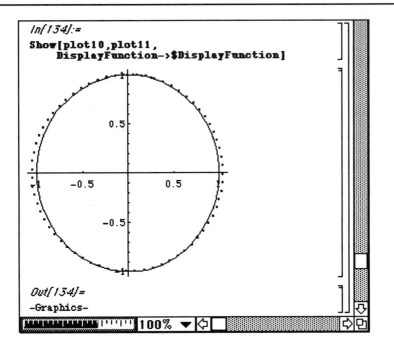

```
In[134]:=
Show[plot10,plot11,
    DisplayFunction->$DisplayFunction]
```

```
Out[134]=
-Graphics-
```

100% ▼

Error Analysis

An important aspect of numerical approximation is understanding the amount of error involved with a particular method. There are two types of error associated with any numerical method, round-off error and formula (truncation) error. The round-off error is due to the fact that the computer can hold only a finite number of significant digits while the formula error is due to the use of only a finite number of terms in the Taylor expansion. Consider the formula used in Euler's method for solving the initial value problem:

$$\frac{dy}{dx} = f(x,y), \ y(x_0) = y_0$$

which is

$$y_{n+1} = y_n + f(x_n, y_n)h \ , \ n = 0, \ 1, \ 2, \ \dots \ .$$

Hence, the difference between the approximation to the solution at $x = x_1$ obtained with Euler's method and the value obtained with the infinite Taylor series is

$$E_1 = \left(y_0 + hf(x_0, y_0) + \frac{h^2}{2!}\frac{df}{dx}(x_0, y_0) + \cdots \right) - \left(y_0 + hf(x_0, y_0) \right) = \frac{h^2}{2!}\frac{df}{dx}(x_0, y_0) + \cdots$$

so we have the following upper bound on the error at the first step:

$$|E_1| \leq \frac{\max\limits_{x_0 \leq x \leq x_1} \left|\frac{d\mathbf{f}}{dx}(x,y)\right| h^2}{2!} \leq K_1 h^2.$$

Since a similar error is obtained at each step, an approximation of n steps has a total error of

$$|E_n| \leq nKh^2$$

where K may differ from the constant K_1 above. However, we also have that

$$|E_n| = |y(x_n) - y_n| \leq K(x_n - x_0)h = Ch \quad \text{since} \quad n = \frac{x_n - x_0}{h}.$$

Hence, the total error is proportional to the step-size h.

A similar method can be followed to investigate the round-off error. Let

$\bar{y}_n$ = the actual value of the solution at $x = x_n$.

Then, the total round-off error after n steps is

$r_n = \bar{y}_n - y_n$.

A bound on the combined error (round-off and formula) is determined in the following manner:

$$|y(x_n) - \bar{y}_n| = |y(x_n) - y_n + y_n - \bar{y}_n| \leq |y(x_n) - y_n| + |y_n - \bar{y}_n| \leq |E_n| + |r_n|.$$

For a more precise discussion of the analysis of error associated with various numerical methods, refer to a numerical analysis text.

Glossary

Abs

Abs[z] yields the absolute value of **z**.

Apart

Apart[expression] computes the partial fraction decomposition of **expression**.

AppendTo

AppendTo[list,element] appends **element** to **list** and names the result **list**.

AspectRatio

AspectRatio is an option used with graphics commands like **Plot**, **ParametricPlot**, and **Show** which specifies the ratio of the height to the width of a plot. The default value is **AspectRatio->1/GoldenRatio**.

Axes

Axes is an option used with graphics commands which specifies whether axes are drawn or not. **Axes->True** specifies that axes are drawn; **Axes->False** specifies that no axes are drawn.

AxesOrigin

AxesOrigin->{xcoord,ycoord} is an option used with two-dimensional graphics commands which specifies that the given axes cross at the point (**xcoord,ycoord**).

BesselJ

BesselJ[n,x] yields the Bessel function of the first kind of order n, $J_n(x)$, which is regular at x=0 for every value of n.

Cancel

Cancel[expression] cancels common factors from the numerator and denominator of **expression**.

CharacteristicPolynomial

CharacteristicPolynomial[m,x] yields the characteristic polynomial of the matrix **m** in terms of the variable **x**.

Clear

Clear[symbol] clears all definitions of symbol.

Coefficient

Coefficient[poly,form] yields the coefficient of **form** in **poly**.

Collect

Collect[expression,variable] collects the terms in **expression** involving the same power of **variable**.

ColumnForm

ColumnForm[list] displays **list** as a column.

Compile

Compile[{x1,x2,...},expression] creates a compiled function which evaluates **expression** for the values given in **{x1,x2,...}**.

ComplexExpand

ComplexExpand[expression] expands **expression** assuming that all variables are real.

ComplexToTrig

ComplexToTrig is contained in the package **Trigonometry** located in the **Algebra** folder (or directory).
ComplexToTrig[expression] writes complex exponential terms in **expression** using trigonometric functions.

ContourPlot

ContourPlot[f[x,y],{x,xmin,xmax},-{y,ymin,ymax}] produces a graph of several contour levels for the two-dimensional function **f[x,y]** over the region [xmin,xmax]× [ymin,ymax].

Contours

Contours->n is an option used with **ContourPlot** to specify that **n** contours are to be plotted; **Contours->{z1,z2,...}** is an option used with **ContourPlot** to specify that the contours corresponding to the numbers **z1**, **z2**, ... are graphed.

ContourShading

ContourShading is an option used with **ContourPlot** to specify whether the space between contours is shaded or not. The default is **ContourShading->True**. If **ContourShading->False** is included, the space between contours is not shaded.

Cos

Cos[z] yields the cosine of **z** where **z** has radian units.

Cosh

Cohs[z] yields the hyperbolic cosine of **z**.

D

D[f,x] computes the derivative of **f** with respect to **x**; **D[f,{x,n}]** computes the **n**th derivative of **f** with respect to **n**; and **D[f,x,y]** yields the partial derivative of **f** with respect to **y** then **x**.

Dashing

Dashing is a two- and three-dimensional graphics directive. **Dashing[{r1,r2,...}]** specifies that subsequent line segments in a plot have lengths **r1**, **r2**, ...; this sequence repeats itself cyclically.

Delta

Delta[t] represent the Dirac delta function with spike at **t**=0. Similarly, **Delta[a+bt]** represents the function with spike at **t=-a/b**. This command is located in the package **LaplaceTransform** in the **Calculus** directory.

Det

Det[matrix] computes the determinant of the square matrix **matrix**.

DisplayFunction

DisplayFunction is an option for graphics functions like **Show**, **Plot**, and **Plot3D** which indicates how the resulting graphics should be displayed. The default setting, **DisplayFunction->$DisplayFunction**, indicates that the graphics objects produced should be displayed; **DisplayFunction->Identity** generates no display.

Do

Do[expression,
{i,imin,imax,istep}] computes
expression for values of the variable **i** from
i=imin to **i=imax** in steps of **istep**. If **istep**
is not included, the default stepsize is 1.

Drop

Drop[list,n] deletes the first **n** elements of
list and returns the resulting list.

DSolve

DSolve[equation,y[x],x] attempts to
solve the differential equation equation for the
dependent variable **y[x]** in terms of the
independent variable **x**.
DSolve[{eq1,eq2,...},
{y1[x],y2[x],...},x] attempts to solve
the system of ordinary differential equations
defined in the list **{eq1,eq2,...}** for the
dependent variables **{y1[x],y2[x],...}** in
terms of **x**.

Eigensystem

Eigensystem[mat] yields a list of the
eigenvalues and eigenvectors of the n × n matrix
mat. These lists are given in corresponding order
if **mat** has an equal number of eigenvalues as
eigenvectors. However, if the number of
eigenvalues exceeds the number of eigenvectors,
then a zero vector is given in the list of eigenvectors
to correspond to each additional eigenvalue.
Numerical values for the eigenvalues and
eigenvectors are given if **mat** contains approximate
real numbers. Output is given the following form:
{{eval1,eval2,...},{evec1,evec2,.-
..}} where each eigenvector is a list of
components.

Eigenvalues

Eigenvalues[mat] determines the
eigenvalues of the n × n matrix **mat**. An
eigenvalue of the n × n matrix A is a number λ such
that $Ax=\lambda x$ for a non-zero vector x. If **mat**
involves approximate real numbers then the result
of **Eigenvalues[mat]** is a list of numerical
eigenvalues. If an eigenvalue is repeated, then it
appears in the list with the appropriate multiplicity.

Eigenvectors

Eigenvectors[mat] computes the
eigenvectors of the n × n matrix **mat**. The output
list consists of the linearly independent
eigenvectors of **mat** and the correct number of zero
vectors so that the length of the list is n. An
eigenvector of the n × n matrix A is a non-zero
vector x such that $Ax=\lambda x$ for the corresponding
eigenvalue λ. **Eigenvectors[mat]** yields
numerical eigenvectors if **mat** contains
approximate real numbers.

Evaluate

Evaluate[expression] causes
Mathematica to evaluate **expression**
immediately even if the command is the argument
of a function the attributes of which indicate that it
should be held unevaluated. This command is
particularly useful when working with lists of
functions.

Expand

Expand is used to expand expressions. This can be accomplished in several ways:

Expand[expression] carries out the multiplication of products and positive integer powers in **expression**. The resulting expression is the sum of terms.

Expand[epression,form] expands only those terms in **expression** which do not match the pattern defined in **form**. When **expression** involves fractions, **Expand[expression]** expands the numerator and divides the denominator into each term in **expression**. **Expand** has the option **Trig->True** which causes trigonometric functions to be treated as exponential rational functions so that they can be expanded.

ExpandAll

ExpandAll[expression] completely expands the numerators and denominators in **expression**.

ExpandAll[expression,form] does not apply **ExpandAll** to terms in **expression** which match the pattern defined by **form**. **ExpandAll** also has the option **Trig->True** as defined above in **Expand**.

ExpandNumerator

ExpandNumerator[expression] multiplies out the terms in **expression** which appear as numerators. Therefore, **ExpandNumerator** applies only to terms having positive integer exponents.

Exponent

Exponent[expression,variable] yields the highest power of **variable** which appears in **expression** where **expression** is a single term or a sum of terms. This command can also be stated in the form

Exponent[expression,variable, function] to apply **function** to the set of exponents which appear in **expression**.

Factor

Factor[poly] completely factors the polynomial **poly** over the integers. Factorization can also be accomplished modulo a prime number **p** with

Factor[poly,Modulus->p]. Other options include **Trig->True** which interprets trigonometric functions as exponential rational functions and factors them correctly, **GaussianIntegers->True** which allows for Gaussian integer coefficients in the factorization, and **Variables->{x1,x2,...}** which indicates an ordering on the variables in **poly**. This ordering may affect the time needed to complete the factorization.

FindRoot

FindRoot is used to approximate the numerical solution of an equation or a system of equations. This command is entered as

FindRoot[lhs==rhs,{x,x0}] where **x0** is the initial guess of the solution to the equation **lhs==rhs**. However, if the derivatives of the equation can not be computed symbolically, then the command

FindRoot[lhs==rhs,{x,x0,x1}] may be used to find the approximate solution. In this case, the first two values of **x** are taken to be **x0** and **x1**. The **FindRoot** command of the form

FindRoot[lhs==rhs,{x,x0}] is based on

Newton's method while that of
FindRoot[lhs==rhs,{x,x0,x1}] is a
variation of the secant method. The options of
FindRoot include **AccuracyGoal** to indicate
the desired accuracy of the function at the root,
Compiled to instruct whether or not the function
should be compiled, **DampingFactor** to specify
the damping factor in Newton's method,
Jacobian to enter the Jacobian of the system of
equations, **MaxIterations** to specify the
maximum number of iterations to be taken in the
search for the solution, and **WorkingPrecision**
to indicate the number of digits to be used in
internal computations. Some of the default settings
include **DampingFactor->1** and
MaxIterations->15. The default settings for
AccuracyGoal is 10 digits fewer than that of
WorkingPrecision.

Flatten

Flatten is used to flatten out nested lists. This
command can be entered in several forms to obtain
different results. **Flatten[list]** flattens out all
levels of **list**. Flattening out unravels all nested
lists by omitting the inner braces {}.
Flatten[list,n] only flattens the first **n**
levels in **list**.

Graphics

Graphics[primitives,options] denotes
the graphical image of a two-dimensional object.
These objects are viewed with **Show**. The
primitives which may be used include **Circle**,
Point, **Polygon**, **Rectangle**, and **Line** while
the lengthy list of options includes
AspectRatio, **Axes**, **DisplayFunction**,
Frame, **PlotRange**, and **Ticks**. Sound
primitives such as
SampledSoundList and
SampledSoundFunction can also be

considered with **Graphics**. **Graphics**
directives such as **GrayLevel**, **RGBColor**, and
Thickness may also be included in a **Graphics**
command.

GraphicsArray

GraphicsArray is used to represent an array
of graphics objects. This command can be entered
in several ways.
GraphicsArray[{graph1,graph2,...}
] represents a row of objects while
GraphicsArray[{
  {graph11,graph12,...},
  {graph21,graph22,...},
...}] yields the array with rows
{graph11,graph12,...},
{graph21,graph22,...}, ... The objects
contained within **GraphicsArray** are displayed
in identical rectangular regions with **Show**. The
same options which are available to **Graphics**
can also be used with **GraphicsArray** with the
exception of **StringConversion**,
ColorOutput, and **DisplayFunction**. Also,
with **GraphicsArray**, the default setting for
Ticks and **FrameTicks** is **None**. In addition to
these options, **GraphicsArray** has the option
GraphicsSpacing which is used to control the
spacing between the rectangular display regions.
The default setting for **GraphicsSpacing** is
0.1.

GrayLevel

GrayLevel[intensity] is a directive used
with graphics functions to indicate the level of gray
to be used when displaying graphics objects. The
value of
intensity is between 0.0 and 1.0 where a level
of 0.0 denotes black and 1.0 white. This directive
is most useful when graphing several plots
simultaneously.

Head

Head[expression] yields the head of the given expression. The head of an expression can be an operator such as **Plus**, a function name, a type of number such as **Real**, or another *Mathematica* object.

HermiteH

HermiteH[n, x] represents the Hermite polynomial $H_n(x)$ which satisfies the ordinary differential equation $\dfrac{d^2y}{dx^2} - 2x\dfrac{dy}{dx} + 2ny = 0$. The Hermite polynomials are orthogonal with weight function $w(x) = e^{-x^2}$.

IdentityMatrix

IdentityMatrix[n] yields the $n \times n$ identity matrix.

If

If is a conditional block similar to those of programming languages.
If[condition,then,else] performs then if condition is **True** and else if condition is **False**. This block can be entered as
If[condition,then,else,other] if condition is neither **True** nor **False**. In this case, the result is **other**.

ImplicitPlot

ImplicitPlot is used to plot curves which are given implicitly as solutions to equations. This can be accomplished in several ways.
ImplicitPlot[eq,{x,x0,x1}] determines the solution to **eq** using **Solve** and then plots this solution over the indicated interval for **x**.
ImplicitPlot[eq,{x,x0,r1,r2,..., x1}] performs a similar

task but does not use the list of x-coordinates **r1**, **r2**, ... **ImplicitPlot[eq,{x,xo,x1}, {y,y0,y1}]** uses a method similar to **ContourPlot** to generate the graph over the indicated **x** and **y** intervals. Finally, **ImplicitPlot[{eq1,eq2,...},cints]** plots several curves simultaneously using the coordinate intervals specified in **cints**. This command is located in the **Graphics** package directory.

Integer

Integer represents a type of number in *Mathematica*. Integers can be of any length and can be entered in base **b** with the command **b^^digits**. The largest allowable base is 35.

Integrate

Integrate is used to compute the exact value of definite and indefinite integrals. Integrals may be single or multiple. Syntax for these commands is given below.

`Integrate[f, x]` determines the indefinite integral $\int f \, dx$.

`Integrate[f, {x, x1, x2}]` evaluates the definite integral $\int_{x1}^{x2} f \, dx$. For multiple integrals, the command

`Integrate[f, {x, x1, x2}, {y, y1, y2}]`

calculates $\int_{x1}^{x2} dx \int_{y1}^{y2} dy \, f$. Notice that the first variable listed within the command is the variable involved in the outermost integral.

`Integrate` is applicable to those integrals which depend on exponential, logarithmic, trigonometric, and inverse trigonometric functions if the outcome also involves these functions.

Inverse

`Inverse[m]` computes the inverse of the $n \times n$ matrix **m**. *Mathematica* prints a warning message if **m** is not invertible. If the matrix **m** involves approximate real or complex numbers, then the accuracy used in the computation of the inverse can be acquired with `Accuracy`. `Inverse` has the option `ZeroTest->test` which applies `test` to the elements of the matrix to determine whether these elements are zero. The default setting for `ZeroTest` is `ZeroTest->(Together[#]==0)&`.
`Inverse` also has the option `Modulus->p` which computes the inverse of the matrix modulo **p**.

LaguerreL

`LaguerreL[n, a, x]` represents the Laguerre polynomial which is denoted $L_n^a(x)$. These polynomials satisfy the ordinary differential equation $x \dfrac{d^2y}{dx^2} + (a + 1 - x)\dfrac{dy}{dx} + ny = 0$ and are orthogonal with respect to the weight function $w(x) = x^a e^{-x}$.

LaplaceTransform

`LaplaceTransform[func, t, s]` calculates the Laplace transform of `func` (which is treated as a function of `t`). Hence, the result is a function of `s`. This command is located in the `LaplaceTransform` package in the `Calculus` directory.

LegendreP

`LegendreP[n, x]` represents the Legendre polynomial $P_n(x)$ while
`LegendreP[n, m, x]` is the associated Legendre polynomial $P_n^m(x)$. The Legendre polynomials are orthogonal with respect to the weight function $w(x) = 1$ and are solutions to the ordinary differential equation

$$\left(1 - x^2\right)\frac{d^2y}{dx^2} - 2x\frac{dy}{dx} + n(n + 1)y = 0.$$ The associated Legendre polynomials are given by $P_n^m(x) = (-1)^m (1 - x^2)^{m/2} \dfrac{d^m}{dx^m} P_n(x)$.

Length

Length[list] yields the number of elements in **list**. This command can be used to determine the number of terms in a polynomial as well with **Length[poly]**. If **expression** cannot be divided into subexpressions, then **Length[expression]** = 0.

Limit

Limit[expression,x->x0] determines the limit of **expression** as **x** approaches **x0**. An option used with **Limit** is that of **Direction**. **Direction->1** causes *Mathematica* to take the limit as **x** approaches **x0** from below, and **Direction->-1** forces the limit to be evaluated as **x** approaches **x0** from above. In some cases, the result of **Limit** is given in the form **RealInterval[{a,b}]** when **Limit** encounters an uncertain value which is located on the interval {**a,b**}. **Limit** leaves expressions which depend on an arbitrary function unevaluated. **Analytic** is an option which can be used with **Limit**.

Line

Line[{point1,point2,...}] represents the graphics primitive for a line in two or three dimensions which connects the points listed in {**point1,point2,...**}. Hence, the points can be entered as {**x,y**} or {**x,y,z**}. They can be scaled as well with **Scaled[{x,y}]** and **Scaled[{x,y,z}]**. Several options are available for displaying the line. **Thickness** or **AbsoluteThickness** affects the thickness of the line. **Dashing** or **AbsoluteDashing** controls the dashing of the line. Finally, **CMYKColor**, **GrayLevel**, **Hue**, and **RGBColor** monitor the coloring of the line.

ListPlot

ListPlot is used to plot a list of points. This can be accomplished in one of two ways. **ListPlot[{f1,f2,...}]** plots the list of points {1, **f1**}, {2, **f2**}, ... On the other hand, the points {x,y} in two dimensions to be plotted can be entered within the command using **ListPlot[{x1,y1},{x2,y2},...]**. Since the plot which results is a **Graphics** object, **ListPlot** has the same available options as **Graphics** along with the additional options **PlotJoined** and **PlotStyle**.

Log

Log[z] represents the natural logarithm (base e) of **z** while **Log[b,z]** yields the logarithm of **z** to base **b**. Exact values result from these functions whenever possible.

LogicalExpand

LogicalExpand[expression] expands out expressions in **expression** which involve the logical connectives **&&** and **||** by applying appropriate distributive properties.

Map

Map is used with functions and expressions to apply a function to particular elements of an expression. This can be done with **Map[f,expression]** to apply the function **f** to each element on the first level of **expression** or with **Map[f,expression,lev]** to apply the function **f** to each element on the level **lev** of **expression**. Note that the command **Map[f,expression]** can be entered as **f/@expression**.

MatrixForm

MatrixForm[{{a11,a12,...},{a21,a-22,...},...}] is used to display the rectangular array given in the usual form of a matrix where each element **{aj1,aj2,...}** becomes a row. **MatrixForm** only affects the printing of the array. It does not alter any calculations. The options used with **TableForm** are used with **MatrixForm** as well.

MatrixPower

MatrixPower[a,n] takes the product of the square matrix **a** with itself **n** times. Note that if **n** is negative, then *Mathematica* computes the powers of the inverse of **a**.

Method

Method is an option used with numerical functions such as **Solve** and **NIntegrate** to indicate the algorithm to be used to determine the solution. Two examples of settings used for this option with **NIntegrate** are **GaussKronrod** and **MultiDimensional**.

Module

Module is a means by which local variables are defined in *Mathematica*.
Module[{x,y,...},body] defines the local variables **{x,y,...}** used in the commands given in **body**. Even if a variable in the list **{x,y,...}** has the same name as a global variable, changes in the value of that local variable do not affect the value of the corresponding global variable. Initial values can be assigned to local variables with **Module[{x=x0,y=y0,...},body]**. The initial values are evaluated before the module is executed. **Module** can be used in defining a function with a condition attached. This is accomplished with

lhs:=Module[variables,
                       rhs/;condition].
This allows for the sharing of the local variables listed in **variables** with **rhs**. Symbols created within **Module** have the attribute **Temporary** while **Module** itself carries the attribute **HoldAll**.

N

N[expression] yields the numerical value of **expression**. The number of digits of precision can be requested with **N[expression,n]**. With this command *Mathematica* performs computations with **n**-digit precision numbers. However, results may involve fewer than **n** digits.

NDSolve

NDSolve is used to numerically solve ordinary differential equations (i.e., those depending on one independent variable). A single equation is solved with
NDSolve[{equation,y[x0]==y0},y, {x,xmin,xmax}] where equation is solved for **y[x]** and the solution is valid over the interval **{xmin,xmax}**. Systems of equations can be solved   with
NDSolve[{eq1,eq2,...,y1[x0]==a, y2[x0]==b,...},{y1,y2,...},{x,xmin,xmax}]. Note that enough initial conditions must be given to determine the solution. Hence, there are no arbitrary variables in the solution which results from
NDSolve. Each component of the result is given in the form
InterpolatingFunction[{xmin,xmax}-,<>]. These solutions can be plotted in two dimensions with **ParametricPlot** or in three dimensions with
ParametricPlot3D. The options of **NDSolve** are **AccuracyGoal**, **MaxSteps**,

PrecisionGoal, **StartingStepSize**, and **WorkingPrecision**. The default setting for **MaxSteps** is 500 while that for **AccuracyGoal** and **PrecisionGoal** are each 10 digits fewer than that of **WorkingPrecision**. The default for **WorkingPrecision** is **$WorkingPrecision.**

NIntegrate

NIntegrate[f,{x,xmin,xmax}] numerically computes the definite integral of **f** from **x** = **xmin** to **x** = **xmax**. Multidimensional integrals can be considered using the syntax for multidimensional integrals given for **Integrate** and the **Method->MultiDimensional** option. The options for **NIntegrate** along with their default values are **AccuracyGoal->Infinity**, **Compiled->True**, **GaussPoints->Floor[WorkingPrecision/3]**, **MaxRecursion->6**, **MinRecursion->0**, **PrecisionGoal->Automatic**, **SingularDepth->4**, and **WorkingPrecision->$MachinePrecision**. **Integrate** continues to take subdivisions until the specified value of **AccuracyGoal** or **PrecisionGoal** is met. The default setting for **PrecisionGoal** is ten digits fewer than the setting for **WorkingPrecision**. **NIntegrate** has the attribute **HoldAll**.

Normal

Normal[expression] changes **expression** to a normal expression. If **expression** is a power series, then this is accomplished by truncating the series by removing the higher order terms.

Numerator

Numerator[expression] yields the numerator of **expression** by selecting the terms of **expression** which do not have exponents having a negative number as a factor. The argument of **Numerator** can be a rational expression.

ParametricPlot

ParametricPlot is used to plot parametric curves in two dimensions. **ParametricPlot[{x[t],y[t]}, {t,tmin,tmax}]** plots the curve given by **x=x[t]** and **y=y[t]** from **t** = **tmin** to **t** = **tmax**. More than one such curve is generated with **ParametricPlot[{{x1[t],y1[t]}, {x2[t],y2[t]},...}, {t,tmin,tmax}]**. The options used with **Plot** are available to **ParametricPlot**. The default setting **Axes->True** is different, however. The result of **ParametricPlot** is a **Graphics** object.

ParametricPlot3D

ParametricPlot3D is used to plot parametric curves in three dimensions. **ParametricPlot3D[{x[t],y[t],z[t]} ,{t,tmin,tmax}]** generates the three-dimensional curve defined by **x** = **x[t]**, **y** = **y[t]**, **z** = **z[t]** for **t=tmin** to **t=tmax**. If the coordinates depend on two parameters, then **ParametricPlot3D[{x[t,u],y[t,u],z[t,u]}, {t,tmin,tmax},{u,umin,umax}]** plots the surface which results. The command **ParametricPlot3D[{x[t,u],y[t,u],z[t,u],shade}, {t,tmin,tmax},{u,umin,umax}]** shades the surface using the colors defined by **shade**. A

simultaneous plot is generated with
`ParametricPlot3D[{x1,y2,z2},`

`{x2,y2,y3},...]`. `ParametricPlot3D`
uses the same options as `Graphics3D` with the
addition of `Compiled` and `PlotPoints`. If the
default setting `PlotPoints->Automatic` is
used, then the setting `PlotPoints->75` is used
for curves and `PlotPoints->{15,15}` is used
for surfaces.

Part ([[...]])

Part is used to extract particular parts of
expressions. `Part[expr,i]` or `expr[[i]]`
yields the `i`th part of expression. `expr[[-i]]`
counts the `i`th part from the end. `expr[[0]]`
yields the head of `expr`.
`Part[expr,{i,j,..}]` or
`expr[[i,j,..]]` yields the part `{i,j,..}` of
`expr`. Several parts are given with
`Part[expr,i,j,..]`.

Partition

Partition is used to partition lists into
sublists. `Partition[list,n]` forms `n` sublists
from `list`. `Partition[list,n,d]` also
creates n sublists with each sublist offset by `d`
elements. Also,
`Partition[list,{n1,n2,...},`

`                {d1,d2,...}]`
partitions successive levels of list into sublists of
length `ni` using offsets `di`.

Plot

Plot is used to plot functions of one variable.
`Plot[f[x],{x,xmin,xmax}]` plots the
function `f[x]` over the interval `x=xmin` to
`x=xmax`.
`Plot[{f1[x],f2[x],...},`

`{x,xmin,xmax}]` plots several functions
simultaneously. The options used with `Graphics`

are also available to `Plot` along with the addition
of several more. These additional options as well
as their default setting are `Complied->True`,
`MaxBend->10`, `PlotDivision->20`,
`PlotPoints->25`, and
`PlotStyle->Automatic`. The setting for
`PlotPoints` is used to initially select equally
spaced sample points. *Mathematica* then selects
other sample points in an attempt to minimize
`MaxBend`. The `PlotDivision` setting limits the
number of subdivisions taken on any interval. Note
that since *Mathematica* samples only a finite
number of points, the plot which is produced by
`Plot` may be inaccurate. Plot returns a
`Graphics` object.

PlotPoints

PlotPoints is a plotting function option used
to indicate the total number of
equally spaced sample points to be used. This
setting can be made in several ways.
`PlotPoints->n` indicates that `n` points be used
while with two variables `PlotPoints->n`
implies that `n` points be selected in the direction of
both coordinates. If different numbers are to be
used in the two directions, then
`PlotPoints->{nx,ny}` is used.

PlotRange

PlotRange is an option used with graphics
functions to indicate the function values to be
included in the plot. If the function is
one-dimensional, then `PlotRange` limits the
y-values. However, if the function is
two-dimensional, then the z-coordinate is restricted.
The setting `PlotRange->All` indicates that all
points are to be included while
`PlotRange->Automatic` implies that outlying
points be omitted. Also, the setting
`PlotRange->{min,max}` places definite limits

on the y-values or z-values to be displayed in the plot depending on the number of dimensions present. In addition, this option can be used as **PlotRange->{{xmin,xmax},{ymin,ymax}}** to indicate translation from original to translation coordinates. Note also that **Automatic** can be used as one component of a setting of the form **{min,max}**.

PlotStyle

PlotStyle is an option used with **Plot** and **ListPlot** to indicate the manner in which lines or points be displayed. The setting **PlotStyle->style** implies that all lines or points be drawn using the graphics directive **style**. Such directives include **RGBColor**, **Thickness**, **Hue**, and **GrayLevel**. **PlotStyle** is often useful in multiple plots. The setting **PlotStyle->{{style1,style2,...}}** indicates that successive lines or points be rendered with the listed directives. Styles must be enclosed in lists as they are used cyclically.

PlotVectorField

PlotVectorField[{fx,fy}, {x,x0,x1},{y,y0,y1}] generates the two-dimensional vector field given by the vector function **{fx,fy}** using the indicated intervals for **x** and **y**. This command is located in the **PlotField** package in the **Graphics** directory.

PlotVectorField3D

PlotVectorField3D[{fx,fy,fz}, {x,x0,x1},{y,y0,y1},{z,z0,z1}] generates the three-dimensional vector field given by the vector function **{fx,fy,fz}** using the indicated intervals for **x**, **y**, and **z**. This command is located in the **PlotField3D** package in the **Graphics** directory.

Plot3D

Plot3D is used to plot functions of two variables. **Plot3D[f[x,y], {x,xmin,xmax},{y,ymin,ymax}]** yields a three-dimensional plot of the function **f[x,y]** over the region defined by **{xmin,xmax}** × **{ymin,ymax}** in the xy-plane. Also, **Plot3D[{f[x,y],shade}, {x,xmin,xmax},{y,ymin,ymax}]** yields a three- dimensional plot with shading according to the settings of **shade**. The possible directives used to define shade are **RGBColor**, **GrayLevel**, and **Hue**. The same options which are used with **SurfaceGraphics** are used with **Plot3D** as well, but there are two additional options for use with **Plot3D**. These are **Compiled** and **PlotPoints**. Note that the default setting for these options are **Compiled->True** and **PlotPoints->15**. Hence, *Mathematica* automatically selects 15 sample points in each direction. If no shading option is included and if **Lighting->False**, then shading is done according to height. The object returned by **Plot3D** is a **SurfaceGraphics** object which includes a setting for the **MeshRange** option.

Point

Point[coords] is the graphics primitive for a point in two or three dimensions. The coordinates given in **coords** can be expressed as **{x,y}** or **{x,y,z}** or they can be scaled with **Scaled[{x,y}]** or **Scaled[{x,y,z}]**. Points are rendered as circular regions and may be shaded or colored using **GrayLevel**, **Hue**, **RGBColor**, or **CMYKColor**. The size of the point is controlled by the option **PointSize**.

PointSize

PointSize[r] indicates that all **Point** elements be drawn as circles of radius **r** in the graphics object. The radius **r** is measured as a fraction of the width of the entire plot. The default setting in two dimensions is **PointSize[0.008]** while it is **PointSize[0.01]** for three dimensions.

PowerExpand

PowerExpand[expression] expands out terms of the form **(a b)^c** and **(a^b)^c** which appear in **expression**. Unlike **Power**, this expansion is carried out even if **c** is not in the form of an integer.

PrependTo

PrependTo[list,element] is equivalent to **list=Prepend[list,element]** in that **element** is prepended to **list** and the value of **list** is reset to the prepended list.

Prolog

Prolog is a two- and three-dimensional graphics function option which indicates graphics directives to use before a plot is started. The graphics primitives given in Prolog are displayed after axes, boxes, and frames are rendered. Primitives used in two dimensions can also be used for three.

ReplaceAll (/.)

ReplaceAll[exp,rules], symbolized by **exp/.rules**, is used to apply a rule or list of rules, **rules**, to every part of an expression **exp**. Each listed rule is applied to every subexpression of **exp**. When a rule applies, then the result that is obtained after the appropriate transformation is returned. If none of the rules apply, then **exp** is returned by *Mathematica*.
ReplaceAll[exp,rules] is equivalent to

MapAll[Replace[#,rules]&,exp].

Roots

Roots[lhs==rhs,variable] produces a set of equations of the form
eq1||eq2||... which represents the solution to the polynomial equation
lhs==rhs. When such functions as **Solve** cannot determine the solution to an equation, it returns **{ToRules[Roots[eqn,var]]}** the approximate solutions of which can be determined numerically with **N**. Roots has many options. These include along with their default settings **Cubics->True**, **EquatedTo->Null**, **Modulus->Infinity**, **Multiplicity->1**, **Quartics->True**, and **Using->True**. Roots uses **Decompose** and **Factor** in attempting to find the roots.

Series

Series is used to obtain power series expansions for functions of one and two variables. **Series[f[x],{x,a,n}]** yields the power series expansion for the function **f** about **x=a** to order $(x-a)^n$. Note that this power series is given by the formula
$$\sum_{k=0}^{\infty} \frac{f^{(k)}(a)}{k!}(x-a)^n.$$ If f is a function of two variables, then **Series[f[x,y],{x,a,nx},{y,b,ny}]** successively determines the power series of **f** with respect to y and then with respect to x.

Short

Short is used to give output in a shortened form. **Short[expression,n]** prints **expression** on at most **n** lines while **Short[expression]** prints **expression** in its usual form minus one line. The omitted sequences of k elements of **expression** are given as **<<k>>**. If entire sequences are omitted, then they are represented as **Skeleton** objects. **Short** is not only useful with **OutputForm**. It can be used with **InputForm** as well. This function only affects the printing of expressions. It does not alter any calculations.

Show

Show is used to display two- and three-dimensional graphics objects. Therefore, it is useful with **ContourGraphics**, **DensityGraphics**, **Graphics**, **Graphics3D**, **GraphicsArray**, and **SurfaceGraphics**. The command **Show[graphics,options]** displays graphics using the indicated options. Several graphics objects can be shown at once with **Show[graphics1,graphics2,...]**. Note that all options included in the **Show** command override options used in the original graphics commands. The option **DisplayFunction** specifies whether graphs are to be shown or not. The setting **DisplayFunction->$DisplayFunction** causes the graph to be shown while **DisplayFunction->Identity** causes the display to be suppressed. **Show** is automatically used with such functions as **Plot** and **Plot3D** to display graphics.

Simplify

Simplify[expression] tries to determine the simplest form of **expression** through a sequence of algebraic transformations. If **expression** is complicated, then *Mathematica* can spend a great deal of time testing various transformations in an attempt to determine the simplest form of **expression**. In this case, commands like **Expand** and **Factor** are useful

prior to application of **Simplify**. **Simplify** has the option **Trig** where **Trig->False** does not apply trigonometric identities in the simplification process.

Sin

Sin[x] represents the trigonometric sine function of **x** where **x** is given in radian measurement. Multiplication by **Degree** converts degrees to radian measure.

Sinh

Sinh[x] represents the hyperbolic sine function.

Solve

Solve is used to solve equations or systems of equations where all equations must include a double equals sign, **==**. Systems of equations may be entered in the form of a list or be connected by the logical connective **&&**.
Solve[{lhs1==rhs1,lhs2==rhs2,..},
{var1,var2,...}] solves the given equations for the variables listed. The commands
Solve[
{lhs1,lhs2,...}==
{rhs1,rhs2,...},
{var1,var2,...}] and
Solve[lhs1==rhs1&&lhs2==rhs2&&
...,{var1,var2,...}] also yield the same results. The output of **Solve** is rendered in the form of the list {{var1->a},...} if there are several variables or several solutions. The result appears in the form var1->a if there is only one. *Mathematica* indicates the multiplicity of roots by listing a repeated root the appropriate number of times.
Solve[eqn,var,elim] solves the given equation or system of equations for the variable(s) given in **var** by eliminating the variable(s) given in

elim. This allows for certain variables to be given in terms of parameters. **Solve** has several options. One of these options is **Mode**. **Mode->Modular** allows for a weaker form of equality in which two integer expressions are equal modulo a fixed integer. This integer can be explicitly given with **Modulus==p**. Otherwise, *Mathematica* tries to determine an integer for which the given equations can be satisfied. Another option is **InverseFunctions** which indicates whether or not inverse functions should be used in the solution process. The default setting for **InverseFunctions** is **Automatic** which indicates that inverses are employed. If this is the case, a warning message is printed in the output. If there are no solutions, then **Solve** yields the symbol { }. **Solve** gives solutions which cannot be solved for explicitly in terms of **Roots**. **Solve** is most useful when dealing with polynomial equations.

SolveAlways

SolveAlways[equations,variables] yields parameter values for which the given equation or system of equations which depend on a set of parameters is valid for all variable values. As with **Solve**, systems of equations can be entered in the form of a list or through the use of logical connectives **&&**. Note that equations must include a double equals sign, **==**. The results of **SolveAlways** are given in the form of a list if there is more than one parameter.
Solve[[!Eliminate[!equations,variables]] can be used to accomplish the same results as **SolveAlways**. **SolveAlways** is most useful when dealing with polynomial equations.

Sum

Sum computes finite sums of single and multiple sums.
Sum[a[i],{i,imin,imax}] yields the value of

$$\sum_{i=imin}^{imax} a[i]$$ while a step size other than one unit is

indicated with **Sum[a[i],{i,imin,imax,ip}]**

The multiple sum $$\sum_{i=imin}^{imax} \sum_{j=jmin}^{jmax} a[i,j]$$

is computed with **Sum[a[i,j],**
{i,imin,imax},{j,jmin,jmax}]. Note that the order of summation is the same as that used with **Integrate** for multiple integrals so the syntax for summation of more than a double sum is easily extended.

Table

Table is used in several forms to create tables.
Table[expression,{i}] generates n copies of expression.
Table[expression,{i,imax}] creates a list of the values of expression from **i** = 1 to **i** = **imax**. A minimum value of **i** other than **1** is indicatedwith**Table[expression,{i,imin-,imax}]** while a step size other than one unit is defined with
Table[expression,
{i,imin,imax,istep}].
Finally, **Table[expression,**
{i,imax,imax},{j,jmin,jmax},...]
creates a nested list where the outermost list is associated with **i**. Matrices, vectors, and tensors are all represented as tables in *Mathematica*.

TableForm

TableForm[list] prints **list** in the form of a rectangular array of cells. Note that the size of the elements of **list** are not required to be the same as they are with **MatrixForm**. The width of each column and the height of each row is determined by the size of the largest element in the respective row or column. The options used with **TableForm** are **TableAlignments**, **TableDepth**, **TableDirections**, **TableHeadings**, and **TableSpacing**. **TableForm** only affects the appearance of **list**. It does not change later calculations involving **list**.

Ticks

Ticks is a two- and three-dimensional graphics function option which indicates how tick marks should be placed on axes. **Ticks** can be specified in several ways. **Ticks->None** implies that no marks be drawn, **Ticks->Automatic** displays tick marks automatically by placing the marks at points whose coordinates have the fewest number of decimal digits, and **Ticks->{xaxis,yaxis,...}** which states tick marks for each axis. In specifying individual axes specifications, many settings for **xaxis**, **yaxis**, ... are possible. These include **None** and **Automatic** which were previously described as well as **{m1,m2,...}** which lists the tick mark positions, **{{m1,label1},{m2,label2},...}** which lists the mark positions and corresponding label, **{{m1,label1,length1},...}** which also indicates a scaled length for each mark, **{{m1,length1, {poslength1,neglength1}},...}** which states lengths in the positive and negative directions, and **{{m1,length1,length1,style1},...}**

which gives a specified style to use to draw the tick marks as well. These styles are indicated with such directives as **Thickness**, **GrayLevel**, and **RGBColor**. Labels which accompany tick marks appear in **OutputForm**.

Together

Together[expression] combines the terms of **expression** by adding them over a common denominator and simplifying the result. **Trig** is an option used with **Together**. The setting **Trig->True** causes trigonometric functions to be interpreted as rational functions of exponentials.

Transpose

Transpose is used to transpose both matrices and tensors. **Tranpose[m]** transposes the matrix **m** by interchanging the rows and columns of m. Also, **Tensor[t,{k1,k2,...}]** yields the tensor **t'** such that the **i**-th index becomes the **ki**-th.

Trig

Trig is an option used with functions such as **Expand** and **Factor** to indicate whether or not to consider trigonometric functions as rational functions of exponentials. **Trig->False** is the default setting of all functions except **Simplify**. With this setting, *Mathematica* treats trigonometric functions as indivisible objects. With the setting **Trig->True**, these functions are converted to exponential rational functions so that they can be manipulated.

Variables

Variables[polynomial] lists the independent variables in the given polynomial.

VectorHeads

VectorHeads->setting is an option used with the commands located in the **PlotField3D** package in the **Graphics** folder (or directory) which specifies whether heads should be placed on the vectors which are drawn. The default setting is **VectorHeads->False.**

Selected References

Abell, Martha L. and Braselton, James P., The *Mathematica* Handbook, Academic Press, 1992.

Abell, Martha L. and Braselton, James P., *Mathematica* by Example, Academic Press, 1992.

Blachman, Nancy, *Mathematica: Quick Reference*, Version 2, Addison-Wesley, 1992.

Blachman, Nancy, *Mathematica: A Practical Approach*, Prentice-Hall, 1992.

Boyce, William E. and DiPrima, Richard C., Elementary Differential Equations and Boundary Value Problems, Fourth Edition, John Wiley & Sons, 1986.

Bronson, Richard, Matrix Methods: An Introduction, Academic Press, 1970.

Campbell, Robert, The *Mathematica* Help Stack Version 2, Variable Symbols, 1992.

Churchill, Ruel V., Operational Mathematics, Third Edition, McGraw-Hill, 1972.

Conway, John B., Functions of One Complex Variable, Second Edition, Springer-Verlag, 1978.

Crandall, Richard E., *Mathematica* for the Sciences, Addison-Wesley, 1991.

Gray, Theodore W. and Glynn, Jerry, Exploring Mathematics with *Mathematica*, Addison-Wesley, 1991.

Gutterman, Martin M. and Nitecki, Zbigniew H., Differential Equations: A First Course, Third Edition, Saunders College Publishing, 1991.

Jordan, D. W. and Smith, P., Nonlinear Ordinary Differential Equations, Second Edition, Oxford University Press, 1987.

Kreyszig, Erwin, Advanced Engineering Mathematics, Sixth Edition, John Wiley & Sons, 1988.

Maeder, Roman, Programming in *Mathematica*, Second Edition, Addison-Wesley, 1992.

Myint-U, Tyn, Partial Differential Equations of Mathematical Physics, Second Edition, North Holland, 1980.

Saff, E. B. and Snider, A. D, <u>Fundamentals of Complex Analysis for Mathematics, Science, and Engineering</u>, Prentice-Hall, 1976.

Strang, Gilbert, <u>Introduction to Applied Mathematics</u>, Wellesley-Cambridge Press, 1986.

Strauss, Walter A., <u>Partial Differential Equations: An Introduction</u>, John Wiley & Sons, 1992.

Uhl, J., Porta, H., and Brown, D., <u>Calculus & *Mathematica*</u>, Addison-Wesley, 1991.

Wagon, Stan, *Mathematica* <u>in Action</u>, W. H. Freeman and Company, 1991.

Wolfram, Stephen, *Mathematica*: <u>A System for Doing Mathematics by Computer</u>, Second Edition, Addison Wesley, 1991.

Zill, Dennis, <u>A First Course in Differential Equations</u>, Second Edition, Prindle, Weber & Schmidt, 1982.

Zwillinger, Daniel, <u>Handbook of Differential Equations</u>, Second Edition, Academic Press, 1992.

Index